AF899566

Superatoms

Superatoms

Principles, Synthesis and Applications

Edited by

Purusottam (Puru) Jena
Virginia Commonwealth University,
Richmond, VA, USA

Qiang Sun
Peking University,
Beijing, China

WILEY

This edition first published 2022
© 2022 John Wiley & Sons Ltd

All rights reserved. No part of this publication may be reproduced, stored in a retrieval system, or transmitted, in any form or by any means, electronic, mechanical, photocopying, recording or otherwise, except as permitted by law. Advice on how to obtain permission to reuse material from this title is available at http://www.wiley.com/go/permissions.

The right of Purusottam (Puru) Jena and Qiang Sun to be identified as the authors of the editorial material in this work has been asserted in accordance with law.

Registered Offices
John Wiley & Sons, Inc., 111 River Street, Hoboken, NJ 07030, USA
John Wiley & Sons, Ltd., The Atrium, Southern Gate, Chichester, West Sussex, PO19 8SQ, UK

Editorial Office
The Atrium, Southern Gate, Chichester, West Sussex, PO19 8SQ, UK

For details of our global editorial offices, customer services, and more information about Wiley products visit us at www.wiley.com.

Wiley also publishes its books in a variety of electronic formats and by print-on-demand. Some content that appears in standard print versions of this book may not be available in other formats.

Limit of Liability/Disclaimer of Warranty
In view of ongoing research, equipment modifications, changes in governmental regulations, and the constant flow of information relating to the use of experimental reagents, equipment, and devices, the reader is urged to review and evaluate the information provided in the package insert or instructions for each chemical, piece of equipment, reagent, or device for, among other things, any changes in the instructions or indication of usage and for added warnings and precautions. While the publisher and authors have used their best efforts in preparing this work, they make no representations or warranties with respect to the accuracy or completeness of the contents of this work and specifically disclaim all warranties, including without limitation any implied warranties of merchantability or fitness for a particular purpose. No warranty may be created or extended by sales representatives, written sales materials, or promotional statements for this work. The fact that an organization, website, or product is referred to in this work as a citation and/or potential source of further information does not mean that the publisher and authors endorse the information or services the organization, website, or product may provide or recommendations it may make. This work is sold with the understanding that the publisher is not engaged in rendering professional services. The advice and strategies contained herein may not be suitable for your situation. You should consult with a specialist where appropriate. Further, readers should be aware that websites listed in this work may have changed or disappeared between when this work was written and when it is read. Neither the publisher nor authors shall be liable for any loss of profit or any other commercial damages, including but not limited to special, incidental, consequential, or other damages.

Library of Congress Cataloging-in-Publication data applied for:
HB ISBN: 9781119619529

Cover Design: Wiley
Cover Image: © American Chemical Society

Set in 9.5/12.5pt STIXTwoText by Straive, Pondicherry, India

Contents

Preface

One of chemistry's biggest contribution to science and society is the Mendeleev's periodic table of elements. Created in 1869 before the discovery of the electron and the knowledge of quantum mechanics, the periodic table accounts for the chemistry of elements in terms of their valence electrons. At present, the periodic table contains 118 elements, 94 of which occur in nature. Atoms of these naturally occurring elements are the building blocks of all matter. For ages, alchemists have tried to alter the chemistry of elements with the hope of transforming, for example, base metals to gold, but to no avail. With the knowledge we have gained over more than a century, we now ask the question: Can we achieve an alchemist's dream? Progress in nanoscience over the past few decades has provided some hope. As matter is reduced to nanometer or even sub-nanometer length scale, surface atoms dominate over bulk atoms. Combined with quantum confinement and reduced coordination, nanomaterials possess properties not seen in their bulk phase.

Atomic clusters, composed of a few to a few hundred atoms, are the ultimate nanoparticles where every atom counts. Research over the past 50 years has shown that properties of clusters are size- and composition-specific and can be tailored such that they mimic the chemistry of elements in the periodic table, even when none of their constituents belong to that group. Such clusters, termed as *superatoms*, can be used to build a three-dimensional periodic table where superatoms constitute the third dimension. Because there are unlimited ways of designing these superatoms, the three-dimensional periodic table will have no limits, at least in principle. Can properties of matter be fundamentally altered if we change their building blocks from individual atoms to these superatomic clusters? For example, can a superatomic cluster containing only nonmagnetic atoms be magnetic? Can a cluster composed of metal atoms behave like a semiconductor? Can noble gas atoms and inert molecules form chemical bonds with tailored superatoms? Can a cluster of gold atoms be reactive? This book examines these possibilities.

Although the field of atomic clusters can be traced to 1930s, it became attractive to the physics and chemistry community in 1970s when clusters of atoms and molecules with specific size and composition could be produced in the gas phase and studied both experimentally and theoretically. The motivation for studying clusters at the dawn of this field was to understand how properties of matter evolve, one atom at a time. It was soon realized that the structure and properties of atomic clusters not only are very different from their bulk matter but also evolve non-monotonically, often varying unpredictably and abruptly with the addition of a single atom. In contrast to conventional materials composed of individual atoms, new matter with tailored properties can be created by using these superatoms as the building blocks, primarily because of their unique size, shape, composition, and electronic structure. For example, lattice constants of cluster-assembled crystals would be larger than those of conventional crystals. The energy bands of cluster-assembled crystals formed by the overlap of superatomic orbitals would be different from the conventional energy band structure formed by the overlap of atomic orbitals. The nonspherical geometry of clusters would introduce novel features as their rotational degree of freedom can affect the energy landscape. Similarly, intra-cluster and inter-cluster vibrations would further affect the electron–phonon interaction. This paradigm shift in materials design and synthesis would open new doors for focused discovery of materials with unprecedented properties.

This book contains 13 chapters covering the recent advances in our understanding of superatoms and their assemblies. A group of international experts discuss the design of superatoms, study properties that make these unique, and explore their potential as building blocks of a new class of materials with applications in devices and energy-related technologies. Chapter 1 highlights some seminal papers and reviews on atomic clusters that made cluster science an emerging field bridging physics, chemistry, and materials science. Chapter 2 shows how superatoms mimicking the properties of group 1, 15, 16, and 17 elements can be rationally designed using electron-counting rules. Also discussed in this chapter is how different electron-counting rules can be used simultaneously to design multiply charged superatoms that neither fragment nor spontaneously eject the added electrons. Such superatoms can be used to promote unusual reactions enabling noble gas atoms such as argon to form chemical bonds at room temperature. Chapter 3 provides a comprehensive review of superhalogens that mimic the chemistry of halogens but possess electron affinities far exceeding those of halogen atoms. The design, stability, and structure-property relationships of endohedral clusters where metal atoms are encapsulated inside cage clusters composed of Si, Ge, Sn, and coinage metal atoms are discussed in Chapter 4. Also discussed in this chapter are assemblies of these metal encapsulated clusters. Chapter 5 shows how magnetic superatoms that mimic the properties of transition and rare earth metal atoms can be created by tailoring the role of *s* and *p* electrons that account for bonding while *d* and *f* electrons carry the magnetic moment. Chapters 6 and 7 cover the experimental and theoretical aspects of cluster-assembled materials focusing on atomically precise ligand-protected clusters composed of noble metal atoms. Two-dimensional materials created with superatoms as the building blocks are presented in Chapter 8. Potential applications of superatom-based materials in ferroelectrics, solar cells, Li-ion batteries, hydrogen storage, and capture and conversion of CO_2 are covered in Chapters 9–12. The challenges and opportunities that lie ahead are summarized in the concluding chapter. Cluster science in general and superatoms in particular are expected to remain a vibrant field for years to come, leading the way to new developments in materials design, synthesis, and applications.

Puru Jena
Qiang Sun

List of Contributors

Hong Fang
Department of Physics, Virginia Commonwealth University, Richmond, VA, USA

Nicola Gaston
The MacDiarmid Institute for Advanced Materials and Nanotechnology, Department of Physics, The University of Auckland, Auckland, New Zealand

Hannu Häkkinen
Departments of Physics and Chemistry, Nanoscience Center, University of Jyväskylä, Jyväskylä, Finland

Purusottam (Puru) Jena
Department of Physics, Virginia Commonwealth University, Richmond, VA, USA

Vijay Kumar
Center for Informatics, School of Natural Sciences, Shiv Nadar University, Tehsil Dadri, Uttar Pradesh, India
Dr. Vijay Kumar Foundation, Gurgaon, Haryana, India

Tingwei Li
School of Materials Science and Engineering, Peking University, Beijing, China
Center for Applied Physics and Technology, Peking University, Beijing, China

Zhifeng Liu
School of Physical Science and Technology, Inner Mongolia University, Hohhot, Inner Mongolia, China

Haoming Shen
School of Materials Science and Engineering, Peking University, Beijing, China
Center for Applied Physics and Technology, Peking University, Beijing, China

Piotr Skurski
Laboratory of Quantum Chemistry, Faculty of Chemistry, University of Gdańsk, Gdańsk, Poland
Henry Eyring Center for Theoretical Chemistry, Department of Chemistry, University of Utah, Salt Lake City, UT, USA

Qiang Sun
School of Materials Science and Engineering, Peking University, Beijing, China
Center for Applied Physics and Technology, Peking University, Beijing, China

Shinjiro Takano
Department of Chemistry, School of Science, The University of Tokyo, Tokyo, Japan

Tatsuya Tsukuda
Department of Chemistry, School of Science, The University of Tokyo, Tokyo, Japan

Elements Strategy Initiative for Catalysts and Batteries (ESICB), Kyoto University, Kyoto, Japan

Menghao Wu
School of Physics, Huazhong University of Science and Technology, Wuhan, Hubei, China

1

Introduction

Puru Jena[1] *and Qiang Sun*[2,3]

[1] *Physics Department, Virginia Commonwealth University, Richmond, Virginia, USA*
[2] *School of Materials Science and Engineering, Peking University, Beijing, China*
[3] *Center for Applied Physics and Technology, Peking University, Beijing, China*

Atomic clusters are a group of homo- or hetero-nuclear atoms that form in the gas phase when a hot plume of atoms cool through collisions with noble gas atoms. Once mass is isolated in a quadrupole or time of flight mass spectrometer, these clusters are studied with atomic precision. Although both molecules and atomic clusters are composed of a group of atoms, the similarity ends there. Molecules are found in nature and are stable under ambient conditions, while clusters are produced in the laboratory and are stable only when held in isolation. In addition, unlike molecules, clusters of any size and composition, in principle, can be formed. The early motivation for studying clusters was to understand how properties of matter evolve, one atom at a time. It was soon realized that the structure and properties of clusters are unique and do not resemble their bulk properties until their sizes are very large. Equally important, their properties vary nonmonotonically and unpredictably with size, shape, and composition. This realization gave clusters their own identity, and they came to represent a new phase of matter intermediate between the atoms and their bulk matter. The field of cluster science, born over half a century ago, has attracted scientists from different disciplines to study, understand, and manipulate their properties by changing size, shape, composition, charge, and environment [1–136]. More than 200 000 papers have appeared in this field with about 10 000 papers appearing per year over the past decade. Clusters now bridge disciplines, and their role in the design and synthesis of novel materials with tailored properties is the subject of this book.

In 1992 in an article in *Physical Review Letters*, Shiv Khanna and Puru Jena showed that a cluster with specific size and composition could mimic the properties of an element in the periodic table and named such a cluster as a "superatom" [137]. They further suggested that these superatoms could be used to build a three-dimensional periodic table with superatoms forming the third dimension [138]. Furthermore, a new class of materials, called "cluster-assembled materials" [138], can be formed by using superatomic clusters as building blocks, much as conventional materials are formed by using atoms as building blocks. This concept originated from a seminal experiment in the field of cluster science by Walter

Superatoms: Principles, Synthesis and Applications, First Edition. Edited by Purusottam (Puru) Jena and Qiang Sun.
© 2022 John Wiley & Sons Ltd. Published 2022 by John Wiley & Sons Ltd.

Knight and his group in 1984 [139]. These authors measured the mass spectra of Na clusters and observed that clusters containing 2, 8, 20, 40,... atoms are more abundant than their neighbors (see Figure 1.1). They noted that the nuclei with the same number of nucleons were already known to be very stable. In analogy with the nuclear physics, the authors termed these very stable clusters as magic clusters and showed that the origin of the magic numbers in Na clusters is due to electronic shell closure, just as the magic numbers in nuclei are due to nuclear shell closure.

Knight et al. described the electronic shell structure of Na clusters using the jellium model where the clusters were assumed to have spherical symmetry with the charges of the positive ion core distributed uniformly inside the sphere (see Figure 1.2) [140]. They assumed that valence electrons in Na clusters would behave like free electrons, similar to the conduction electrons in a Na metal. In the jellium model of a cluster, the charge density $n^{+}(r)$ of the positive ions is given by,

$$n^{+}(r) = n_0\,\Theta(r - R)$$

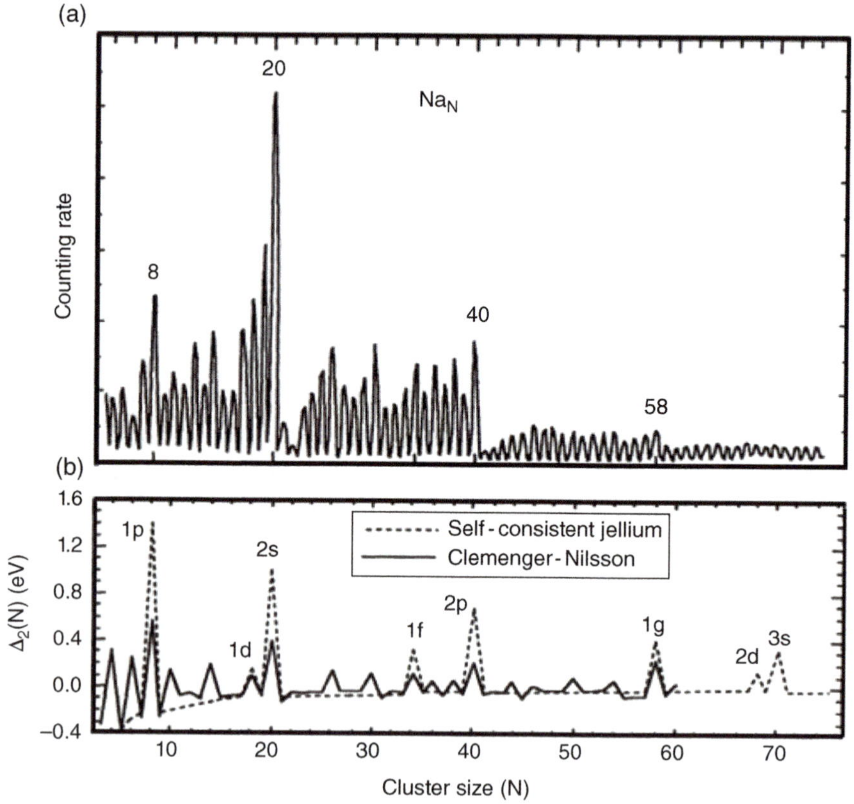

Figure 1.1 Sodium cluster abundance spectrum, (a) experimental data of Knight et al. (b) Second derivative of the total energy, dashed line using Woods-Saxon potential; solid line using the ellipsoidal shell (Clemenger-Nilsson model). *Source:* Adapted with permission from Ref. [130]. Copyright 1993 American Physical Society.

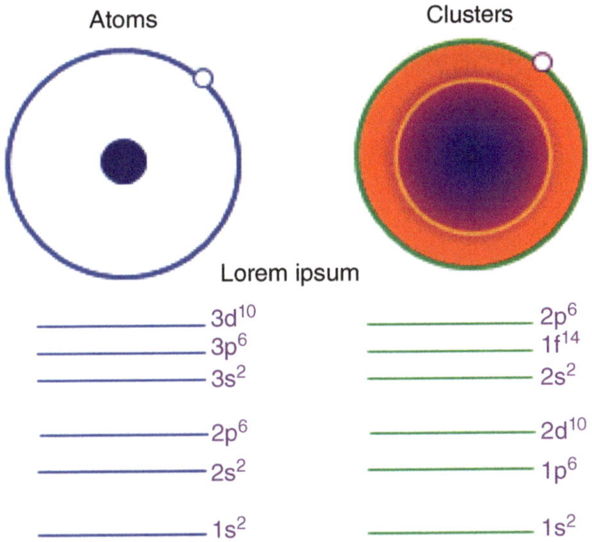

Figure 1.2 Schematic diagram of an atom and atomic orbitals (left panel) where the positively charged nucleus is localized at a point, and jellium model of a cluster where the positive charge is smeared over a sphere of finite radius with corresponding electronic orbitals (right panel). *Source:* Adapted with permission from Ref. [140]. Copyright 2013 American Chemical Society.

where n_0 is the density of the positive charge and R is the radius of the sphere. $\Theta(r - R) = 1$ for $r \leq R$ and $\Theta(r - R) = 0$ for $r > R$. The electrons responding to this charge distribution occupy orbitals similar to those in the atoms. The major difference is that in this case the orbitals of clusters are characterized as 1S 1P 1D 2S 1F 2P.... shells. Because the number of free electrons in a Na cluster is equal to the number of Na atoms, the magic numbers observed in Na clusters, namely 2, 8, 20, 40,... correspond to electronic shell closure of $1S^2$, $1S^2\,1P^6$, $1S^2\,1P^6\,1D^{10}\,2S^2$, $1S^2\,1P^6\,1D^{10}\,2S^2\,1F^{14}\,2P^6$,.... orbitals, respectively. The validity of the jellium model has since been established in simple as well as noble metals. Of particular note is the Al_{13}^- cluster, which contains 40 electrons. Will Castleman and coworkers noted that Al_{13}^- with a conspicuous peak in the mass spectra is indeed a very stable cluster [141] and is much less reactive toward oxygen than the neighboring clusters (Figure 1.3). In addition, Al_{13}, having an icosahedral geometry, also represents an atomically shell closed structure. The extraordinary stability of Al_{13}^-, therefore, derives from both electronically and atomically closed shell system.

The above findings led Khanna and Jena to propose that neutral Al_{13} cluster, with 39 electrons should behave the same way as a halogen atom, as both need one extra electron to close their electronic shells, the former being consistent with the jellium rule and the latter being consistent with the octet rule [137]. Indeed, calculations and experiments showed that the electron affinity of Al_{13}, namely 3.57 eV, is identical to that of Cl [142, 143]. That Al_{13} mimics the chemistry of halogens that was later shown theoretically [144] by studying its interaction with K. KAl_{13} was shown to be an ionically bonded cluster just like KCl. Experimental confirmation that KAl_{13} is indeed ionically bonded validates the superatom concept [145].

For the sake of history, it should be mentioned that properties of clusters have been linked to atoms as early as 1981 when Gutsev and Boldyrev [146] showed that the electron affinity of a cluster with composition MX_{k+1}, where M is a metal atom with valence k and X is a halogen atom, can exceed that of a halogen atom. Note that MX_{k+1} cluster, just like a halogen atom, would need an extra electron to close its electronic shell. Because the size

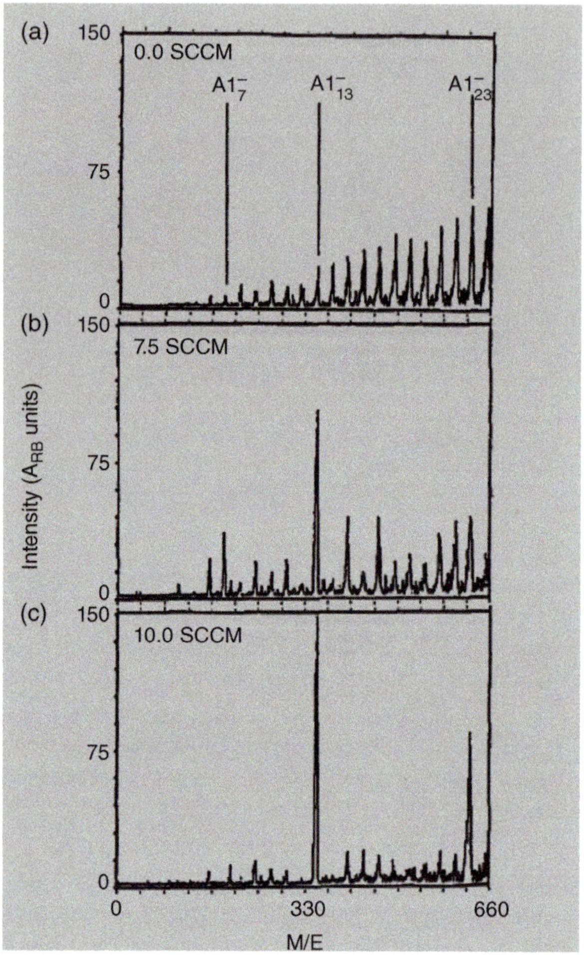

Figure 1.3 Series of mass spectra showing progression of the etching reaction of Al anions with oxygen in 0.0 SCCM (a), 7.5 SCCM (b), and 10.0 SCCM s(c). *Source:* Leuchtner, et al. [141]. © AIP Publishing.

of MX_{k+1} cluster is larger than that of a halogen atom, the Coulomb repulsion experienced by the extra electron decreases as the size increases. Hence, the electron affinity of MX_{k+1} is larger than that of X. The authors coined the word "superhalogen" to describe these clusters. In a subsequent publication, Gutsev and Boldyrev also showed that the ionization potential of a cluster with composition $M_{k+1}X$ is less than that of the alkali atom, M where k is the valence of atom X, and coined the word "superalkali" to describe these clusters. Thus, superhalogens and superalkalis that mimic the chemistry of alkali and halogen atoms in the periodic table, respectively, can be regarded as the first demonstration of what is now commonly termed as "superatoms." Figure 1.4 shows an example of a three-dimensional periodic table where superhalogens and superalkalis constitute the third dimension representing Group 1 and Group 17 elements [147].

A few years later, Saito and Ohnishi [148] noted that Na_8 satisfying electronic shell closure ($1S^2\ 1P^6$) according to the jellium model should be chemically inert like the noble gas atoms, which have their outermost s and p shells closed. Similarly, Na_{19} lacks an electron

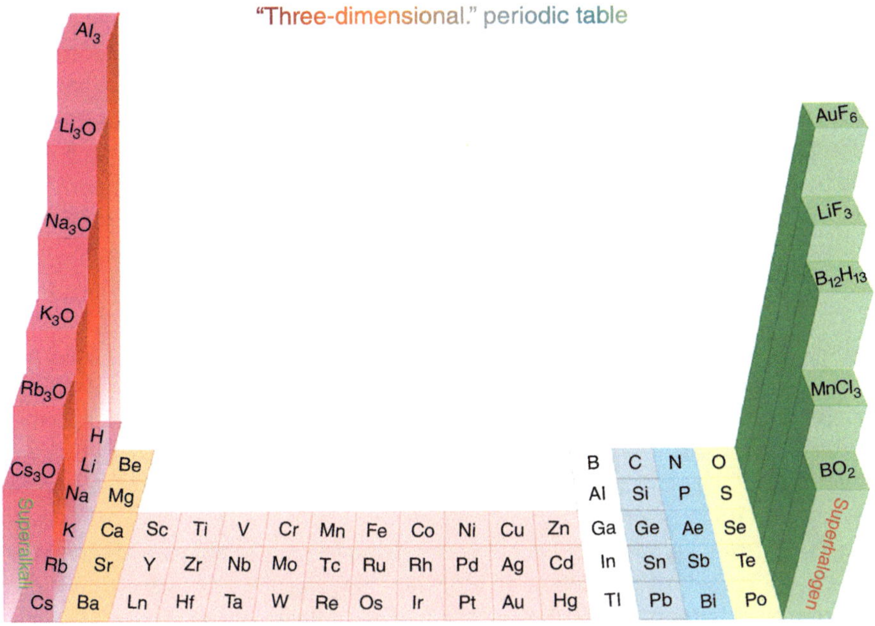

Figure 1.4 Three-dimensional periodic table with superatoms mimicking the chemistry of halogens and alkalis. Examples are those of superalkalis designed using the octet rule (Li_3O, Na_3O, K_3O, Rb_3O, Cs_3O) and jellium rule (Al_3) and superhalogens designed using the octet rule (AuF_6, LiF_3, $MnCl_3$, BO_2) and Wade-Mingos rule ($B_{12}H_{13}$).

to achieve electronic shell closure and should behave like a halogen atom. The authors termed these as "giant atoms." However, atomic clusters neither have spherical geometries nor do their electrons behave as if they are free electrons. On the contrary, the electrons are confined within the cluster. A later calculation that took explicit account of the cluster geometry showed that Na_8 retains its geometry up to 600 K on a NaCl substrate, but it spontaneously collapses forming an epitaxial layer on a Na (110) surface [149]. When two Na_8 clusters interact, instead of remaining as individual clusters, they coalesce and form a Na_{16} cluster. Similarly, no evidence exists to demonstrate that Na_{19} can form a salt-like molecule, analogous to KAl_{13}, while interacting with an alkali atom.

In later years, superatoms have been described in terms of their molecular orbital structures specific to their optimized geometries. Instead of using the jellium model and identifying superatoms as they fill the jellium orbitals, superatoms are regarded as a single unit with their molecular orbitals filled much the same way as electrons fill orbitals of a single atom. The chemistry of the superatoms is then determined by the outer molecular orbitals. As an example, consider $Al_{13}I_2^-$ [150]. Since Al_{13} behaves as a halogen, one could regard $Al_{13}I_2^-$ mimicking a triiodide I_3^- ion. Indeed, the outer electronic orbitals of $Al_{13}I_2^-$ and I_3^- ion have similar features. Similarly, $Al_{14}I^-$ can be viewed as $Al_{14}^{2+}.3I^-$, where Al_{14}^{2+} behaves like an alkaline-earth element. Considerable research over the past couple of decades has led to the design of superatoms using electron counting rules such as the octet rule, the 18-electron rule, Hückel's aromatic rule, and the Wade-Mingos rule [147].

"Magnetic superatoms," the name coined by Vijay Kumar and Yoshi Kawazoe [151], is another example of how a cluster could possess a magnetic moment just as transition and rare earth metal atoms do. An early realization of this concept dates back to the work of Rao et al. [152] who showed a Li_4 cluster confined to a tetrahedral geometry has a total spin 1 while its rhombus ground state structure has a total spin 0. That the geometry of a cluster can be linked to its underlying spin structure gives scientists an unprecedented opportunity to design magnetic superatoms where the constituent atoms are not even magnetic. Similarly, clusters of antiferromagnetic elements such as Mn can be ferromagnetic while the magnetic moments of transition metal clusters can far exceed that of their bulk value [153].

Knowing that superatoms can have properties different from the atoms, one can envision a new class of materials where these superatomic clusters are used as building blocks. One of the basic requirements for this purpose is that the superatomic clusters must be stable when assembled to a form a bulk material. This class of materials is called "cluster-assembled materials." The advantage of cluster-assembled materials over atom-assembled materials is that their underlying electronic structure is different from the electronic structure of atoms. In Figure 1.5 we present the electronic orbitals of an Al atom and compare it to that of an Al_{13} superatom [140]. We note that the energy spacing between molecular orbitals and atomic orbitals as well as their degeneracies and energy gap between highest occupied and lowest unoccupied molecular orbitals (HOMO–LUMO) are different. As these energy levels overlap and broaden when individual atoms and superatoms are brought together, the resulting energy bands would be very different. Thus, the electronic structure of cluster-assembled materials would be different from that of atom-assembled materials, even though the cluster is composed of the same elemental atoms. A good example is crystals made of carbon atoms and that of C_{60} fullerene. Graphite, the ground state of

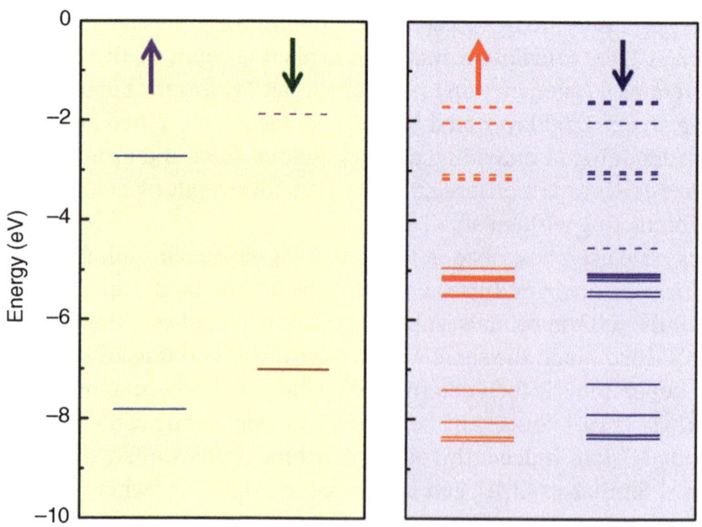

Figure 1.5 Spin polarized electron orbitals of Al atom (left panel) and the Al_{13} cluster (right panel). *Source:* Jena [140]. © American Chemical Society.

atom-assembled carbon is metallic and made of honeycomb arrangements of layers coupled weakly with each other. Fulleride, a crystal of C_{60}, on the other hand, forms an insulating fcc lattice with weakly bonded C_{60} clusters. Alkalization of fulleride crystal gives rise to a superconductor while intercalation of graphite by alkali atoms does not. In addition to the different electronic structure between atoms and superatoms, there are other features that contribute to the unique properties of cluster-assembled materials. For example, in a conventional crystal there is only one length scale, namely the lattice constant, while in a cluster-assembled material, there are two length scales – the intra-cluster distance and the inter-cluster distance. Clusters due to their nonspherical shape can influence the potential energy surface due to their rotational degree of freedom, while in a conventional crystal, the atoms, being spherical, do not have that freedom. The phonons generated from the lattice vibration and their coupling with electrons can also render unusual properties depending upon whether the crystal has atoms *vs* superatoms as building blocks.

The central question then is: how to ensure that the superatoms retain their geometry after assembly? This can be accomplished in a number of ways: (i) The superatoms should be very stable (e.g., C_{60}) and must not coalesce or deform as they come together to form a crystal. Electron counting rules as well as atomic shell closure rules can be used to identify such superatoms. Stability of clusters satisfying the jellium shell closure rule is one such scheme that is discussed in the above. However, stable superatoms can also be designed by satisfying other electron counting rules such as the octet rule for simple elements (s^2 p^6), the 18-electron rule for transition metal elements (s^2 p^6 d^{10}), 32-electron rule for rare earth elements (s^2 p^6 d^{10} f^{14}), the aromatic rule for organic molecules, and the Wade-Mingos rule for boron-based and Zintl clusters. (ii) Endohedral doping of metal atoms can also be used as an effective strategy to stabilize clusters. (iii) Atomic clusters can be soft-landed on a substrate and kept apart by limiting their density or (iv) coated with ligands that protect the core when assembled. In the latter two cases, it is likely that the substrate and the ligands can interact with the atomic clusters and can affect both their geometry and properties. Instead of viewing such interactions as undesired, they can be used to tailor the properties of atomic clusters by choosing the right substrate and the ligands.

In the following 11 chapters, various authors discuss how to design superatoms by using simple electron counting rules, how to stabilize them by endohedral doping of metal atoms, and how to protect them from coalescing with each other by coating them with suitable ligands, or soft-landing them on a chosen substrate to form cluster-based thin films. Cluster-assembled materials and how their properties can be tailored to produce novel catalysts, magnetic materials, and materials for energy production, storage, and conversion are also discussed. The concluding chapter describes outstanding problems and provides an insight into the future developments.

References

1 Becker, E.W., Bier, K., and Henkes, W. (1956). Strahlen aus kondensierten atomen und Molekeln im hochvakuum. *Eur. Phys. J. A* 146: 333–338.

2 Kubo, R. (1962). Electronic properties of metallic fine particles. I. *J. Phys. Soc. Jpn.* 17: 975–986.

3 Davenas, J. and Rabette, P. (1982). Contribution of clusters physics to materials science and technology. *Proceedings of the NATO Advanced Study Institute on Impact of Clusters Physics in Materials Science and Technology*, Cap d'Agde, France (1986).

4 Jena, P., Rao, B., and Khanna, S. (1987). *Physics and Chemistry of Small Clusters*. Richmond, VA: Virginia Commonwealth University.

5 Jena, P., Khanna, S., and Rao, B. (1992). *Physics and Chemistry of Finite Systems: From Clusters to Crystals*. New York: Springer Science & Business Media.

6 Sugano, S. (1991). *Microcluster Physics*. New York: Springer.

7 Sattler, K. (1996). *Cluster Assembled Materials*. New York: CRC Press.

8 Duncan, M.A. (1998). *Advances in Metal and Semiconductor Clusters: Cluster Materials*. New York: Elsevier.

9 Castleman, A.W. and Khanna, S.N. (2003). *Quantum Phenomena in Clusters and Nanostructures*. Berlin, Heidelberg: Springer.

10 Jena, P. and Castleman, A. Jr. (2010). *Nanoclusters–A Bridge Across Disciplines*. New York: Elsevier.

11 Chattaraj, P.K. (2010). *Aromaticity and Metal Clusters*. New York: CRC Press.

12 Campbell, E.E.B. (2011). *Proceedings of Nobel Symposium*. Sweden: World Scientific Publishing Co.

13 Alonso, J.A. (2012). *Structure and Properties of Atomic Nanoclusters*, 2e. River Edge, NJ: World Scientific.

14 Jellinek, J. (2012). *Theory of Atomic and Molecular Clusters: With a Glimpse at Experiments*. New York: Springer Science & Business Media.

15 Meiwes-Broer, K.-H. (2012). *Metal Clusters at Surfaces: Structure, Quantum Properties, Physical Chemistry*. New York: Springer Science & Business Media.

16 Milani, P. and Iannotta, S. (2012). *Cluster Beam Synthesis of Nanostructured Materials*. New York: Springer Science & Business Media.

17 Ciobanu, C.V., Wang, C.-Z., and Ho, K.-M. (2013). *Atomic Structure Prediction of Nanostructures, Clusters and Surfaces*. New York: John Wiley & Sons.

18 Kawazoe, Y., Kondow, T., and Ohno, K. (2013). *Clusters and Nanomaterials: Theory and Experiment*. New York: Springer Science & Business Media.

19 Heiles, S. and Schafer, R. (2014). *Dielectric Properties of Isolated Clusters: Beam Deflection Studies*. New York: Springer.

20 Tsukuda, T. and Hakkinen, H. (2015). *Protected Metal Clusters: From Fundamentals to Applications*. New York: Elsevier.

21 Chapon, C., Gillet, M.F., and Henry, C.R. (1989). *Small Particles and Inorganic Clusters: Proceedings of the Fourth International Meeting on Small Particles and Inorganic Clusters University Aix-Marseille III Aix-en-Provence, France, 5–9 July 1988*. New York: Springer Science & Business Media.

22 Jena, P. and Behera, S.N. (1996). *Clusters and Nanostructured Materials*. New York: Nova Publishers.

23 Jena, P., Khanna, S., and Rao, B. (1996). *Proceedings of the Science and Technology of Atomically Engineered Materials: Richmond, Virginia, USA, Oct. 30–Nov. 4*. Singapore: World Scientific.

24 Jena, P., Khanna, S., and Rao, B. (2000). *Cluster and Nanostructure Interfaces*. Singapore: World Scientific.

25 Jena, P., Khanna, S., and Rao, B. (2005). *Clusters and Nano-Assemblies: Physical and Biological Systems*. Singapore: World Scientific.

26 Sahu, S., Choudhury, R., and Jena, P. (2006). *Nano-Scale Materials: From Science to Technology*. New York: Nova Publishers.

27 Band, E. and Muetterties, E.L. (1978). Mechanistic features of metal cluster rearrangements. *Chem. Rev.* 78: 639–658.

28 Muetterties, E.L., Rhodin, T.N., Band, E. et al. (1979). Clusters and surfaces. *Chem. Rev.* 79: 91–137.

29 Beuhler, R. and Friedman, L. (1986). Larger cluster ion impact phenomena. *Chem. Rev.* 86: 521–537.

30 Castleman, A.W. and Keesee, R.G. (1986). Ionic clusters. *Chem. Rev.* 86: 589–618.

31 Koutecky, J. and Fantucci, P. (1986). Theoretical aspects of metal atom clusters. *Chem. Rev.* 86: 539–587.

32 Bonacic-Koutecky, V., Fantucci, P., and Koutecky, J. (1991). Quantum chemistry of small clusters of elements of groups Ia, Ib, and IIa: fundamental concepts, predictions, and interpretation of experiments. *Chem. Rev.* 91: 1035–1108.

33 Bonačić-Koutecký, V. and Mitrić, R. (2005). Theoretical exploration of ultrafast dynamics in atomic clusters: analysis and control. *Chem. Rev.* 105: 11–66.

34 Berry, R.S. (1993). Potential surfaces and dynamics: what clusters tell us. *Chem. Rev.* 93: 2379–2394.

35 Morse, M.D. (1986). Clusters of transition-metal atoms. *Chem. Rev.* 86: 1049–1109.

36 Fendler, J.H. (1987). Atomic and molecular clusters in membrane mimetic chemistry. *Chem. Rev.* 87: 877–899.

37 Jelski, D.A. and George, T.F. (1988). Clusters: link between molecules and solids. *J. Chem. Educ.* 65: 879–883.

38 Kappes, M.M. (1988). Experimental studies of gas-phase main-group metal clusters. *Chem. Rev.* 88: 369–389.

39 Weltner, W. and Van Zee, R.J. (1989). Carbon molecules, ions, and clusters. *Chem. Rev.* 89: 1713–1747.

40 Van Orden, A. and Saykally, R.J. (1998). Small carbon clusters: spectroscopy, structure, and energetics. *Chem. Rev.* 98: 2313–2358.

41 Kong, X.-J., Long, L.-S., Zheng, Z. et al. (2010). Keeping the ball rolling: fullerene-like molecular clusters. *Acc. Chem. Res.* 43: 201–209.

42 Adams, R.D. (1989). Metal cluster complexes containing heteroatom-substituted carbene. ligands. *Chem. Rev.* 89: 1703–1712.

43 Mingos, D.M.P., Slee, T., and Zhenyang, L. (1990). Bonding models for ligated and bare clusters. *Chem. Rev.* 90: 383–402.

44 Leutwyler, S. and Boesiger, J. (1990). Rare-gas solvent clusters: spectra, structures, and order-disorder transitions. *Chem. Rev.* 90: 489–507.

45 Chalasinski, G. and Szczesniak, M.M. (1994). Origins of structure and energetics of van der Waals clusters from ab Initio calculations. *Chem. Rev.* 94: 1723–1765.

46 Shang, Q.Y. and Bernstein, E.R. (1994). Energetics, dynamics, and reactions of Rydberg state molecules in van der Waals clusters. *Chem. Rev.* 94: 2015–2025.

47 Garvey, J.F., Herron, W.J., and Vaidyanathan, G. (1994). Probing the structure and reactivity of hydrogen-bonded clusters of the type {M}n{H2O}H+, via the observation of magic numbers. *Chem. Rev.* 94: 1999–2014.

48 Plesek, J. (1992). Potential applications of the boron cluster compounds. *Chem. Rev.* 92: 269–278.
49 Hawthorne, M.F. and Maderna, A. (1999). Applications of radiolabeled boron clusters to the diagnosis and treatment of cancer. *Chem. Rev.* 99: 3421–3434.
50 Schmid, G. (1992). Large clusters and colloids: metals in the embryonic state. *Chem. Rev.* 92: 1709–1727.
51 Parent, D.C. and Anderson, S.L. (1992). Chemistry of metal and semimetal cluster ions. *Chem. Rev.* 92: 1541–1565.
52 Brutschy, B. (1992). Ion-molecule reactions within molecular clusters. *Chem. Rev.* 92: 1567–1587.
53 Illenberger, E. (1992). Electron-attachment reactions in molecular clusters. *Chem. Rev.* 92: 1589–1609.
54 Hobza, P., Selzle, H.L., and Schlag, E.W. (1994). Structure and properties of benzene-containing molecular clusters: nonempirical ab initio calculations and experiments. *Chem. Rev.* 94: 1767–1785.
55 Kim, K.S., Tarakeshwar, P., and Lee, J.Y. (2000). Molecular clusters of π-systems: theoretical studies of structures, spectra, and origin of interaction energies. *Chem. Rev.* 100: 4145–4186.
56 Sun, T. and Seff, K. (1994). Silver clusters and chemistry in zeolites. *Chem. Rev.* 94: 857–870.
57 Braga, D., Dyson, P.J., Grepioni, F., and Johnson, B.F.G. (1994). Arene clusters. *Chem. Rev.* 94: 1585–1620.
58 Gates, B.C. (1995). Supported metal clusters: synthesis, structure, and catalysis. *Chem. Rev.* 95: 511–522.
59 Alexeev, O.S. and Gates, B.C. (2003). Supported bimetallic cluster catalysts. *Ind. Eng. Chem. Res.* 42: 1571–1587.
60 Yachandra, V.K., Sauer, K., and Klein, M.P. (1996). Manganese cluster in photosynthesis: where plants oxidize water to dioxygen. *Chem. Rev.* 96: 2927–2950.
61 Bačić, Z. and Miller, R.E. (1996). Molecular clusters: structure and dynamics of weakly bound systems. *J. Phys. Chem.* 100: 12945–12959.
62 Ogino, H., Inomata, S., and Tobita, H. (1998). Abiological iron-sulfur clusters. *Chem. Rev.* 98: 2093–2122.
63 Henderson, R.A. (2005). Mechanistic studies on synthetic Fe—S-based clusters and their relevance to the action of nitrogenases. *Chem. Rev.* 105: 2365–2438.
64 Lee, S.C., Lo, W., and Holm, R.H. (2014). Developments in the biomimetic chemistry of cubane-type and higher nuclearity iron-sulfur clusters. *Chem. Rev.* 114: 3579–3600.
65 Desfrançois, C., Carles, S., and Schermann, J.P. (2000). Weakly bound clusters of biological interest. *Chem. Rev.* 100: 3943–3962.
66 Buck, U. and Huisken, F. (2000). Infrared spectroscopy of size-selected water and methanol clusters. *Chem. Rev.* 100: 3863–3890.
67 Rohmer, M.-M., Bénard, M., and Poblet, J.-M. (2000). Structure, reactivity, and growth pathways of metallocarbohedrenes M8C12 and transition metal/carbon clusters and nanocrystals: a challenge to computational chemistry. *Chem. Rev.* 100: 495–542.
68 Dedonder-Lardeux, C., Grégoire, G., Jouvet, C. et al. (2000). Charge separation in molecular clusters: dissolution of a salt in a salt—(solvent)n cluster. *Chem. Rev.* 100: 4023–4038.

69 Niedner-Schatteburg, G. and Bondybey, V.E. (2000). FT-ICR studies of solvation effects in ionic water cluster reactions. *Chem. Rev.* 100: 4059–4086.

70 Zhong, Q. and Castleman, A.W. (2000). An ultrafast glimpse of cluster solvation effects on reaction dynamics. *Chem. Rev.* 100: 4039–4058.

71 Stace, A.J. (2002). Metal ion solvation in the gas phase: the quest for higher oxidation states. *J. Phys. Chem. A* 106: 7993–8005.

72 Bieske, E.J. and Dopfer, O. (2000). High-resolution spectroscopy of cluster ions. *Chem. Rev.* 100: 3963–3998.

73 Bieske, E.J. and Maier, J.P. (1993). Spectroscopic studies of ionic complexes and clusters. *Chem. Rev.* 93: 2603–2621.

74 Celii, F.G. and Janda, K.C. (1986). Vibrational spectroscopy, photochemistry, and photophysics of molecular clusters. *Chem. Rev.* 86: 507–520.

75 Neusser, H.J. and Siglow, K. (2000). High-resolution ultraviolet spectroscopy of neutral and ionic clusters: hydrogen bonding and the external heavy atom effect. *Chem. Rev.* 100: 3921–3942.

76 Alonso, J.A. (2000). Electronic and atomic structure, and magnetism of transition-metal clusters. *Chem. Rev.* 100: 637–678.

77 Boldyrev, A.I. and Wang, L.-S. (2001). Beyond classical stoichiometry: experiment and theory. *J. Phys. Chem. A* 105: 10759–10775.

78 Gabriel, J.-C.P., Boubekeur, K., Uriel, S., and Batail, P. (2001). Chemistry of hexanuclear rhenium chalcohalide clusters. *Chem. Rev.* 101: 2037–2066.

79 Lombardi, J.R. and Davis, B. (2002). Periodic properties of force constants of small transition-metal and lanthanide clusters. *Chem. Rev.* 102: 2431–2460.

80 Lee, S.C. and Holm, R.H. (2004). The clusters of nitrogenase: synthetic methodology in the construction of weak-field clusters. *Chem. Rev.* 104: 1135–1158.

81 Dermota, T.E., Zhong, Q., and Castleman, A.W. (2004). Ultrafast dynamics in cluster systems. *Chem. Rev.* 104: 1861–1886.

82 Chisholm, M.H. and Macintosh, A.M. (2005). Linking multiple bonds between metal atoms: clusters, dimers of “dimers”, and higher ordered assemblies. *Chem. Rev.* 105: 2949–2976.

83 Sevov, S.C. and Goicoechea, J.M. (2006). Chemistry of deltahedral Zintl ions. *Organometallics* 25: 5678–5692.

84 Ferrando, R., Jellinek, J., and Johnston, R.L. (2008). Nanoalloys: from theory to applications of alloy clusters and nanoparticles. *Chem. Rev.* 108: 845–910.

85 Erdemir, D., Lee, A.Y., and Myerson, A.S. (2009). Nucleation of crystals from solution: classical and two-step models. *Acc. Chem. Res.* 42: 621–629.

86 Schnöckel, H. (2010). Structures and properties of metalloid Al and Ga clusters open our eyes to the diversity and complexity of fundamental chemical and physical processes during formation and dissolution of metals. *Chem. Rev.* 110: 4125–4163.

87 Qiu, J. and Burns, P.C. (2013). Clusters of actinides with oxide, peroxide, or hydroxide bridges. *Chem. Rev.* 113: 1097–1120.

88 Yano, J. and Yachandra, V. (2014). Mn4Ca cluster in photosynthesis: where and how water is oxidized to dioxygen. *Chem. Rev.* 114: 4175–4205.

89 Cruz-Olvera, D., de la Trinidad Vasquez, A., Geudtner, G. et al. (2015). Transition-state searches in metal clusters by first-principle methods. *J. Phys. Chem. A* 119: 1494–1501.

90 Fernando, A., Weerawardene, K.L.D.M., Karimova, N.V., and Aikens, C.M. (2015). Quantum mechanical studies of large metal, metal oxide, and metal chalcogenide nanoparticles and clusters. *Chem. Rev.* 115: 6112–6216.

91 Tomalia, D.A. and Khanna, S.N. (2016). A systematic framework and nanoperiodic concept for unifying nanoscience: hard/soft nanoelements, superatoms, meta-atoms, new emerging properties, periodic property patterns, and predictive mendeleev-like nanoperiodic tables. *Chem. Rev.* 116: 2705–2774.

92 Gillespie, R.J. (1979). Nyholm memorial lecture. Ring, cage, and cluster compounds of the main group elements. *Chem. Soc. Rev.* 8: 315–352.

93 Holm, R.H. (1981). Centenary lecture. Metal clusters in biology: quest for a synthetic representation of the catalytic site of nitrogenase. *Chem. Soc. Rev.* 10: 455–490.

94 Greenwood, N.N. (1984). Liversidge lecture. Molecular tectonics: the construction of polyhedral clusters. *Chem. Soc. Rev.* 13: 353–374.

95 Mingos, D.M.P. (1986). Bonding in molecular clusters and their relationship to bulk metals. *Chem. Soc. Rev.* 15: 31–61.

96 Coolbaugh, M.T. and Garvey, J.F. (1992). Magic numbers in molecular clusters: a probe for chemical reactivity. *Chem. Soc. Rev.* 21: 163–169.

97 Imhof, D. and Venazi, L.M. (1994). Trimetallic units as building blocks in cluster chemistry. *Chem. Soc. Rev.* 23: 185–193.

98 Housecroft, C.E. (1995). Denuding the boron atom of B-H interactions in transition metal-boron clusters. *Chem. Soc. Rev.* 24: 215–222.

99 Gatteschi, D., Caneschi, A., Sessoli, R., and Cornia, A. (1996). Magnetism of large iron-oxo clusters. *Chem. Soc. Rev.* 25: 101–109.

100 Schnepf, A. (2007). Metalloid group 14 cluster compounds: an introduction and perspectives to this novel group of cluster compounds. *Chem. Soc. Rev.* 36: 745–758.

101 Hakkinen, H. (2008). Atomic and electronic structure of gold clusters: understanding flakes, cages and superatoms from simple concepts. *Chem. Soc. Rev.* 37: 1847–1859.

102 Schmid, G. (2008). The relevance of shape and size of Au55 clusters. *Chem. Soc. Rev.* 37: 1909–1930.

103 Verlet, J.R.R. (2008). Femtosecond spectroscopy of cluster anions: insights into condensed-phase phenomena from the gas-phase. *Chem. Soc. Rev.* 37: 505–517.

104 Fischer, I. (2003). Time-resolved photoionization of radicals, clusters and biomolecules: relevant model systems. *Chem. Soc. Rev.* 32: 59–69.

105 Kostakis, G.E., Ako, A.M., and Powell, A.K. (2010). Structural motifs and topological representation of Mn coordination clusters. *Chem. Soc. Rev.* 39: 2238–2271.

106 Schubert, U. (2011). Cluster-based inorganic-organic hybrid materials. *Chem. Soc. Rev.* 40: 575–582.

107 Long, D.-L., Burkholder, E., and Cronin, L. (2007). Polyoxometalate clusters, nanostructures and materials: from self-assembly to designer materials and devices. *Chem. Soc. Rev.* 36: 105–121.

108 Rozes, L. and Sanchez, C. (2011). Titanium oxo-clusters: precursors for a Lego-like construction of nanostructured hybrid materials. *Chem. Soc. Rev.* 40: 1006–1030.

109 Cabeza, J.A. and Garcia-Alvarez, P. (2011). The N-heterocyclic carbene chemistry of transition-metal carbonyl clusters. *Chem. Soc. Rev.* 40: 5389–5405.

110 Sculfort, S. and Braunstein, P. (2011). Intramolecular d10-d10 interactions in heterometallic clusters of the transition metals. *Chem. Soc. Rev.* 40: 2741–2760.

111 Lu, Z. and Yin, Y. (2012). Colloidal nanoparticle clusters: functional materials by design. *Chem. Soc. Rev.* 41: 6874–6887.

112 Lu, Y. and Chen, W. (2012). Sub-nanometre sized metal clusters: from synthetic challenges to the unique property discoveries. *Chem. Soc. Rev.* 41: 3594–3623.

113 Olid, D., Nunez, R., Vinas, C., and Teixidor, F. (2013). Methods to produce B-C, B-P, B-N and B-S bonds in boron clusters. *Chem. Soc. Rev.* 42: 3318–3336.

114 Fuhr, O., Dehnen, S., and Fenske, D. (2013). Chalcogenide clusters of copper and silver from silylated chalcogenide sources. *Chem. Soc. Rev.* 42: 1871–1906.

115 Sterenberg, B.T., Scoles, L., and Carty, A.J. (2002). Synthesis, structure, bonding and reactivity in clusters of the lower phosphorus oxides. *Coord. Chem. Rev.* 231: 183–197.

116 LaViolette, R.A. and Benson, M.T. (2012). Structure and thermodynamics of phosphorus oxide caged clusters. *J. Struct. Chem.* 53: 48–54.

117 Sokolov, M.N. and Fedin, V.P. (2004). Chalcogenide clusters of vanadium, niobium and tantalum. *Coord. Chem. Rev.* 248: 925–944.

118 Thompson, L.K., Waldmann, O., and Xu, Z. (2005). Polynuclear manganese grids and clusters—a magnetic perspective. *Coord. Chem. Rev.* 249: 2677–2690.

119 Alexandrova, A.N., Boldyrev, A.I., Zhai, H.-J., and Wang, L.-S. (2006). Allboron aromatic clusters as potential new inorganic ligands and building blocks in chemistry. *Coord. Chem. Rev.* 250: 2811–2866.

120 Armelao, L., Barreca, D., Bottaro, G. et al. (2006). Recent trends on nanocomposites based on Cu, Ag and Au clusters: a closer look. *Coord. Chem. Rev.* 250: 1294–1314.

121 Huang, Y.-G., Jiang, F.-L., and Hong, M.-C. (2009). Magnetic lanthanide–transition-metal organic–inorganic hybrid materials: from discrete clusters to extended frameworks. *Coord. Chem. Rev.* 253: 2814–2834.

122 Llusar, R. and Vicent, C. (2010). Trinuclear molybdenum cluster sulfides coordinated to dithiolene ligands and their use in the development of molecular conductors. *Coord. Chem. Rev.* 254: 1534–1548.

123 Glover, S.D., Goeltz, J.C., Lear, B.J., and Kubiak, C.P. (2010). Inter- or intramolecular electron transfer between triruthenium clusters: we'll cross that bridge when we come to it. *Coord. Chem. Rev.* 254: 331–345.

124 Shieh, M., Miu, C.-Y., Chu, Y.-Y., and Lin, C.-N. (2012). Recent progress in the chemistry of anionic groups 6–8 carbonyl chalcogenide clusters. *Coord. Chem. Rev.* 256: 637–694.

125 Kostakis, G.E., Perlepes, S.P., Blatov, V.A. et al. (2012). High-nuclearity cobalt coordination clusters: synthetic, topological and magnetic aspects. *Coord. Chem. Rev.* 256: 1246–1278.

126 Mayasree, O., Sankar, C.R., Kleinke, K.M., and Kleinke, H. (2012). Cu clusters and chalcogenchalcogen bonds in various copper polychalcogenides. *Coord. Chem. Rev.* 256: 1377–1383.

127 Sokolov, M.N. and Abramov, P.A. (2012). Chalcogenide clusters of Groups 8–10 noble metals. *Coord. Chem. Rev.* 256: 1972–1991.

128 Hong, K. and Chun, H. (2013). Nonporous titanium–oxo molecular clusters that reversibly and selectively adsorb carbon dioxide. *Inorg. Chem.* 52: 9705–9707.

129 Brack, M. (1993). The physics of simple metal clusters: self-consistent jellium model and semiclassical approaches. *Rev. Mod. Phys.* 65: 677–732.

130 de Heer, W.A. (1993). The physics of simple metal clusters: experimental aspects and simple models. *Rev. Mod. Phys.* 65: 611–676.

131 Jensen, P. (1999). Growth of nanostructures by cluster deposition: Experiments and simple models. *Rev. Mod. Phys.* 71: 1695–1735.

132 Herschbach, D. (1999). Chemical physics: molecular clouds, clusters, and corrals. *Rev. Mod. Phys.* 71: S411–S418.

133 Maier, T., Jarrell, M., Pruschke, T., and Hettler, M.H. (2005). Quantum cluster theories. *Rev. Mod. Phys.* 77: 1027–1080.

134 Fennel, T., Meiwes-Broer, K.H., Tiggesbäumker, J. et al. (2010). Laser-driven nonlinear cluster dynamics. *Rev. Mod. Phys.* 82: 1793–1842.

135 Einax, M., Dieterich, W., and Maass, P. (2013). Colloquium: cluster growth on surfaces: densities, size distributions, and morphologies. *Rev. Mod. Phys.* 85: 921–939.

136 Furrer, A. and Waldmann, O. (2013). Magnetic cluster excitations. *Rev. Mod. Phys.* 85: 367–420.

137 Khanna, S.N. and Jena, P. (1992). Assembling crystals from clusters. *Phys. Rev. Lett.* 69: 1664–1667.

138 Khanna, S.N. and Jena, P. (1995). Atomic clusters: building blocks for a class of solids. *Phys. Rev. B* 51: 13705–13716.

139 Knight, W.D., Clemenger, K., de Heer, W.A. et al. (1984). Electronic shell structure and abundances of sodium clusters. *Phys. Rev. Lett.* 52: 2141–2144.

140 Jena, P. (2013). Beyond the periodic table of elements: the role of superatoms. *J. Phys. Chem. Lett.* 4: 1432.

141 Leuchtner, R.E., Harms, A.C., and Castleman, A.W. Jr. (1989). Thermal metal cluster anion reactions: behavior of aluminum clusters with oxygen. *J. Chem. Phys.* 91: 2753.

142 Li, X., Wu, H., Wang, X.B., and Wang, L.S. (1998). s–p Hybridization and electron shell structures in aluminum clusters: a photoelectron spectroscopic study. *Phys. Rev. Lett.* 81: 1909–1912.

143 Rao, B.K. and Jena, P. (1999). Evolution of the electronic structure and properties of neutral and charged aluminum clusters: a comprehensive analysis. *J. Chem. Phys.* 111: 1890.

144 Khanna, S.N. and Jena, P. (1994). Designing ionic solids from metallic clusters. *Chem. Phys. Lett.* 219: 479–483.

145 Zheng, W.-J., Thomas, O.C., Lippa, T.P. et al. (2006). The ionic KAl_{13} molecule: a stepping stone to cluster assembled materials. *J. Chem. Phys.* 124: 144304–144305.

146 Gutsev, G.L. and Boldyrev, A.I. (1981). DVM-Xα calculations on the ionization potentials of MXk_{+1}–complex anions and the electron affinities of MXk_{+1} "superhalogens". *Chem. Phys.* 56: 277–283.

147 Jena, P. and Sun, Q. (2018). Super atomic clusters: design rules and potential for building blocks of materials. *Chem. Rev.* 118: 5755–5870.

148 Saito, S. and Ohnishi, S. (1987). Stable $(Na_{19})_2$ as a giant alkali-metal- atom dimer. *Phys. Rev. Lett.* 59: 190–193.

149 Hakkinen, H. and Manninen, M. (1996). How "magic" is a magic cluster? *Phys. Rev. Lett.* 76: 1599–1602.

150 Bergeron, D.E., Castleman, A.W. Jr., Morisato, T., and Khanna, S.N. (2004). Formation of $Al_{13}I^-$: evidence for the superhalogen character of Al_{13}. *Science* 304: 84–87.

151 Kumar, V. and Kawazoe, Y. (2003). Metal-doped magic clusters of Si, Ge, and Sn: The finding of a magnetic superatom. *Appl. Phys. Lett.* 83: 2677.

152 Rao, B.K., Jena, P., and Manninen, M. (1985). Relationship between topological and magnetic order in small metal clusters. *Phys. Rev. B* 32: 477.

153 Nayak, S.K. and Jena, P. (1998). Anomalous magnetism of small Mn clusters. *Chem. Phys. Lett.* 289: 473.

2

Rational Design of Superatoms Using Electron-Counting Rules

Puru Jena[1], *Hong Fang*[1], *and Qiang Sun*[2,3]

[1] *Physics Department, Virginia Commonwealth University, Richmond, Virginia, USA*
[2] *School of Materials Science and Engineering, Peking University, Beijing, China*
[3] *Center for Applied Physics and Technology, Peking University, Beijing, China*

2.1 Introduction

The periodic table of elements developed by Mendeleev and presented at a meeting of the Russian Chemical Society on 6 March 1869, was based purely on the chemical properties of the then known elements. However, a fundamental understanding of the chemistry of the elements and the sites they occupy in the periodic table had to wait until the discovery of the electron in 1897 and the development of quantum mechanics in the early part of the twentieth century. Central to this understanding is the knowledge of the atomic orbitals and the manner in which they are filled as one moves along the columns and the rows of the periodic table. The elements belonging to the same column of the periodic table have similar chemistry. As one moves along the rows, the electrons continue to fill the successive atomic orbitals in keeping with the Pauli's principle and the chemistry changes accordingly. Consider, for example, the Group 18 elements, which have an outer electron configuration of $ns^2\ np^6$. Because these orbitals are full, the energy cost to remove an electron is high and the energy gain in adding an electron is negligibly small. Thus, these elements do not participate in chemical reactions, justifying their name as the noble gas atoms. They are very stable and their bonding is characterized by weak van der Waals forces. The origin of their stability is, therefore, attributed to the electronic shell closure, which is referred to as the "octet rule" where $2+6 = 8$ electrons are enough to fill the s and p orbitals. On the other hand, the group 1 alkali elements are characterized by an outer electron orbital occupation, namely ns^1. The energy cost to remove this electron, i.e., the ionization potential, is smaller than elements belonging to the same row in the periodic table and hence in a chemical reaction they tend to donate that electron, leaving behind an ionic core with closed electronic shells. Similarly, group 17 halogen atoms have an outer electron configuration of $ns^2\ np^5$ and need an extra electron to satisfy the octet rule. Consequently, they have large electron affinities and gain energy by accepting an electron in a chemical reaction. Thus, when an alkali atom and a halogen atom approach each other, electron transfer from the alkali to the halogen atom satisfies

Superatoms: Principles, Synthesis and Applications, First Edition. Edited by Purusottam (Puru) Jena and Qiang Sun.
© 2022 John Wiley & Sons Ltd. Published 2022 by John Wiley & Sons Ltd.

the octet rule of both the atoms, resulting in the formation of a salt and an ionic bond between the cations and anions. In other cases, such as H_2, O_2, and N_2, electron shell closure is achieved not by transferring electrons from one atom to the other but rather by sharing electrons. This leads to a covalent bond, which, in general, is stronger than an ionic bond. A fourth kind of bonding appears as atoms from groups 2–13 and some elements in the higher groups upto 16 come together to form a crystal. Here, each atom contributes its outer valence electrons to a common pool. These electrons move "freely" and collectively, forming a metal. Note that, the ionic cores of the metal also have electronic shell closure.

The periodic table (see Figure 2.1) currently consists of 118 elements among which 94 occur in nature. All materials are created by combining atoms from one or more groups. Among the 94 naturally occurring elements, some are expensive while some others occur in trace quantities. Some of these elements are even toxic. Is it possible to replace the expensive elements by earth abundant materials and the toxic elements by nontoxic ones? This has been the dream of alchemists for centuries. With the birth of cluster science, we have arrived at a stage where this dream may not be as farfetched as it once seemed.

Atomic clusters are groups of atoms whose size and composition can be varied by design, one atom at a time. More than half a century of research has made it clear that the properties of clusters are very different from any other form of matter [1]. Because their properties are size-, composition-, and shape-specific, clusters can be tailored with atomic precision. In 1992, Khanna and Jena [2] coined the word "superatom" to describe a cluster that has the same chemistry as an atom in the periodic table and suggested that these superatoms can be used as the building blocks of a new three-dimensional periodic table, with superatoms forming the third dimension [3]. If these superatoms can retain their geometry and properties

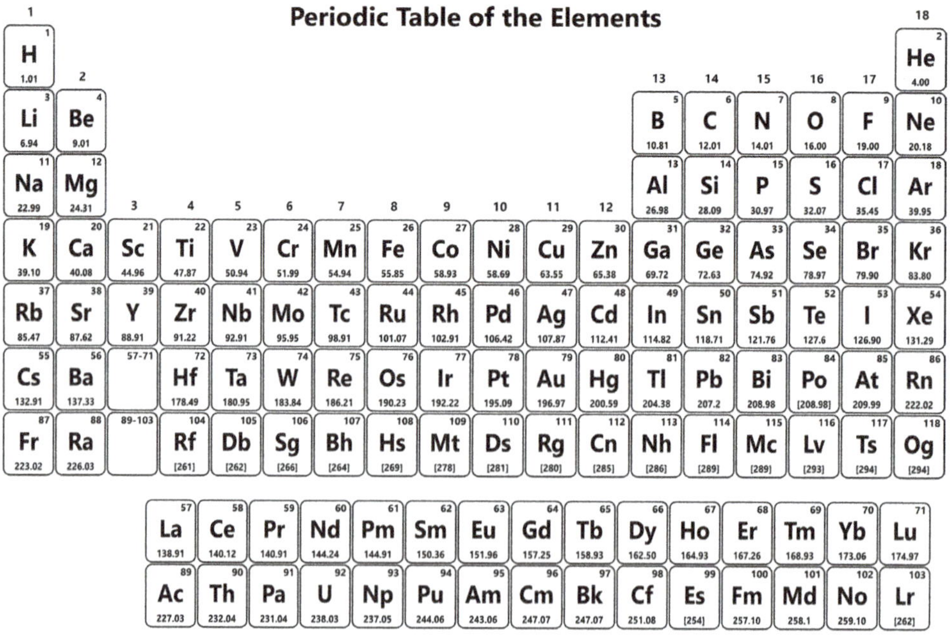

Figure 2.1 The periodic table of elements.

when assembled, a new class of cluster-assembled materials with tailored properties can be formed. A classic example of such a crystal is based on C_{60} fullerene, which was discovered by Smalley and coworkers in the gas phase in 1985 [4] and later synthesized in bulk quantities by Kratschmer et al. [5]. Once assembled, C_{60} fullerenes retain their shape, but the property of the fulleride crystal is very different from that of graphite and diamond. The former is the ground state of carbon while the latter is metastable but protected by a very large energy barrier. Note that the discovery of C_{60} was not the result of a rational design approach. The question is: can other clusters like C_{60} be rationally designed by using some prescribed rules?

In this chapter we examine if the electron-counting rules, known to explain the stability of atoms and compounds, can be used to rationally design stable clusters. The first glimpse of such a possibility came from the experiment of Knight and collaborators in 1984 [6]. The authors observed conspicuous peaks in the mass spectra of Na clusters containing 2, 8, 20, 40,... atoms. Realizing that similar observation was made in nuclear physics where nuclei with 2, 8, 20, 40,... nucleons were found to be very stable, Knight et al. suggested an electronic shell model, analogous to the nuclear shell model [7], to explain the magic numbers in Na clusters. They used the jellium model where free electrons move in a uniform distribution of positive ion charge. Assuming that a Na cluster has a spherical geometry and the charges on the positive ion cores are distributed uniformly, they showed that the stability of the magic Na clusters is due to shell closures of their electronic orbitals such as $1S^2$, $1S^2\ 1P^6$, $1S^2\ 1P^6\ 1D^{10}\ 2S^2$, $1S^2\ 1P^6\ 1D^{10}\ 2S^2\ 1F^{14}\ 2P^6$,....

In the following, we first study clusters of simple metals whose stability can be well explained by the jellium model and see if magic clusters can be assembled to make a bulk material. Next, we explore a number of other electron-counting rules such as the octet rule for *sp* elements [8–10], 18-electron rule for transition metal elements [11], 32-electron rule for rare earth elements, Hückel's aromaticity rule for organic molecules [12, 13], and Wade-Mingos rule [14–17] for boron-based clusters and Zintl ions [18, 19]. We focus not only on neutral but also on charged clusters that can be stabilized by using any one of the above rules and combinations thereof.

2.2 Electron-Counting Rules

2.2.1 Jellium Rule

One of the early works using the jellium rule to study clusters as "giant atoms" was due to Saito and Ohnishi [20]. The authors studied if a closed shell Na cluster will interact weakly as noble gas atoms do and if an open shell Na cluster will be reactive. They showed that two Na_8 clusters interact weakly just as two noble gas atoms do, thus implying that Na_8 clusters with $1S^2\ 1P^6$ closed electronic shells are chemically inert. In a similar fashion, they showed that a Na_{19} cluster with electronic configuration of $1S^2\ 1P^6\ 1D^{10}\ 2S^1$ can be viewed as an alkali atom as both need one extra electron to close the s-shell. In Figure 2.2 we show the binding energy of two Na_{19} clusters as a function of distance computed by these authors. Note that there is an initial attraction leading to the formation of a Na_{19} dimer with the centers of the Na_{19} clusters 17 a. u. apart. As the distance between the two Na_{19} jellium clusters is further reduced, the clusters face a significant energy barrier and eventually

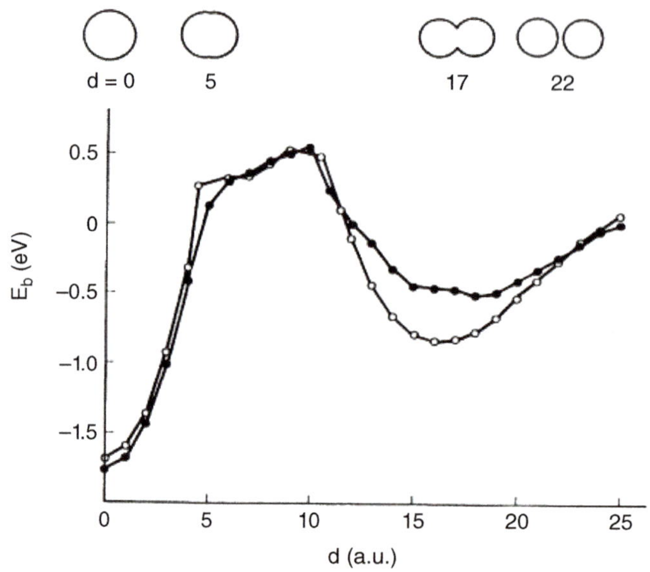

Figure 2.2 Binding-energy curves of $(Na_{19})_2$ for two different electronic configurations, ($N_\uparrow$, $N_\downarrow$ = 20, 18) (filled circles) and ($N_\uparrow$, $N_\downarrow$ = 19,19) (open circles). For several inter-jellium distances, schematic pictures for positive background are shown. *Source:* Saito and Ohnishi [20]. © American Physical Society.

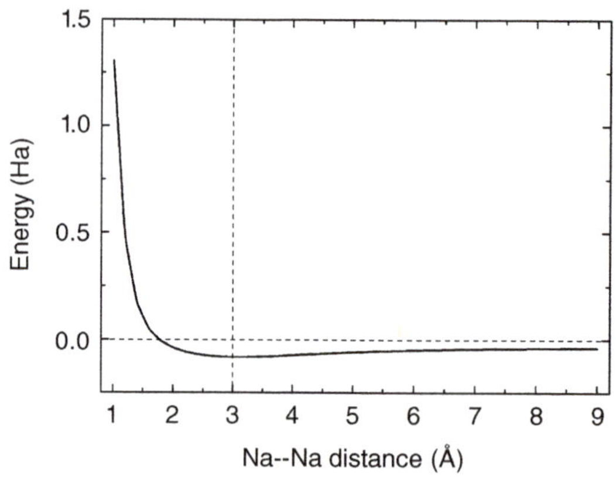

Figure 2.3 Binding energy as a function of distance between two Na atoms. The computed bond length (3.0 Å) of the Na_2 dimer agrees well with the experimental value of 3.08 Å.

coalesce to form a Na_{38} jellium cluster that is magnetic with two unpaired spins. But, does Na_{19} cluster mimic the chemistry of a Na atom? To understand this, we plot in Figure 2.3 the binding energy as function of distance between two Na atoms using the Gaussian 16 code [21] and the density functional theory with the Perdew, Burke, Ernzerhof (PBE) form for the generalized gradient approximation [22] and 6-31+G* basis function. As can be seen, the energy profile in Figure 2.3 is very different from that in Figure 2.2. Clearly, Na_{19} cannot be regarded as a superatom mimicking the chemistry of a Na atom.

To understand the effect of geometry of a cluster on its electronic structure, we focus on Na_{20}, which is a closed shell cluster in the jellium model. In Figure 2.4 we show the ground state geometry of Na_{20} calculated by Sun et al. [23]. Clearly, its geometry is not spherical. However, the molecular orbitals of Na_{20} (Figure 2.5) show strong resemblance with that in

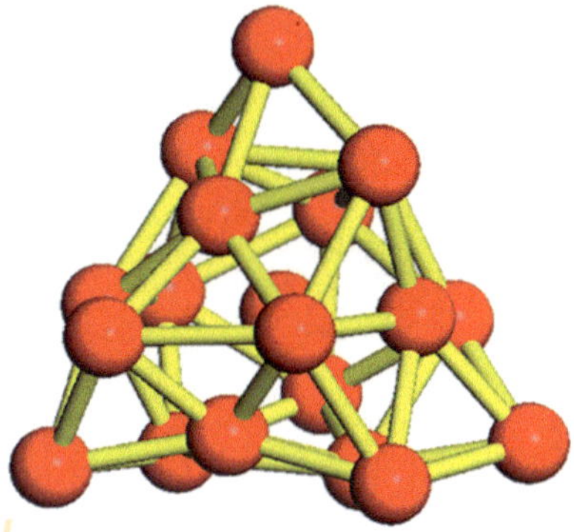

Figure 2.4 Ground-state geometry of Na_{20}. *Source:* Sun et al. [23]. © American Chemical Society.

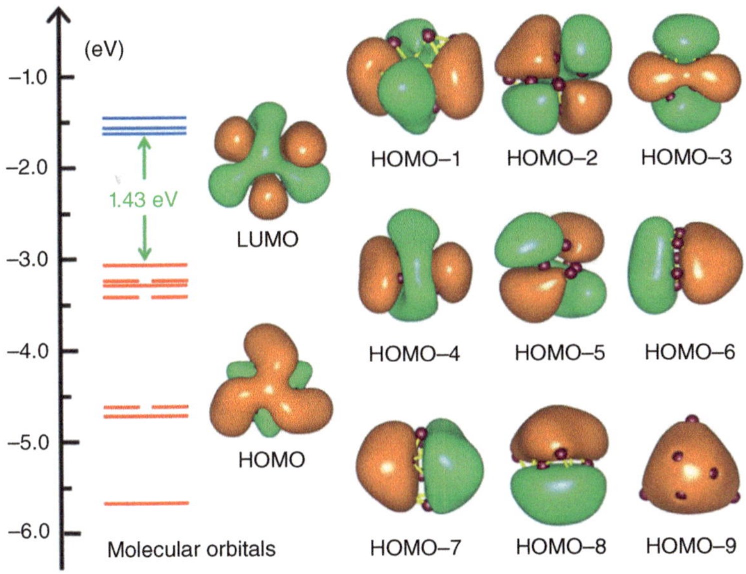

Figure 2.5 Molecular orbitals and energy levels of neutral Na_{20} cluster. The HOMO–LUMO energy gap is indicated (in green). *Source:* Sun et al. [23]. © American Chemical Society.

the jellium model. The nondegenerate highest occupied molecular orbital (HOMO) is primarily a 2S orbital and HOMO-q ($q = 1–5$) are d-type, $q = 6–8$ are p-type, and $q = 9$ is s-type, just as the case in the jellium model. In addition, a HOMO–lowest unoccupied molecular orbital (LUMO) gap of 1.43 eV is indicative of a chemically inert behavior of Na_{20} cluster.

To what extent can a jellium model describe the interaction between two *real* clusters was further investigated by Hakkinen and Manninen [24] by taking into account the geometries and electronic structure of clusters, explicitly. Using molecular dynamics and density functional theory, they considered a Na_8 cluster in a variety of surroundings. In the gas phase, Na_8 cluster was found to retain its geometry even up to 600 K. But, when two Na_8 clusters are brought together (see Figure 2.6), they collapse forming a deformed Na_{16} cluster and the electronic shell structure is destroyed. They further showed that Na_8 cluster forms an epitaxial layer (Figure 2.7) when supported on a Na (100) surface. This shows that Na_8 is a magic cluster only when it is held in isolation.

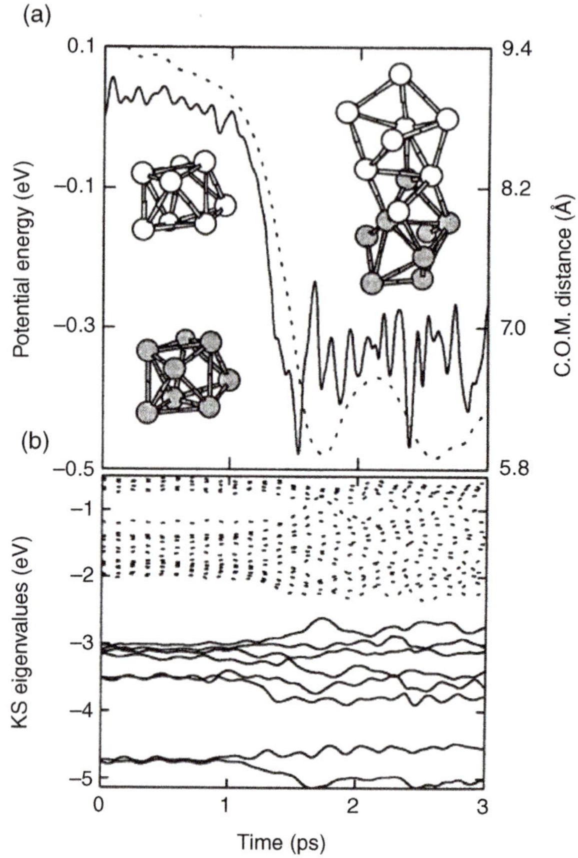

Figure 2.6 Reaction between two Na_8 clusters in vacuum. (a) Time evolution of the potential energy relative to its value in the initial configuration (solid curve, scale on the left) and the center-of-mass distance of the two clusters (dotted curve, scale on the right). The two snapshots indicate the initial configuration (left) and the configuration at 2.6 ps (right). (b) Time evolution of the Kohn-Sham eigen values. The dotted curves indicate empty states. *Source:* Hakkinen and Manninen [24]. © American Physical Society.

While Na_8 was found to see its geometry destroyed when interacting with another Na_8 cluster or when supported on a Na (100) surface, the result for Au_{20} is different. Note that according to the jellium model, Au_{20} is also a closed shell cluster. Although it has a pyramidal geometry (Figure 2.8), its molecular orbitals show the same pattern as in the jellium model [25]. As shown in Figure 2.9, Au_{20} maintains its pyramidal structure when deposited on a carbon substrate [26]. However, it is not clear if Au_{20} would continue to maintain its gas phase geometry when interacting with each other or when supported on an Au substrate? No studies have yet been done to make any conclusion. However, based on extensive studies of Al_{13}, another free electron meal cluster [27–36], it is unlikely that Au_{20} would maintain its virgin geometry in the above situation.

With 39 valence electrons and an electronic configuration of $1S^2\ 1P^6\ 1D^{10}\ 2S^2\ 1F^{14}\ 2P^5$, Al_{13} is known to mimic the chemistry of a halogen atom. Indeed, its electron affinity of

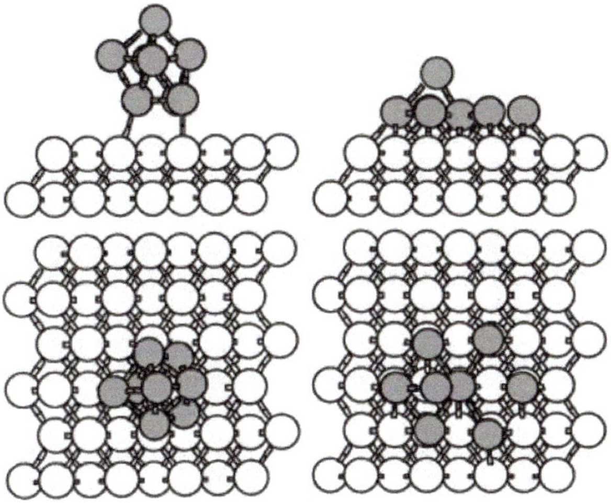

Figure 2.7 The initial (left) and the final (right) configuration of the collapse of Na_8 on Na (100). Note the epitaxial arrangement of the adatoms at the end of the run (at 2.8 ps). Both side and top views of the two configurations are shown. *Source:* Hakkinen and Manninen [24]. © American Physical Society.

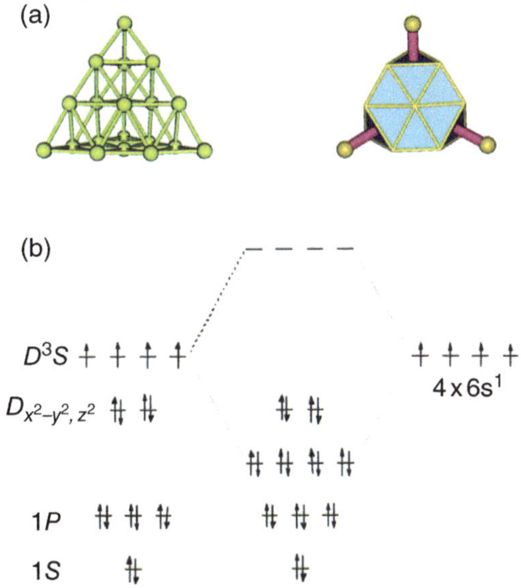

Figure 2.8 (a) Structure and super atomic-molecule models of Au_{20} (TAu4). (b) Schematic representation for the superatom–atom D^3S–s bonding of Au_{20} (TAu4). *Source:* Cheng et al. [25]. © Royal Society of Chemistry.

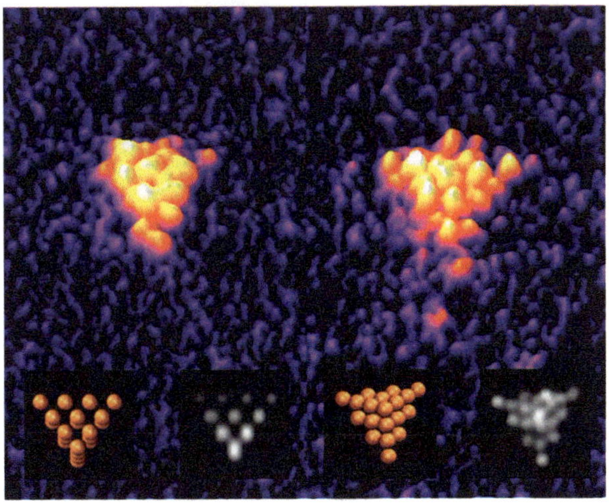

Figure 2.9 Direct atomic imaging and dynamical fluctuations of the tetrahedral Au20 cluster soft-landed on amorphous carbon substrate. *Source:* Adapted with permission from Wang et al. [26]. © Royal Society of Chemistry.

3.57 eV is almost identical to that of the Cl atom. It was theoretically predicted [30] and experimentally verified [31] that KAl_{13} is an ionically bonded cluster where an electron is transferred from K to Al_{13}. Evidence that Al_{13} behaves like a halogen also came from an experiment of Bergeron et al. who showed that $Al_{13}I_2^-$ can be viewed as $Al_{13}^-.2I$, making it look like a triiodide (I_3^-) ion [32]. Similarly, $Al_{14}I_3^-$ can be viewed as $Al_{14}^{2+}.3I^-$ with Al_{14} behaving like an alkaline earth element. From the above results, the authors concluded that Al_{13} and Al_{14} exhibit a new form of superatom chemistry in which superatoms behave like atoms when they react with other atoms/molecules. However, a different conclusion was reached by Han and Jung who examined whether Al*n* clusters exhibit multiple atomic characteristics depending upon *n* by studying halogenated Al*n* ($n = 11$–15) complexes and plotted the charge (Q) distribution in MX and MX_2 systems ($M = Al_{11}$–Al_{15}, X = F, Cl, Br, I) *vs* electronegativity, η of X [33, 34]. The results are presented in Figure 2.10. Noting that the

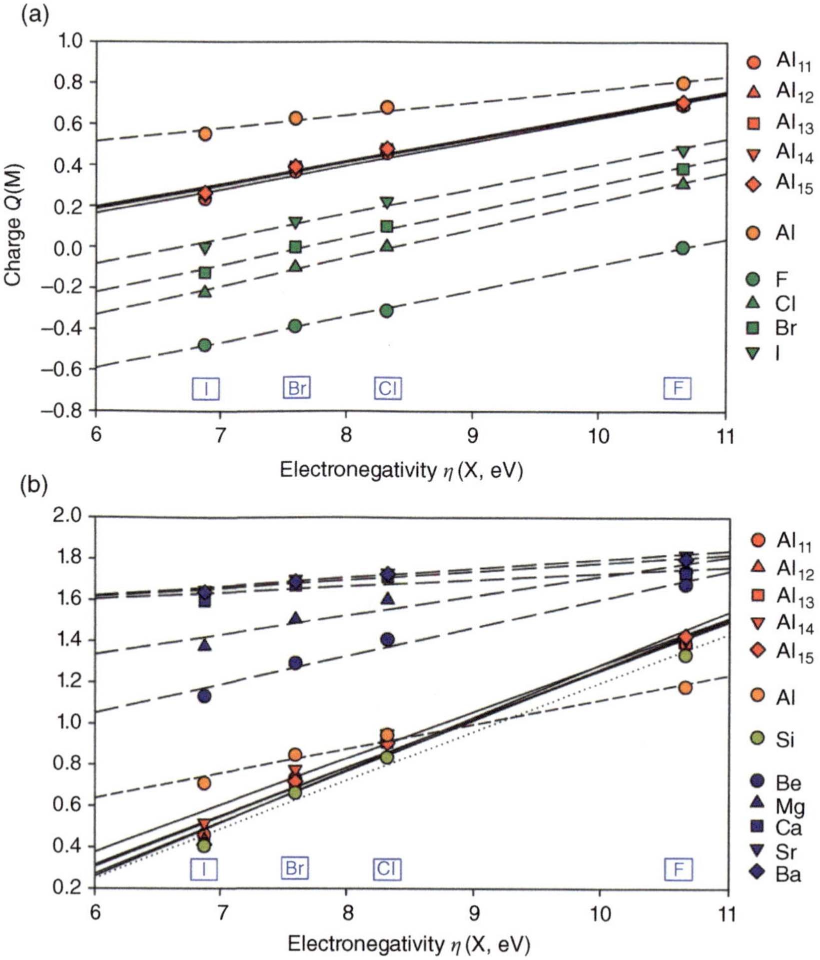

Figure 2.10 Q(M) versus η (X = F, Cl, Br, I) for (a) MX ($M = Al_{11}$–Al_{15}, Al, halogen atoms) and (b) MX_2 ($M = Al_{11}$–Al_{15}, Al, Si, alkaline earth atoms). The data of Al_{11} through Al_{15} basically coincide without revealing any exceptions for Al_{13} or Al_{14}. *Source:* Han and Jung [33]. © American Chemical Society.

charge transfer Q(M) is nearly independent of n in Aln clusters in both the systems, the authors concluded that "there is no evidence of an alkaline earth superatom in the Al_{14} clusters" and that "there are no theoretical grounds to regard $Al_{13}I_2^-$ as Al_{13}^-.2I."

For superatoms to be used as building blocks of cluster-assembled materials, it is important that not only they be stable and mimic the chemistry of atoms but also they should remain in their virgin form when forming a crystal. Liu et al. [35] studied the stability of a KAl_{13} crystal confined to the CsCl structure. The hypothesis was that Al_{13}^- clusters will stay apart from each other due to the negative charge they carry. On the contrary, they found that the Al atoms in neighboring Al_{13} clusters interact and KAl_{13} as a crystal was unstable. To determine whether changing the cation from a metal to a nonmetallic one would result in stabilizing the Al_{13}^- icosahedral geometry, Huang et al. [36] recently studied the stability of $[(CH_3)_4N^+][Al_{13}^-]$ crystal. Note that the ionization potential of $(CH_3)_4N^+$ is 3.27 eV, which is even smaller than that of a K atom, namely, 4.34 eV. In addition, the diameter of $(CH_3)_4N^+$ is 4.2 Å, which is comparable to the diameter of Al_{13}, namely 5.3 Å. The binding energy of $[(CH_3)_4N^+][Al_{13}^-]$ cluster, namely 2.68 eV, is also larger than that of KAl_{13} cluster, which is 2.49 eV. Expecting that a crystal of $[(CH_3)_4N^+][Al_{13}^-]$ may be stable with both the cation and the anion maintaining their individual geometry, Huang et al. confined the initial crystal structure to three forms – (i) body-centered-cubic, (ii) rock salt, and (iii) zinc-blende phases, which are common crystal structures of binary salts (e.g., CsCl, NaCl, and ZnO) (see Figure 2.11). After optimization, the results show that while the $(CH_3)_4N$ maintains its structure in all of the above systems, Al_{13} clusters coalesce, ceasing to remain as individual clusters. It is, thus, safe to conclude that stable metallic

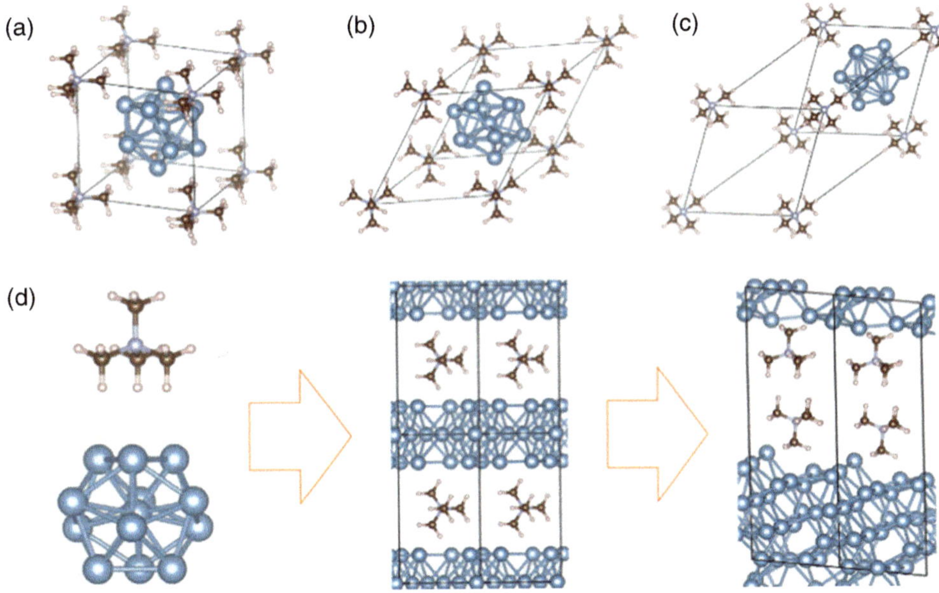

Figure 2.11 Initial crystal structure for (a) body-centered-cubic (bcc), (b) rock salt (rs), and (c) zinc-blende (zb) phases of $(CH_3)_4N^+Al_{13}^-$ bulk. (d) Optimized structures of $(CH_3)_4N^+Al_{13}^-$ show the coalescence of Al_{13} clusters when forming a bulk material. *Source:* Huang [36]. © American Chemical Society.

clusters with electronic closed shells are not good candidates for cluster-assembled materials.

On the contrary, jellium shell closure rule has been effectively used to explain the stability of ligated metal clusters, particularly ligated Cu, Ag, and Au clusters. As the number of atoms in the metal core as well as the number and type of ligands can be varied independently, one is gifted with considerable flexibility to design ligated metal clusters as superatoms. Consider, for example a metal core consisting of N_C number of core atoms and N_S number of surface atoms. The surface atoms are the ones that bind to the ligands forming either an ionic or a covalent bond. If the number of valence electrons in the core atoms correspond to shell closure, it is likely that such a cluster can gain unusual stability. Hakkinen and coworkers [37, 38] used this approach to explain the unusual stability of some ligated Al and Au clusters. For example, the unusual stability of $Al_{50}Cp^*_{12}$ can be explained by realizing that 12 Cp* ligands withdraw 12 electrons from the Al_{50} core, thus leaving $50 \times 3 - 12 = 138$ electrons. This is just enough electrons to close the 1I superatom jellium shell. Similarly, the authors showed that ligand-protected Au clusters, satisfying electron shell closure at 8, 34, and 58, can indeed be synthesized.

That stable ligated metal clusters consistent with the jellium shell closure rule can be synthesized has been verified by experiments. Some of these clusters include $Au_{25}(SR)_{18}^-$ [39] and As_7 and As_{11}, with cryptated alkali atoms [40, 41]. However, it has been pointed out that the strength of the ligand interaction and its effect on the geometry on the metal core play an important role in the electron counting [33, 42, 43]. Examples of some ligated clusters consistent with the jellium model are given in Table 2.1.

2.2.2 Octet Rule

The octet rule was developed in early 1900s to account for the chemistry of low atomic number (<20) elements [8–10]. The noble gas atoms, with their outer electron configuration of $ns^2\ np^6$, have closed electronic shells and hence, a large HOMO–LUMO gap, high ionization potential, and low electron affinity. Thus, they are very stable and chemically

Table 2.1 Examples of ligated clusters with effective number (n_e) of valence electrons for jellium shell closure.

n_e	Examples
2	$Ag_{14}(SR)_{12}(PR)_8$, $Ag_{16}(SR)_{14}(dppe)_4$, $\{Au_{34}[Fe(CO)_3]_6\ [Fe(CO)_4]_8\}^{8-}$
8	$Au_{11}X_3(PR_3)_7$, $[Au_{13}Cl_2(PR_3)_{10}]^{3+}$, $[Au_{25}(PET)_{18}]^-$, $Au_{28}(SR)_{20}$, $Al_4(C_5Me_5)_4$, $Al_4[SiC(CH_3)_3]_4$, $[Au_{13}Cu_2(PR)_6(SPy)_6]^+$, $[Au_{13}Cu_4(PR_2Py)_4(SR)_8]^+$, $[Au_{13}Cu_8(SPy)_{12}]^+$, $[Au_{11}(dpdp)_6]^{3+}$, $\{Ag_{21}[S_2P(O/Pr)_2]_{12}\}^+$
18	$[Ag_{44}(SR)_{30}]^{-4}$, $[Ag_{44}(SR)_{30}]^{-4}$, $[Au_{12}Ag_{32}(SR)_{30}]^{-4}$, $[Au_{12+n}Cu_{32}(SR)_{30+n}]^{-4}$ ($n = 0, 2, 4, 6$), $Au_{24}Ag_{20}(SR)_4(C_2Ph)_{20}Cl_2$
34	$[Au_{39}(PR_3)_{14}Cl_6]^-$, $Au_{68}(SR)_{34}$, $[Au_{67}(SR)_{35}]^{-2}$, $Au_{39}Cl_6(PH_3)_{14}$
40	$\{Ge_9[Si(SiMe_3)_3]_3\}^-$, $SiAl_{14}(C_5Me_5)_6$
58	$Au_{102}(p\text{-}MBA)_{44}$, $Au_{102}(SMe)_{44}$, $\{GaGa_{11}[GaN(SiMe_3)_2]_{11}\}$
138	$Al_{50}(C_5Me_5)_{12}$

inert. A variety of superatomic clusters containing elements with low atomic number can be designed such that constituent atoms satisfy the octet shell closure. We begin with the description of Group 1 alkali and Group 17 halogen atoms.

2.2.2.1 Superalkalis and Superhalogens

Alkali atoms, with outer electronic configuration of ns^1, have one excess electron while halogen atoms, with outer electronic configuration of $ns^2\ np^5$, need one extra electron to achieve the octet shell closure of their respective ionic cores. As a result, both the atoms are reactive. Gutsev and Boldyrev [44, 45] were the first ones to use the octet rule to design superalkali and superhalogen clusters that not only mimic the chemistry of alkali and halogen atoms, respectively, but also surpass their properties. The ionization potentials of superalkalis are lower than those of the alkali atoms while the electron affinities of superhalogens are higher than those of the halogen atoms. The composition of superalkali is $M_{k+1}X$, where M is an alkali atom and X is an atom with valence k. An example of a superalkali is Li_3O whose ionization potential of 3.54 eV [46] is smaller than that of the Li atom, namely, 5.39 eV. The composition of a superhalogen, on the other hand, is MY_{k+1}, where Y is a halogen atom and M is a metal atom with valence k. A typical example of a superhalogen is LiF_2, which has an electron affinity of 5.45 eV [47]. This is much larger than the electron affinity of F, namely 3.4 eV. The reason for these superior properties is inherent in the nature of the distribution of electrons. Note that the phase space occupied by the outer electrons increases with cluster size. In superalkalis, this makes it easier to remove an electron, hence leading to a lower ionization potential. In a superhalogen, on the other hand, the increased phase space for the electron distribution causes a reduction in electron–electron repulsion; hence, leading to a higher electron affinity. That superhalogens can promote unusual reactions was already realized by Bartlett in 1962, long before Gutsev and Boldyrev coined the word. Bartlett and coworkers showed that O_2 and noble gas atoms such as Xe can be ionized by using PtF_6 and estimated its electron affinity as 6.8 eV [48, 49]. The fact that clusters can mimic the chemistry of alkali and halogen atoms with superior properties provides new opportunities to design supersalts with superalkalis and superhalogens as building blocks [50].

Numerous studies of superalkalis and superhalogens have been carried out over the past 20 years, and Chapter 3 of this book covers the advancements in superhalogen design and synthesis. For the purpose of illustration, we show in Figure 2.12 the electron affinities of fluorinated coinage metal clusters [51]. Note that the electron affinities are higher than that of F once the number of F atoms is greater than 2 and reach a value as high as 8.6 eV in AuF_6.

The octet rule can also be applied to nonmetallic elements to design superalkalis and superhalogens. Consider, for example molecules such as BH_4, BO_2, CN, and NO_3. Each of these molecules needs one extra electron for electronic shell closure. Indeed, with electron affinities of 4.42, 4.32, 3.86, and 4.03 eV, respectively, these molecules are superhalogens. Following the guidelines of Gutsev and Boldyrev, one could imagine that a new class of clusters with electron affinities even higher than those of superhalogens can be designed. Consider, for example, a cluster with composition MZ_{k+1}, where Z is a superhalogen. This new class of clusters, named hyperhalogens, was discovered by Willis et al. [52] during the study of the interaction of Au with BO_2 molecules. The electron affinities of $Au(BO_2)_n$

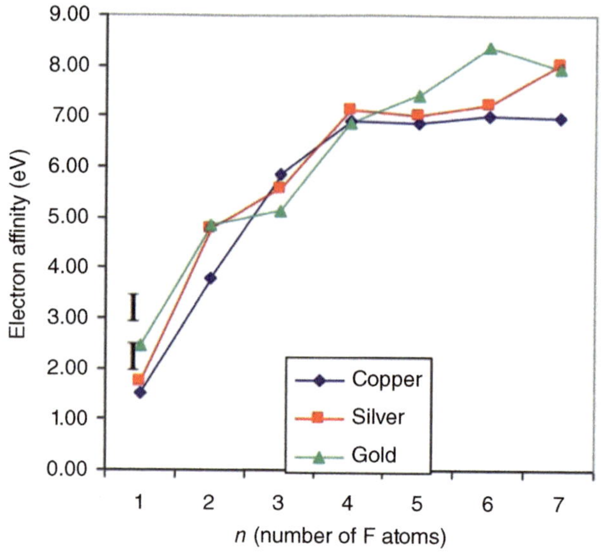

Figure 2.12 Electron affinity of coinage metal atoms decorated with F. *Source:* Koirala et al. [51]. © American Chemical Society.

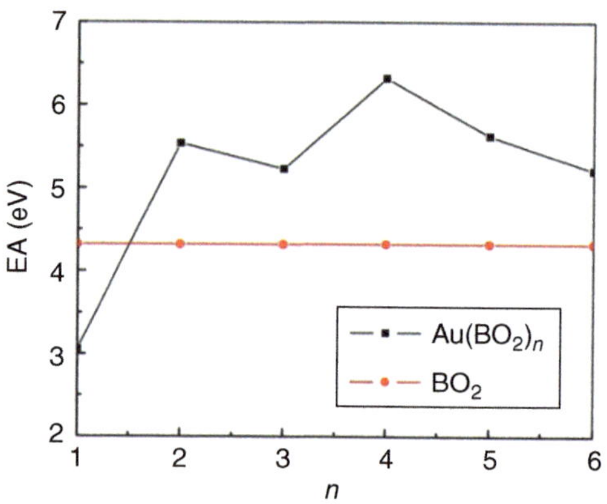

Figure 2.13 Electron affinity (EA) of $Au(BO_2)n$ as a function of n (black line). *Source:* Adapted with permission from Ref. [52]. © John Wiley & Sons.

($n = 1$–6) is compared with that of BO_2 in Figure 2.13. Note that for $n \geq 2$, the electron affinities are larger than that of BO_2. This provides a way to design species with ever increasing electron affinities. We will discuss later how the synthesis of such highly electronegative species can be used to promote unusual reactions.

Superhalogens and hyperhalogens can be used in the design of novel salts for applications in solar cells, batteries, and hydrogen storage materials. While we discuss these applications in detail in Chapter 10, in the following chapter we show how hyperhalogen concept led to the synthesis of a hypersalt, $KAl(BH_4)_4$. Note that $Al(BH_4)_3$ is a volatile pyrophoric liquid. Although it contains 16.8 wt % hydrogen, it cannot be used as a hydrogen storage material because of safety concerns. However, by adding one more BH_4 unit to $Al(BH_4)_3$, an $Al(BH_4)_4$ hyperhalogen can be formed. By combining it with a K cation, Knight et al. [53] synthesized $KAl(BH_4)_4$ hypersalt, which is solid and nonvolatile under ambient conditions (Figure 2.14).

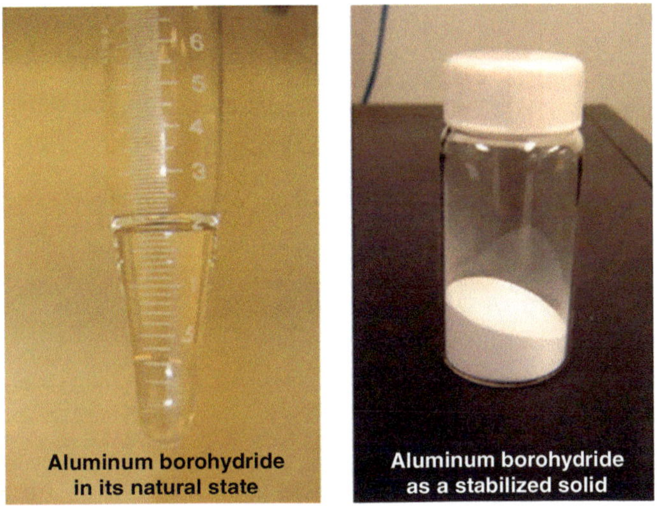

Figure 2.14 $Al(BH_4)_3$ (left panel) and $KAl(BH_4)_4$ (right panel). *Source:* Adapted with permission from Knight et al. [53]. © American Chemical Society (courtesy of D. Knight and R. Zidan, private communication).

2.2.2.2 Superchalcogens

Atoms in the Group 16 of the periodic table require two extra electrons to satisfy the octet rule. When isolated, these atoms cannot retain both the electrons due to electron–electron repulsion. However, an atomic cluster could be stable as a dianion if it is large enough to reduce electron–electron repulsion. The question is: how small a cluster has to be so that it can retain two extra electrons without fragmenting or ejecting the second electron spontaneously? Such a cluster could be viewed as a superchalcogen that is stable, yet mimics the chemistry of Group 16 elements. Chen et al. [54] studied this possibility by focusing on $M(CN)_4$ clusters where M is a divalent alkaline earth metal atom (Be, Mg, Ca, Zn, Cd), which contributes two electrons while each CN molecule would need one electron to satisfy the octet rule. The authors calculated the equilibrium geometries and total energies of neutral, monoanionic, and dianionic $M(CN)_4$ clusters using density functional theory. The results are presented in Figure 2.15. The energy gains in adding the first (second) electron to $M(CN)_4$ clusters are 3.13, 2.94, 2.89, 2.78, and 2.59 eV (0.32, 0.97, 1.21, 0.83, and 0.56 eV), respectively, for M = Be, Mg, Ca, Zn, and Cd. The stability of the $M(CN)_4^{2-}$ indicates that the octet rule can be effectively used to rationally design doubly charged species that are stable in the gas phase.

It is interesting to compare the relative robustness of clusters obeying the jellium and octet shell closure rules in the rational design of cluster-assembled materials. It was discussed earlier that Al_{13}^- cluster, in spite of its being stable and chemically "inert," coalesces when crystals of KAl_{13} and $[(CH_3)_4N^+][Al_{13}^-]$ are formed. Note that $(CH_3)_4N^+$, obeying the octet rule, retains its geometry [36]. Huang et al. [36] studied the stability of crystals composed of $(CH_3)_4N^+$ cation and $B(CN)_4^-$ anion. Note that both molecules satisfy the octet rule. The authors found $[(CH_3)_4N^+][B(CN)_4^-]$ to be a stable charge-transfer transparent salt with a band gap of 6.5 eV and having a diverse range of structural phases. In Figure 2.16 we show the geometry of the isolated $[(CH_3)_4N^+]$ $[B(CN)_4^-]$ cluster and the optimized geometry of the $[(CH_3)_4N^+][Al_{13}^-]$ crystal having

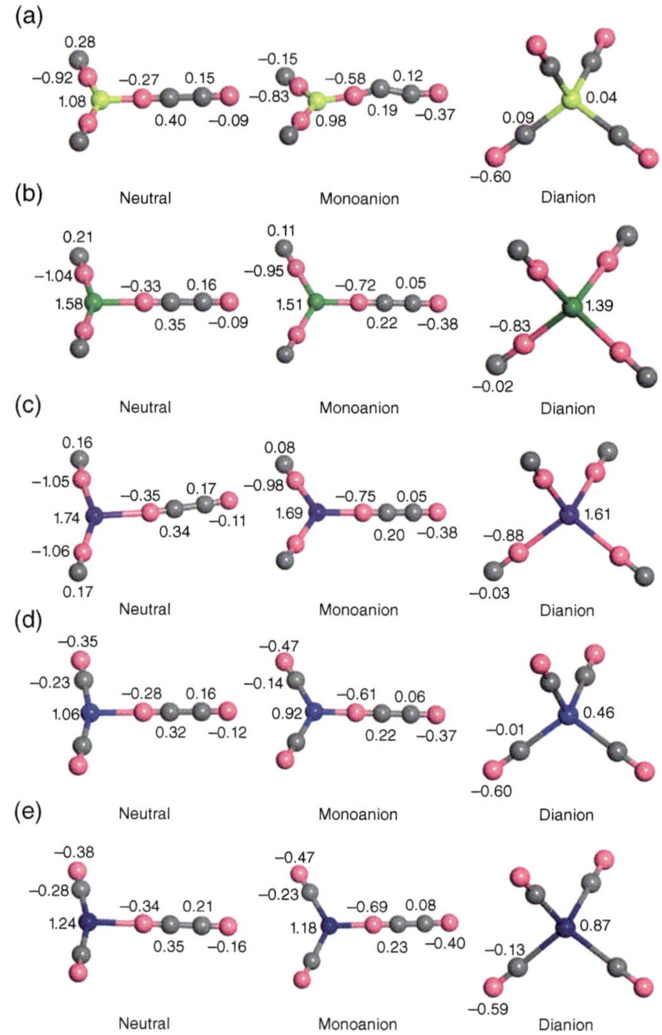

Figure 2.15 (a)–(e) are the globally optimized geometries for $M(CN)_4{}^{0,1-,2-}$ (M = Be, Mg, Ca, Zn, and Cd) clusters, respectively. Gray, pink, light green, dark green, purple, light blue, and dark blue spheres stand for C, N, Be, Mg, Ca, Zn, and Cd atoms, respectively. The charges on selected atoms are also given. *Source:* Chen et al. [54]. © American Chemical Society.

body-centered cubic structure and T_d symmetry. The band gap of the crystal phase is close to that of the HOMO–LUMO gap of the isolated cluster, implying that the electronic structure of the crystal is guided by the properties of the individual cluster building blocks.

In spite of the success of the octet rule accounting for the stability of clusters composed of light elements, like the jellium model, it has limitations; stable clusters exist even though they do not satisfy the octet rule. These include, for example, NO (which has an odd number of valence electrons), BH_3 and BF_3 (which are electron deficient), and PCl_5, SF_4, and SF_6 (which are electron rich).

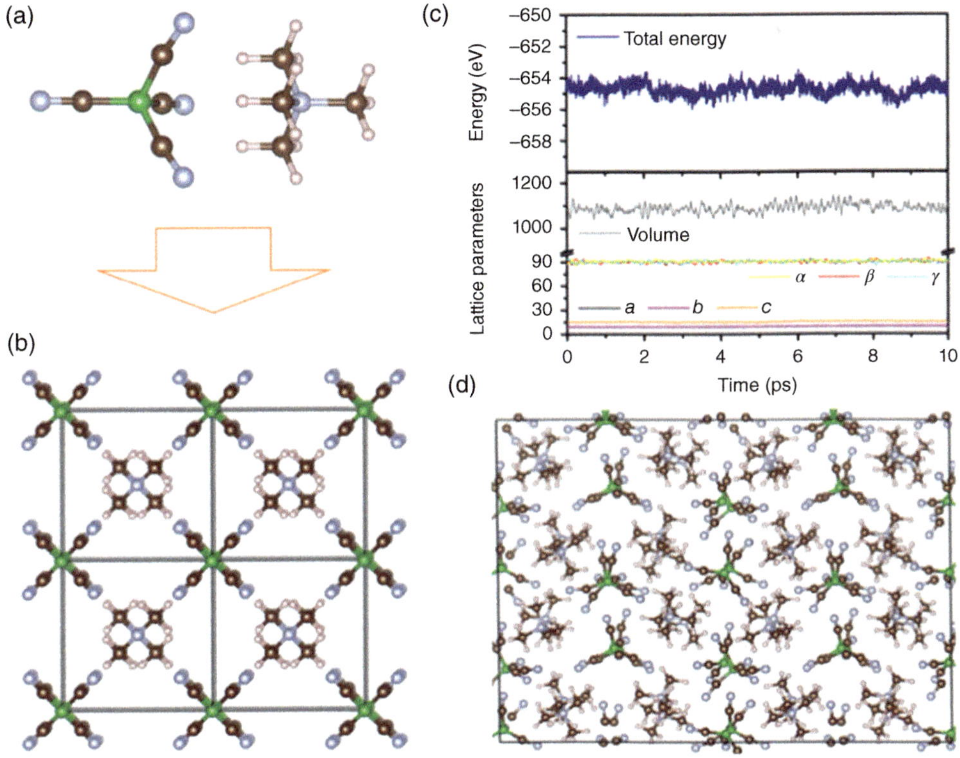

Figure 2.16 (a) Isolated $(CH_3)_4N^+Al_{13}^-$ cluster. (b) Optimized body-centered-cubic phase of $(CH_3)_4N^+Al_{13}^-$ crystal with T_d symmetry (same as the molecular point-group of the respective ions). (c) Total energy and lattice parameter (lattice constants a, b, and c in Å, interaxial angles α, β, and γ in degrees, and volume in Å^3) fluctuations during the ab initio molecular dynamics simulation. (d) Snapshot of the structure obtained from the ab initio molecular dynamics (AIMD) simulation for 10 ps after 4 ps to allow the system to reach thermal equilibrium. *Source:* Huang et al. [36]. © American Chemical Society.

2.2.3 18-Electron Rule

Stability of transition metal compounds can be accounted for by using the 18-electron rule where 18 electrons are needed to fill $ns^2\ np^6$ and $(n-1)d^{10}$ orbitals [11]. Classic examples of complexes stabilized by the 18-electron rule are chromium bisbenzene [$Cr(C_6H_6)_2$] and ferrocene [$Fe(C_5H_5)_2$]. C_6H_6 and C_5H_5 have six and five π electrons each. As Cr and Fe have outer electronic configuration of $3d^5\ 4s^1$ and $3d^6\ 4s^2$, respectively, one can see that both $Cr(C_6H_6)_2$ and $Fe(C_5H_5)_2$ are 18-electron systems.

Pykko and Runenberg [55] showed that an all-metal cluster, $Au_{12}W$, with a HOMO–LUMO gap of 3.0 eV is very stable due to the 18-electron rule. Here, 12 Au atoms contribute 12 electrons while W atom ($3d^5\ 4s^1$) contributes 6 electrons. This prediction was later verified in photoelectron spectroscopy experiment by Wang and collaborators [56]. A further proof of the 18-electron rule can also be seen by measuring the electron affinity of Ta@Au_{12}. Note that with 17-valence electrons, Ta@Au_{12} needs one extra electron to satisfy the 18-electron shell closure rule. Indeed, the measured electron affinity of 3.76 eV makes Ta@

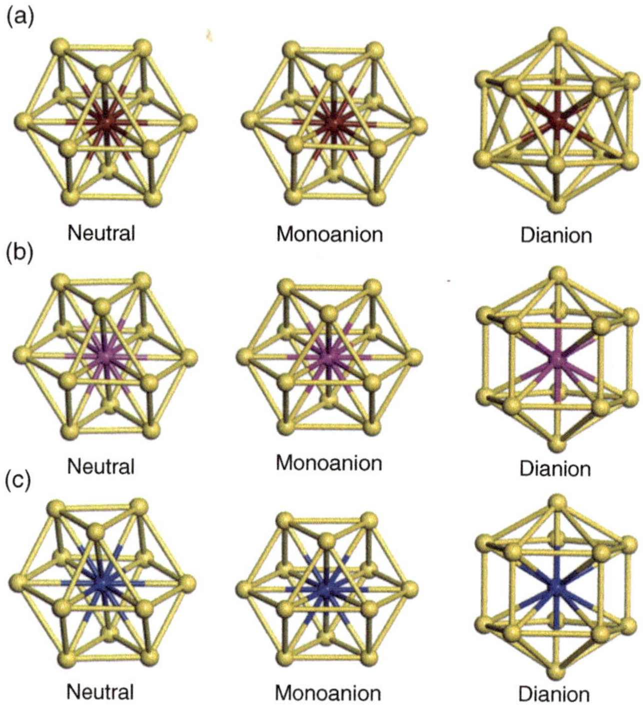

Figure 2.17 (a)–(c) are the optimized geometries for $MAu_{12}{}^{0,1-,2-}$ (M = Ti, Zr, and Hf) clusters, respectively. Yellow, dark red, purple, and blue spheres stand for Au, Ti, Zr, and Hf atoms, respectively. *Source:* Chen et al. [54]. © American Chemical Society.

Au_{12} an all-metal superhalogen [57]. Chen et al. calculated the structure and stability of $M@Au_{12}{}^{2-}$ (M = Ti, Zr, Hf) to see if these clusters can be stable and thus can be regarded as superchalcogens. The results are given in Figure 2.17. Note that all these structures are dynamically stable. However, $Ti@Au_{12}{}^{2-}$ is unstable against an electron loss by 0.23 eV while $M@Au_{12}{}^{2-}$ (M = Zr, Hf) dianions are stable against the second electron loss by 0.05 eV, due to their increased size.

2.2.4 32-Electron Rule

The 32-electron rule applies to clusters containing early 5f elements where complete shell closure of $s^2\ p^6\ d^{10}\ f^{14}$ orbitals give them stability and chemical inertness. The discovery of empty icosahedral Zintl ions such as $Pb_{12}{}^{2-}$ and $Sn_{12}{}^{2-}$ [58, 59] motivated Dognon et al. [60] to study the stability of endohedral clusters $Pu@Pb_{12}$. With an electronic configuration of [Rn] $5f^6\ 7s^2$ for Pu and [Xe] $4f^{14}\ 5d^{10}\ 6s^2\ 6p^2$ for Pb, $Pu@Pb_{12}$ constitutes a 32-electron system. The calculated HOMO–LUMO gap of 1.93 eV and the binding energy of 22.17 eV measured against the ionic dissociation limit confirm that the stability of $Pu@Pb_{12}$ arises due to the 32-electron shell closure. In Figure 2.18 we compare the orbital energies of $Pu@Pb_{12}$ with that of $Pb_{12}{}^{2-}$. One can see the strong participation of the central atom orbitals in bonding.

A few years later, Ghanty and coworkers [61] showed that the stability of Pu embedded in a C_{24} cage also follows the 32-electron rule, with 8 electrons contributed by Pu and 24 π electrons contributed by the C_{24} fullerene. The authors found that the C_2 symmetry of the

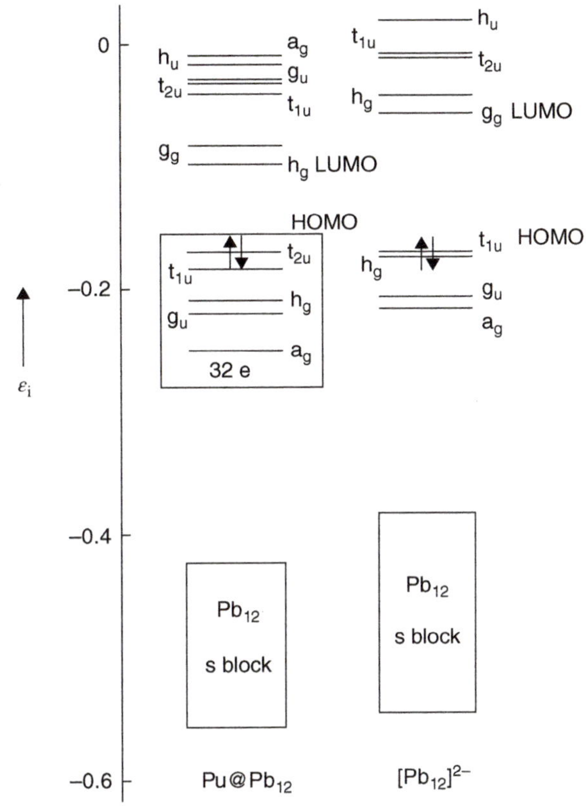

Figure 2.18 Orbital energies of $Pu@Pb_{12}$ and Pb_{12}^{2-}. The latter have been shifted to make the HOMOs equal. *Source:* Dognon et al. [60]. © John Wiley & Sons.

empty C_{24} fullerene transforms to D_{6d} symmetry, once encapsulated with the Pu atom. The HOMO–LUMO gap of 1.83 eV of the bare C_{24} cage changes to 3.26 eV following Pu encapsulation. The binding energy of the $Pu@C_{24}$ clusters measured with respect to atomic fragments is 6.77 eV. Other 32-electron systems studied recently include $An@C_{28}$ [62], $Pu@Sn_{12}$ [63], $(U@Si_{20})^{6-}$ [64], and actinide-encapsulated fullerene systems [65], $U@C_{28}$ [66], $Ln(CO)_8^-$ (Ln = Tm, Yb, Lu) [67], and superatomic CBe_8H_{12} cluster [68].

2.2.5 Aromaticity Rule

Aromaticity rule was developed by Huckel [12, 13, 69] to account for the stability of planar conjugated molecules such as C_6H_6. Based on a molecular orbital theory, it was shown that planar conjugated monocyclic polyene that has $(4n + 2)$ $(n = 0, 1, 2, \ldots)$ π or nonbonding electrons will be aromatic and stable. For benzene, $n = 1$. Because of the high stability, the aromatic molecules have low electron affinities, which seldom exceed 1.17 eV. Indeed, the electron affinity of C_6H_6 is −1.29 eV [70]. Jena and coworkers [71, 72] showed that the aromaticity rule can be used to design superatoms with electron affinities that can even exceed the electron affinity of Cl. This is accomplished by replacing H atoms in C_6H_6 by more electronegative atoms as well as by changing composition of the hexagonal core. In Figures 2.19 and 2.20, we show the globally optimized geometries of $C_6H_{6-x}F_x$ and

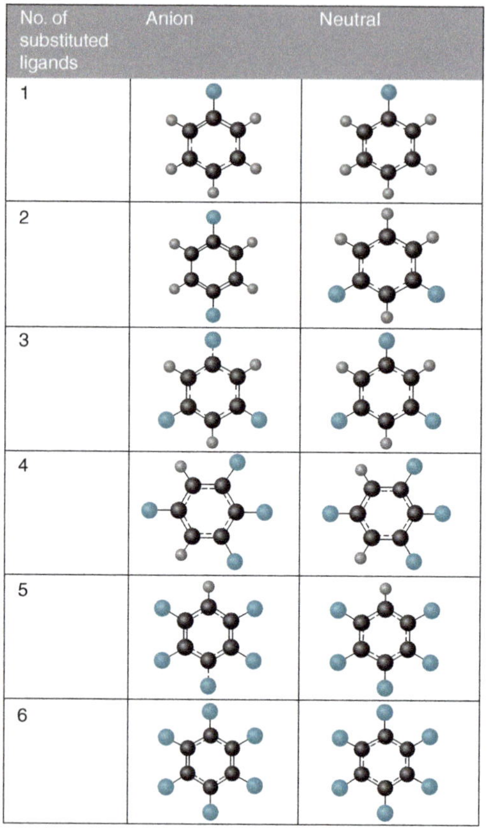

Figure 2.19 Ground state geometries of neutral and anionic $C_6H_{6-x}F_x$. The gray, white, and blue spheres correspond to carbon, hydrogen, and fluorine, respectively. *Source:* Driver and Jena [73]. © John Wiley & Sons.

$BC_5H_{6-x}F_x$ (x = 1–6) computed by Driver and Jena [73] using density functional theory. The corresponding electron affinities are given in Table 2.2. Note that the electron affinities of $C_6H_{6-x}F_x$ and $BC_5H_{6-x}F_x$ steadily increase with x. The computed electron affinity, 0.75 eV of C_6F_6 agree well with the experimental value of 0.86 ± 0.03 eV [74, 75]. Realizing that the electron affinity of C_5H_5, which lacks one electron to be aromatic, is +1.80 eV, Jena and coworkers replaced one of the C atoms in C_6H_6 by a B atom. The resulting BC_5H_6 lacks one electron to be aromatic, and its electronic affinity of 2.31 eV is indeed high. Further replacement of H atoms by F atoms allows the electron affinity to rise systematically, reaching a value of 3.24 eV for BC_5F_6. Thus, an aromatic molecule can mimic the chemistry of a halogen atom by suitable tailoring of its core and/or the ligand atoms.

The aromaticity rule has been extended to inorganic systems, providing further opportunities to design a new class of superatoms. One of these is an all-metal cluster, MAl_4^- (M = Li, Na, Cu). Using experiment and ab initio calculations, Li et al. [76] compared the measured photoelectron spectra with theory. The computed geometries of the MAl_4^- clusters (see Figure 2.21) are found to be pyramidal with the metal atom M at the apex carrying +1 charge and interacting with a square Al_4^{2-} moiety. Note that the geometry of an isolated Al_4^{2-} is also a square, although it is unstable against spontaneous emission of the second

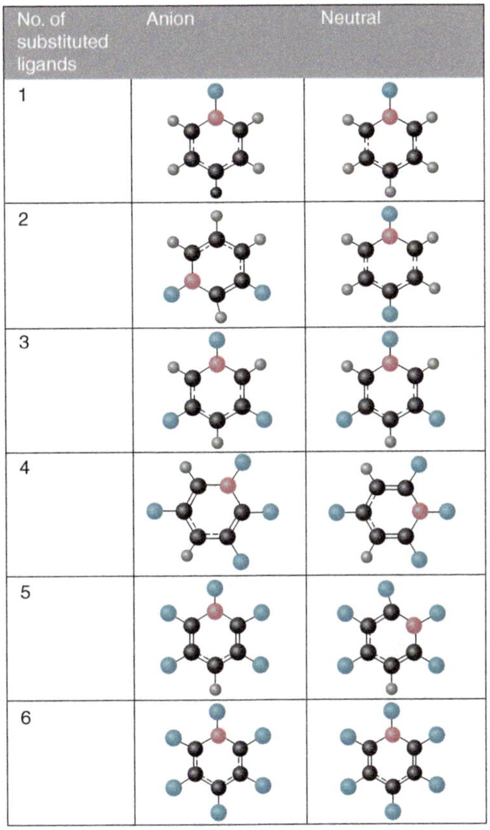

Figure 2.20 Ground state geometries of neutral and anionic $C_5BH_{6-x}F_x$. The gray, white, pink, and blue spheres correspond to carbon, hydrogen, boron, and fluorine, respectively. *Source:* Driver and Jena [73]. © John Wiley & Sons.

Table 2.2 Electron affinities (in eV) of ground state molecules.

x	$C_6H_{6-x}F_x$	$BC_5H_{6-x}F_x$	$C_6H_{6-x}(BO_2)_x$	$BC_5H_{6-x}(BO_2)_x$	$C_6H_{6-x}(CN)_x$
0	−1.29	2.31	−1.15	2.31	−1.29
1	−0.93	2.28	−0.75	2.78	0.09
2	−0.62	2.56	−0.38	2.72	1.06
3	−0.42	2.93	0.79	3.24	1.70
4	0.01	2.91	1.22	3.12	2.44
5	0.45	3.15	1.67	3.93	3.11
6	0.75	3.24	1.80	3.65	3.49

electron. The HOMO of MAl_4^-, as well as that of Al_4^{2-}, was found to be doubly occupied and composed of a delocalized π orbital, with other molecular orbitals being either σ-type or lone pairs. The presence of two valence electrons is characteristic of Huckel aromaticity, with $n = 0$.

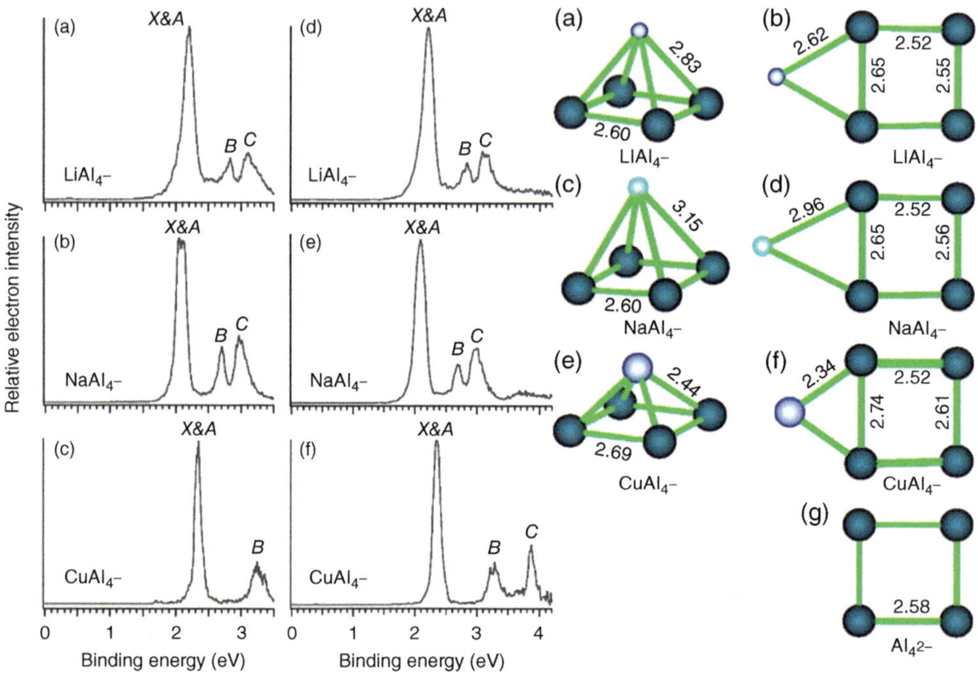

Figure 2.21 Left panel: photoelectron spectra at 355 nm (3.496 eV) for (a) $LiAl_4^-$, (b) $NaAl_4^-$, and (c) $CuAl_4^-$ and at 266 nm (4.661 eV) for (d) $LiAl_4^-$, (e) $NaAl_4^-$, (f) $CuAl_4^-$, and (g) square planar Al_4^{2-} cluster. Right panel: optimized structures of $LiAl_4^-$, $NaAl_4^-$, Al_4^{2-} [at the CCSD(T)/6-311+G* level of theory], and $CuAl_4^-$. *Source:* Li et al. [76]. © American Association for the Advancement of Science.

2.2.6 Wade-Mingos Rule

Polyhedral skeletal electron pair theory (PSEPT), commonly known as the Wade-Mingos rule, was developed by Wade [14, 15] and Mingos [16, 17] to account for the structure and bonding of polyhedral borane clusters (B_nH_n). Because of electron deficiency, B cannot form conventional covalent bonds between adjacent pairs of atoms. Instead, it forms multicenter bonds with the valence electrons separated into external and skeletal electrons. The former forms traditional covalent bonds with ligands while the latter contributes to cage bonding. Consider a borane polyhedron, B_nH_n with n being the number of vertices. According to the Wade-Mingos rule, $2n+2$, $2n+4$, and $2n+8$ electrons are needed to form *closo*, *nido*, and *archano* boranes, respectively. $B_{12}H_{12}^{2-}$ is a well-known example of a *closo*-borane having an icosahedral geometry (Figure 2.22). Here, $n = 12$. Of the three valence electrons of B, one is involved in the covalent bonding with H while the other two are contributed to the cage bonding. Because $2n+2 = 2\times12+2 = 26$ electrons are needed to stabilize the cage, $B_{12}H_{12}^{2-}$ is stable as a dianion. Indeed, the electron affinities of the first and second electron are 4.56 and 0.9 eV, respectively.

If one of the B atoms in B_nH_n is replaced by a C atom, according to the Wade-Mingos rule only one extra electron will be needed to satisfy the stability criterion. Hence, $CB_{n-1}H_n^-$ should be a stable cluster. Indeed, this is the case and mono-carboranes with $n = 7$–12 have been isolated and structurally characterized [77–80]. Pathak et al. [81] hypothesized that $CB_{11}H_{12}$ should also behave like a halogen atom. Using density functional theory, the authors

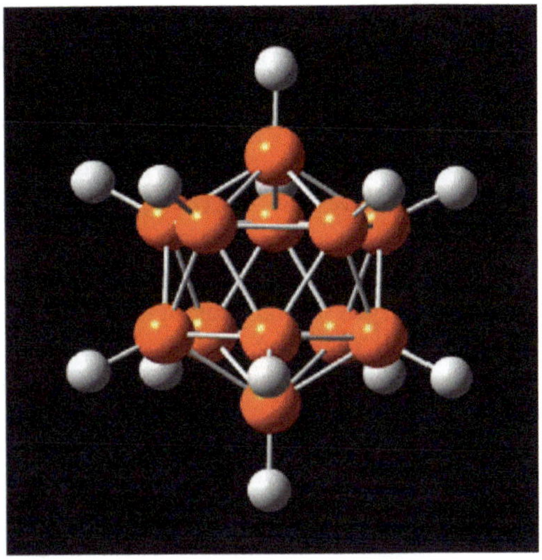

Figure 2.22 Geometry of $B_{12}H_{12}^{2-}$. *Source:* Adapted with permission from Jena and Sun [1]. © American Chemical Society.

calculated the structure and found the electron affinity of $CB_{11}H_{12}$ to be 5.39 eV, establishing it as a superhalogen. They also showed that $MB_{12}H_{12}$ (M = Li, Na, K, Rb, and Cs) clusters are superhalogens as they only need one electron to satisfy the Wade-Mingos rule. In addition, they showed that $CB_{11}H_{12}$ can be used as the building blocks of hyperhalogens.

Al and B belong to the same column in the periodic table, yet they have different chemistry. While Al is a metal, B is not. In addition, Al is not electron deficient like B and its valence electrons are nearly free in the bulk phase. Al does not form the same range of complexes with H as B does. Consequently, Wade-Mingos rule was known not to apply to Al. Bowen and coworkers [106] studied the possibility that small clusters of Al may have properties similar to that of B. Using the pulsed arc cluster ion source (PACIS), the authors reacted Al clusters with hydrogen and observed a vast range of previously unknown Al_nH_m clusters. In Figure 2.23 we show the mass spectra of Al_4H_m clusters, with Al_4H_6 being particularly stable. Photoelectron spectroscopy experiment revealed a large

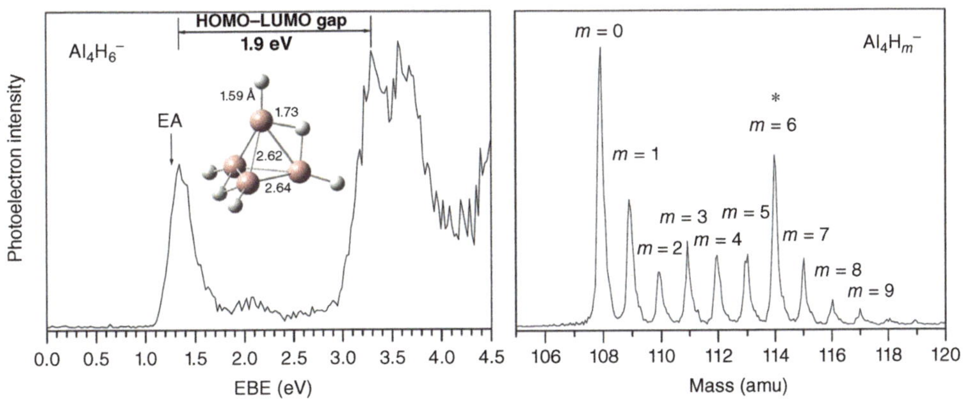

Figure 2.23 PES spectrum of Al_4H_6 anion (left panel) and mass spectra of Al_4Hm^-. The inset shows the geometry of $Al_4H_6^-$. *Source:* X. Li et al. [106]. © American Chemical Society.

HOMO–LUMO gap of 1.9 eV, which further confirmed the unusual stability of Al_4H_6. To understand the origin of this stability, the authors computed the geometry of Al_4H_6, which consists of an Al_4 tetrahedron with four radially bonded and two bridge-bonded H atoms. An examination of the electronic structure revealed that two bridge-bonded H atoms and the four Al-H pairs contribute 10 electrons. Thus, Al_4H_6 can be viewed as $Al_4H_4^{2-}$, which satisfies the Wade-Mingos $2(n+1)$ rule with $n = 4$. Note that $n = 6$ is the smallest value of n for which B_nH_n clusters satisfy the Wade-Mingos rule.

Wade-Mingos rule can also be used to rationalize the bonding in Zintl ions, which are multiply charged clusters belonging to the post-transition metal elements in groups 13, 14, and 15. Consider, for example, Sn_{12}^{2-}. While studying the semiconductor to metal transition, Wang and coworkers [59] observed a very simple photoelectron spectrum (PES) of Sn_{12}^-, which was different from that of its isoelectronic cousin, Ge_{12}^-. The optimized geometry of Sn_{12}^- was found to be a distorted cage with C_{5v} symmetry. Adding an extra electron resulted in closed shell Sn_{12}^{2-} cluster with icosahedral (I_h) symmetry (Figure 2.24). The PES of Sn_{12}^{2-} was found to be similar to that KSn_{12}^-, which can be viewed as $K^+Sn_{12}^{2-}$. Named stannaspherene, Sn_{12}^{2-} has a diameter of 6.1 Å, which is large enough to accommodate a metal atom. Indeed, four years prior to the above experiment Kumar and Kawazoe [82] had predicted M@Sn_{12} (M = Zn, Cd) to be a perfect endohedral cluster with I_h symmetry and a large HOMO–LUMO gap. Cui et al. [83] have subsequently synthesized a series of endohedral clusters with the composition M@Sn_{12} (M = Ti, V, Cr, Fe, Co, Ni, Cu, Y, Nb, Gd, Hf, Ta, Pt, Au). Kandalam et al. [84] showed that Mn@Sn_{12} is a stable cluster with a magnetic moment of 5 μ_B and can be regarded as a magnetic superatom. The bonding pattern of Sn_{12}^{2-} was found to be similar to that of $B_{12}H_{12}^{2-}$ where the 12 $5s^2$ localized electron pairs in Sn_{12} are equivalent to replacing 12 B-H bonds [59].

Pb_{12}^{2-}, which is isoelectronic with Sn_{12}^{2-}, was also found [58] to be a cage cluster whose stability can similarly be justified in terms of the Wade-Mingos rule. Named "plumbaspherene," Pb_{12}^{2-} cage has a diameter of 6.29 Å and can accommodate a metal atom inside it. This is consistent with an earlier experiment where an extremely stable $AlPb_{12}^+$ cluster was synthesized in the gas phase. $AlPb_{12}^+$ can be viewed as Al_3^+: Pb_{12}^{2-} just as KSn_{12}^- was viewed as K^+: Sn_{12}^{2-}. Note that a series of endohedral cage compounds, M@Pb_{12}^{2-} (M = Ni, Pd, Pt), stabilized by K^+ counterions, have been synthesized in solution and in crystalline form. Both of their icosahedral symmetries have been confirmed [85, 86] by X-ray diffraction and nuclear magnetic resonance (NMR) experiments.

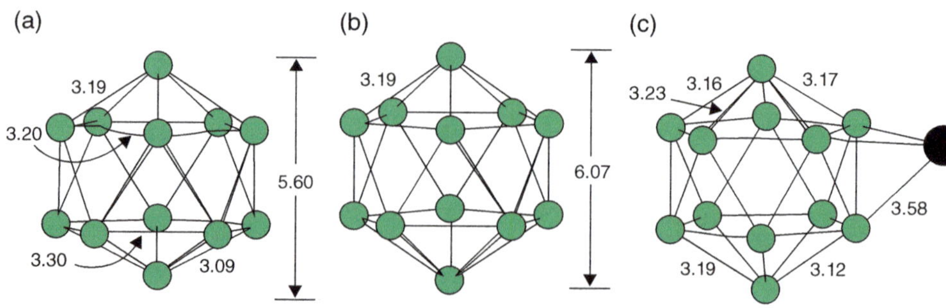

Figure 2.24 Optimized structure of (a) Sn_{12}^-, (b) Sn_{12}^{2-}, and (c) KSn_{12}^-. The bond distances and cage diameters are in angstroms. *Source:* Cui et al. [59]. © American Chemical Society.

Zintl compounds composed of alkali and alkaline earth cations and Zintl anions such as Sn_5^{2-}, Pb_5^{2-}, Pb_9^{4-}, Sb_7^{3-}, and Bi_4^{2-} have been known since 1930s [19, 87]. The entry of Zintl ions into cluster science dates back to the work of Recknagel [88], Duncan [89], Castleman [90], and their coworkers. For example, the unusual stability of $[Na_4Bi_3]^+$ in the mass spectra of $Na_nBi_m^+$ clusters was attributed to the presence of a Bi_3^{3-} Zintl ion. However, the stability of Zintl ions cannot be always understood in terms of the Wade-Mingos rule. The exceptions to this rule include $Ni_2Sn_7Bi_5^{3-}$ [91], $[Ge_9(\eta^4\text{-}Ni(CO))]^{3-}$ [92], and $[Ge_9(\eta^4\text{-}Pd(PPh3))]^{3-}$ [93].

2.3 Stabilizing Negative Ions Using Multiple Electron-Counting Rules

In the preceding section, we discussed how stable clusters can be designed by satisfying any one of the electron-counting rules such as the jellium rule for free-electron systems; the octet rule for low atomic number elements; the 18- and 32-electron rule for clusters containing transition and rare earth metal atoms, respectively; the aromatic rule for organic molecules; and the Wade-Mingos rule for boron-based clusters and Zintl ions. These rules can also be used to design reactive clusters by requiring that they either contain less or more electrons than needed for shell closure. In particular, we discussed the design of superalkalis and superhalogens. Over the past five years, Jena and his group [94, 95] have been studying systematically how unusually stable or reactive clusters can be designed by using *multiple* electron-counting rules simultaneously and how such clusters can be used to promote reactions and properties otherwise unthinkable. Consider, for example, multiply charged negative ions. We know that most atoms in the periodic table have positive electron affinity, i.e., they gain energy as an extra electron is attached. However, an atom with two extra electrons will spontaneously eject the second electron due to Coulomb repulsion. On the other hand, a cluster or a molecule in the dianionic form can be stable if they satisfy one of two criteria; either they are large enough so that the electron–electron repulsion is smaller than the binding energy or addition of electrons satisfies electronic shell closure. As we have pointed out before, $B_{12}H_{12}^{2-}$ is a classic example of a stable dianion in the gas phase. Its stability stems from the fact that the second electron satisfies the Wade-Mingos rule. A lot of work has been done to understand the stability of multiply charged ions. While there are examples of stable dianions, clusters carrying three or more extra electrons are rare. In the following we discuss how the use of multiple electron-counting rules can be used to design clusters with large electron affinities as well as clusters capable of carrying three or more extra electrons.

2.3.1 Monoanions

In previous sections we have discussed the design of stable monoanions by using a single electron-counting rule. Here, we discuss how the stability of monoanions can be further enhanced by using multiple electron-counting rules. Recall that C_6H_6 has a negative electron affinity, which can be systematically increased by replacing H by F (see Table 2.2). Driver and Jena [73] have systematically replaced H atoms with a BO_2 moiety. Note that

BO_2^- is isoelectronic with CO_2 and is a very stable molecule. Indeed, it is a superhalogen with an electron affinity of 4.46 ± 0.03 eV [96]. In Figure 2.25 we show the equilibrium geometries of neutral and anionic $C_6H_{6-x}(BO_2)_x$, $x = 1$–6. Note that in most situations BO_2 molecules dimerize and bind to the carbon atoms. From Table 2.2 we see that the electron affinity increases systemically as H is replaced by F and BO_2 and reaches a value of 1.80 eV in $C_6(BO_2)_6$. Note that $C_6(BO_2)_6$ satisfies both the aromatic rule and the octet rule. The electron affinity can be further increased by replacing one of the C atoms with B. The equilibrium geometries of neutral and anionic $BC_5H_{6-x}(BO_2)_x$, $x = 1$–6 are shown in Figure 2.26,

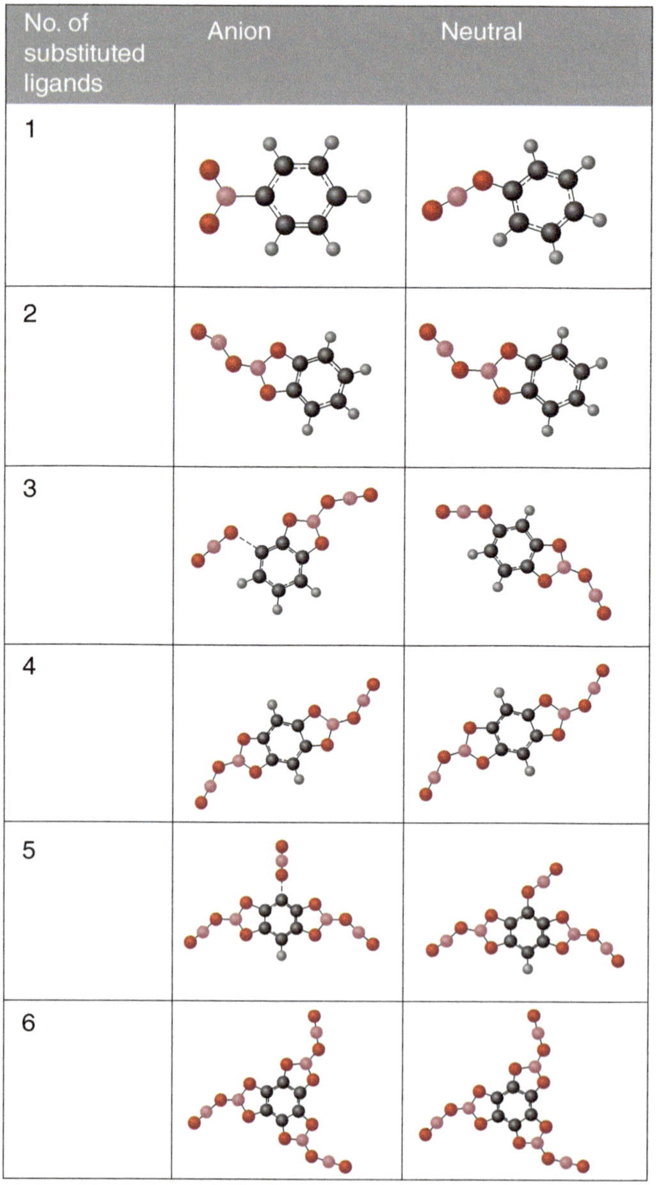

Figure 2.25 Ground state geometries of neutral and anionic $C_6H_{6-x}(BO_2)_x$. The gray, white, pink, and red spheres correspond to carbon, hydrogen, boron, and oxygen, respectively. *Source:* Driver [73].

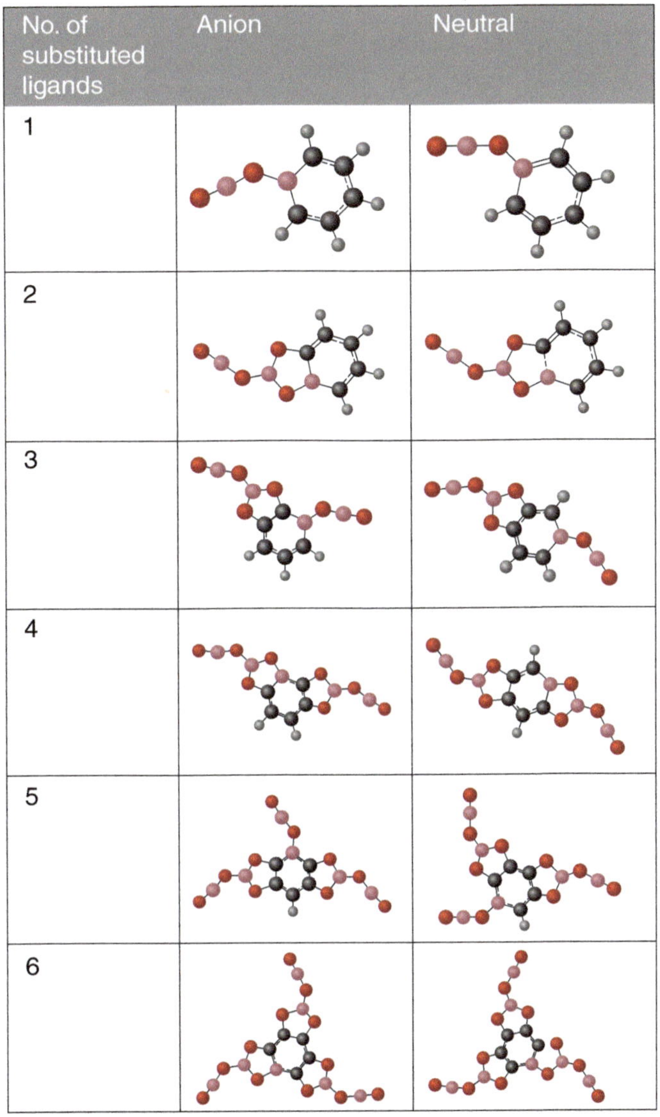

Figure 2.26 Ground state geometries of neutral and anionic $C_5BH_{6-x}(BO_2)_x$. The gray, white, pink, and red spheres correspond to carbon, hydrogen, boron, and oxygen atoms, respectively.

and the corresponding electron affinities are given in Table 2.2. Thus, a benzene molecule can be made into a superhalogen by suitable modification, and the electron affinity of $BC_5H(BO_2)_5$ surpasses that of Cl. In general, the electron affinity of a cluster can be increased by choosing the composition of a cluster that satisfies multiple electron-counting rules. This can be seen from several recent publications [97, 98].

Another example where the use of both octet rule and aromaticity rule can simultaneously give rise to stable monoanions is $BC_5H_{6-x}(CN)_x$, $x = 1$–6. Here, the H atoms in C_6H_6 are sequentially replaced by superhalogen CN. The electron affinities of these clusters calculated using density functional theory and B3LYP hybrid exchange-correlation potential with 6-31 + G* basis set implemented in the Gaussian 16 code are given in Figure 2.27. Note

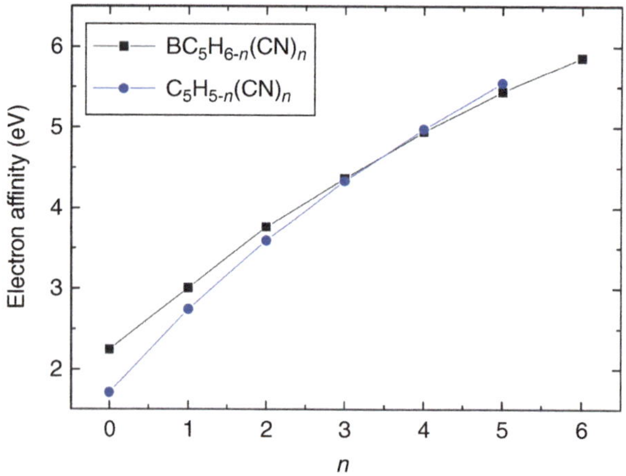

Figure 2.27 Electron affinities of $BC_5H_{6-x}(CN)_x$, $x = 1-6$ and their isoelectronic $C_5H_{5-x}(CN)_x$, $x = 1-5$.

that electron affinity of $BC_5(CN)_6$ is close to 6 eV, making it a hyperhaolgen. It is interesting to compare the electron affinities of these clusters with isoelectronic $C_5H_{5-x}(CN)_x$, $x = 1-5$. The results are compared with the electron affinities $BC_5H_{6-x}(CN)_x$ in Figure 2.27. As the number of CN ligands increases, the electron affinities of both these two classes of clusters become identical, indicating that the ligands as well as total number of "valence" electrons play a role in stabilizing the monoanions.

Unusually stable clusters can also be created by satisfying three electron-counting rules, simultaneously. Consider, for example, $Mn[BC_5(CN)_6]_2^-$ that satisfies the aromaticity rule, the octet rule, and the 18-electron rule, simultaneously. The geometry of neutral and anionic $Mn[BC_5(CN)_6]_2^-$ cluster computed by Giri et al. [94] is shown in Figure 2.28. Here, Mn is sandwiched between two $BC_5(CN)_6$ molecules. Another consequence of the electron shell closure is that the magnetic moment of the Mn atom is quenched. The octet shell closure enables CN to have an electron affinity of 3.86 eV. The electron affinity of $BC_5(CN)_6$, which satisfies both the octet rule and aromaticity rule is 5.87 eV. $Mn[BC_5(CN)_6]_2$, satisfying three electron-counting rules simultaneously, has an electron affinity of 6.40 eV. This provides a recipe for designing highly stable negative ions.

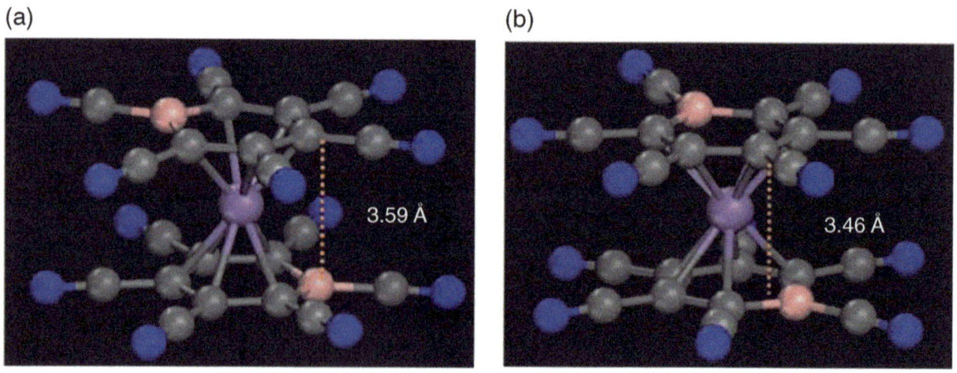

Figure 2.28 Equilibrium geometries of (a) neutral and (b) anionic $Mn[BC_5(CN)_6]_2$ cluster. *Source:* Adapted with permission from Giri et al. [94]. © Royal Society of Chemistry.

2.3.2 Dianions

Unusual stability of $Mn[BC_5(CN)_6]_2^-$ suggests that if Mn is replaced by Cr, one extra electron will be needed to satisfy the octet, aromaticity, and 18-electron rule, simultaneously. Thus, $Cr[BC_5(CN)_6]_2^{2-}$ should be a stable cluster. Giri et al. [94] calculated the second electron affinity of $Cr[BC_5(CN)_6]_2^{2-}$ to be 2.58 eV, which is substantially larger than the 0.9 eV value of $B_{12}H_{12}^{2-}$. Molecular dynamics simulation of $Cr[BC_5(CN)_6]_2^{2-}$ at 300 K confirmed that it is thermally stable. From the snapshots of the geometries taken at different time intervals in Figure 2.29, we see that there is very little change in its geometry.

Another example of an ultra-stable dianion designed by satisfying multiple electron-counting rules is $B_{12}(CN)_{12}^{2-}$ where the H atoms of $B_{12}H_{12}^{2-}$ are replaced by CN molecules. Recall that CN needs an extra electron to satisfy the octet rule. Thus, $B_{12}(CN)_{12}^{2-}$ satisfies both the octet and Wade-Mingos rules. Zhao et al. [99] found the optimized geometry of $B_{12}(CN)_{12}^{2-}$ to be a perfect icosahedron (see Figure 2.30) with the second electron bound by

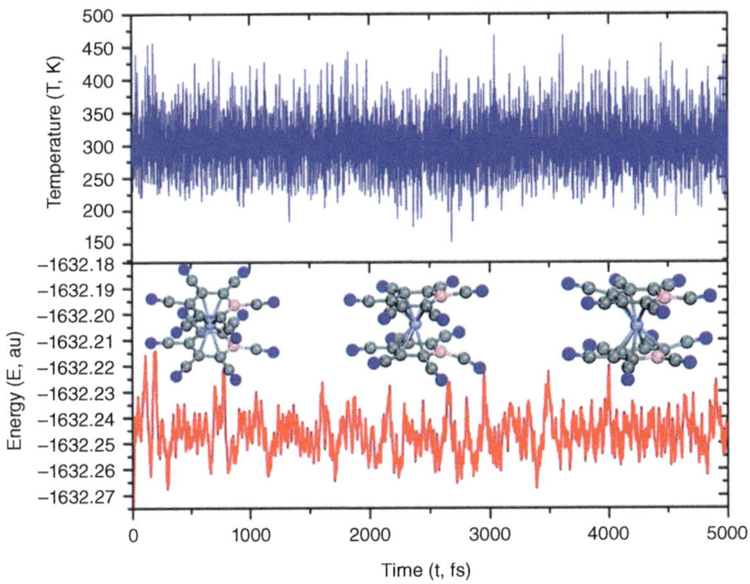

Figure 2.29 Molecular dynamics simulation of $Cr[BC_5(CN)_6]_2^{2-}$ cluster. *Source:* Giri et al. [94]. © Royal Society of Chemistry.

Figure 2.30 Geometry of $B_{12}(CN)_{12}^{2-}$ with I_h symmetry. *Source:* Jena and Sun [1]. © American Chemical Society.

5.3 eV. This colossal stability of $B_{12}(CN)_{12}^{2-}$ has been recently verified experimentally by Mayer et al. [100] and the measured electron affinity of 5.55 eV agrees well with predicted value. To date, $B_{12}(CN)_{12}^{2-}$ is the most stable dianion known in the gas phase.

Dianions of organic molecules such as C_6H_6 and C_8H_8 can also be stabilized by replacing H with BO and CN superhalogens and two of the carbon atoms by boron. The resulting $B_2C_4X_6^{2-}$ (X = H, BO, CN) clusters satisfy both the aromaticity rule and octet rule. Chen et al. [54] computed the geometries of neutral, monoanion, and dianions of $B_2C_4X_6$ (X = H, BO_2, CN) clusters. These are given in Figure 2.31; the electron affinity and vertical

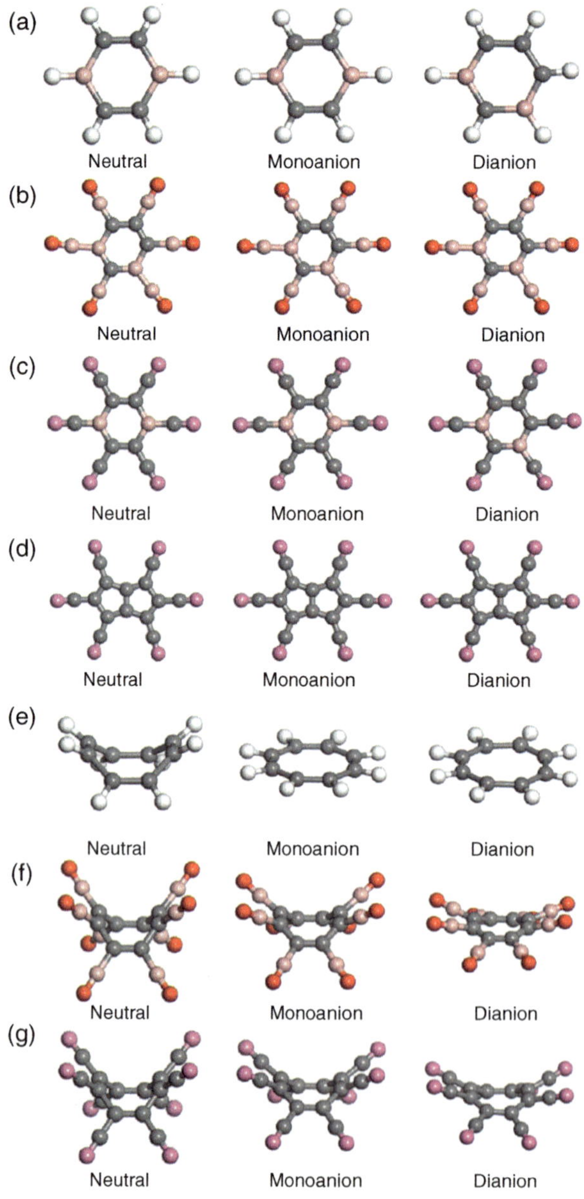

Figure 2.31 Optimized geometries of (a) $B_2C_4H_6$ $^{0,1-,2-}$, (b) $B_2C_4(BO)_6$ $^{0,1-,2-}$, (c) $B_2C_4(CN)_6$ $^{0,1-,2-}$, (d) $C_8(CN)_6$ $^{0,1-,2-}$, (e) C_8H_8 $^{0,1-,2-}$, (f) $C_8(BO)_8$ $^{0,1-,2-}$, and (g) $C_8(CN)_8$ $^{0,1-,2-}$ clusters. Gray, pink, brown, and red spheres stand for C, N, B, and O atoms, respectively. *Source:* Chen et al. [54]. © American Chemical Society.

Table 2.3 First and second electron affinity (EAs) and vertical detachment energy (VDEs) of the studied aromatic molecules.

Clusters	FVDE	FEA	SVDE	SEA
$B_2C_4H_6$	2.01	1.90	−2.68	−3.27
$B_2C_4(BO)_6$	6.93	6.11	1.71	1.58
$B_2C_4(CN)_6$	5.47	5.38	1.59	1.12
$C_8(CN)_6$	5.41	5.19	0.83	0.81
C_8H_8	1.47	0.82	−3.35	−3.49
$C_8(BO)_8$	5.03	4.53	1.70	1.21
$C_8(CN)_8$	5.00	4.64	1.54	1.21

FEA, first electron affinity; SEA, second electron affinity; and SVDE, second vertical detachment energy.

detachment energies associated with the addition of the first and second electron are given in Table 2.3. Note that the dianions of all molecules are stable, once H is replaced by CN and BO. Similarly, $TiC_{12}(CN)_{12}{}^{2-}$ that satisfies both the 18-electron rule and aromaticity rule is also stable with the second electron bound by 0.76 eV.

In some cases, the ligands can play a significant role that can even dominate over the electron-counting rule. Consider, for example, manganocene, $Mn(C_5H_5)_2$. It has 17 electrons with Mn contributing 7 electrons due to its $3d^5\,4s^2$ configuration and each C_5H_5 contributing 5 electrons. Thus, $Mn(C_5H_5)_2{}^-$ should be very stable due to the 18-electron rule and its electron affinity should be high. Instead, its electron affinity is only 0.28 eV [101]. Furthermore, $Mn(C_5H_5)_2{}^-$ should be unstable against electron emission not only because it is a 19-electron system but also because the two extra electrons will repel. This is indeed the case. However, if H is replaced by CN or BO, the corresponding $Mn[C_5(BO)_5]_2$ and $Mn[C_5(CN)_5]_2{}^{2-}$ are superhalogens with electron affinities of 4.85 and 4.78 eV, respectively. In addition, $Mn[C_5(BO)_5]_2{}^{2-}$ and $Mn[C_5(CN)_5]_2{}^{2-}$, which are 19-electron systems, are stable with a second electron affinity of 0.38 and 0.7 eV, respectively.

2.3.3 Trianions

Unlike the studies of dianions, work on trianions is rather scarce. One of the early studies of the trianions was due to Compton and coworkers who reported the mass spectra of $(C_{60})_2(CN)_5{}^{3-}$ and $(C_{60})_2(CN)_7{}^{3-}$ [102]. Note that observation of species in mass spectra does not necessarily mean that the trianions are stable, but simply that they exist only within experimental conditions. Many metastable multiply charged ions are observed due to repulsive Coulomb barrier. Indeed, $(C_{60})_2(CN)_5{}^{3-}$ is metastable against autodetachment of the third electron. Cederbaum and coworkers [103] examined the stability of a number of trianions and found that the best candidate, $B(C_2CO_2)_3{}^{3-}$, is thermodynamically unstable against electron emission by −0.4 eV.

Following the discovery of colossal stability of the $B_{12}(CN)_{12}{}^{2-}$, Zhao et al. [104] calculated the optimized geometry and total energy of $BeB_{11}(CN)_{12}$ in neutral, monoanion, dianion, and trianion form. Note that with Be replacing a B atom, an additional electron will be

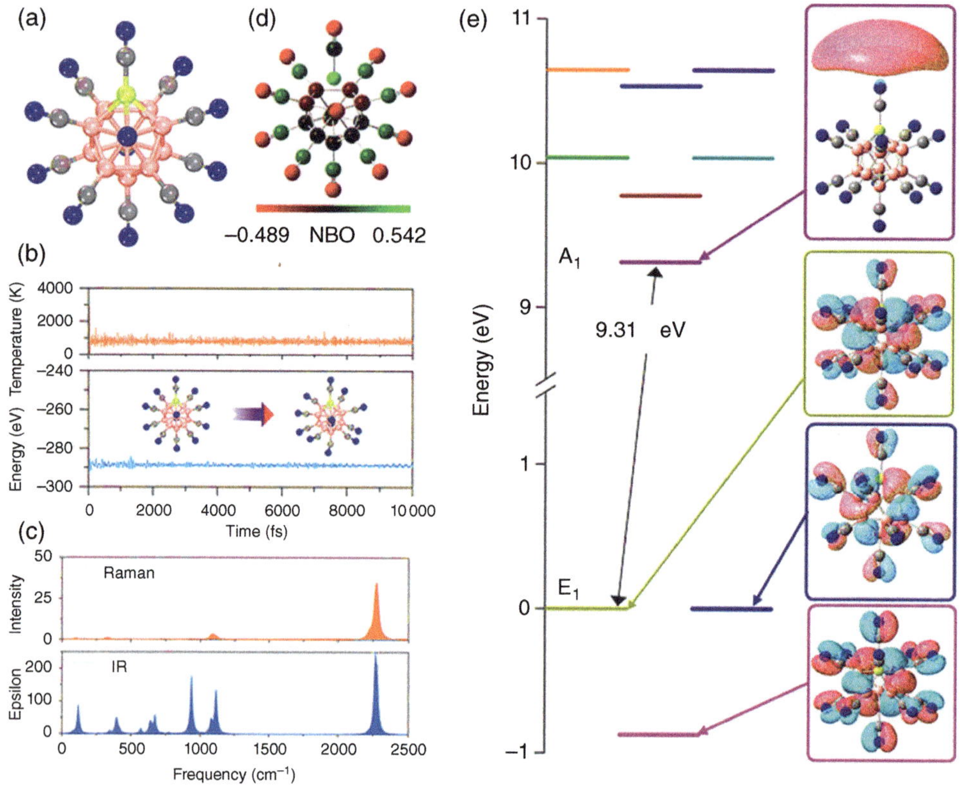

Figure 2.32 $BeB_{11}(CN)_{12}^{3-}$ (a) geometry, (b) AIMD simulation as a function of temperature and total energy fluctuation, (c) Raman and infra-red (IR) simulation spectra, (d) natural bond orbital (NBO) charge distribution, and (e) energy diagram and frontier orbitals. *Source:* Zaho et al. [104]. © John Wiley & Sons.

required to satisfy the Wade-Mingos rule and the octet rule, simultaneously. In Figure 2.32 we show the geometry, thermal stability at 800K, and electronic structure of $BeB_{11}(CN)_{12}^{3-}$, which is stable against auto-ejection of the third electron by 2.65 eV. When CN is replaced by BO or SCN, the third electron affinities of the resulting trianions $BeB_{11}(BO)_{12}^{3-}$ and $BeB_{11}(SCN)_{12}^{3-}$ are, respectively, 1.30 and 0.59 eV.

2.3.4 Tetra-Anions and Beyond

Recently Hong and Jena [105] developed a universal model to examine the stability of multiply charged anions capable of carrying more than three extra electrons by tailoring their proximity to the closed shell as well as the size and the electron affinity of the terminal groups. They discovered several thermodynamically stable tetra- and penta-anions containing as few as 50 and 80 atoms, respectively. The starting point is to realize that the extra electrons in a stable multiply charged cluster must reside in a set of bound states. Assuming a spherical potential well with a given depth V and radius r, the number of bound states depends on a positive coefficient a where $V \geq a/r^2$. The larger the coefficient a, the greater are the number of bound states. A deeper or wider potential well can hold more bound states.

To form a stable tetra-anion, the authors chose two stable dianions. Recall that $M(CN)_4^{2-}$ (M = Mg, Ca, Sr, Ba) and $BeB_{11}(CN)_{12}^{3-}$ are known to be very stable dianion and trianion, respectively. By removing a CN^- from $BeB_{11}(CN)_{12}^{3-}$, Hong and Jena combined $BeB_{11}(CN)_{11}^{2-}$ with a series of $M(CN)_4^{2-}$ clusters and optimized the resulting geometries (Figure 2.33). $BeB_{11}(CN)_{11}M(CN)_4^{4-}$ (M = Ca, Sr, Ba) clusters were found to be stable with the fourth electron bound by 0.79, 0.20, and 0.25 eV, respectively. Similarly, the fourth electron affinity of $Be_2B_{22}(CN)_{23}^{4-}$ was found to be 1.48 eV. A similar procedure led to the

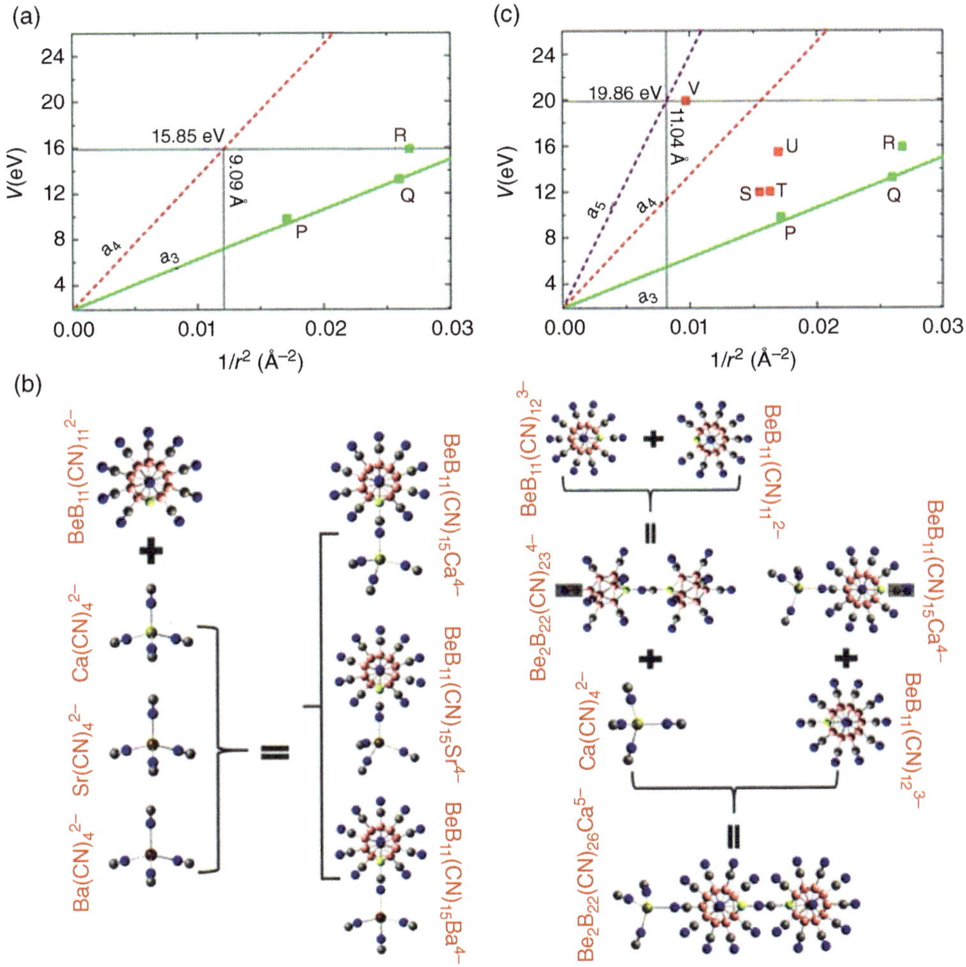

Figure 2.33 (a) Evaluation of the size (r = 9.09 Å) needed for a stable tetra-anion using the estimated a_4 and the maximal V (15.85 eV) of the known trianions (in green). The red dashed line is the estimated threshold line for tetra-anions. The green solid line is the threshold line for trianions. (b) Demonstration of using the known multiply charged clusters with proper sizes to form new clusters with higher negative charge states. To form the penta-anion, certain CN^- terminal group, as indicated under shades, needs to be knocked off. (c) Evaluation of the size (r = 11.04 Å) needed for a stable penta-anion using the estimated a_5 and the maximal V (19.86 eV) of the newly found tetra-anions (in red). The purple dashed line is the estimated threshold line for penta-anions. Boron is in pink, beryllium in light yellow, carbon in gray, nitrogen in blue, calcium in dark yellow, strontium in green yellow, and barium in brown. *Source:* Fang and Jena [105]. © John Wiley & Sons.

discovery of $Be_2B_{22}(CN)_{23}Ca(CN)_4^{5-}$ penta-ion with a fifth electron affinity of 30 meV. The level of theory used to predict the composition, geometry, and electron affinity of the above tetra- and penta-anions is same as that was used to predict the electron affinity of $B_{12}(CN)_{12}^{2-}$. As mentioned above, this prediction has now been experimentally verified. Thus, we believe that the predicted stable tetra- and penta-anions can be found and the synergy between theory and experiment can lead to the focused discovery of other multiply charged anions, potentially opening a new chapter in materials chemistry.

2.4 Conclusions

In this chapter we have focused on the design of superatoms, which are atomic clusters of specific size and composition and whose chemistry mimics that of the atoms in the periodic table. The concept of superatoms is that their free electrons occupy a new set of orbitals that are defined by the entire group of atoms in the cluster, instead of by each atom separately. If these superatomic orbitals have the same symmetry as that of the atoms, one can think of a new class of materials with superatoms as the building blocks. Because of their specific size, cluster-assembled materials may have properties different from those of the corresponding atom-assembled materials. A classic example is C_{60} fullerene-based material *vs* diamond and graphite. All are made of only carbon atoms but their properties are very different because of the way carbon atoms are arranged.

The central question is how to design these superatoms so that they are stable and maintain their structure when assembled. We discussed various electron-counting rules that have been used for nearly a century to explain the stability of atoms and the compounds they form. Just as superatomic orbitals mimic atomic orbitals, it is expected the same electron-counting rules that apply to atoms may also apply to superatoms. Indeed, we demonstrated how the octet rule for low atomic number species, the 18-electron rule for transition metals, the 32-electron rule for rare earth metals, the Wade-Mingos rule for boron-based systems, and the aromaticity rules for organic molecules can be used to design not only stable neutral but anionic species. More importantly, we demonstrated how multiple electron-counting rules can be used simultaneously to design negative ions carrying up to five extra electrons that are stable against fragmentation or auto-ejection of the electron.

In Chapter 10, we will discuss how the superatoms can be used to promote unusual chemistry such as making noble gas atoms form chemical bonds at room temperature and accessing high oxidation states of metal atoms. The potential of cluster-assembled materials in storing hydrogen, catalyzing reactions, serving as building blocks of super- and hyper-salts, electrolytes in Li- and Na-ion batteries, and moisture-insensitive hybrid perovskite-based solar cells is also discussed.

References

1 Jena, P. and Sun, Q. (2018). Super atomic clusters: design rules and potential for building blocks of materials. *Chem. Rev.* 118: 5755–5870. See also references 1-146 in Chapter 1.

2 Khanna, S.N. and Jena, P. (1992). Assembling crystals from clusters. *Phys. Rev. Lett.* 69: 1664–1667.

3 Khanna, S.N. and Jena, P. (1995). Atomic clusters: building blocks for a class of solids. *Phys. Rev. B* 51: 13705–13716.
4 Kroto, H.W., Heath, J.R., O'Brien, S.C. et al. (1985). C60: buckminsterfullerene. *Nature* 318: 162–163.
5 Kratschmer, W., Lamb, L.D., Fostiropoulos, K., and Huffman, D.R. (1990). Solid C60: a new form of carbon. *Nature* 347: 354–358.
6 Knight, W.D., Clemenger, K., de Heer, W.A. et al. (1984). Electronic shell structure and abundances of sodium clusters. *Phys. Rev. Lett.* 52: 2141–2144.
7 Mayer, M.G. (1948). On closed shells in nuclei. *Phys. Rev.* 74: 235–239.
8 Abegg, R. (1904). Die Valenz und das Periodische System. Versuch einer Theorie der Molekularverbindungen (the valency and the periodical system - attempt on a theory of molecular compound). *Z. Anorg. Chem.* 39: 330–380.
9 Lewis, G.N. (1916). The atom and the molecule. *J. Am. Chem. Soc.* 38: 762–785.
10 Langmuir, I. (1919). The arrangement of electrons in atoms and molecules. *J. Am. Chem. Soc.* 41: 868–934.
11 Langmuir, I. (1921). Types of valence. *Science* 54: 59–67.
12 Hückel, E. (1931). Quantentheoretische beiträge zum benzolproblem. *Eur. Phys. J. A* 70: 204–286.
13 Hückel, E. (1930). Zur quantentheorie der doppelbindung. *Eur. Phys. J. A* 60: 423–456.
14 Wade, K. (1971). The structural significance of the number of skeletal bonding electron-pairs in carboranes, the higher boranes and borane anions, and various transition-metal carbonyl cluster compounds. *Chem. Commun.*: 792–793.
15 Wade, K. (1976). Structural and bonding patterns in cluster chemistry. *Adv. Inorg. Chem. Radiochem.* 18: 1–66.
16 Mingos, D.M.P. (1972). A general theory for cluster and ring compounds of the main group and transition elements. *Nat. Phys. Sci.* 236: 99–102.
17 Mingos, D.M.P. (1984). Polyhedral skeletal electron pair approach. *Acc. Chem. Res.* 17: 311–319.
18 Zintl, E., Goubeau, J., and Dullenkopf, W. (1931). Metals and alloys. I. Saltlikecompounds and intermetallic phases of sodium in liquid ammonia. *Z. Phys. Chem.* 154: 1–46.
19 Zintl, E. and Kaiser, H. (1933). Metals and alloys. VI. Ability of elements to form negative ions. *Z. Anorg. Allgem. Chem.* 211: 113–131.
20 Saito, S. and Ohnishi, S. (1987). Stable $(Na_{19})_2$ as a giant alkali-metal- atom dimer. *Phys. Rev. Lett.* 59: 190–193.
21 Gaussian 16 (2016). *Revision C.01*. Wallingford, CT: Gaussian, Inc.
22 Perdew, J.P., Burke, K., and Ernzerhof, M. (1996). Generalized gradient approximation made simple. *Phys. Rev. Lett.* **77**: 3865.
23 Sun, W.G., Wang, J.J., Lu, C. et al. (2017). Evolution of the structural and electronic properties of medium-sized sodium clusters: a honeycomb-like Na20 cluster. *Inorg. Chem.* 56: 1241–1248.
24 Hakkinen, H. and Manninen, M. (1996). How "magic" is a magic cluster? *Phys. Rev. Lett.* 76: 1599–1602.
25 Cheng, L., Zhang, X., Jin, B., and Yang, J. (2014). Superatom-atom superbonding in metallic clusters: a new look to the mystery of an Au_{20} pyramid. *Nanoscale* 6: 12440–12444.
26 Wang, Z.W. and Palmer, R.E. (2012). Direct atomic imaging and dynamical fluctuations of the tetrahedral Au_{20} cluster. *Nanoscale* 4: 4947–4949.

27 Leuchtner, R.E., Harms, A.C., and Castleman, A.W. Jr. (1989). Thermal metal cluster anion reactions: behavior of aluminum clusters with oxygen. *J. Chem. Phys.* 91: 2753–2754.

28 Li, X., Wu, H., Wang, X.B., and Wang, L.S. (1998). s–p hybridization and electron shell structures in aluminum clusters: a photoelectron spectroscopic study. *Phys. Rev. Lett.* 81: 1909–1912.

29 Rao, B.K. and Jena, P. (1999). Evolution of the electronic structure and properties of neutral and charged aluminum clusters: a comprehensive analysis. *J. Chem. Phys.* 111: 1890.

30 Khanna, S.N. and Jena, P. (1994). Designing ionic solids from metallic clusters. *Chem. Phys. Lett.* 219: 479–483.

31 Zheng, W.-J., Thomas, O.C., Lippa, T.P. et al. (2006). The ionic KAl13 molecule: a stepping stone to cluster assembled materials. *J. Chem. Phys.* 124: 144304–144305.

32 Bergeron, D.E., Castleman, A.W. Jr., Morisato, T., and Khanna, S.N. (2004). Formation of $Al_{13}I^-$: evidence for the superhalogen character of Al_{13}. *Science* 304: 84–87.

33 Han, Y.K. and Jung, J. (2008). Does the "superatom" exist in halogenated aluminum clusters? *J. Am. Chem. Soc.* 130: 2–3.

34 Jung, J., Kim, H., and Han, Y.K. (2011). Can an electron-shell closing model explain the structure and stability of ligand-stabilized metal clusters? *J. Am. Chem. Soc.* 133: 6090–6095.

35 Liu, F., Mostoller, M., Kaplan, T. et al. (1996). Evidence for a new class of solids: first principles study of $K(Al_{13})$. *Chem. Phys. Lett.* 248: 213.

36 Huang, C., Fang, H., Whetten, R., and Jena, P. (2020). Robustness of superatoms and their potential as building blocks of materials: Al_{13}^- vs $B(CN)_4^-$. *J. Phys. Chem. C* 124: 6435–6440.

37 Clayborne, P.A., Lopez-Acevedo, O., Whetten, R.L. et al. (2011). The Al50Cp*12 cluster – A 138-electron closed shell (L = 6) superatom. *Eur. J. Inorg. Chem.* 2011: 2649–2652.

38 Walter, M., Akola, J., Lopez-Acevedo, O. et al. (2008). A unified view of ligand-protected gold clusters as superatom complexes. *Proc. Natl. Acad. Sci. U. S. A.* 105: 9157–9162.

39 Jadzinsky, P.D., Calero, G., Ackerson, C.J. et al. (2007). Structure of a thiol monolayer-protected gold nanoparticle at 1.1 Å resolution. *Science* 318: 430–433.

40 Castleman, A.W. and Khanna, S.N. (2009). Clusters, superatoms, and building blocks of new materials. *J. Phys. Chem. C* 113: 2664–2675.

41 Claridge, S.A., Castleman, A.W., Khanna, S.N. et al. (2009). Cluster-assembled materials. *ACS Nano* 3: 244–255.

42 Shafai, G., Hong, S., Bertino, M., and Rahman, T.S. (2009). Effect of ligands on the geometric and electronic structure of Au_{13} clusters. *J. Phys. Chem. C* 113: 12072–12078.

43 Zhang, Z.-G., Xu, H.-G., Feng, Y., and Zheng, W. (2010). Communications: investigation of the superatomic character of Al_{13} via its interaction with sulfur atoms. *J. Chem. Phys.* 132: 161103.

44 Gutsev, G.L. and Boldyrev, A.I. (1981). DVM-Xα calculations on the ionization potentials of $M X_{k+1}$ – complex anions and the electron affinities of $M X_{k+1}$ "superhalogens". *Chem. Phys.* 56: 277–283.

45 Gutsev, G.L. and Boldyrev, A.I. (1982). DVM Xα calculations on the electronic structure of "superalkali" cations. *Chem. Phys. Lett.* 92: 262–266.

46 Lievens, P., Thoen, P., Bouckaert, S. et al. (1999). Ionization potentials of Li_nO $(2 < n < 70)$ clusters: experiment and theory. *J. Chem. Phys.* 110: 10316–10329.

47 Gutsev, G.L., Bartlett, R.J., Boldyrev, A.I., and Simons, J. (1997). Adiabatic electron affinities of small superhalogens: LiF_2, $LiCl_2$, NaF_2, and $NaCl_2$. *J. Chem. Phys.* 107: 3867–3875.

48 Bartlett, N. and Lohmann, D.H. (1962). Dioxygenyl hexafluoroplatinate (V), O2 + [PtF6] −. *Proc. Chem. Soc., London* 3: 115–116.

49 Bartlett, N. (1962). Xenon hexafluoroplatinate (V) Xe+ [PtF6] −. *Proc. Chem. Soc., London* 6: 218.

50 Giri, S., Behera, S., and Jena, P. (2014). Superalkalis and superhalogens as building blocks of supersalts. *J. Phys. Chem. A* 118: 638–645.

51 Koirala, P., Willis, M., Kiran, B. et al. (2010). Superhalogen properties of fluorinated coinage metal clusters. *J. Phys. Chem. C* 114: 16018–16024.

52 Willis, M., Götz, M., Kandalam, A.K. et al. (2010). Hyperhalogens: discovery of a new cass of highly electronegative species. *Angew. Chem. Int. Ed.* 49: 8966–8970.

53 Knight, D.A., Zidan, R., Lascola, R. et al. (2013). Synthesis, characterization, and atomistic modeling of stabilized highly pyrophoric $Al(BH_4)_3$ via the formation of the hypersalt $K[Al(BH_4)_4]$. *J. Phys. Chem. C* 117: 19905–19915.

54 Chen, G., Zhao, T., Wang, Q., and Jena, P. (2019). Rational design of stable dianions and the concept of superchalcogens. *J. Phys. Chem. A* 123: 5753–5761.

55 Pyykkö, P. and Runeberg, N. (2002). Icosahedral Wau12: a predicted closed-shell species, stabilized by aurophilic attraction and relativity and in accord with the 18-electron rule. *Angew. Chem. Int. Ed.* 41: 2174–2176.

56 Li, X., Kiran, B., Li, J. et al. (2002). Experimental observation and confirmation of icosahedral W@Au12 and Mo@Au12 molecules. *Angew. Chem. Int. Ed.* 41: 4786–4789.

57 Zhai, H.-J., Li, J., and Wang, L.-S. (2004). Icosahedral gold cage clusters: $M@Au_{12}^-$ (M = V, Nb, and Ta). *J. Chem. Phys.* 121: 8369–8374.

58 Cui, L.-F., Huang, X., Wang, L.-M. et al. (2006). Pb_{12}^{2-}: Plumbaspherene. *J. Phys. Chem. A* 110: 10169–10172.

59 Cui, L.-F., Huang, X., Wang, L.-M. et al. (2006). Sn_{12}^{2-}: Stannaspherene. *J. Am. Chem. Soc.* 128: 8390–8391.

60 Dognon, J., Clavaguera, C., and Pyykko, P. (2007). Towards a 32-electron principle: Pu@ Pb_{12} and related systems. *Angew. Chem. Int. Ed.* 46: 1427–1430.

61 Manna, D., Sirohiwal, A., and Ghanty, T.K. (2014). $Pu@C_{24}$: a new example satisfying the 32-electron principle. *J. Phys. Chem. C* 118: 7211–7221.

62 Dognon, J.P., Clavaguéra, C., and Pyykkö, P. (2009). A predicted organometallic series following a 32-electron principle: $An@C_{28}$ (An = Th, Pa^+, U^{2+}, Pu^{4+}). *J. Am. Chem. Soc.* 131: 238–243.

63 Dognon, J.P., Clavaguéra, C., and Pyykkö, P. (2010). Chemical properties of the predicted 32-electron systems $Pu@Sn_{12}$ and $Pu@Pb_{12}$. *C. R. Chim.* 13: 884–888.

64 Dognon, J.P., Clavaguéra, C., and Pyykkö, P. (2012). A new, centered 32-electron system: the predicted $[U@Si_{20}]^{6-}$-like isoelectronic series. *Chem. Sci.* 3: 2843–2848.

65 Manna, D. and Ghanty, T.K. (2012). Prediction of a new series of thermodynamically stable actinide encapsulated fullerene systems fulfilling the 32-electron principle. *J. Phys. Chem. C* 116: 25630–25641.

66 Dai, X., Yang Gao, Y., Jiang, W. et al. (2015). U@C28: the electronic structure induced by the 32-electron principle. *Phys. Chem. Chem. Phys.* 17: 23308–23311.

67 Jin, J., Pan, S., Jin, X. et al. (2019). Octacarbonyl anion complexes of the late lanthanides $Ln(CO)_8^-$ (Ln=Tm, Yb, Lu) and the 32-electron rule. *Chem. Eur. J.* 25: 3229–3234.

68 Guo, J.-C., Feng, L.-Y., Dong, C., and Zhai, H.-J. (2020). A designer 32-electron superatomic CBe_8H_{12} cluster: core–shell geometry, octacoordinated carbon, and cubic aromaticity. *New J. Chem.* 44: 7286–7292.

69 Schleyer, P.V.R. (2001). Introduction: aromaticity. *Chem. Rev.* 101: 1115–1117.

70 Nenner, I. and Schulz, G.J. (1975). Temporary negative ions and electron affinities of benzene and N-heterocyclic molecules: pyridine, pyridazine, pyrimidine, pyrazine, and s-triazine. *J. Chem. Phys.* 62: 1747–1758.

71 Child, B.Z., Giri, S., Gronert, S., and Jena, P. (2014). Aromatic superhalogens. *Chem. Eur. J.* 20: 4736–4745.

72 Giri, S., Child, B.Z., and Jena, P. (2014). Organic superhalogens. *ChemPhysChem* 15: 2903–2908.

73 Driver, N. and Jena, P. (2018). Electron affinity of modified benzene. *Int. J. Quantum Chem.* 118: e25504.

74 Wentworth, W.E., Limero, T., and Chen, E.C.M. (1987). Electron affinities of hexafluorobenzene and pentafluorobenzene. *J. Phys. Chem.* 91: 241.

75 Chen, E.C.M. and Wentworth, W.E. (1975). A comparison of experimental determinations of electron affinities of pi charge transfer complex acceptors. *J. Chem. Phys.* 63: 3183.

76 Li, X., Kuznetsov, A.E., Zhang, H.-F. et al. (2001). Observation of all-metal aromatic molecules. *Science* 291: 859–861.

77 Plesek, J., Jelinek, T., Stibr, B., and Hermanek, S. (1988). New monocarbaboranes from stepwise degradation of lcosahedral 1,2-and 1,7-$C_2B_{10}H_{12}$ dicarbaboranes. *J. Chem. Soc. Chem. Commun.*: 348–349.

78 Jelinek, T., Stibr, B., Plesek, J. et al. Eight-vertex polyhedral monocarbaborane chemistry. Three closo anions, $[CB_7H_8]^-$, $[CB_7H_7I]^-$ and $[CB_7H_6I_2]^-$. Preparation and structural studies. *J. Chem. Soc. Dalton Trans.* 995: 431–437.

79 Jelinek, T., Stibr, B., Holub, J. et al. (2001). Monocarbaborane chemistry. Preparation and characterisation of $[4\text{-}CB_8H_9]^-$, the 'missing' closo-carbaborane anion. *Chem. Commun.*: 1756–1757.

80 Stíbr, B., Tok, O.L., Milius, W. et al. (2002). The [*closo*-2-CB_6H7]$^-$ ion: the first representative of the 7-vertex monocarbaborane series. *Angew. Chem. Int. Ed.* 41: 2126–2128.

81 Pathak, B., Samanta, D., Ahuja, R., and Jena, P. (2011). Borane derivatives: a new class of super- and hyperhalogens. *ChemPhysChem* 12: 2423–2428.

82 Kumar, V. and Kawazoe, Y. (2002). Metal-encapsulated icosahedral superatoms of germanium and tin with large gaps: $Zn@Ge_{12}$ and $Cd@Sn_{12}$. *Appl. Phys. Lett.* 80: 859–861.

83 Cui, L.-F., Huang, X., Wang, L.-M. et al. (2007). Endohedral stannaspherenes $M@Sn_{12}^-$: a rich class of stable molecular cage clusters. *Angew. Chem. Int. Ed.* 46: 742–745.

84 Kandalam, A.K., Chen, G., and Jena, P. (2008). Unique magnetic coupling between Mn doped stannaspherenes $Mn@Sn_{12}$. *Appl. Phys. Lett.* 92: 143109.

85 Neukermans, S., Janssens, E., Chen, Z.F. et al. (2004). Extremely stable metal-encapsulated $AlPb_{10}^+$ and $AlPb_{12}^+$ clusters: mass-spectrometric discovery and density functional theory study. *Phys. Rev. Lett.* 92: 163401.

86 Esenturk, E.N., Fettinger, J., Lam, Y.-F., and Eichhorn, B. (2004). $[Pt@\ Pb_{12}]^{2-}$. *Angew. Chem. Int. Ed.* 43: 2132–2134.

87 Zintl, E., Goubeau, J., and Dullenkopf, W. (1931). Metals and alloys. I. Saltlike compounds and intermetallic phases of sodium in liquid ammonia. *Z. Phys. Chem.* 154: 1–46.

88 Schild, D., Pflaum, R., Sattler, K., and Recknagel, E. (1987). Stability of free intermetallic compound clusters: lead/antimony and bismuth/ antimony. *J. Phys. Chem.* 91: 2649–2653.

89 Wheeler, R., LaiHing, K., Wilson, W., and Duncan, M. (1988). Growth patterns in binary clusters of Group IV and V metals. *J. Chem. Phys.* 88: 2831–2839.

90 Farley, R.W. and Castleman, A.W. (1989). Observation of gas phase anionic bismuth zintl ions. *J. Am. Chem. Soc.* 111: 2734–2735.

91 Lips, F. and Dehnen, S. (2011). Neither electron precise nor in accordance with Wade–Mingos rules: the ternary cluster anion $[Ni_2Sn_7Bi_5]^{3-}$. *Angew. Chem. Int. Ed.* 50: 955–959.

92 Goicoechea, J.M. and Sevov, S.C. (2006). Deltahedral germanium clusters: insertion of transition-metal atoms and addition of organometallic fragments. *J. Am. Chem. Soc.* 128: 4155–4161.

93 Sun, Z.-M., Zhao, Y.-F., Li, J., and Wang, L.-S. (2009). Diversity of functionalized germanium Zintl clusters: syntheses and theoretical studies of $[Ge_9 PdPPh_3]^{3-}$ and $[Ni@(Ge_9 PdPPh3)]^{2-}$. *J. Clust. Sci.* 20: 601–609.

94 Giri, S., Child, B.Z., Zhou, J., and Jena, P. (2015). Unusual stability of multiply charged organo-metallic complexes. *RSC Adv.* 5: 44003–44008.

95 Zhao, H., Zhou, J., and Jena, P. (2016). Stability of $B_{12}(CN)_{12}{}^{2-}$: implications for lithium and magnesium ion batteries. *Angew. Chem. Int. Ed.* 55: 3704–3708.

96 Zhai, H.-J., Wang, L.-M., Li, S.-D., and Wang, L.-S. (2007). Vibrationally resolved photoelectron spectroscopy of BO- and BO2-: a joint experimental and theoretical study. *J. Phys. Chem. A* 111: 1030.

97 Zhao, H., Zhou, J., Fang, H., and Jena, P. (2016). From halogen to superhalogen behavior of organic molecules created by functionalizing benzene. *Chem. Phys. Chem.* **17**: 184.

98 Zhong, M.-M., Zhou, J., Fang, H., and Jena, P. (2017). Role of ligands on the stability of B*n*X*n* and CB*n*$_{-1}$X*n* (n = 5–10; X = H, F, CN) and their potential as building blocks of electrolytes in lithium ion batteries. *Phys. Chem. Chem. Phys.* **19**: 17937.

99 Zhao, H., Zhou, J., and Jena, P. (2016). Stability of $B_{12}(C\ N)_{12}{}^{2-}$: implications for lithium and magnesium ion batteries. *Angew. Chem. Int. Ed.* 55: 3704–3708.

100 Mayer, M., van Lessen, V., Rohdenburg, M. et al. (2019). Rational design of an argon-binding superelectrophilic anion. *Proc. Natl. Acad. Sci. U. S. A.* 116: 8167–8173.

101 Yadav, M., Fang, H., Santanab Giri, G., and Jena, P. (2019). Ligand stabilization of manganocene dianions – in defiance of the 18-electron rule. *Phys. Chem. Chem. Phys.* 21: 24300–24307.

102 Tuinman, A.A. and Compton, R.N. (1998). Structures of gas-phase $(C60)_n(CN)_m$ trianions from reactions of C_{60} with NaCN in solution. *J. Phys. Chem. A* 102: 9791–9796.

103 Feuerbacher, S. and Cederbaum, L.S.J. (2005). Stable and long-lived trianions in the gas phase. *Phys. Chem. A* 109: 11401–11406.

104 Zhao, T., Zhou, J., Wang, Q., and Jena, P. (2017). Colossal stability of gas-phase trianions: super-pnictogens. *Angew. Chem. Int. Ed.* 56: 13421–13425.

105 Fang, H. and Jena, P. (2019). Stable tetra- and penta-anions in the gas phase. *Angew. Chem. Int. Ed.* 58: 11248–11252.

106 Li, X., Grubisic, A., Stokes, S. T., Gentefor, G. F., Bowen, K. H., Boggavarapu, K., Willis, M., Jena, P., Burgert, R., and Schnockel, H. (2007). Unexpected Stability of Al4H6: A Borane Analog? *Science* 315, 356.

3

Superhalogens – Enormously Strong Electron Acceptors

Piotr Skurski

Laboratory of Quantum Chemistry, Faculty of Chemistry, University of Gdańsk, Gdańsk, Poland
Henry Eyring Center for Theoretical Chemistry, Department of Chemistry, University of Utah, Salt Lake City, Utah, USA

3.1 Superhalogen Concept

3.1.1 Early Studies

In the 1970s very few molecules exhibiting high electron affinity (EA) were known; hence, it was generally assumed that the EA values of neutral systems cannot exceed 4 eV. However, the relentless efforts leading to the improvement of the experimental methods enabling the EA measurements were being undertaken [1], which led to the first reports describing the molecules with high EA values [2, 3]. Although some of those predicted EAs seemed surprisingly significant (e.g. EA of 8.6–9.3 estimated for PtF_6 and EA = 9.5–10 eV predicted for AuF_6) [3, 4], their correctness and accuracy were questionable [5, 6] (the startling examples include a clearly incorrect estimation of electronic stability of $(PF_6)^-$ anion reported in 1978 by Barber et al. as equal to 13.4 eV) [7].

In 1981, Gutsev and Boldyrev published a brilliant paper describing several compounds whose theoretically evaluated EAs were expected to exceed those of the halogen atoms [8]. Without detracting from the importance of this milestone paper, it needs to be pointed out that the possible existence of certain neutral systems involving main-group elements in high oxidation states was postulated four years earlier by Sidorov who suggested that such compounds as BeF_3 and AlF_4 could be synthesized [9]. However, according to Kiselev and Tretiyakov, more recent studies demonstrated that various attempts to obtain compounds containing Be and Al in their oxidation states "higher than the extreme values" (i.e. +2 for beryllium and +3 for aluminum) in chemical processes failed [10]. Undaunted by these discouraging reports, Gutsev and Boldyrev performed in 1981 a theoretical investigation on several anions matching the $(MF_{k+1})^-$ formula and a few $(MO_{(k+1)/2})^-$ anions (where M is main-group central atom of maximal valence k), which resulted in reporting very large excess electron-binding energies thereof. In particular, they studied the following negatively

Superatoms: Principles, Synthesis and Applications, First Edition. Edited by Purusottam (Puru) Jena and Qiang Sun.
© 2022 John Wiley & Sons Ltd. Published 2022 by John Wiley & Sons Ltd.

charged systems: $(BeF_3)^-$, $(MgF_3)^-$, $(BF_4)^-$, $(AlF_4)^-$, $(SiF_5)^-$, $(PF_6)^-$, $(BO_2)^-$, $(AlO_2)^-$, $(NO_3)^-$, $(PO_3)^-$, and $(ClO_4)^-$. The vertical electron detachment energies (VDEs) of these species were calculated using a self-consistent discrete variational Xα method (SCF DVM-Xα) and basis sets of double-zeta quality. In order to have a better insight into the results obtained in the course of those calculations, one must realize that all the systems considered represented closed-shell valence-bound anions whose geometrical structures were assumed (based on the literature) to correspond to high-symmetry configurations – namely, linear $D_{\infty h}$-symmetry for $(BO_2)^-$ and $(AlO_2)^-$, triangular D_{3h}-symmetry for tetratomic systems, tetrahedral T_d-symmetry for penta-atomic species, trigonal bipyramidal D_{3h}-symmetry for $(SiF_5)^-$, and octahedral O_h-symmetry for $(PF_6)^-$. The resulting VDEs for these anions were found to exceed 3.62 eV (i.e. the EA of chlorine atom) [11] although the authors realized that the precision of their evaluations was likely insufficient (0.5–1.0 eV error margin was assumed). In particular, the following VDEs were predicted for $(MF_{k+1})^-$ systems: 4.5 eV $(BeF_3)^-$, 3.8 eV $(MgF_3)^-$, 5.5 eV $(BF_4)^-$, 5.6 eV $(AlF_4)^-$, 6.3 eV $(SiF_5)^-$, and 6.6 eV $(PF_6)^-$, whereas the corresponding values determined for $(MO_{(k+1)/2})^-$ anions were found to span the 3.6–6.3 eV range [8]. According to Gutsev and Boldyrev, two factors were primarily responsible for such large electron-binding energies: (i) effective delocalization of an excess negative charge among relatively large number of electronegative ligands and (ii) the absence of destabilizing antibonding ligand-central atom interactions in highest occupied molecular orbital (HOMO). Indeed, it was confirmed that the VDE values of $(MF_{k+1})^-$ anions increase when k develops from two to five. Also, the analysis of HOMOs revealed their nonbonding character with respect to ligand-central atom interactions (Figure 3.1).

Although 40 years have passed since Gutsev and Boldyrev published their first results on the strongly electronically bound anions [8], the importance of that milestone paper is not to be underestimated as it paved the way to further exploration of numerous molecular

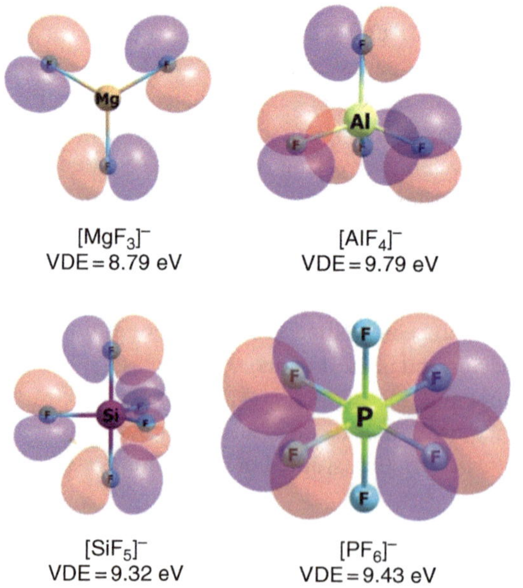

Figure 3.1 HOMOs for $(MgF_3)^-$, $(AlF_4)^-$, $(SiF_5)^-$, and $(PF_6)^-$ superhalogen anions showing nonbonding character with respect to central atom–ligand interactions. The reliable estimates of the VDE values are taken from Refs. [12–15]. HOMO for $(SiF_5)^-$ and $(PF_6)^-$ is doubly and triply degenerated, respectively. *Sources:* Adapted from Anusiewicz et al. [12], Sikorska et al. [13], Marchaj et al. [14], and Sobczyk et al. [15]. © John Wiley & Sons.

anions exhibiting large excess electron-binding energies. In addition, it enabled establishing and consolidating the term *superhalogens*, which has been used to describe the whole family of such molecular systems ever since.

3.1.2 Further Research (until 1999)

In the years 1981–1999, many other theoretical [16–45] and experimental [46–64] efforts have been undertaken to estimate the excess electron-binding energies of various anions having superhalogens as their neutral parents. In fact, certain experimental attempts to measure the electronic stability of such anions were made even earlier [2–5, 7, 9, 65–81]; however, one has to keep in mind that all such determinations performed before 1999 were related to condensed phases only (i.e. the gas-phase electron detachment energies had not been measured).

One of the issues addressed in those investigations was the estimation of adiabatic EAs of neutral superhalogen molecules. As mentioned above, superhalogen anions matching the $(MX_{k+1})^-$ formula (where X represents halogen atom) are closed-shell systems having singlet ground electronic states. Hence, the vertical excess electron-binding energy of $(MX_{k+1})^-$ (which is often referred to as vertical electron detachment energy, VDE) corresponds to the energy amount required to detach an electron from the anion at its equilibrium structure. Neutral superhalogen molecules matching the MX_{k+1} formula, by contrast, are open-shell systems having one unpaired electron and thus doublet ground electronic states. The adiabatic EA of a given compound is commonly defined as the decrease of its energy upon attachment of an excess electron [23]. Hence, the adiabatic EA of a neutral molecule represents the energy difference between the parent neutral and daughter anionic systems both corresponding to their ground electronic states and equilibrium geometries (see the references [23, 82] for the precise definitions of EA and VDE). Therefore, the knowledge of both neutral and anionic equilibrium structures is required for any theoretical evaluation of adiabatic EA. In the case of superhalogens, the first attempts were focused exclusively on the negatively charged species, which resulted in providing the VDE values only [8]. In the following years, partial geometry optimizations of the neutral superhalogens were performed, however, using symmetry constrains that limited the potential energy surface explorations to high-symmetry structures [23]. For example, tetrahedral T_d-symmetry was imposed while optimizing BF_4 and AlF_4 neutral molecules (likely due to the limited computer resources available then) [23]. Clearly, such partial geometry optimizations could not lead to the proper equilibrium structures of the neutral superhalogens as they significantly differ from those of their corresponding anions [44, 45]. This in turn rendered those early EA estimations inaccurate. More advanced calculations performed later for $LiX_2/(LiX_2)^-$ and $NaX_2/(NaX_2)^-$ (X = F, Cl) revealed that the adiabatic EAs are in fact much smaller (by ca. 1 eV) than the corresponding VDE values [44]. Moreover, it was found that the neutral structures of LiF_2, $LiCl_2$, NaF_2, and $NaCl_2$ are strongly bent [44, 83–85], whereas those of their corresponding anions are linear [44]. Substantial differences between the equilibrium structures of neutral and anionic superhalogens (and hence between the EAs of the former and VDEs of the latter systems) were also confirmed for structurally larger $MX_{k+1}/(MX_{k+1})^-$ neutral/anion pairs – namely, $BF_4/(BF_4)^-$ and $AlF_4/(AlF_4)^-$. According to the studies performed by Gutsev, Jena, and Bartlett [45], the lowest

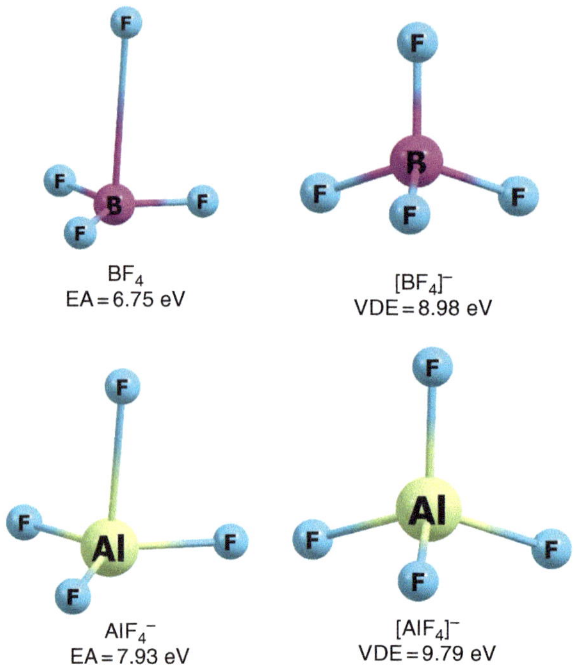

Figure 3.2 The equilibrium structures of neutral BF_4 and AlF_4 superhalogens and their corresponding $(BF_4)^-$ and $(AlF_4)^-$ anions. The EA values were taken from Ref. [45], whereas the VDEs were taken from Ref. [13]. *Source:* Adapted from Gutsev et al. [45] and Sikorska et al. [13]. © John Wiley & Sons.

energy structures of these superhalogen anions are tetrahedral and contain four equivalent (by symmetry) M-F bonds, whereas their corresponding neutral structures resemble F···MF_3 adduct-type systems (yet stable against the fragmentation yielding MF_3+F) composed of nearly planar triangular MF_3 fragment and a fluorine atom weakly tethered to it (see Figure 3.2 where the EA and VDE values are also provided). As for other superhalogens, the comparison of adiabatic EAs of those neutral compounds and VDEs of their corresponding anions revealed significant differences.

Among other issues concerning superhalogens and raised shortly after their possible existence and gas-phase stability were postulated, most urgent and challenging included: (i) designing structurally more complex compounds of superhalogen nature that would allow to increase the number of electronegative ligands in the molecule; (ii) searching for the neutral and anionic systems exhibiting very large EAs and VDEs, respectively; (iii) establishing the excess electron-binding energy limit that a molecular system can be characterized with; (iv) verifying whether various nonmetal elements may play the central atom role in superhalogens; (v) verifying whether halogen ligands can be replaced with nonhalogen elements or groups of atoms; and (vi) exploring the usefulness of both neutral superhalogens and superhalogen anions in synthesis, material science, and other areas. Having these topics in mind and taking into account that nowadays the design of novel molecules is being done primarily by theoretical chemists, it seemed necessary to establish the level of computational treatment that could be used to provide reliable results concerning superhalogens. Nevertheless, addressing all the above-mentioned issues required the involvement of many independent research groups, which could in turn be conceivable only in contemporaneous response to a scientific breakthrough in that area. In 1999, the experimental confirmation of superhalogens' existence in gas phase made it possible and triggered various researchers to study these systems more extensively.

3.1.3 First Measurement of Gas-Phase Experimental Electron Detachment Energies

Although the stability of isolated superhalogens and their anions was suggested by theoretical calculations performed in 1981, their existence in gas phase has not been experimentally confirmed until 1999 when the Wang group measured gas-phase electron detachment energies of $(MX_2)^-$ (M = Li, Na; X = Cl, Br, I) systems [86]. As indicated in the preceding sections, all earlier reports on such anions were related to condensed phases only, primarily due to the low stability of their neutral counterparts, which made the Knudsen-type equilibrium experiments almost unmanageable. Also, performing direct photodetachment experiments (using photoelectron spectroscopy, PES) was very difficult because of extremely large electron detachment energies of superhalogen anions. Fortunately, at that time the Wang group managed to develop an innovative experimental facility that couples a magnetic-bottle time-of-flight PES device and an electrospray ionization (ESI) source [87]. Since the ESI source turned out to be suitable for producing superhalogen anions while the high sensitivity of the magnetic-bottle PES technique allowed high photon energy experiments to be appropriately performed, the very first measurement of gas-phase electron detachment energies of those species became possible. The desired $(MX_2)^-$ anions were generated using the solutions of their typical MX salts in water/methanol (2/98) mixed solvent, which were sprayed through a syringe needle at ambient atmosphere and at −2.2 kV voltage bias. The resulting highly charged liquid droplets were introduced into a desolvation capillary. The anions emerging from the capillary were guided into an ion trap, accumulated there, and then pushed into the extraction zone of a time-of-flight mass spectrometer. After being mass selected and decelerated, the $(MX_2)^-$ anions were intercepted by a laser beam in the detachment zone of the magnetic-bottle photoelectron analyzer. Due to the large excess electron-binding energies of superhalogen anions, photodetachment at 193 nm from ArF excimer laser was used. The photoelectron spectra were collected for each $(MX_2)^-$ anion, converted to kinetic energy spectra, and calibrated. Finally, the binding energy spectra were obtained by subtracting the kinetic energy spectra from the photon energy (i.e. 193 nm (6.424 eV)) [86].

The analysis of the experimental photoelectron spectra of $(LiX_2)^-$ and $(NaX_2)^-$ systems (X = Cl, Br, I) confirmed large electron-binding energies of these anions (Figure 3.3). Since the measurements performed by the Wang group were supported by the advanced ab initio calculations executed by Boldyrev and Simons, a detailed assignment of the spectral features was possible. Moreover, it turned out that the VDEs elucidated from the photoelectron spectra were in excellent agreement with the values predicted by theoretical calculations [86]. Namely, the first VDEs of superhalogen anions were measured to be 5.92 ± 0.04 $(LiCl_2)^-$, 5.42 ± 0.03 $(LiBr_2)^-$, 4.88 ± 0.03 $(LiI_2)^-$, 5.86 ± 0.06 $(NaCl_2)^-$, 5.36 ± 0.06 $(NaBr_2)^-$, and 4.84 ± 0.06 eV $(NaI_2)^-$, whereas the corresponding VDEs obtained at the outer valence green function OVGF/6-311+G(2df) theory level (with the spin-orbit corrections) were calculated to be 5.90 $(LiCl_2)^-$, 5.48 $(LiBr_2)^-$, 4.57 $(LiI_2)^-$, 5.81 $(NaCl_2)^-$, 5.43 $(NaBr_2)^-$, and 4.50 eV $(NaI_2)^-$.

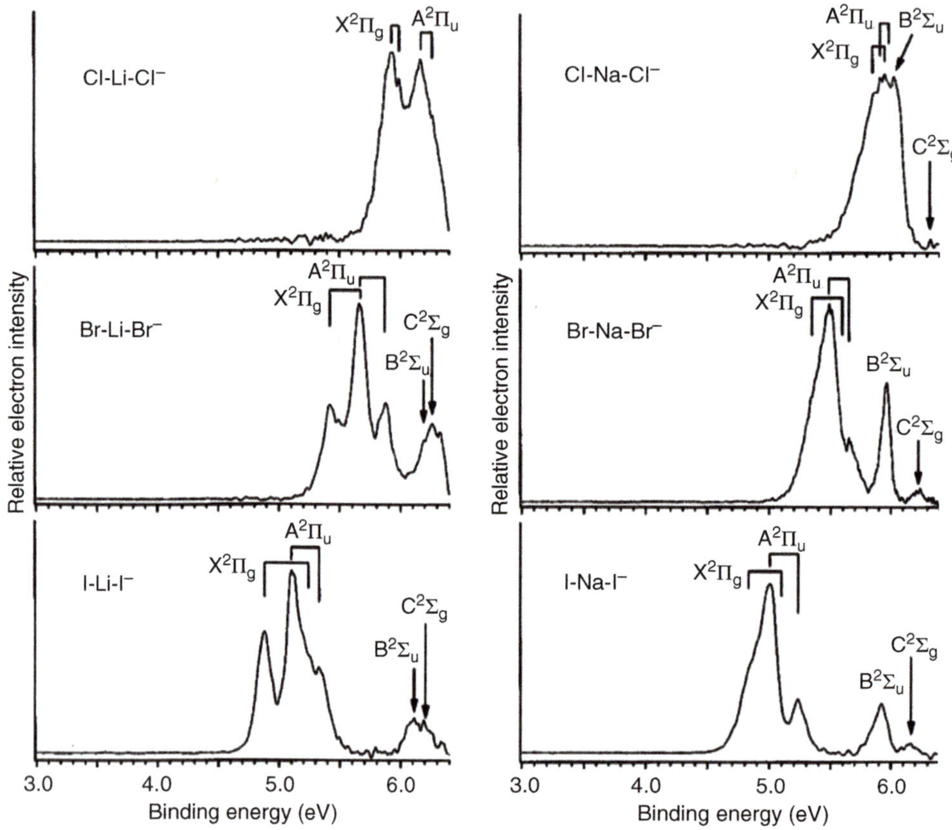

Figure 3.3 Photoelectron spectra of $(LiX_2)^-$ anions (left) and $(NaX_2)^-$ anions (right) (X = Cl, Br, I) at 193 nm. *Source:* Wang et al. [86]. © AIP Publishing.

This combined experimental and theoretical study of selected superhalogen anions was in fact a milestone achievement, which not only confirmed the existence and stability of such species in gas phase but also demonstrated the usefulness of certain computational ab initio techniques to predict their structures and excess electron-binding energies. Since the appearance of this paper, many research groups turned their attention to superhalogens that resulted in designing numerous new compounds of that type in the following years.

3.1.4 The Performance of Theoretical Treatments in Estimating VDEs

Despite the correctness of the advanced ab initio approaches in reproducing the electron-binding energies of superhalogen anions has been confirmed, little was known about the usefulness of more modest theoretical treatments. The comparison of the earlier results obtained by employing less-reliable methods (such as SCF DVM-Xα or DFT(B3LYP)) [8, 23, 88] and smaller basis sets (such as STO-nG or 6-31G(d,p)) to those determined experimentally or evaluated at much more advanced theory levels revealed substantial differences in predicted VDE values. In particular, it became clear that the electron-binding energies calculated at insufficient level are usually underestimated [86]. On the other hand, it was established that the use of either the OVGF or the coupled-cluster method with the single, double, and noniterative triple excitations (CCSD(T)) method (the former allows to evaluate the VDE in a direct manner, whereas the latter enables the prediction of VDE via a supermolecular approach) together with triple-zeta basis sets supplemented with additional sets of polarization and diffuse functions permits to achieve correct VDEs. Indeed, such obtained electron detachment energies of small superhalogen anions were found to match the experimentally measured VDEs in a sense that the differences between calculated and measured values are usually smaller than reported experimental inaccuracies. Such an advanced theoretical treatment, however, although correct and appropriate, seems computationally demanding. Hence, it limits the size of a superhalogen compound that might be the subject of theoretical studies to relatively small molecules. Taking into account the potential importance of significantly larger superhalogen species (e.g. in material science), it became clear that the usefulness of various theoretical treatments needed to be tested and verified.

The first systematic study on this subject appeared in 2011 with the aim of testing the performance of selected ab initio methods and one electron basis sets in reproducing the electron detachment energies of superhalogen anions [89]. In the course of this investigation, the VDEs of four relatively small negatively charged systems (i.e. $(LiCl_2)^-$, $(NaCl_2)^-$, $(MgCl_3)^-$, and $(BCl_4)^-$ whose experimentally measured electron detachment energies were available) were calculated by applying various methods (i.e. Hartree-Fock (HF), MPn ($n = 2$–4), CCSD(T), and OVGF) and basis sets (i.e. 3-21G, 3-21G*, 3-21+G, 3-21+G*, 6-31G, 6-31G*, 6-31+G, 6-31+G*, 6-311G, 6-311G*, 6-311+G, 6-311+G*, 6-311+G(3df), cc-pVDZ, aug-cc-pVDZ, cc-pVTZ, and aug-cc-pVTZ). The representative results for $(NaCl_2)^-$ are shown in Table 3.1 and Figure 3.4.

Not surprisingly, it was found that the HF method (the HF results are not shown in Figure 3.4) is definitely not recommended for estimating the VDEs of superhalogen anions since the resulting deviations may reach 10% (even with large basis sets supplemented with polarization and diffuse functions, see Table 3.1). The use of the Møller-Plesset perturbational methods, however, lead to much better results when applied with selected basis sets. In particular, the relative deviations (calculated as $|VDE^{est}–VDE^{exp}|$ divided by VDE^{exp}, where VDE^{est} stands for the VDE evaluated at a given level while VDE^{exp} is the experimentally measured value) produced by the MP2 method with either aug-cc-pVTZ or 6-311+G(3df) basis never exceed 3%. More computationally demanding MP3 and MP4 treatments were found to generate only slightly improved VDEs with respect to those obtained with the MP2 approach. Surprisingly, even the use of the CCSD(T) method led to the VDEs of similar accuracy to those estimated with the MP2 treatment, see Table 3.1. The OVGF technique was found to perform well as it produced relatively accurate results (within 2%) when applied together with the aug-cc-pVTZ or 6-311+G(3df) basis sets (although the predicted VDEs were slightly shifted up).

As far as the basis sets are concerned, it was found that the use of Pople split-valence basis sets of double-zeta quality may lead to the deviations of 8% for all methods considered except the OVGF method for which the deviations were smaller (up to 5%). Employing the Pople triple-zeta split-valence basis set improves the results as it produces the deviations not exceeding 7%. However, a significant improvement is observed when the 6-311G basis

Table 3.1 The VDEs (in eV) for $(NaCl_2)^-$.

	HF	MP2	MP3	MP4	CCSD(T)	OVGF
3-21G	5.295	5.105	5.188	5.156	5.125	5.283
3-21G*	5.227	5.149	5.217	5.152	5.131	5.247
3-21+G	5.551	5.445	5.512	5.489	5.457	5.677
3-21+G*	5.484	5.476	5.529	5.468	5.450	5.602
6-31G	5.354	5.181	5.261	5.232	5.203	5.377
6-31G*	5.279	5.219	5.286	5.224	5.203	5.326
6-31+G	5.510	5.404	5.468	5.450	5.415	5.655
6-31+G*	5.438	5.439	5.491	5.433	5.413	5.585
6-311G	5.594	5.461	5.531	5.496	5.472	5.644
6-311G*	5.496	5.466	5.524	5.456	5.441	5.550
6-311+G	5.527	5.423	5.484	5.457	5.428	5.636
6-311+G*	5.447	5.455	5.504	5.441	5.424	5.565
6-311+G(3df)	5.383	5.744	5.809	5.757	5.737	5.886
cc-pVDZ	5.248	5.262	5.324	5.256	5.233	5.362
cc-pVTZ	5.304	5.610	5.682	5.621	5.604	5.719
aug-cc-pVDZ	5.409	5.693	5.746	5.696	5.671	5.868
aug-cc-pVTZ	5.365	5.770	5.831	5.786	5.764	5.917

Experimental value: 5.86 ± 0.06 eV [86]. *Source:* Sikorska et al. [89]. © World Scientific Publishing.

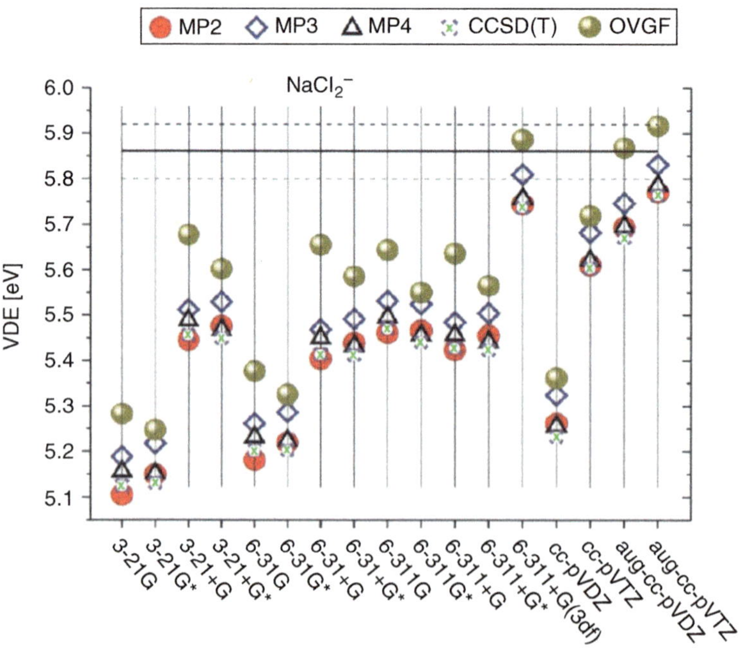

Figure 3.4 Electron detachment energies for the $(NaCl_2)^-$ superhalogen anion predicted with various ab initio methods and basis sets. The solid horizontal line at VDE = 5.86 eV corresponds to the VDE measured experimentally with the error margin indicated by the horizontal dashed lines.

is supplemented with a standard diffuse set, and the 3df set of polarization functions on heavy atoms (leading to the 6-311+G(3df) basis) as the electron-binding energies are then reproduced with the accuracy of 3%. The use of Dunning's correlation-consistent basis sets of double- and triple-zeta quality leads to quite reliable results but only when these basis sets are supplemented with the standard sets of diffuse functions; see Figure 3.4. In particular, the deviations produced by the use of the aug-cc-pVDZ basis set were found to span the 1–3% range while those obtained specifically at the OVGF/aug–cc–pVDZ level never exceeded 1%. Interestingly, the use of the aug-cc-pVTZ basis set generated the results of similar accuracy to those obtained with the aug-cc-pVDZ basis and even led to slightly less accurate VDEs in some cases.

In general, it was concluded that the MP2/6-311+G(3df) and MP2/aug-cc-pVnZ ($n = 2$, 3) treatments represent reasonable choices for predicting the VDEs of superhalogen anions, whereas the MPn ($n = 3$, 4) and CCSD(T) methods are not recommended (as computationally too demanding and not leading to relevant improvements). The use of the OVGF method with a medium-size basis set (e.g. 6-31+G*) was advised as a reasonable choice with the expected accuracy of ca. 5% for either preliminary estimates (for smaller systems) or final predictions (for structurally larger superhalogen anions). Finally, the OVGF/aug-cc-pVDZ and OVGF/6-311+G(3df) treatments were recommended (for accurate evaluation of VDE) as leading to the deviations never exceeding 1% [89].

The conclusions and recommendations gathered in Ref. 89, albeit useful for computational chemists investigating the electronic structure of superhalogens, are not sufficient for those who are focused primarily on their applications. Most of the recent studies concerning such systems seem to be moving toward very large species and molecular clusters containing sizeable numbers of heavy atoms. Therefore, there is a strong need for a new systematic study on the performance of theoretical treatments in estimating VDEs of superhalogen anions that would also cover various density functional theory (DFT) functionals, effective core potentials (ECP), and perhaps other approaches that enable performing calculations even for nanoparticles.

3.2 Alternative Superhalogens

Even though the precise definition of superhalogens is difficult to formulate, these species are usually described as the compounds whose EA exceeds the EA of chlorine atom (3.62 eV) [11]. In fact, such a definition was originally suggested by Gutsev and Boldyrev, and it covers various molecular systems regardless their composition and stoichiometry [8]. The early design of superhalogens made use of the octet electron-counting rule [90, 91]; however, Jena and Sun pointed out that other electron-counting rules (i.e. jellium rule, 18-electron rule, and Wade-Mingos rule) [92] may also be used for this purpose. It should also be emphasized that many well-known systems meet the above-mentioned criterion and thus can be termed superhalogens (e.g. CN or C_6F_5 (pentafluorophenyl) radicals whose EAs are equal to 3.86 [93] and 4.5 eV [94, 95], respectively). Therefore, any general formula covering all superhalogens cannot be provided. Nevertheless, there exists a simple approach that enables to recognize a superhalogen on the basis of two fundamental factors: (i) the EA of a system has to exceed the 3.62 eV threshold, and (ii) its chemistry must mimic that of halogen atoms [92].

Taking into account the absence of any general chemical formula covering all possible superhalogens, we decided to limit our description provided in this section to the compounds matching the $MX_{k+1}/(MX_{k+1})^-$ formula proposed by Gutsev and Boldyrev in their original paper published in 1981 [8]. In fact, these *Gutsev-Boldyrev superhalogens* remain the most extensively studied compounds in the whole superhalogen family of molecular systems. The original MX_{k+1} recipe assumed that a metal atom M (having its maximal valence given by k) decorated with halogen atoms X whose number exceeds the maximal valence of M by 1 would correspond to a superhalogen. Hence, this formula seemed to limit the central atoms M to metal atoms and the ligands X to halogens, but it was soon revealed that these limitations do not apply if M and X fulfill certain criteria.

3.2.1 Nonmetal Central Atoms

The possibility of using nonmetal atoms as central atoms was actually mentioned already in 1981 as the PF_6 compound was one of the systems studied [8]. In the following years, more neutral superhalogens containing nonmetal or metalloid central atoms were proposed, such as BF_4, PCl_6, SiF_5, $SiCl_5$, AsF_6, and $AsCl_6$, although their EAs were anticipated using relatively modest theoretical approaches only [18, 23]. Later predictions provided more accurate electron-binding energies of those systems and enabled a thorough insight into their properties and electronic structure [13–15, 96–100]. It turned out that the VDEs of such superhalogen anions far exceed the 3.62 eV threshold value (e.g. 9.32 eV for $(SiF_5)^-$, 9.74 eV for $(GeF_5)^-$, and 9.43 eV for $(PF_6)^-$; see Figures 3.1 and 3.5) [14, 15]. It was also recognized that even hydrogen atom may play a central atom role in superhalogens [101]. Indeed, the VDEs pre-

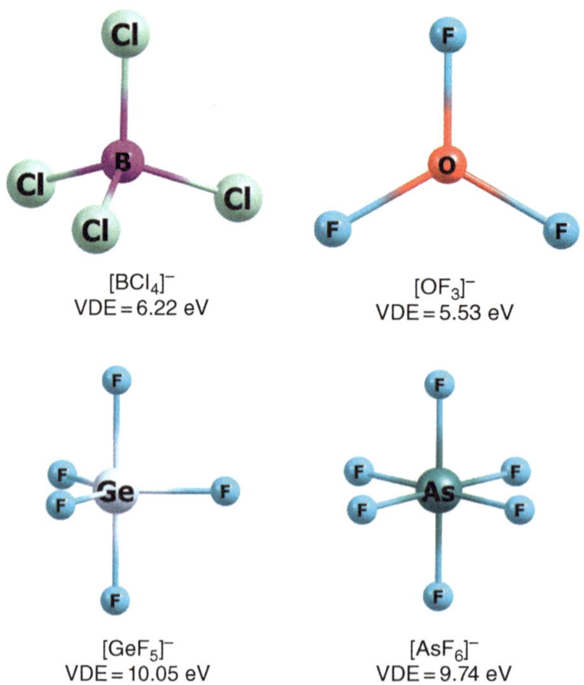

Figure 3.5 Equilibrium structures of selected alternative superhalogen anions utilizing nonmetal or metalloid central atoms.

dicted for $D_{\infty h}$-symmetry structures of $(FHF)^-$, $(ClHCl)^-$, and $(BrHBr)^-$ anions (and also for the polynuclear $(H_nF_{n+1})^-$ (n = 2–5, 7, 9, 12) anions, which will be discussed in the following sections) were found to be 6.453, 5.080, and 4.663 eV, respectively [101]. Next, it is worth noting that halogen atoms may also act as central atoms in superhalogens, and the examples of such species include the bromine-centered $ClBrF_n/(ClBrF_n)^-$ (n = 2–5) systems (whose EAs and VDEs were predicted to span the 4.41–5.62 and 5.56–6.49 eV range, respectively) [102], the chlorine-centered $BrClF_n/(BrClF_n)^-$ (n = 2–5) compounds (EA = 4.83–6.12 eV/ VDE = 5.87–7.13 eV) [102], and $(ClO_4)^-$ anion having its VDE = 5.25 eV [8, 103, 104] (the latter system matches a superhalogen formula of $MO_{(k+1)/2}$, which is equivalent to MX_{k+1} when the divalent oxygen atoms act as ligands X) [8]. Finally, we provide a very untypical example of a superhalogen system that obeys the MX_{k+1} formula and contains oxygen as the central atom. Namely, the $(OF_3)^-$ anion was found to adopt D_{3h}-symmetry triangular equilibrium structure (see Figure 3.5) while its VDE was estimated to be 5.53 eV [105].

3.2.2 Nonhalogen Ligands

On the other hand, the possibility of utilizing nonhalogen ligands has been extensively explored likewise. First, it seemed natural to verify whether halogenoids (such as CN, SCN, or OCN) could play the ligand role in superhalogens. It turned out that the $(MX_{k+1})^-$ anions comprising Li, Na, Be, Mg, B, and Al central atoms (M) and CN, SCN, and OCN ligands (X) correspond to thermodynamically stable systems whose VDEs far exceed the 3.62 eV threshold [106]. In particular, the VDEs of the anions involving the -CN (or -NC) ligands were found to span the 7–9 eV range, whereas the VDEs of the anions comprising the remaining halogenoids were predicted to span the 5–8 eV range (for -OCN or -NCO) and the 5–6.5 eV range (for -SCN or -NCS) [106]. The trend in the VDEs of the resulting superhalogen anions was recognized as being related to the electron-binding energies of the isolated ligands (the CN radical possesses the largest EA of 3.862 ± 0.004 eV while the EAs of OCN and SCN radicals are equal to 3.609 ± 0.005 and 3.537 ± 0.005 eV, respectively) [93].

It soon became clear that not only halogenoids (whose similarity to halogens is evident) but also other radical fragments and functional groups may serve as ligands in superhalogens. It was found that various electrophilic groups (e.g. NO_2, CF_3, CCl_3, HSO_3, and COOH) represent convenient ligands, and their use leads to substantial electron-binding energies even for the systems containing Na or Mg central atoms [107]. Certain acidic functional groups (i.e. ClO_4, ClO_3, ClO_2, ClO, NO_3, PO_3, H_2PO_4, HSO_4, HCO_3, and HS) [108] were also recognized as capable of serving as ligands in superhalogen anions as all of the negatively charged $(NaX_2)^-$ and $(MgX_3)^-$ species containing these groups have proven to possess electron-binding energies exceeding 4.2 and approaching 9 eV in some cases. For example, the VDEs evaluated for $(Mg(ClO_4)_3)^-$, $(Mg(PO_3)_3)^-$, $(Mg(HSO_4)_3)^-$, $(Mg(NO_3)_3)^-$, and $(Mg(HCO_3)_3)^-$ are equal to 8.914, 8.676, 8.606, 8.293, and 7.379 eV, respectively [108]. Interestingly, the superhalogen anions comprising acidic functional groups exhibit the dependence of their excess electron-binding energies on the strength of the underlying HX acids that these groups were derived from. Namely, the largest VDE values were found for the anions containing the ClO_4 ligands ($HClO_4$ was the strongest acid among the systems considered), whereas the smallest VDE values were found for the anions involving HS groups (H_2S was the weakest acid considered) [108].

Other molecular fragments were also identified as potentially capable of serving as ligands in superhalogen anions, for example, the OH and C_6F_5 radicals [109, 110]; see Figure 3.6. Indeed, the $(M(OH)_{k+1})^-$ anions (M = Li, Na, K, Be, Mg, Ca, B, Al, and Ga) were found to be thermodynamically stable, and their VDEs were predicted in the 4.27–6.21 eV range (with the largest excess electron-binding energies of 6.07 and 6.21 eV for S_4-symmetry $(Al(OH)_4)^-$ and $(Ga(OH)_4)^-$ systems, respectively) [110]. It seems important to notice that the OH fragment, unlike other previously described radicals, does not represent a strong electron acceptor (EA = 1.825 ± 0.002 eV) [111], yet it acts as an effective ligand in superhalogens. Hence, it turned out that even common radicals exhibiting relatively modest EAs may play the ligand role in such systems.

Another important finding coming from that report [110] was the confirmation of the fact that the formation of the $(MX_{k+1})^-$ superhalogen anions (by combining the MX_k and X^- systems) should proceed spontaneously in the gas phase since the corresponding reactions involving $(M(OH)_{k+1})^-$ anions (M = Li, Na, K, Be, Mg, Ca, B, Al, and Ga) were found to be barrier-free and thermodynamically favorable; see the representative energy profile shown in Figure 3.7.

The verification of thermodynamic stability is required while designing the alternative superhalogen anions. In some cases, the promising compounds exhibiting large electron-binding energies turn out to be susceptible to fragmentation and thus impractical as such. The $(M(OOH)_{k+1})^-$ anions (M = Li, Na, K, Be, Mg, Ca, B, Al, and Ga) represent the examples of the systems whose predicted VDEs are large (3.7–6.3 eV), but their fragmentations leading to the O_2 loss are plausible [112].

Establishing a large body of work concerning superhalogens involving various ligands enabled recognizing that the fluorine atoms lead to the largest VDEs when used as substituents X in the $(MX_{k+1})^-$ systems. Therefore, the alternative nine-electron ligands (i.e. isoelectronic with fluorine atom) were investigated to verify their usefulness in designing superhalogen anions [113]. The ligands considered consisted of various possible combinations of atoms containing nine electrons when brought together, which led to the following

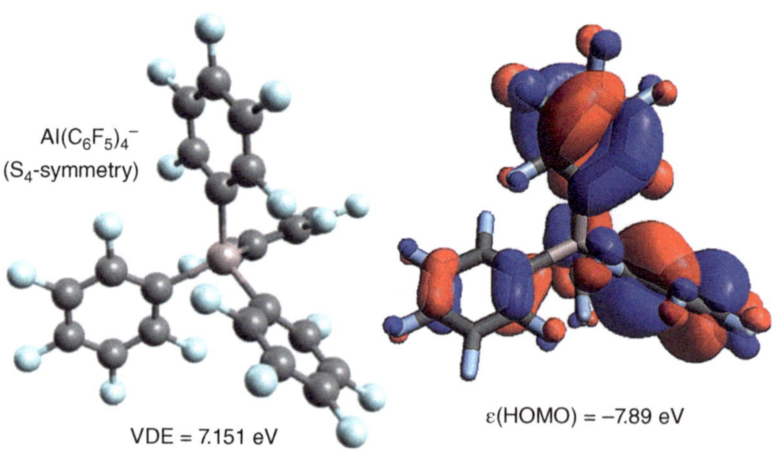

Figure 3.6 Equilibrium structure of the $(Al(C_6F_5)_4)^-$ anion and its doubly degenerate HOMO. *Source:* Sikorska and Skurski [109]. © Taylor & Francis.

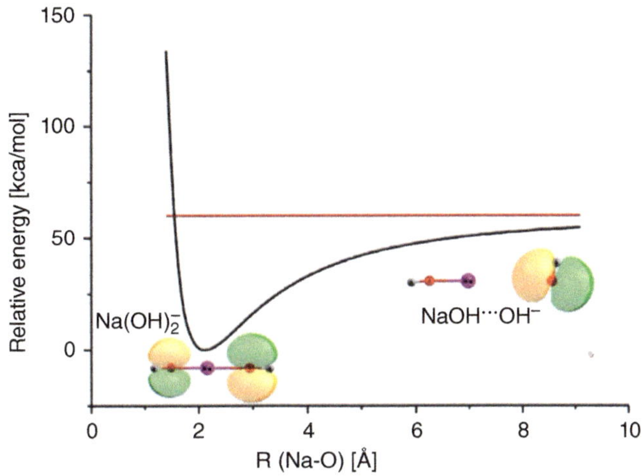

Figure 3.7 The energy profile for the formation of the $(Na(OH)_2)^-$ anion according to the $NaOH + (OH)^- \rightarrow (Na(OH)_2)^-$ reaction. The HOMOs holding the excess electron are depicted for the structures corresponding to $R = 2.102$ Å (equilibrium geometry) and $R = 4.100$ Å. *Source:* Świerszcz and Anusiewicz [110]. © Elsevier.

radicals: Li_2H_3, OH, H_2BeLi, BeB, NH_2, HBLi, CH_3, Be_2H, and CLi. Among these molecular fragments only Li_2H_3, OH, and NH_2 were found suitable for serving as ligands in superhalogen anions due to the thermodynamic stability and large values of electron-binding energy exhibited by the resulting $(MX_{k+1})^-$ systems. In particular, the neutral Li_2H_3 radical (whose EA was estimated as equal to 4.00 eV) appeared to be potentially useful because the VDEs of certain anions containing such ligands are relatively large (i.e. 4.30 eV for $(Na(Li_2H_3)_2)^-$ and 5.78 eV for $(Al(Li_2H_3)_4)^-$) [113].

The use of electronegative systems (atoms or molecular fragments) as substituents in superhalogen anions seems natural as the excess negative charge in $(MX_{k+1})^-$ is expected to delocalize among $k+1$ ligands, which in turn increases the electronic stability of the anion. The significance of this effect was demonstrated [114, 115] by a successive replacement of the fluorine ligands in $(BF_4)^-$ and $(PF_6)^-$ superhalogen anions with nonelectronegative alkyl ligands (CH_3 and C_2H_5). Namely, it was found that subsequent replacement of the fluorine ligands with alkyl R groups in the $(BF_4)^-$ or $(PF_6)^-$ anion results in a substantial electronic stability decrease (by 1.13–3.42 and 2.38–6.74 eV, respectively). The VDE of the $(BF_4)^-$ anion (8.975 eV) decreases by ca. 3 eV and then by ca. 1 eV when the first and the second F atom is replaced with CH_3 group, whereas the replacement of the third and fourth F ligand causes only small VDE change (ca. 0.1 eV); see Table 3.2. These changes are similar to those observed when the F ligands in $(BF_4)^-$ are being replaced with the ethyl (C_2H_5) functional groups, although the effect of replacing the first fluorine ligand is more profound as it leads to VDE decrease by 3.42 eV [114]. Despite this, even the fully substituted $(B(CH_3)_4)^-$ and $(B(C_2H_5)_4)^-$ anions remain stable and relatively strongly bound (by ca. 4.3–4.4 eV) negatively charged systems, which can be considered superhalogen anions as their excess electron-binding energies exceed 4 eV. Recalling that the EAs of the isolated CH_3 and C_2H_5 radicals are either negligibly small (0.08 ± 0.03 eV for CH_3) [116] or negative (-0.263 ± 0.089 eV for C_2H_5) [117], it was concluded that the presence of four substituents bound to the boron atom is the most important factor that enables the stability of the $(BF_{4-n}R_n)^-$ anions and causes large electron detachment energies thereof [114]. Clearly, the alkyl groups serving as ligands cannot be regarded as electronegative substituents that

Table 3.2 The VDEs (in eV) of $(BF_{4-n}R_n)^-$ anions (n = 0–4, R=CH_3, C_2H_5).

Species	VDE	Species	VDE
$(BF_4)^-$	8.975	$(BF_4)^-$	8.975
$(BF_3CH_3)^-$	5.900	$(BF_3C_2H_5)^-$	5.550
$(BF_2(CH_3)_2)^-$	4.629	$(BF_2(C_2H_5)_2)^-$	4.425
$(BF(CH_3)_3)^-$	4.508	$(BF(C_2H_5)_3)^-$	4.335
$(B(CH_3)_4)^-$	4.422	$(B(C_2H_5)_4)^-$	4.272

Source: Sieradzan and Anusiewicz [114]. © Elseiver.

would efficiently contribute to excess electron binding. However, it turns out that even such fragments may play the ligand role in superhalogen anions if their number is adequate (i.e. if the resulting compound matches the MX_{k+1} formula). Analogous conclusions have been formulated while studying the subsequent replacement of the F ligands in $(PF_6)^-$ anion with either methyl or ethyl functional groups [115]. Again, the presence of six (i.e. $k+1$ for $k = 5$) substituents bound to the phosphorus atom was identified as the most important factor that enables the stability of the $(PF_{6-n}R_n)^-$ anions (n = 0–6, R = CH_3, C_2H_5) and assures their substantial VDEs [115]. The studies performed for $(BF_{4-n}R_n)^-$ and $(PF_{6-n}R_n)^-$ anions not only confirmed the usefulness of the original MX_{k+1} formula and the significance of selecting electronegative substituents as ligands (when large electron-binding energies are desired) but also revealed the importance of symmetry breaking in superhalogen anions (see Ref. [118] in which the symmetry issue is discussed).

Among many other atoms and fragments whose ability to serve as ligands in superhalogen anions has been verified through the years, the hydrogen atoms seem particularly interesting (mostly because of their simple electronic structure and the fact that the EA of H is rather small (0.75 eV)) [119]. According to the theoretical predictions, the systems with hydrogen ligands (i.e. the $(MH_{k+1})^-$ anions) possess surprisingly high electron-binding energies and some of them can indeed be classified as superhalogen anions (e.g. $(BeH_3)^-$, $(MgH_3)^-$, $(BH_4)^-$, $(AlH_4)^-$, $(TiH_5)^-$, and $(VH_6)^-$ whose VDEs were estimated to be in the 3.74–4.75 eV range) [29, 120]. Recently, the existence and stability of alanate $(AlH_4)^-$ anion was confirmed experimentally; namely, the $(AlH_4)^-$ system was generated in the gas phase using a pulsed arc cluster ionization source and its photoelectron spectrum was then measured with 193 nm photons [121]. The VDE of $(AlH_4)^-$ was measured to be 4.4 eV, and this value was in agreement with the results obtained by theoretical calculations [121].

3.2.3 Beyond the MX_{k+1} Formula

Since the large excess electron-binding energies of superhalogen anions matching the $(MX_{k+1})^-$ formula have been attributed mostly to the fact that the excess negative charge can be delocalized over $k+1$ ligands [8, 23], the attempts to increase the number of ligands have been undertaken. In 2011, Lo and Hopkinson studied [122] the anions in which the central atom M is decorated with halogen atoms X whose number exceeds the maximal valence of M by 2 (rather than by 1, as in *Gutsev-Boldyrev superhalogens*) [8]. It was found that such $(MX_{k+2})^-$ anions might be both stable and strongly electronically bound. In particular, the existence of the $(KF_3)^-$ anion was proposed and its VDE was estimated to be 6.83 eV. Since this value appeared to be significantly larger (by 1.2 eV) [34] than the VDE of $(KF_2)^-$ anion, it was suggested that the $(KF_3)^-$ might be a representative example of novel $(MX_{k+2})^-$ anions having extremely large electronic stabilities [122]. Nevertheless, more thorough studies covering large number of $(MF_{k+2})^-$ systems (M = Li, Na, K, Be, Mg, Ca, B, Al, Ga) revealed that although the most stable isomers of the $(LiF_3)^-$, $(NaF_3)^-$, $(KF_3)^-$, $(BeF_4)^-$, $(MgF_4)^-$, and $(CaF_4)^-$ anions correspond to the structures having all fluorine ligands bound to the metal central atom (with no bonding interactions among the F atoms present), the structures corresponding to the global minima of the $(BF_5)^-$, $(AlF_5)^-$, and $(GaF_5)^-$ anions resemble quasi-tetrahedral MF_4 subunits with the fifth fluorine atom tethered to them via the elongated F–F bonds [123]; see Figure 3.8. Hence, the $(MF_5)^-$ species (M = B, Al, Ga) should be treated as $(MF_4\cdots F)^-$ complex anions rather than $(MF_5)^-$ anions. In addition, it was found [123] that some of the $(MF_{k+2})^-$ anions (e.g. $(BeF_4)^-$ and $(BF_5)^-$) are thermodynamically unstable (i.e. susceptible to the fluorine atom loss according to the $(MF_{k+2})^- \rightarrow (MF_{k+1})^- + F$ fragmentation process) leading to the formation of the corresponding *Gutsev-Boldyrev superhalogen anions*. Hence, taking also into account that the VDEs of the $(MF_{k+2})^-$ systems turned out to be usually only slightly larger (by ca. 0.05–0.16 eV) than the VDEs of their corresponding $(MF_{k+1})^-$ *Gutsev-Boldyrev superhalogen anions* (although in certain cases may be larger by ~1 eV, e.g. for $(BeF_4)^-$ VDE = 8.76 eV vs. $(BeF_3)^-$ VDE = 7.63 eV) [123], the systems matching the $(MF_{k+2})^-$ formula do not appear to be promising strongly bound molecular anions in general, mostly because of their either small thermodynamic stability or instability. Also, the VDE increase resulting from the presence of the additional ligand is not substantial for the majority of $(MF_{k+2})^-$ systems examined [123].

Finally, it may be anticipated that a further increase of the ligand number would likely lead to even more unstable compounds (due to enhancement of steric hindrance and destabilizing valence repulsion effects). Therefore, the well-established $(MX_{k+1})^-$ formula remains the best choice for designing strongly electronically bound and thermodynamically stable superhalogen anions.

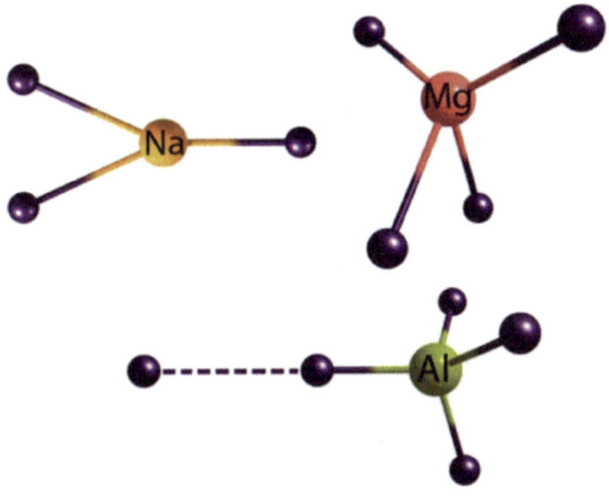

Figure 3.8 The lowest energy isomers of $(NaF_3)^-$ (C_{2v}-symmetry, VDE = 7.465 eV), $(MgF_4)^-$ (C_{2v}-symmetry, VDE = 8.949 eV), and $(AlF_5)^-$ (C_{3v}-symmetry, VDE = 9.851 eV) anions matching the $(MX_{k+2})^-$ formula. *Source:* Marchaj et al. [123]. © Elsevier.

3.2.4 Superhalogens as Ligands

Since the use of electronegative substituents as ligands in superhalogens has been proven successful, variously directed efforts have been undertaken to design novel compounds exhibiting large excess electron-binding energies. In 2010, the idea of using nonmodified superhalogens as ligands in order to increase the EA of the resulting systems has been proposed [124, 125]. This concept was motivated mostly by the observation that the EA of a MF_n cluster (M = Cu, Ag, Au; $n = 1–6$) increases when n develops from one to six (i.e. when the number of fluorine ligands bound to the central atom M increases) [126, 127]. In a $(MX_{k+1})^-$ superhalogen anion, the extra electron is delocalized over $k+1$ halogen atoms X, hence the ability of ligands X to accept the excess negative charge seems to be crucial for the effectiveness of this process. Namely, one may anticipate that the use of ligands X having large EA should lead to the enhancement of the EA exhibited by the resulting MX_{k+1} superhalogen molecule. In fact, such a prediction is known as a collective-effect concept, which is based on a simple assumption that each X ligand contributes to the total EA of MX_{k+1} system. However, it has to be stressed that this assumption is not always correct. For example, the EAs of Cl atom and C_6F_5 radical are larger than the EA of fluorine atom, yet the VDEs predicted for $(MF_{k+1})^-$ anions are always (i.e. for various central atoms M) larger than those estimated for the corresponding $(MCl_{k+1})^-$ or $(M(C_6F_5)_{k+1})^-$ anions. Nevertheless, the idea of utilizing strong electron acceptors as ligands in superhalogen anions is generally correct, and it prompted the researchers to investigate whether the superhalogens themselves might serve as ligands X in the MX_{k+1}/$(MX_{k+1})^-$ systems [125].

According to theoretical evaluation, the EA of $Au(BO_2)_2$ (5.54 eV) is nearly two times larger than that of AuO_2 while the EA of $AuO(BO_2)$ (4.21 eV) lies between that of AuO_2 and $Au(BO_2)_2$. Since these results were validated by PES measurements [125], it was concluded that the EAs of superhalogens can likely be further enhanced by modifying the building blocks, namely, by utilizing superhalogens themselves (e.g. BO_2) to serve as ligands. This concept is believed to offer a promising path for designing highly electronegative species by suitable choosing of the composition of these compounds (i.e. the superhalogen ligands and central atoms). The systems designed in such a way have been termed *hyperhalogens* because their EAs are significantly increased upon replacing the conventional atomic ligands (e.g. F, Cl, O) with superhalogen moieties (simply put, hyperhalogens are the systems comprising the central atom decorated with superhalogen ligands and having their EA larger than the EA of those ligands) [125]. The possibility of enhancing the EA by the use of superhalogen moieties was originally demonstrated for the molecules containing Au as a central atom. Namely, the comparison of the EA values for AuF_2, $AuCl_2$, $AuBr_2$, and AuI_2 (4.84, 4.63, 4.46, and 4.38 eV, respectively) to the EA of $Au(BO_2)_2$ (5.54 eV) revealed a substantial EA increase of the AuX_2 system caused by replacing the halogen atoms with superhalogen BO_2 fragments [125].

Since 2010 when the hyperhalogen concept was introduced, many novel compounds comprising superhalogen ligands have been proposed and investigated. Although most of them were proposed by the Jena group, other research groups also contributed to the field. The BO_2 superhalogen was used as a building block in designing $(CuBO_2)^-$, $(Cu(BO_2)_2)^-$,

$(Cu_2BO_2)^-$, and $(Cu_2(BO_2)_2)^-$ anions whose properties were investigated using PES. This study revealed that the neutral $Cu(BO_2)_2$ system can be classified as a hyperhalogen as its experimentally predicted EA of 5.07 eV exceeds that of its BO_2 superhalogen building block (4.46 eV) [128]. It was soon recognized that designing compounds that utilize superhalogens as ligands does not always lead to expected results, which was demonstrated for AlO_2 and VO_3 superhalogen moieties. Namely, it was found that neither $Na(AlO_2)_2$ nor $Na(VO_3)_2$ corresponds to a hyperhalogen as their EAs (3.19 and 4.22 eV for $Na(AlO_2)_2$ and $Na(VO_3)_2$, respectively) do not exceed those of their superhalogen AlO_2 and VO_3 building blocks, likely due to the large binding energies of $(AlO_2)_2$ and $(VO_3)_2$ dimers [88]. On the other hand, the use of $MnCl_3$ superhalogen ligands enabled the design of $(Mn(MnCl_3)_3)^-$ hyperhalogen anion whose VDE was predicted to be 6.76 eV [129]. Moreover, it was found that even higher VDE values could be achieved when the $Mn(MnCl_3)_3$ is used as a ligand (e.g. the VDE of 7.26 eV was estimated for the $(Li(Mn(MnCl_3)_3)_2)^-$ anion) [129]. Analogous enhancement of excess electron-binding energy was predicted when $Na(BF_4)_2$ (EA = 8.33 eV) and $Li(BF_4)_2$ (EA = 8.51 eV) systems were used as ligands in $Na(Na(BF_4)_2)_2$ and $Li(Li(BF_4)_2)_2$ whose EAs were estimated as equal to 9.18 and 9.01 eV, respectively [130]. Also, using the $Mg(BH_4)_3$ moieties as building blocks led to designing the $Mg(Mg(BH_4)_3)_3$ compound exhibiting hyperhalogen nature [131]; see Figure 3.9. Although the authors of this procedure (i.e. the enhancement of EA by using more and more structurally complex superhalogen building blocks) considered it to be the recipe that provides a systematic way for "creating species with ever increasing EAs," such an expectation seems rather too optimistic because numerous investigations on molecular systems (designed using this technique) performed in the last decade led to the largest EAs of ~10–11 eV [132], whereas the values exceeding 12 eV remain unattainable.

Further experimental and theoretical studies on various systems involving superhalogens as ligands included the systems gathered in Table 3.3.

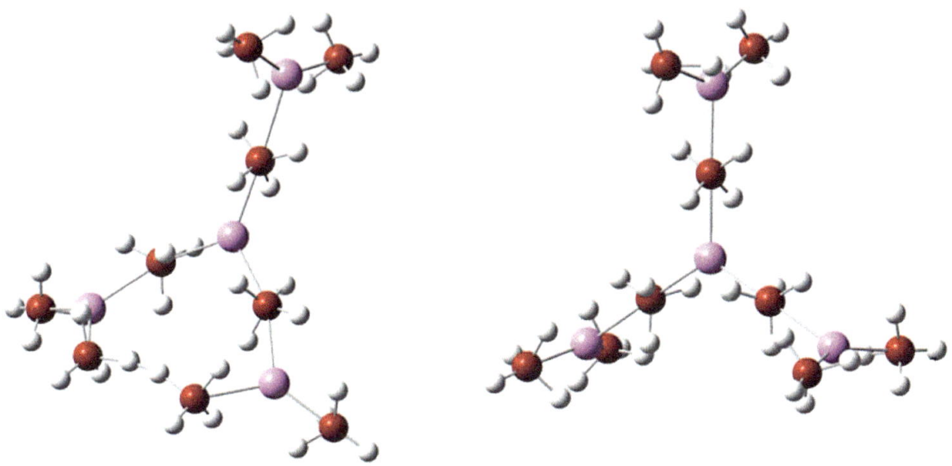

Figure 3.9 The equilibrium structure of neutral $Mg(Mg(BH_4)_3)_3$ hyperhalogen (left, EA = 6.18 eV) and its $(Mg(Mg(BH_4)_3)_3)^-$ anion (right, VDE = 7.60 eV). *Source:* Paduani and Jena [131]. © Elsevier.

Table 3.3 EAs (in eV) and VDEs (in eV) of neutral and anionic systems containing superhalogen ligands, respectively.

System	EA/VDE (eV)	System	EA/VDE (eV)
$(H(IO_3)_2)^-$	6.25 [133]	$(Li(IO_3)_2)^-$	6.57 [133]
$(Na(IO_3)_2)^-$	6.60 [133]	$(K(IO_3)_2)^-$	6.51 [133]
$Gd(AlH_4)_4$	4.95 [134]	$Gd(AlF_4)_4$	8.42 [134]
$(Gd(AlH_4)_4)^-$	6.23 [134]	$(Gd(AlF_4)_4)^-$	9.66 [134]
$Gd(BO_2)_4$	6.67 [135]	$(Gd(BCl_4)_4)^-$	7.12 [136]
$(Gd(BO_2)_4)^-$	7.02 [135]	$(Gd(GdCl_4)_4)^-$	7.70 [136]
$(BH_{4-n}(BH_4)_n)^-$ ($n = 1$–4)	5.52–7.28 [137]	$AuF_{4-n}(PO_3)_n$ ($n = 1$–4)	6.16–7.27 [138]
$Mg(BF_4)_3$	9.24 [139]	$(AuF_{4-n}(PO_3)_n)^-$ ($n = 1$–4)	7.16–8.20 [138]
$(Mg(BF_4)_3)^-$	9.33 [139]	$Gd(BF_4)_4$	9.70 [139]
AuB_2O_3	6.05 ± 0.08 [140]	$(Gd(BF_4)_4)^-$	10.22 [139]
$(AuB_2O_3)^-$	6.17 ± 0.08 [140]	$(Li(N_5)_2)^-$	6.87 [141]
$(Be(N_5)_3)^-$	7.42 [141]	$(B(N_5)_4)^-$	7.86 [141]
$(Li_2(N_5)_3)^-$	7.17 [141]	$(Be_2(N_5)_5)^-$	7.67 [141]
$(Sc(BO_2)_4)^-$	9.08 [142]	$(Ti(BO_2)_n)^-$ ($n = 4, 5$)	6.66–9.23 [142]
$Ag(BO_2)_2$	6.20 [143]	$(V(BO_2)_n)^-$ ($n = 3$–6)	5.18–9.44 [142]
$(Ag(BO_2)_2)^-$	6.28 [143]	$(Li(C_8F_7)_2)^-$	5.74 [144]
$(Na(C_8F_7)_2)^-$	5.37 [144]	$(K(C_8F_7)_2)^-$	5.11 [144]
$(Be(C_8F_7)_3)^-$	6.45 [144]	$(Mg(C_8F_7)_3)^-$	6.37 [144]
$Al(BH_4)_4$	5.43 [145]	$(Ca(C_8F_7)_3)^-$	6.12 [144]
$(Al(BH_4)_4)^-$	7.18 [145]	$Al(BF_4)_4$	8.86 [145]
$Mn(BO_2)_3$	5.71 [146]	$(Al(BF_4)_4)^-$	9.96 [145]
$(Mn(BO_2)_3)^-$	6.33 [146]	$Mn(BO_2)_4$	6.28 [146]
$Fe(BO_2)_3$	5.73 [146]	$(Mn(BO_2)_4)^-$	6.83 [146]
$(Fe(BO_2)_3)^-$	5.75 [146]	$Fe(BO_2)_4$	6.37 [146]
$(Fe(BO_2)_4)^-$	6.73 [146]		

3.3 Polynuclear Systems and the Search for EA and VDE Limits

The possible existence and stability of superhalogens containing more than one central atom was suggested in the early 1980s by Gutsev and Boldyrev who predicted the equilibrium structures and electron-binding energies of $(Al_2F_7)^-$, $(P_2F_{11})^-$, $(As_2F_{11})^-$, and $(V_2F_{11})^-$ anions [20, 147]. It was then recognized that the mononuclear MX_{k+1} formula could be extended to $M_nX_{n\times k+1}$ (where n stands for the number of central atoms) to cover the polynuclear superhalogens and their corresponding $(M_nX_{n\times k+1})^-$ anions. In addition, it was found that the lowest energy isomers of $(M_nX_{n\times k+1})^-$ systems correspond to the structures containing the central atoms M connected to one another via the bridging halogen ligands (i.e. comprising the M-X-M linking fragments). Although several such species were proposed back in the 1980s [20, 23, 147], little was known about the precise values of their electron-binding energies. The studies that followed provided a deeper insight into their electronic structure.

3.3.1 Polynuclear Superhalogens

The pioneering investigations on polynuclear superhalogens and their corresponding anions have been performed using rather modest computational methods and led to likely approximate values of EAs (e.g. Al_2F_7 (7.5 eV), P_2F_{11} (8.6 eV), As_2F_{11} (8.6 eV), V_2F_{11} (8.0 eV)) [20, 23]; however, a systematic increase of EA by 1–2 eV was demonstrated while going from the mononuclear MX_{k+1} to their corresponding dinuclear M_2X_{2k+1} systems [23]. It was then anticipated that even much larger EA values could be obtained by designing the $M_nX_{n\times k+1}$ superhalogens for $n > 2$, but such investigations were not possible due to limited computer resources available at that time.

In the course of more recent studies, previously described polynuclear superhalogens were reexamined and various novel $M_nX_{n\times k+1}$ systems were proposed. Due to questionable thermodynamic stability of the neutral $M_nX_{n\times k+1}$ compounds (which are often vulnerable to fragmentation), most of the research was focused on the negatively charged $(M_nX_{n\times k+1})^-$ systems. These investigations led to determination of the equilibrium structures of numerous polynuclear superhalogens and/or their anions and evaluation of their EA/VDE values (see Table 3.4).

Among the polynuclear systems studied, dinuclear and trinuclear anions involving magnesium as central atoms have attracted particular attention. The global minimum D_{3h}-symmetry structure of the $(Mg_2F_5)^-$ anion (VDE = 9.38 eV) was described as the system in

Table 3.4 EAs (in eV) and VDEs (in eV) of neutral and anionic polynuclear superhalogens, respectively.

System	EA/VDE (eV)	System	EA/VDE (eV)
$(Al_2F_7)^-$	11.16 [15]	$(Al_2Cl_7)^-$	7.75 [15]
$(Al_2Br_7)^-$	7.08 [15]	$(P_2F_{11})^-$	10.95 [15]
$(V_2F_{11})^-$	10.98 [15]	$(As_2F_{11})^-$	11.43 [15]
$(Ta_2F_{11})^-$	11.84 [15]	$(As_3F_{16})^-$	12.20 [15]
$(Ta_3F_{16})^-$	12.63 [15]	$(Ca_2(CN)_5)^-$	9.70 [148]
$Fe_2(CN)_5$	6.23 [149]	$Fe_3(CN)_7$	6.02 [149]
$(Fe_2(CN)_5)^-$	6.71 [149]	$(Fe_3(CN)_7)^-$	6.35 [149]
$(Au_2F_{11})^-$	11.24 [150]	$(Au_3F_{16})^-$	11.96 [150]
Pt_2Cl_6	5.41 [151]	Pt_2Cl_7	5.43 [151]
Pt_2Cl_8	5.74 [151]	Pt_2Cl_9	6.55 [151]
Pt_2Cl_{10}	6.10 [151]		

which two central magnesium atoms are linked via symmetrical triangle formed by three fluorine atoms [152]. Replacing the fluorine ligands with chlorine atoms leads to an analogous lowest energy structure of $(Mg_2Cl_5)^-$ anion although its VDE was found to be substantially smaller (7.70 eV) [153]. The most stable isomer of $(Mg_3Cl_7)^-$ anion seems particularly interesting although its VDE (8.09 eV) exceeds that of $(Mg_2Cl_5)^-$ by only 0.4 eV. Namely, its C_{3v}-symmetry equilibrium structure (predicted in gas phase) resembles that observed in the chloromagnesite crystal lattice, where chlorine atoms are linked to three Mg atoms, forming the umbrella-like fragments (see Figure 3.10) [153].

The $(Mg_2X_5)^-$ (X = F, Cl, CN) and $(Mg_3X_7)^-$ (X = F, Cl) anions were also used as testing compounds while verifying whether certain theoretical treatments could be considered adequate for evaluating VDEs of such species [154]. In addition, the studies performed for Mg-containing polynuclear superhalogen anions include the $(Mg_2(CN)_5)^-$ anion whose VDE was predicted to be 8.67 eV [155–158] and the $(Mg_2X_5)^-$ (X = OH, OOH, OF) anions whose VDE values were all remarkably larger than 3.6 eV (the highest VDE of 7.30 eV was predicted for a triple-bridged structure based on the OF ligands) [159]. Finally, the $(RMg_2F_4)^-$ and $(RMg_2Cl_4)^-$ (R = C_2H_3, C_2H_5, C_6H_5) anions were found to exhibit substantial VDEs of 6.00 eV $(C_2H_3Mg_2F_4)^-$, 4.77 eV $(C_2H_5Mg_2F_4)^-$, 5.58 eV $(C_6H_5Mg_2F_4)^-$, 6.22 eV $(C_2H_3Mg_2Cl_4)^-$, 5.01 eV $(C_2H_5Mg_2Cl_4)^-$, and 5.75 eV $(C_6H_5Mg_2Cl_4)^-$ [160].

The investigation on the $(H_nF_{n+1})^-$ (n = 1–5, 7, 9, 12) anions (utilizing hydrogen and fluorine atoms as central atoms and ligands, respectively) revealed remarkably large VDEs thereof (see Table 3.5) [161].

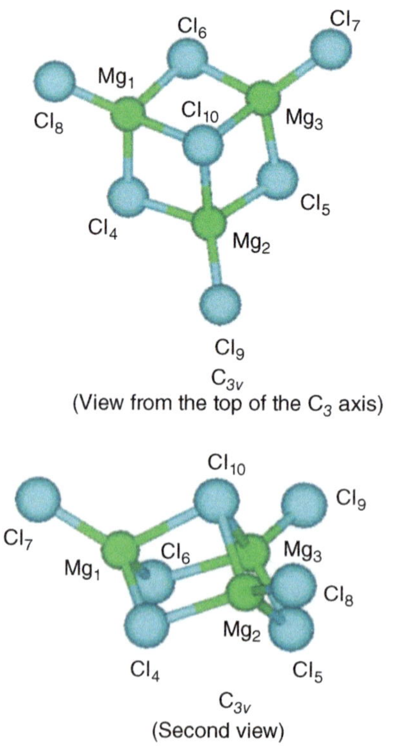

Figure 3.10 The lowest energy structure of the $(Mg_3Cl_7)^-$ anion. *Source:* Anusiewicz [153]. © CSIRO.

Table 3.5 VDEs predicted for the $(H_nF_{n+1})^-$ (n = 1–5, 7, 9, 12) anions.

System	Symmetry	VDE (eV)
$(HF_2)^-$	$D_{\infty h}$	6.98
$(H_2F_3)^-$	C_{2v}	8.47
$(H_3F_4)^-$	D_{3h}	9.70
$(H_4F_5)^-$	T_d	10.66
$(H_5F_6)^-$	C_{2v}	11.29
$(H_7F_8)^-$	D_{3d}	12.32
$(H_9F_{10})^-$	C_3	12.97
$(H_{12}F_{13})^-$	C_{3h}	13.87

Source: Freza and Skurski [161]. © Elseiver.

Also, a crude estimation (at the Koopmans' Theorem level) of the VDE for even larger $(H_{25}F_{26})^-$ system was provided (~14 eV) [161]. It is worth noting that the equilibrium structures of the $(H_nF_{n+1})^-$ anions were found to adopt relatively high symmetry configurations (see Figure 3.11).

The studies on other polynuclear superhalogen anions resulted in predicting the structures and electron-binding energies of systems comprising various central atoms and halogen ligands, such as $(Li_nF_{n+1})^-$ (n = 2–5) [162], $(Na_nCl_{n+1})^-$ (n = 2–4) [163], $(Al_nF_{3n+1})^-$ (n = 2–5) [164], $(Ti_nF_{4n+1})^-$ (n = 2, 3) [165], $(Ge_nF_{4n+1})^-$ (n = 2, 3) [165], $(In_nF_{3n+1})^-$ (n = 2, 3) [166], $(Sn_nF_{4n+1})^-$ (n = 2, 3) [166], $(Sb_nF_{5n+1})^-$ (n = 2, 3) [166], $(AlBF_7)^-$ [167], and $(AsSbF_{11})^-$ [167]; see Table 3.6 where the corresponding results are gathered.

Finally, it should be pointed out that there exist other polynuclear anions that do not match the extended Gutsev-Boldyrev $(M_nX_{n\times k+1})^-$ formula, yet they can be considered

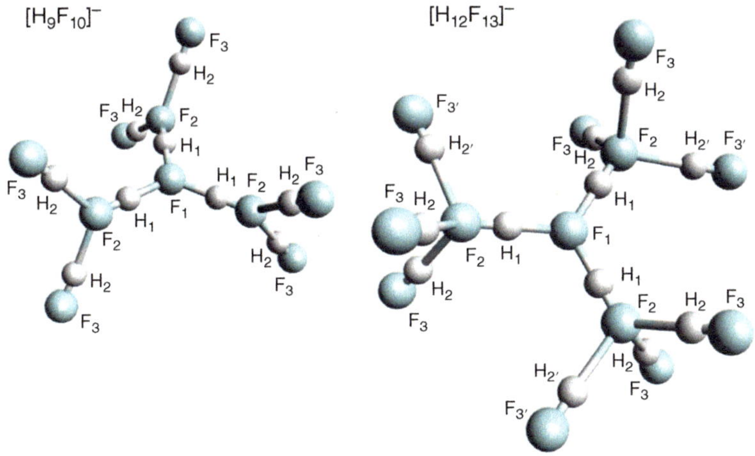

Figure 3.11 The equilibrium structures of C_3-symmetry $(H_9F_{10})^-$ and C_{3h}-symmetry $(H_{12}F_{13})^-$ anions. *Source:* Freza and Skurski [161]. © Elsevier.

Table 3.6 VDEs for the most stable isomers of various polynuclear superhalogen anions (symmetry point group is given in parenthesis).

System	VDE (eV)	System	VDE (eV)
$(Li_2F_3)^-$ ($D_{\infty h}$) [162]	8.51	$(Na_2Cl_3)^-$ ($D_{\infty h}$) [163]	6.46 ± 0.04/6.49[a]
$(Li_3F_4)^-$ (C_{3v}) [162]	8.49	$(Na_3Cl_4)^-$ (C_{3v}) [163]	6.3/6.68[a]
$(Li_4F_5)^-$ (C_{4v}) [162]	9.36	$(Na_4Cl_5)^-$ (C_{4v}) [163]	7.0/7.28[a]
$(Li_5F_6)^-$ (D_{5h}) [162]	10.18	$(Ti_2F_9)^-$ (D_{3h}) [165]	11.04
$(Al_2F_7)^-$ (D_{3d}) [164]	11.36	$(Ti_3F_{13})^-$ (C_{3v}) [165]	12.06
$(Al_3F_{10})^-$ (C_{2v}) [164]	12.04	$(Ge_2F_9)^-$ (D_{3h}) [165]	11.45
$(Al_4F_{13})^-$ (C_2) [164]	12.40	$(Ge_3F_{13})^-$ (C_1) [165]	11.97
$(Al_5F_{16})^-$ (C_1) [164]	11.09	$(Sn_2F_9)^-$ (S_2) [166]	12.01
$(In_2F_7)^-$ (C_{2h}) [166]	11.40	$(Sn_3F_{13})^-$ (C_{2v}) [166]	12.56
$(In_3F_{10})^-$ (C_{2v}) [166]	11.98	$(AlBF_7)^-$ (C_s) [167]	10.96
$(Sb_2F_{11})^-$ (C_2) [166]	12.54	$(AsSbF_{11})^-$ (C_s) [167]	12.38
$(Sb_3F_{16})^-$ (S_2) [166]	13.27		

[a] Experimental/theoretical result.

superhalogen anions due to their significant VDEs (e.g. the icosahedral $(CHB_{11}X_{11})^-$ (X = H, F, Cl, CN) anions [168–170] formed by deprotonation of carborane superacids [171, 172]).

3.3.2 Search for EA and VDE Limits

The excess electron-binding energy is one of the important characteristics of chemical compounds. It is commonly discussed in terms of adiabatic EA (for neutral molecules) and VDE (for anions). The compounds exhibiting maximal possible EA/VDE are of special interest because a wide range of new materials (such as chemical compounds of inert gases, organic superconductors, organic metals, ionic liquids, etc.) could be designed and synthesized on their base. Certainly, the search for maximal possible electron-binding energy has been focused on superhalogens as they represent the species having larger EAs than other commonly known molecules. Since the electronic stability of a singly negatively charged system (i.e. monoanion) strongly depends on the ability of excess charge delocalization over the molecular framework, it seemed natural to assume that superhalogen compounds containing large number of ligands would exhibit significant electron-binding energies. Indeed, such a dependence was confirmed back in the 1980s for a series of MX_{k+1} molecules comprising central atoms (M) of developing maximal valences (k) [8, 18–23]. Several extrapolations based on various theoretical models have led to the estimations of maximal possible EA in the 10–12 eV [173] (for mononuclear superhalogens) or even 15–17 eV [20] range. Although the existence of the molecules having such large EAs had been anticipated, the actual examples were neither provided nor suggested.

In 1984, Gutsev and Boldyrev discussed the directions that could be pursued to find the systems with the highest possible EA and suggested that a promising way to increase the

EA is to resort to polynuclear complexes with a large number of electronegative ligands [20]. The argumentation was based on the earlier suggestion presented by Lowe who recognized that the smaller the extra fraction of an electron to be added to an atom, the easier for an atom to hold this portion [174]. Thus, it was expected that the binding of an extra electron in the $(MX_{k+1})^-$ anion should be strengthened with the increase of k. Indeed, the largest VDE of a mononuclear anion (i.e. 11.9 eV for $(TeF_7)^-$) was predicted for the system containing seven fluorine ligands [33] (interestingly, that result confirms the limit of ~12 eV suggested earlier for superhalogen anions consisting of one central atom and $k+1$ halogen ligands) [173]. Further considerations [15, 20, 23, 82, 161, 173] were based on the following line of reasoning: assuming that an excess charge of $-1\,e$ is being evenly delocalized over p ligands, each ligand holds the partial charge of $-1/p\ e$. When the number of ligands (p) increases, then the partial charge on each ligand ($-1/p\ e$) decreases. In particular, when p goes to infinity, then the partial charge goes to zero, which in turn means that the VDE of an anion goes to the ionization potential (IP) of a neutral molecule. In other words, making the number of ligands larger and larger ($p \to \infty$) leads to smaller and smaller fractional charges ($-1/p\ e \to 0$), and thus the energy required to detach an excess electron from the anion (VDE) approaches the energy required to detach an electron from the neutral molecule (IP). Therefore, it was concluded that the IPs of the neutral molecules that are difficult to ionize (such as CF_4, SiF_4, SF_6 having their IPs in the 15–17 eV range) should be considered as the limiting VDE values [15, 23, 82, 161, 173].

It is understandable that the number of ligands bound to a single central atom cannot be increased beyond certain values (mostly due to destabilizing valence repulsion effects and steric hindrance). Hence, the attention was directed toward the $(M_nX_{n\times k+1})^-$ polynuclear superhalogen anions in which an excess electron density is expected to delocalize over $n\times k+1$ electronegative ligands. Indeed, many negatively charged superhalogens containing several central atoms turned out to exhibit large VDE values (see the preceding section); however, most of them have their VDEs not exceeding 12 eV. The list of anions whose VDE exceeds 12 eV is relatively short and to the best of our knowledge contains $(As_3F_{16})^-$, $(Ta_3F_{16})^-$, $(H_7F_8)^-$, $(H_9F_{10})^-$, $(H_{12}F_{13})^-$, $(Al_3F_{10})^-$, $(Al_4F_{13})^-$, $(Ti_3F_{13})^-$, $(Sn_2F_9)^-$, $(Sn_3F_{13})^-$, $(AsSbF_{11})^-$, $(Sb_2F_{11})^-$, and $(Sb_3F_{16})^-$ (whose VDEs are provided in the preceding section) as well as recently studied $(AlF_4(HF)_m)^-$ (m = 4–8, 12), $(AsF_6(HF)_m)^-$ (m = 5, 6, 12), and $(SbF_6(HF)_m)^-$ (m = 3–6, 12), anionic clusters whose formation in binary superacids was predicted [175]. Even though the latter anions are not polynuclear systems (instead, they correspond to mononuclear $(AlF_4)^-$, $(AsF_6)^-$, or $(AsF_6)^-$ superhalogen anions solvated by various number of HF molecules), their VDEs are remarkably large (see Table 3.7).

It should be emphasized that only few anions whose VDEs approach (or even exceed) 14 eV have been predicted to date: $(H_{12}F_{13})^-$ (VDE = 13.87 eV, see Figure 3.11) [161], $(AlF_4(HF)_{12})^-$ (13.96 eV) [175], $(AsF_6(HF)_{12})^-$ (VDE = 14.03 eV) [175], and $(SbF_6(HF)_{12})^-$ (VDE = 14.03 eV) [175]. These systems represent the most strongly bound anions, which have been described in the literature thus far. Although their VDEs are remarkably large, the predicted limit of 15–17 eV has not been reached yet. Taking into account the slopes of the plots showing the VDE dependence on the number of central atoms (n) for the $(H_nF_{n+1})^-$ anions [161] or on the number of HF molecules (m) solvating the $(AlF_4)^-$, $(AsF_6)^-$, or $(SbF_6)^-$ anions [175] (see Figure 3.12), achieving the 15–17 eV limit might not be easy, unless a different direction of designing molecular anions is proposed.

Table 3.7 VDEs (in eV) of superhalogen anions solvated by various number of HF molecules.

System	VDE (eV)	System	VDE (eV)
$(AlF_4(HF)_4)^-$	12.34	$(AlF_4(HF)_5)^-$	12.59
$(AlF_4(HF)_6)^-$	12.87	$(AlF_4(HF)_7)^-$	13.14
$(AlF_4(HF)_8)^-$	13.56	$(AlF_4(HF)_{12})^-$	13.96
$(AsF_6(HF)_5)^-$	12.10	$(AsF_6(HF)_6)^-$	12.37
$(AsF_6(HF)_{12})^-$	14.03	$(SbF_6(HF)_3)^-$	12.31
$(SbF_6(HF)_4)^-$	12.91	$(SbF_6(HF)_5)^-$	13.19
$(SbF_6(HF)_6)^-$	13.35	$(SbF_6(HF)_{12})^-$	14.03

Source: Modified from Czapla et al. [175].

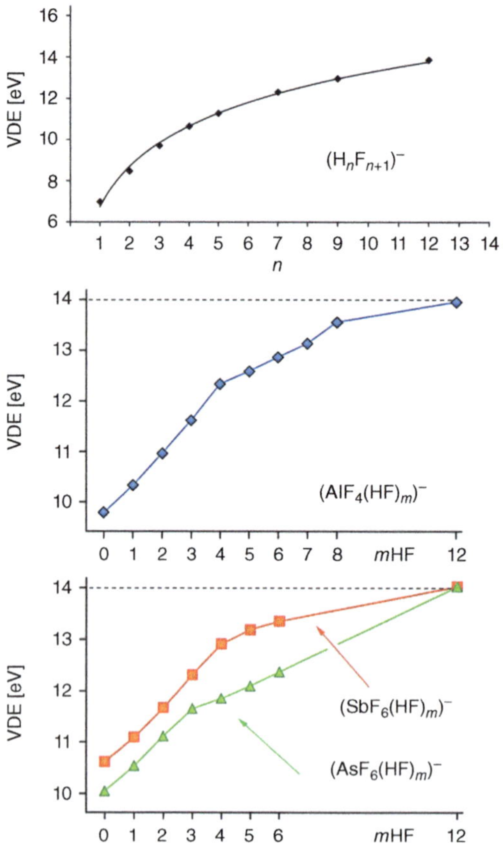

Figure 3.12 The VDE dependence on the number of central atoms (n) for the $(H_nF_{n+1})^-$ anions. *Source:* Taken from Ref. [161] and the VDE dependence on the number of HF molecules (m) solvating the $(AlF_4)^-$, $(AsF_6)^-$, or $(SbF_6)^-$ anions. *Source:* Modified from Ref. [175].

3.3.3 Magnetic Superhalogens

Since the systems called *magnetic superhalogens* (proposed by the Jena group) [176] are essentially polynuclear compounds, it seems appropriate to discuss them in the section devoted to superhalogen systems containing a larger number of central atoms. Taking into account the electronic configuration of Mn ([Ar]$3d^5 4s^2$) and the fact that the outer $4s^2$ electrons are involved in chemical bonding, it is expected that the remaining half-filled $3d^5$ shell should carry a magnetic moment of 5 μ_B. Having this in mind, Jena and coworkers investigated the neutral and anionic Mn_nCl_m systems by measuring their mass spectra and photoelectron spectra and performing theoretical calculations [129, 176]. It was found that in the case of $(Mn_nCl_{2n+1})^-$ ($n \geq 2$), the superhalogen behavior is attained at the +2 oxidation state of Mn (despite the maximum oxidation state of Mn being +7). Namely, the EA = 5.9 ± 0.2 eV and VDE = 6.3 ± 0.1 eV were experimentally measured for Mn_2Cl_5 and $(Mn_2Cl_5)^-$, respectively, while the electron-binding energies (EA = 6.06 eV and VDE = 6.74 eV) for $Mn_3Cl_7/(Mn_3Cl_7)^-$ were predicted theoretically [176]. But more importantly, the Mn_nCl_{2n+1} systems were recognized as magnetic (i.e. exhibiting substantial magnetic moments).

Next, the Yin group studied the $(Mn_2Cl_5)^-$ anion and its isoelectronic analogues ($Fe_2Cl_5^+$ and $MnFeCl_5$) and found that $Fe_2Cl_5^+$ possesses even stronger magnetic coupling as the degree of spin delocalization in $Fe_2Cl_5^+$ is larger than that in $(Mn_2Cl_5)^-$ [177]. These studies paved the way to design new magnetic oxidizing agents and were followed by reporting the stability of magnetic superhalogens composed of one or two iron atoms at the center and surrounded with fluorine atoms serving as ligands. Knowing that Fe atoms have predominant oxidation states of +2 and +3 and taking into account their [Ar]$3d^6 4s^2$ electronic configuration, one may expect each Fe atom to carry a spin magnetic moment of 4μ_B, which should result in ferromagnetic coupling in the bulk. Ding et al. performed a systematic investigation [178] of neutral and anionic Fe_mF_n ($m = 1, 2$; $n = 1$–7) clusters and found that the FeF_n compounds ($n \geq 3$) correspond to superhalogens (as indicated by their large EA values) while the neutral Fe_2F_n and anionic $(Fe_2F_n)^-$ can be classified as superhalogens when $n \geq 5$. Moreover, the calculated spin magnetic moments of neutral and anionic Fe_2F_n ($n = 2$–7) clusters showed that the magnetic moments of the larger clusters ($n \geq 5$) were larger than those of the smaller ones. Also, the magnetic moments of the neutral clusters exhibited an odd-even oscillation between 7 and 9 μ_B. Finally, it was found that the magnetic moments of anionic clusters are consistently 1 μ_B smaller than those of the corresponding neutral systems (excluding the Fe_2F_6 case) [178].

Since the existence of magnetic superhalogens has been proposed, it became clear that such species can serve as the building blocks of bulk materials (e.g. as it was demonstrated for $KMnCl_3$ salt (in which the $(MnCl_3)^-$ anion is neutralized by K^+ counterion) whose total magnetic moment of 5 μ_B is almost entirely located at the manganese site) [176]. These findings seem particularly important as they will likely enable the design of many novel materials with interesting chemistry because polynuclear superhalogen building blocks carrying a net magnetic moment provide an additional degree of freedom that can be used to design and synthesize new magnetic insulators [129, 149, 176, 179].

3.4 Superhalogens' Applications at a Glance

As suggested back in 1981, the salts containing superhalogens (i.e. LMX_{k+1} compounds where L stands for an alkali metal) are ionic and thus their dissociation into L^+ and $(MX_{k+1})^-$ (instead of L and MX_{k+1}) could be anticipated, which in turn may lead to obtaining low-temperature plasmas [8]. Even though one might expect the $LMX_{k+1} \rightarrow LX + MX_k$ fragmentation channel to be competitive, it was demonstrated that the presence of $NaAlCl_4$ in gaseous $AlCl_3$-Al_2Cl_6 mixtures imparts electrical conductance of 10^{-11}–$10^{-6}\,\Omega^{-1}cm^{-1}$ in the range of $T = 500$–900 K [180]. Further studies revealed many other superhalogen applications (e.g. in designing hydrogen storage materials, solar cells, and Li-ion batteries) [181–184].

Superhalogens as the molecules with high EAs continue to be employed in the preparation of many new materials, including organic metals and organic superconductors [185]. Due to their large ability of withdrawing an electron from other molecules, neutral superhalogens can be used in synthesis (e.g. in the oxidation of counterpart systems with high IPs). As demonstrated recently, even various moderately reactive molecules possessing high IPs (such as SiO_2, $CHCl_3$ (chloroform), and CCl_2F_2 (Freon-12)) should form stable and strongly bound ionic compounds when combined with a properly chosen superhalogen system (acting as an oxidizing agent) [186]. Indeed, the resulting compounds (i.e. $[(SiO_2)^+(AlF_4)^-]$, $[(CHCl_3)^+(AlF_4)^-]$, and $[(CCl_2F_2)^+(AlF_4)^-]$) were recognized as ionic complexes characterized by large binding energies of 20–90 kcal/mol, which indicate strong interaction between the constituting fragments [186]. Superhalogens were also found to act as effective oxidizers with respect to water molecules (by inducing the formation of H_3O^+ and OH^- ions) [187]. Similar ability of superhalogens to ionize molecules having large IPs was demonstrated for benzene [188] (IP = 9.24 eV), chloromethane [189] (IP = 11.09 eV), carbon tetrachloride [189] (IP = 11.28 eV), and even carbon dioxide [190] (IP = 13.80 eV). As far as the chemical synthesis is concerned, superhalogens were found capable of serving as the trigger compounds initiating the radical-substitution reactions [189].

A possible use of superhalogens as Lewis-Brønsted superacid precursors [191, 192] and steric shielding agents with respect to selected metal ions [193] was reported in the last decade. In addition, the role played by the superhalogen anions in ionic liquids (in which they serve as negatively charged components) has been examined [194]. As it turned out, certain physical properties of ionic liquids (such as viscosity and electrical conductivity) are anion dependent and thus might be achieved by the proper choice of superhalogen anions (being the ionic liquid (IL)'s negatively charged building blocks). As a consequence, the tailored ionic liquids exhibiting desired properties can be successfully designed [194–196].

Numerous other applications of superhalogens and their anions include the use of superhalogen-like fragments as substituents in molecules (to increase their EAs [197–199] or stabilize certain isomeric structures [200]), the implementations of so-called supersalts [201–205], as well as the exploitation of superhalogens as building blocks of cluster-assembled materials (see the recent review of Jena and Sun in which the potential applications of superhalogens and other "superatoms" are described [92]).

3.5 Final Remarks

Although 40 years have passed since the invention of superhalogen concept, these systems keep attracting attention due to their unusual properties and diversity of applications in synthesis and material science. The exploration of various superhalogen compounds will likely be continued in the future as many issues are yet to be addressed (e.g. designing stable molecules in which He and Ne atoms are involved into chemical bonding or finding the real compounds having their excess electron-binding energies in the predicted 15–17 eV limit). We look forward to the future when superhalogens will become an integral part of material science and other areas.

Acknowledgements

The author dedicates this work to his beloved wife Joanna.

References

1 Massey, H.S. (1976). *Negative Ions*. Cambridge: Cambridge University Press.
2 Bartlett, N. and Lohmann, D.H. (1962). *J. Chem. Soc.*: 5253–5261. https://doi.org/10.1039/JR9620005253.
3 Compton, R.N., Reinhardt, P.W., and Cooper, C.D. (1978). *J. Chem. Phys.* 68: 2023–2036.
4 Boring, M., Wood, J.H., and Moskowitz, J.W. (1974). *J. Chem. Phys.* 61: 3800–3803.
5 Benson, S.W. (1976). *Thermochemical Kinetics*. New York: Wiley.
6 Sidorov, L.N. (1982). *Russ. Chem. Rev.* 51: 356–367.
7 Barber, M., Clark, J.D., and Hinchliffe, H. (1978). *Mol. Phys.* 35: 1745–1751.
8 Gutsev, G.L. and Boldyrev, A.I. (1981). *Chem. Phys.* 56: 277–283.
9 Sidorov, L.N. (1977). *Koord. Chim.* 3: 1128–1133.
10 Kiselev, Y.M. and Tretiyakov, Y.D. (1999). *Russ. Chem. Rev.* 68: 365–379.
11 Hotop, H. and Lineberger, W.C. (1985). *J. Phys. Chem. Ref. Data* 14: 731–750.
12 Anusiewicz, I., Sobczyk, M., Dąbkowska, I., and Skurski, P. (2003). *Chem. Phys.* 291: 171–180.
13 Sikorska, C., Smuczyńska, S., Skurski, P., and Anusiewicz, I. (2008). *Inorg. Chem.* 47: 7348–7354.
14 Marchaj, M., Freza, S., and Skurski, P. (2012). *J. Phys. Chem. A* 116: 1966–1973.
15 Sobczyk, M., Sawicka, A., and Skurski, P. (2003). *Eur. J. Inorg. Chem.*: 3790–3797.
16 Gutsev, G.L. and Boldyrev, A.I. (1985). *Adv. Chem. Phys.* 61: 169–221.
17 Bloor, J.E. and Sherrod, R.E. (1980). *J. Am. Chem. Soc.* 102: 4333–4340.
18 Gutsev, G.L. and Boldyrev, A.I. (1981). *Chem. Phys. Lett.* 84: 352–355.
19 Gutsev, G.L. and Boldyrev, A.I. (1983). *Chem. Phys. Lett.* 101: 441–445.
20 Gutsev, G.L. and Boldyrev, A.I. (1984). *Chem. Phys. Lett.* 108: 250–254.
21 Gutsev, G.L. and Boldyrev, A.I. (1984). *Chem. Phys. Lett.* 108: 255–258.
22 Gutsev, G.L. and Boldyrev, A.I. (1984). *Mol. Phys.* 53: 23–31.
23 Gutsev, G.L. and Boldyrev, A.I. (1987). *Russ. Chem. Rev.* 56: 519–531.
24 Sakai, Y. and Miyoshi, E. (1987). *J. Chem. Phys.* 87: 2885–2892.
25 Miyoshi, E., Sakai, Y., Murakami, A. et al. (1988). *J. Chem. Phys.* 89: 4193–4198.
26 Mota, F., Novoa, J.J., and Ramirez, A.C. (1988). *J. Mol. Struct. (THEOCHEM)* 166: 153–158.
27 Zakrzewski, V.G. and Boldyrev, A.I. (1990). *J. Chem. Phys.* 93: 657–660.

28 Kölmel, C., Palm, G., Ahlrichs, R. et al. (1990). *Chem. Phys. Lett.* 173: 151–156.
29 Boldyrev, A.I. and von Niessen, W. (1991). *Chem. Phys.* 155: 71–78.
30 Weikert, H.-G., Cederbaum, L.S., Tarantelli, F., and Boldyrev, A.I. (1991). *Z. Phys. D* 18: 299–305.
31 Gutsev, G.L. (1991). *Chem. Phys. Lett.* 184: 305–309.
32 Scheller, M.K. and Cederbaum, L.S. (1992). *J. Phys. B* 25: 2257–2265.
33 Boldyrev, A.I. and Simons, J. (1992). *J. Chem. Phys.* 97: 2826–2827.
34 Scheller, M.K. and Cederbaum, L.S. (1993). *J. Chem. Phys.* 99: 441–455.
35 Ortiz, J.V. (1993). *J. Chem. Phys.* 99: 6727–6731.
36 Ortiz, J.V. (1993). *Chem. Phys. Lett.* 214: 467–472.
37 Weikert, H.-G. and Cederbaum, L.S. (1993). *J. Chem. Phys.* 99: 8877–8891.
38 Gutsev, G.L. (1993). *J. Chem. Phys.* 98: 444–452.
39 Gutsev, G.L. (1993). *J. Chem. Phys.* 99: 3906–3912.
40 Scheller, M.K. and Cederbaum, L.S. (1994). *J. Chem. Phys.* 100: 8934–8942.
41 Gutsev, G.L., Leś, A., and Adamowicz, L. (1994). *J. Chem. Phys.* 100: 8925–8933.
42 Weikert, H.-G., Meyer, H.-D., Cederbaum, L.S., and Tarantelli, F. (1996). *J. Chem. Phys.* 104: 7122–7138.
43 Gutowski, M., Boldyrev, A.I., Simons, J. et al. (1996). *J. Am. Chem. Soc.* 118: 1173–1180.
44 Gutsev, G.L., Bartlett, R.J., Boldyrev, A.I., and Simons, J. (1997). *J. Chem. Phys.* 107: 3867–3875.
45 Gutsev, G.L., Jena, P., and Bartlett, R.J. (1998). *Chem. Phys. Lett.* 292: 289–294.
46 Sidorov, L.N., Borshchevsky, A.Y., Rudny, E.B., and Butsky, V.D. (1982). *Chem. Phys.* 71: 145–156.
47 Pyatenko, A.T. and Gorokhov, L.N. (1984). *Chem. Phys. Lett.* 105: 205–208.
48 Pyatenko, A.T., Gusarov, A.V., and Gorokhov, L.N. (1984). *Zhur. Fiz. Khim.* 58: 1–9.
49 Heni, M. and Illenberger, E. (1985). *J. Chem. Phys.* 83: 6056–6057.
50 Igolkina, N.A., Nikitin, M.I., Boltalina, O.V., and Sidorov, L.N. (1986). *High. Temp. Sci.* 21: 111–117.
51 Igolkina, N.A., Nikitin, M.I., Sidorov, L.N., and Boltalina, O.V. (1987). *High. Temp. Sci.* 23: 89–95.
52 Metz, R.B., Kitsopoulos, T., Weaver, A., and Neumark, D.M. (1988). *J. Phys. Chem.* 88: 1463–1465.
53 Weaver, A., Metz, R.B., Bradforth, S.E., and Neumark, D.M. (1988). *J. Phys. Chem.* 92: 5558–5560.
54 Borshchevskii, A.Y., Boltalina, O.V., Sorokin, I.D., and Sidorov, L.N. (1988). *J. Chem. Thermodyn.* 20: 523–537.
55 Sidorov, L.N., Boltalina, O.V., and Borschevskii, A.Y. (1989). *Int. J. Mass Spectrom. Ion Proc.* 87: R1–R3.
56 Korobov, M.V., Kuznetsov, S.V., Sidorov, L.N. et al. (1989). *Int. J. Mass Spectrom. Ion Proc.* 87: 13–27.
57 Boltalina, O.V., Borshchevskii, A.Y., and Sidorov, L.N. (1991). *Russ. J. Phys. Chem.* 65: 466–481.
58 Compton, R.N. (1995). Negative ion formation. In: *Negative Ions* (ed. V.A. Esaulov). Cambridge: Cambridge University Press.
59 Chen, E.C.M., Shuie, L.-R., Desai D'sa, E. et al. (1988). *J. Chem. Phys.* 88: 4711–4719.

60 Huey, L.G., Dunlea, E.J., and Howard, C.J. (1996). *J. Phys. Chem.* 100: 6504–6508.
61 Cha, C.Y., Ganteför, G., and Eberhardt, W. (1994). *J. Chem. Phys.* 100: 995–1010.
62 Wu, H.B., Desai, S.R., and Wang, L.-S. (1996). *J. Am. Chem. Soc.* 118: 7434–7434.
63 Desai, S.R., Wu, H.B., Rohlfing, C.M., and Wang, L.-S. (1997). *J. Chem. Phys.* 106: 1309–1317.
64 Taylor, T.R., Xu, C., and Neumark, D.M. (1998). *J. Chem. Phys.* 108: 10018–10026.
65 Bartlett, N. (1968). *Angew. Chem. Int. Ed. Eng.* 7: 433–439.
66 Jensen, D.E. (1969). *Trans. Faraday Soc.* 65: 2123–2132.
67 Jensen, D.E. and Miller, W.J. (1970). *J. Chem. Phys.* 53: 3287–3292.
68 Ferguson, E.E., Dunkin, D.B., and Feshenfeld, F.C. (1972). *J. Chem. Phys.* 57: 1459–1463.
69 Miller, W.J. (1972). *J. Chem. Phys.* 57: 2354–2358.
70 Burgess, J., Haigh, I., Peacock, R.D., and Taylor, P. (1974). *J. Chem. Soc. Dalton Trans.* 10: 1064–1066.
71 Leffert, C.B., Tang, S.Y., Rothe, E.W., and Cheng, T.C. (1974). *J. Chem. Phys.* 61: 4929–4930.
72 Cooper, C.D. and Compton, R.N. (1974). *Bull. Am. Phys. Soc.* 19: 1106–1106.
73 Tang, S.Y., Reck, G.P., and Rothe, E.W. (1974). *Bull. Am. Phys. Soc.* 19: 1173–1173.
74 Gould, R.K. and Miller, W.J. (1975). *J. Chem. Phys.* 62: 644–649.
75 Beauchamp, J.L. (1976). *J. Chem. Phys.* 64: 929–935.
76 Mathur, B.P., Rothe, E.W., Tang, S.Y. et al. (1976). *J. Chem. Phys.* 64: 1247–1248.
77 Compton, R.N. (1977). *J. Chem. Phys.* 66: 4478–4485.
78 Mathur, B.P., Rothe, E.W., and Reck, G.P. (1977). *J. Chem. Phys.* 67: 377–381.
79 Webb, J.D. and Bernstein, E.R. (1978). *J. Am. Chem. Soc.* 100: 483–485.
80 Sidorov, L.N., Nikitin, M.I., and Korobov, V.M. (1979). *Dokl. Akad. Nauk SSSR* 248: 1387–1395.
81 Sidorov, L.N., Skokan, E.V., Nikitin, M.I., and Sorokin, I.D. (1980). *Int. J. Mass. Spectr. Ion Phys.* 35: 215–224.
82 Kalcher, J. and Sax, A.F. (1994). *Chem. Rev.* 94: 2291–2318.
83 Maessen, B. and Cade, P.E. (1984). *J. Chem. Phys.* 80: 5120–5129.
84 Ju, G. and Davidson, E.R. (1992). *J. Phys. Chem.* 96: 3683–3688.
85 Machado, F.B.C., Ju, G.-Z., and Davidson, E.R. (1993). *J. Phys. Chem.* 97: 5882–5885.
86 Wang, X.-B., Ding, C.-F., Wang, L.-S. et al. (1999). *J. Chem. Phys.* 110: 4763–4771.
87 Wang, L.-S., Ding, C.F., Wang, X.-B., and Barlow, S.E. (1999). *Rev. Sci. Instrum.* 70: 1957–1966.
88 Pradhan, K. and Jena, P. (2011). *J. Chem. Phys.* 135: 144305.
89 Sikorska, C., Ignatowska, D., Freza, S., and Skurski, P. (2011). *J. Theor. Comput. Chem.* 10: 93–110.
90 Lewis, G.N. (1916). *J. Am. Chem. Soc.* 38: 762–785.
91 Langmuir, I. (1919). *J. Am. Chem. Soc.* 41: 868–934.
92 Jena, P. and Sun, Q. (2018). *Chem. Rev.* 118: 5755–5870.
93 Bradforth, S.E., Kim, E.H., Arnold, D.W., and Neumark, D.M. (1993). *J. Chem. Phys.* 98: 800–810.
94 Frazier, J.R., Christophorou, L.G., Carter, J.G., and Schweinler, H.C. (1978). *J. Chem. Phys.* 69: 3807–3818.
95 Tegeder, P. and Illenberger, E. (1998). *J. Radioanal. Nucl. Chem.* 232: 53–62.
96 Xu, W., Li, G., Yu, G. et al. (2003). *J. Phys. Chem. A* 107: 258–266.

97 Gutsev, G.L. (1992). *Chem. Phys.* 166: 57–68.
98 Gutsev, G.L. (1994). *J. Phys. Chem.* 98: 1570–1575.
99 King, R.A., Mastryukov, V.S., and Schaefer, H.F. III (1996). *J. Chem. Phys.* 105: 6880–6886.
100 Li, Q., Li, G., Xu, W. et al. (1999). *J. Chem. Phys.* 111: 7945–7953.
101 Smuczyńska, S. and Skurski, P. (2007). *Chem. Phys. Lett.* 443: 190–193.
102 Ignatyev, I.S. and Schaefer, H.F. III (1999). *J. Am. Chem. Soc.* 121 (29): 6904–6910.
103 Wang, X.-B. and Wang, L.-S. (2000). *J. Chem. Phys.* 113: 10928–10933.
104 Van Huis, T.J. and Schaefer, H.F. III (1997). *J. Chem. Phys.* 106: 4028–4037.
105 Anusiewicz, I., Freza, S., Sikorska, C., and Skurski, P. (2010). *Chem. Phys. Lett.* 493: 234–237.
106 Smuczyńska, S. and Skurski, P. (2009). *Inorg. Chem.* 48: 10231–10238.
107 Anusiewicz, I. (2009). *J. Phys. Chem. A* 113: 6511–6516.
108 Anusiewicz, I.J. (2009). *Phys. Chem. A* 113: 11429–11434.
109 Sikorska, C. and Skurski, P. (2012). *Mol. Phys.* 110: 1447–1452.
110 Świerszcz, I. and Anusiewicz, I. (2011). *Chem. Phys.* 383: 93–100.
111 Hotop, H., Patterson, T.A., and Lineberger, W.C. (1974). *J. Chem. Phys.* 60: 1806–1812.
112 Świerszcz, I. and Anusiewicz, I. (2013). *Mol. Phys.* 111: 377–385.
113 Sikorska, C., Freza, S., Skurski, P., and Anusiewicz, I. (2011). *J. Phys. Chem. A* 115: 2077–2085.
114 Sieradzan, I. and Anusiewicz, I. (2013). *Chem. Phys.* 425: 55–61.
115 Sieradzan, I., Skurski, P., and Anusiewicz, I. (2014). *Mol. Phys.* 112: 2811–2819.
116 Ellison, G.B., Engelking, P.C., and Lineberger, W.C. (1978). *J. Am. Chem. Soc.* 100: 2556–2558.
117 DePuy, C.H., Gronert, S., Barlow, S.E. et al. (1989). *J. Am. Chem. Soc.* 111: 1968–1973.
118 Smuczyńska, S. and Skurski, P. (2008). *Chem. Phys. Lett.* 452: 44–48.
119 Lykke, K.R., Murray, K.K., and Lineberger, W.C. (1991). *Phys. Rev. A* 43: 6104–6107.
120 Boldyrev, A.I. and Simons, J. (1993). *J. Chem. Phys.* 99: 4628–4637.
121 Graham, J.D., Buytendyk, A.M., Zhang, X. et al. (2014). *J. Phys. Chem. A* 118: 8158–8162.
122 Lo, S. and Hopkinson, A.C. (2011). *Comp. Theor. Chem.* 973: 9–12.
123 Marchaj, M., Freza, S., and Skurski, P. (2013). *Chem. Phys. Lett.* 560: 15–21.
124 Götz, M., Willis, M., Kandalam, A.K. et al. (2010). *ChemPhysChem* 11: 853–858.
125 Willis, M., Götz, M., Kandalam, A.K. et al. (2010). *Angew. Chem. Int. Ed. Eng.* 49: 8966–8970.
126 Wang, Q., Sun, Q., and Jena, P. (2009). *J. Chem. Phys.* 131: 124301.
127 Koirala, P., Willis, M., Kiran, B. et al. (2010). *J. Phys. Chem. C* 114: 16018–16024.
128 Feng, Y., Xu, H.-G., Zheng, W. et al. (2011). *J. Chem. Phys.* 134: 094309.
129 Li, Y., Zhang, S., Wang, Q., and Jena, P. (2013). *J. Chem. Phys.* 138: 054309.
130 Paduani, C. and Jena, P. (2012). *J. Phys. Chem. A* 116: 1469–1474.
131 Paduani, C. and Jena, P. (2013). *Chem. Phys. Lett.* 556: 173–177.
132 Kulsha, A.V. and Sharapa, D.I. (2019). *J. Comput. Chem.* 40: 2293–2300.
133 Hou, G.-L., Wu, M.M., Wen, H. et al. (2013). *J. Chem. Phys.* 139: 044312.
134 Paduani, C. (2013). *Chem. Phys.* 417: 1–7.
135 Paduani, C. (2015). *Chem. Phys.* 463: 52–57.
136 Paduani, C. (2016). *Phys. E.* 77: 199–205.
137 Srivastava, A.K. and Misra, N. (2016). *Electrochem. Commun.* 68: 99–103.

138 Yang, Y.-F., Cui, Z.-H., and Ding, Y.-H. (2014). *Mol. Phys.* 112: 1589–1599.

139 Paduani, C. and Jena, P. (2012). *J. Nanopart. Res.* 14: 1035.

140 Tian, W.-J., Xu, H.-G., Kong, X.-Y. et al. (2014). *Phys. Chem. Chem. Phys.* 16: 5129–5136.

141 Sun, W.-M., Hou, D., Wu, D. et al. (2015). *Dalton Trans.* 44: 19901–19908.

142 Liu, J.-Y., Ma, H.-D., Sun, Y.-B. et al. (2017). *RSC Adv.* 7: 47073–47082.

143 Chen, H., Kong, X.-Y., Zheng, W. et al. (2013). *ChemPhysChem* 14: 3303–3308.

144 Sun, W.-M., Li, X.-H., Li, Y. et al. (2016). *ChemPhysChem* 17: 1468–1474.

145 Paduani, C., Wu, M.M., Willis, M., and Jena, P. (2011). *J. Phys. Chem. A* 115: 10237–10243.

146 Koirala, P., Pradhan, K., Kandalam, A.K., and Jena, P. (2013). *J. Phys. Chem. A* 117: 1310–1318.

147 Gutsev, G.L. and Boldyrev, A.I. (1984). *Zhur. Strukt. Khim.* 25: 16–22.

148 Li, J.-F., Sun, Y.-Y., Bai, H. et al. (2015). *AIP Adv.* 5: 067143.

149 Ding, L.-P., Shao, P., Lu, C. et al. (2017). *Inorg. Chem.* 56: 7928–7935.

150 Czapla, M. and Skurski, P. (2018). *Int. J. Quantum Chem.* 118: e25494.

151 Shi, Y., Bian, S., Ma, Y. et al. (2019). *J. Phys. Chem. A* 123: 187–193.

152 Anusiewicz, I. and Skurski, P. (2007). *Chem. Phys. Lett.* 440: 41–44.

153 Anusiewicz, I. (2008). *Aust. J. Chem.* 61: 712–717.

154 Sun, Y.-Y., Li, J.-F., Li, M.-M. et al. (2016). *J. Chem. Phys.* 144: 054303.

155 Yin, B., Li, T., Li, J.-F. et al. (2014). *J. Chem. Phys.* 140: 094301.

156 Li, J.-F., Li, M.-M., Bai, H. et al. (2015). *ChemPhysChem* 16: 3652–3659.

157 Díaz-Tinoco, M. and Ortiz, J.V. (2016). *ChemPhysChem* 17: 2945–2946.

158 Díaz-Tinoco, M. and Ortiz, J.V. (2016). *J. Chem. Phys.* 145: 147101.

159 Zhao, R.-F., Yu, L., Zhou, F.-Q. et al. (2017). *Phys. Chem. Chem. Phys.* 19: 26986–26995.

160 Li, M.-M., Li, J.-F., Bai, H. et al. (2015). *Phys. Chem. Chem. Phys.* 17: 20338–20346.

161 Freza, S. and Skurski, P. (2010). *Chem. Phys. Lett.* 487: 19–23.

162 Wileńska, D., Skurski, P., and Anusiewicz, I. (2014). *J. Fluor. Chem.* 168: 99–104.

163 Alexandrova, A.N., Boldyrev, A.I., Fu, Y.-J. et al. (2004). *J. Chem.Phys.* 121: 5709–5719.

164 Sikorska, C. and Skurski, P. (2012). *Chem. Phys. Lett.* 536: 34–38.

165 Anusiewicz, I., Freza, S., and Skurski, P. (2018). *Polyhedron* 144: 125–130.

166 Czapla, M. and Skurski, P. (2015). *J. Phys. Chem. A* 119: 12868–12875.

167 Cyraniak, A., Skurski, P., and Czapla, M. (2019). *Chem. Phys. Lett.* 717: 77–81.

168 Zhao, R.-F., Zhou, F.-Q., Xu, W.-H. et al. (2018). *Inorg. Chem. Front.* 5: 2934–2947.

169 Díaz-Tinoco, M. and Ortiz, J.V. (2019). *Chem. Phys.* 521: 77–84.

170 Brzeski, J., Czapla, M., and Skurski, P. (2020). *ChemPlusChem* 85: 312–318.

171 Reed, C.A. (2005). *Chem. Commun.*: 1669–1677.

172 Stoyanov, E.S., Hoffmann, S.P., Juhasz, M., and Reed, C.A. (2006). *J. Am. Chem. Soc.* 128: 3160–3161.

173 Gutsev, G.L. and Boldyrev, A.I. (1990). *J. Phys. Chem.* 94: 2256–2259.

174 Lowe, J.P. (1977). *J. Am. Chem. Soc.* 99: 5557–5570.

175 Czapla, M., Ciepła, O., Brzeski, J., and Skurski, P. (2018). *J. Phys. Chem. A* 122: 8539–8548.

176 Wu, M.M., Wang, H., Ko, Y.J. et al. (2011). *Angew. Chem. Int. Ed. Eng.* 50: 2568–2572.

177 Yin, B., Li, J.L., Bai, H.C. et al. (2012). *Phys. Chem. Chem. Phys.* 14: 1121–1130.

178 Ding, L.-P., Kuang, X.-Y., Shao, P. et al. (2013). *J. Chem. Phys.* 139: 104304.

179 Yu, Y., Li, C., Yin, B. et al. (2013). *J. Chem. Phys.* 139: 054305.

180 Dewing, E.W. (1971). *J. Phys. Chem.* 75: 1260–1264.

181 Giri, S., Behera, S., and Jena, P. (2014). *Angew. Chem. Int. Ed. Eng.* 53: 13916–13919.
182 Jena, P. (2015). *J. Phys. Chem. Lett.* 6: 1119–1125.
183 Fang, H. and Jena, P. (2016). *J. Mater. Chem. A* 4: 4728–4737.
184 Fang, H. and Jena, P. (2016). *J. Phys. Chem. Lett.* 7: 1596–1603.
185 Wudl, F. (1984). *Acc. Chem. Res.* 17: 227–232.
186 Sikorska, C. and Skurski, P. (2011). *Inorg. Chem.* 50: 6384–6391.
187 Marchaj, M., Freza, S., Rybacka, O., and Skurski, P. (2013). *Chem. Phys. Lett.* 574: 13–17.
188 Czapla, M., Freza, S., and Skurski, P. (2015). *Chem. Phys. Lett.* 619: 32–35.
189 Marchaj, M., Freza, S., and Skurski, P. (2014). *Chem. Phys. Lett.* 612: 172–176.
190 Czapla, M. and Skurski, P. (2017). *Phys. Chem. Chem. Phys.* 19: 5435–5440.
191 Czapla, M. and Skurski, P. (2015). *Chem. Phys. Lett.* 630: 1–5.
192 Czapla, M., Anusiewicz, I., and Skurski, P. (2016). *Chem. Phys.* 465–466: 46–51.
193 Czapla, M. (2016). *Theor. Chem. Accounts* 135: 231.
194 Wileńska, D., Anusiewicz, I., Freza, S. et al. (2015). *Mol. Phys.* 113: 630–639.
195 Sosnowska, A., Barycki, M., Gajewicz, A. et al. (2016). *ChemPhysChem* 17: 1591–1600.
196 Barycki, M., Sosnowska, A., Gajewicz, A. et al. (2016). *Fluid Phase Equilib.* 427: 9–17.
197 Skurski, P., Gutowski, M., and Simons, J. (1999). *J. Chem. Phys.* 111: 9469–9474.
198 Skurski, P. and Simons, J. (2000). *J. Chem. Phys.* 112: 6563–6570.
199 Świerszcz, I. and Skurski, P. (2012). *Chem. Phys. Lett.* 537: 27–32.
200 Wileńska, D., Skurski, P., and Anusiewicz, I. (2016). *Mol. Phys.* 114: 1494–1501.
201 Rasheed, T., Siddiqui, S.A., Pandey, A.K. et al. (2013). *J. Fluor. Chem.* 146: 59–65.
202 Rasheed, T., Siddiqui, S.A., Pandey, A.K. et al. (2017). *J. Fluor. Chem.* 195: 85–92.
203 Fang, W.-H., Zhang, L., and Zhang, J. (2017). *Chem. Commun.* 53: 3949–3951.
204 Srivastava, A.K., Misra, N., and Tiwari, S.N. (2020). *SN Appl. Sci.* 2: 307.
205 Hou, G.L. and Wang, X.-B. (2020). *Chem. Phys. Lett.* 741: 137094.

4

Endohedrally Doped Superatoms and Assemblies

Vijay Kumar

Center for Informatics, School of Natural Sciences, Shiv Nadar University, Tehsil Dadri, Uttar Pradesh, India
Dr. Vijay Kumar Foundation, Gurgaon, Haryana, India

4.1 Introduction

Past three decades have seen tremendous research on clusters and nanoparticles of a wide variety of materials, including those of elemental metals and semiconductors as well as compounds and alloys [1–7]. Their applications in diverse fields including catalysis, optical and opto-electronic applications, magnetism, electronic devices, bio-applications, and environment, to name a few, have brought further attention. One important aspect that emerged from these studies is the finding of certain clusters and nanoparticles with a specific number of atoms and structures that have high stability and high abundance in the mass spectrum, compared with clusters having one more or one less atom. These, often high symmetry clusters, are referred to as magic clusters [8]. One of the striking examples is the discovery of C_{60} with an empty highest symmetry icosahedral cage that appeared almost exclusively [9] in the mass spectra and later led to the formation of a solid phase [10] of C_{60}. These developments raised interest in assembling such atomically precise clusters also referred to as superatoms [11–13] and nanoparticles of other materials. These constitute a new family of novel designer materials, both in 3D bulk and in lower dimensions. Here, it is of interest to note that before the discovery of solid C_{60}, many cluster compounds have been known for a long time. These are Chevrel phases with Mo_6S_8 clusters in which superconductivity was also found [14], cluster-based materials [15], and Zintl phases [16, 17] with isolated clusters such as Rb_2In_3 that is made of layers [18] of four-bonded *closo* octahedra of indium. The latter are separated by rubidium cations. Similarly, in Cs_4Ge_9 there are isolated nine-atom clusters [19] of group 14. There are also clathrates in which caged clusters are fused [20]. Accordingly, the vast knowledge accumulated on clusters in the past few decades can enrich and lead to new families of cluster-assembled materials in a bottom-up approach.

The discovery of carbon fullerenes with empty cage structures [9] and subsequently those of BN [21, 22] was novel, but empty cage structures of free clusters of other materials are generally not stable. In addition, aggregation of clusters into larger clusters is a

Superatoms: Principles, Synthesis and Applications, First Edition. Edited by Purusottam (Puru) Jena and Qiang Sun.
© 2022 John Wiley & Sons Ltd. Published 2022 by John Wiley & Sons Ltd.

problem. This has led to a slow progress in the synthesis of cluster-assembled materials. Kumar and Kawazoe [23] showed from ab initio calculations that endohedral doping of an atom can lead to a significant increase in the stability of clusters as well as the formation of novel structures different from those of the undoped ones with a specific number of atoms. They predicted Zr@Si_{16} fullerene, Ti@Si_{16} Frank-Kasper (FK) polyhedron shown in Figure 4.1, and other structures to be highly stable. For Hf@Si_{16}, both the fullerene and FK isomers were found to be nearly degenerate, although the fullerene isomer was found to be 0.006 eV/atom lower in energy. Also, in the FK isomer as a triangle of Si rotates [24] by 60° the symmetry changes, but the energy and highest occupied molecular orbital–lowest unoccupied molecular orbital (HOMO–LUMO) gap remain nearly the same as for the FK isomer. We call it a distorted FK isomer. Here, the @ symbol denotes that the metal atom is encapsulated in a Si_{16} cage. These structures are very different from the one known [25] for elemental Si_{16} cluster. Also, interestingly by changing the endohedral atom even in the same column (isoelectronic) in the periodic table led to very different structures due to the slightly different atomic sizes of Ti and Zr atoms. This is due to the strong bonding between the endohedral atom and the cage atoms. It is to be noted that endohedral fullerenes of carbon have been known and synthesized with different cage sizes [26], but the endohedral atom is generally not needed for the stabilization of the cage. Thus, using an endohedral atom to form new forms of clusters of different materials was a new idea. The predictions of M@Si_{16} (M = Ti, Zr, and Hf) were later confirmed by experiments [27–31] with high abundances of $TiSi_{16}$, $ZrSi_{16}$, and $HfSi_{16}$ neutral clusters as shown in Figure 4.2. Also, anions of group 3 metal atom doped clusters as well as cations of group 5 metal atom doped clusters, which are isoelectronic with the neutral group 4 metal atom doped clusters, showed high abundance at 16 silicon atoms. Among these, the abundances of neutral Ti@Si_{16} and cation of Ta@Si_{16} are strikingly high compared with other sizes. The different abundances for the doping of different metal atoms reflect the role of the atomic size. Earlier, Kumar and Kawazoe [32] showed that a slightly bigger size of Ge atom compared with that of Si leads to a different growth behaviour of endohedrally doped Ge cages. In particular Zr@Ge_{16} favours FK polyhedral structure (Figure 4.1b) compared with a

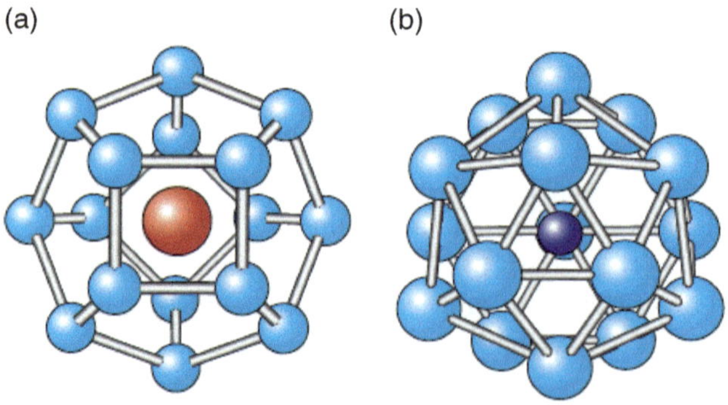

Figure 4.1 (a) Fullerene isomer of Zr@Si_{16} and (b) Frank-Kasper (FK) polyhedral isomer of Ti@Si_{16}. The metal atom is inside the Si_{16} cage. Bonds connecting Si atoms with the metal atom are not shown. The lowest energy structure of Zr@Ge_{16} and some other clusters is also FK type as discussed in the text. *Source:* Adopted from Kumar and Kawazoe [23].

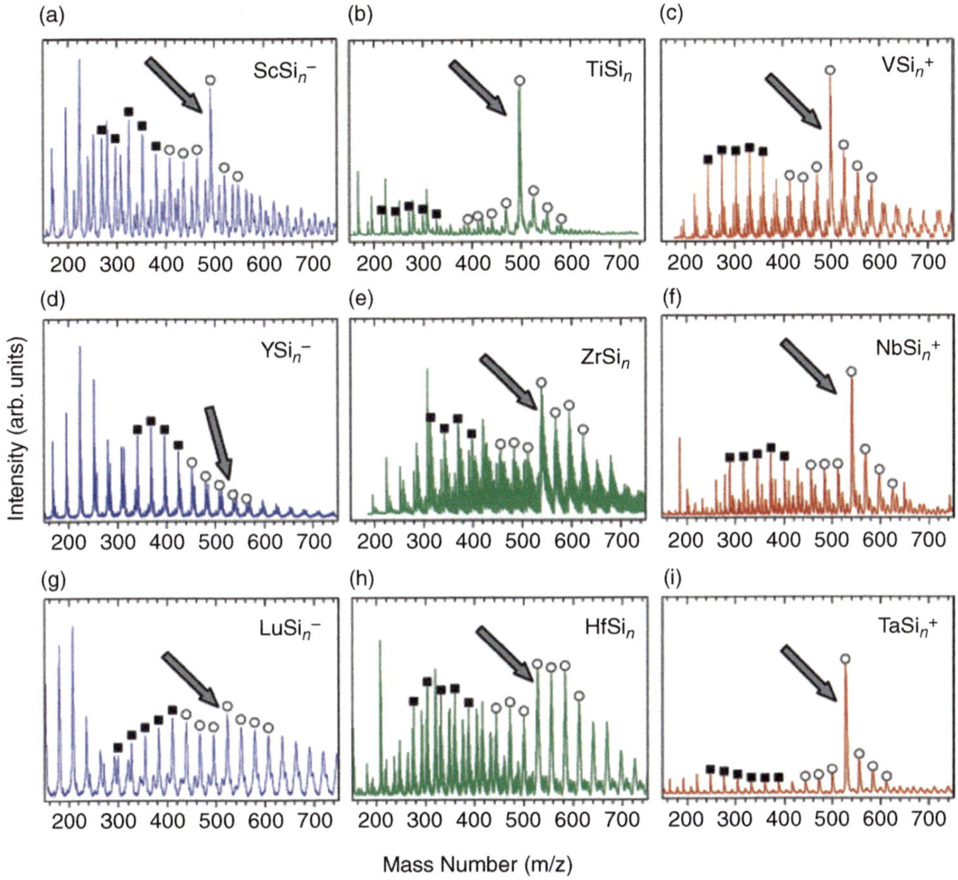

Figure 4.2 Mass spectra of groups 3 (anions), 4 (neutral), and 5 (cations) metal (M) atom doped Si_n clusters (MSi_n): (a) $ScSi_n^-$, (b) $TiSi_n$, (c) VSi_n^+, (d) YSi_n^-, (e) $ZrSi_n$, (f) $NbSi_n^+$, (g) $LuSi_n^-$, (h) $HfSi_n$, and (i) $TaSi_n^+$. The solid squares show distribution similar to that of pure Si_n clusters in their respective charge state while the open circles represent the distribution of M doped Si clusters, which have a prominent peak at MSi_{16}, indicating that species with 68 valence electrons in total are particularly stable. Arrows indicate the mass peaks of the $MSi_{16}^{+/0/-}$ clusters. *Source:* After Koyasu et al. [29].

fullerene structure of Zr@Si_{16}. Thus, by choosing an appropriate combination of the cage atoms and the endohedral atom, a variety of highly stable endohedral clusters with a specific number of atoms and size can be formed [4, 7, 33, 34]. These findings created much interest in the study of endohedral clusters of a variety of elements and compounds in the past two decades, and many experimental and theoretical studies have been performed. These developments have been recently reviewed by Zhao et al. [7]. The particular abundance of 16-atom neutral, cation, and anion metal doped clusters is not specific to silicon and has been also found in the mass abundances in the case of Ge and Sn [4, 32–34]. More recently Ta@Si_{16} and Ti@Si_{16} clusters have been produced in macroscopic quantities [35], thus leading to the possibility of their assemblies. It is to be noted that structures of silicon similar to carbon fullerenes were sought [36] after the discovery of carbon fullerenes, but the weak π bonding in silicon does not favour sp^2 bonding. However, metal encapsulation does wonders and can stabilize smaller fullerenes and other polyhedral structures. The doping of a metal atom can lead to a large gain in energy of about 10 eV or more and totally transform the structure of the undoped clusters leading to the discovery of many novel magic clusters (see Ref. [7]).

There have also been interesting developments using a solution route in different Zintl phases of endohedrally doped Ge, Sn, and Pb clusters [37–43]. Among these, endohedral clusters $[M@E_n]^{q-}$ with E = Ge, Sn, and Pb; $n = 9, 10$, and 12; $q = 2, 3, 4, 5$, and 6; and M a transition metal atom have been found in many cases. Also, crystals of endohedrally doped gold and silver clusters have been synthesized [44–47] (see also Chapter 7). These developments of endohedral clusters and assemblies of such species in bulk have been reviewed by Zhao et al. [7]. The stability of such superatoms can be understood by different electron-counting rules (see also Chapter 2) discussed in this Chapter and by Jena and Sun [6] as well as Zhao et al. [7]. These electron-counting rules are based on e.g. a jellium model [1] applicable to undoped and doped clusters of generally *s*–*p*-bonded metallic materials, 18e or 32e rules [48, 49] as extension of the octet rule, Wade-Mingos rules [50, 51] for polyhedral clusters primarily derived from borane structures, and aromaticity rule [52]. Zhao et al. applied these to study $B_{12}(CN)_{12}{}^{2-}$ [53]. Some clusters may satisfy more than one electron-counting rule giving rise to some special superatoms, which will be discussed (see also Zhao et al. [7]).

In this chapter, we first recapitulate briefly the main considerations for the stability of such superatoms and discuss the discovery of magic endohedral silicon clusters as well as several other interesting examples of highly stable and generally high symmetry and electronically closed shell superatoms of groups 14, 13, and 11. Also, it has been found [54–56] that magnetic superatoms without electronic shell closing can be produced with large magnetic moments similar to atoms, and we discuss some examples here (see also Chapter 5). Further, we discuss possibilities of assembling such superatoms. It is heartening to see that many endohedral cluster materials have been synthesized, and some of these have been found to become superconducting. These include $Mo_8Ga_{41-x}Sn_x$ compound with $Sn@Ga_6$ octahedra and $Mo@Ga_{10}$ polyhedra and $Mo_4Ga_{21-x-\delta}Sn_x$ compound that has $Mo@Ga_9Sn$ clusters, $Ga@Ga_{12}$ cuboctahedra, and Sn_4 squares. Their superconducting transition temperature [57] T_c is 8.7 and 5.85 K, respectively. Another example is of $TaIr_2Ge_2$ compound that has Ta@ Ir_7Ge_4 [58] clusters, and it becomes superconducting at around 3.5 K. The endohedral gallide cluster family includes Mo_8Ga_{41} [59], Mo_4Ga_{21} and Mo_6Ga_{31} [60, 61] compounds with two fused $Mo@Ga_{10}$ clusters. Rh_2Ga_9, Ir_2Ga_9 [62], and $PdGa_5$ [63] compounds have edge-sharing clusters while $ReGa_5$ [64] as well as $PuCoGa_5$ [65] compounds have $Pu@Ga_{12}$ and $Co@Ga_8$ clusters ($T_c = 18.5$ K). The Rh_2Ga_9 compound has the Co_2Al_9 phase with $Co@Al_9$ capped square antiprisms, and it has also been shown to become a superconductor. $Na_{10}NiGa_{10}$ [66] has Ni-centred Ga_{10} clusters. These developments give much hope that more studies would be carried out in the future leading to new families of materials with endohedral clusters.

4.2 Magic Clusters and Their Electronic Stability

For some high symmetry clusters, a sphere is a good approximation to understand their properties. Considering a spherical cluster of free electron-like metals [1]; the electronic states can be well represented by a jellium model according to which the ordering of electronic states is 1S, 1P, 1D, 2S, 1F, 2P, 1G, 2D, 3S, 1H, . . . (see also Chapter 1). Here, the angular momentum character is denoted by capital letters to distinguish them from atoms. Accordingly, clusters with 2, 8, 18, 20, 34, 40, 58, 68, 70, 92, . . . valence electrons have closed electronic shells. Some such clusters have large HOMO–LUMO gaps and exhibit high abundance in the mass spectrum such as icosahedral $Al@Al_{12}^-$, $Si@Al_{12}$, and $B@Al_{12}^-$ clusters, all of which have 40 valence electrons [67]. Such magic clusters or superatoms are similar to atoms with a closed electronic shell, e.g. He or Ar. In realistic cases, there are deviations from the spherical jellium model depending upon the shape of the cluster. However, these results serve as a helpful guide to understand the stability of a variety of clusters. In the case of the endohedrally doped cage clusters, the angular momentum character of the valence orbitals of the endohedral atom should be the same as that of the HOMO of the cage to get a strong hybridization [68]. It is important to find strongly stable endohedral clusters with large HOMO–LUMO gaps as we shall discuss.

The 18e and 32e rules are extensions [48, 49] of the octet rule where *d* and *f* orbitals are filled, respectively. On the other hand, the spherical aromaticity rule [52, 69] is for fullerene-like systems such as icosahedral C_{60} where the p_z orbitals on each C atom are predominantly π bonded. The electronic states of such systems can be described by a spherical potential with high stability of systems having $2(N+1)^2$ electrons where N is a positive integer. Thus, systems with 2, 8, 18, 32, 50, . . . valence electrons are stable. Note that some of the electron numbers are common with other rules. Here, it is worth pointing out that in C_{60} fullerene there are 60 π-bonded valence electrons that do not correspond to one of these electron counts, but it is very stable due to its highly symmetric structure and strong covalent as well as π bonding. This is also a reason that C_{60} can be multiply charged. Therefore, structure and bonding also play an important role in the stability. Very useful rules for the stability of boranes and other polyhedral clusters with a particular number of valence electrons were developed by Wade [50] and Mingos [51]. Considering B–H to represent a group 14 element, one can correlate the stability of clusters such as a three-coordinated polyhedron with $5n_v$ valence electrons, n_v being the number of vertices. Also, a polyhedron with all triangular faces (such as an icosahedron) also referred to as *closo*-deltahedron needs $4n_v + 2$ valence electrons. Out of these $2n_v + 2$ valence electrons are needed for the skeletal bonding molecular orbitals and $2n_v$ electrons in radial covalent bonding with ligands. Incomplete deltahedra, with one (*nido*) and two (*arachno*) vertices missing, need $4n_v + 4$ and $4n_v + 6$ valence electrons, respectively. In the following we refer to these rules to understand the stability of several magic clusters or superatoms.

4.3 Discovery of Silicon Fullerenes and Other Polyhedral Forms

There has been great interest in understanding the properties of nanostructures of silicon for a long time due to the continuous miniaturization of devices. An interesting outcome of research on silicon nanoparticles has been their attractive optical properties with bright light emission unlike bulk Si, which is not a good absorber of light due to its indirect band gap. However, unlike C_{60} fullerene, silicon clusters do not show high abundance of a particular size. Since the HOMO–LUMO gap and other properties of clusters depend on size and shape, it is important to have size selectivity. Some experiments [70] have shown the formation of discrete hydrogenated silicon clusters such as $Si_{29}H_{24}$ of about 1 nm size that were found to emit blue light. Also, in an earlier interesting work, Beck [71, 72] studied the interaction of metals with silicon at the nanoscale. He reported high abundances of 15- and 16-atom silicon clusters with one Cr, Mo, or W atom when $M(CO)_6$ was added to the He career gas during laser vaporization of silicon. He hypothesized that the silicon atoms might have wrapped around the metal atom, but these findings did not attract any attention for quite some time as the understanding of elemental Si clusters was itself evolving. About 10 years later, Jackson and Nellermoe [73] performed ab initio calculations on a Si_{20} dodecahedron fullerene structure similar to that of C_{20}, with a Zr atom inside. They obtained a large binding energy (BE) of the endohedral Zr atom with the Si_{20} cage. Later, Hiura et al. [74] studied interaction of silane gas with different metal monomers and found that there was no H atom attached with $Si_{12}W$ and $Si_{18}W_2$ clusters. They also performed ab initio calculations and suggested a hexagonal prism structure of W@Si_{12} with the W atom inside the Si_{12} prism. Accordingly, two different experiments led to quite different results. Subsequently, Kumar and Kawazoe [23] studied dodecahedral Zr@Si_{20} cluster from ab initio calculations but found the cluster to shrink after optimization. This suggested that 20 silicon atoms were not optimal around a Zr atom. They removed the bulging out silicon atoms and optimized the resulting structure, which ultimately led to a symmetric fullerene-like structure with 16 silicon atoms as shown in Figure 4.1a. This silicon fullerene has eight pentagons and two isolated rhombi (squares) with $D_{4}d$ symmetry, whereas carbon fullerenes have 12 pentagons and different numbers of hexagons such as 20 hexagons in C_{60}. Also, unlike isolated pentagon rule in carbon fullerenes, it was later shown from studies on silicon fullerenes of different sizes such as with 10, 12, 14, 16, and 20 Si atoms that these follow an isolated rhombus rule [4, 34]. Silicon favours pentagonal faces as the bonding in these fullerenes is close to sp^3 type compared to predominantly sp^2 bonding in carbon fullerenes, and rhombi are defects as pentagons are in carbon fullerenes. Si_{20} is ideal as all faces are pentagonal.

Following the similarity of bonding between silicon and germanium in bulk, when a Zr@Ge_{16} fullerene structure was optimized [32], it transformed to a Z16 FK polyhedron with T_d symmetry (Figure 4.1b). This is due to the fact that the atomic radius of a Ge atom is about 4% larger than that of a Si atom. Therefore, bonding in the fullerene isomer of Ge_{16} was not optimal and a compact FK polyhedron with triangular faces (Figure 4.1b) became more favourable. On the other hand, when a Ti atom, which is slightly smaller than a Zr atom, was doped in Si_{16}, the FK isomer became lower in energy than the fullerene isomer. These 16-atom clusters have the highest BE with atomic and electronic shell closing, which supports their high abundance. Calculations with PBE0 hybrid functional in Vienna Ab initio Simulation Package (VASP) code [75] and with projector augmented wave (PAW) pseudopotentials [76] show a large HOMO–LUMO gap of 3.68 eV for FK-Ti@Si_{16} while a smaller gap of 2.768 eV (2.775 eV) for Zr@Si_{16} (Hf@Si_{16}) in the fullerene structure. Photoelectron spectra of anion clusters [29] also showed a large gap between HOMO and HOMO-1 for Ti@Si_{16}^- and a smaller gap for Zr@Si_{16}^- and Hf@Si_{16}^- as predicted. These results demonstrated the subtle dependence of the cage on the size of the endohedral atom as the bonding between the endohedral atom and the cage is very strong. This also quenches the magnetic moments on the transition metal atom. Accordingly, different cages may be stabilized depending upon the combination of the dopant and the cage atoms. Note that none of these cage structures is favoured by elemental Si or Ge clusters. Thus, endohedral doping emerged as a novel way to create new nanostructured materials. These results are different from the endohedral carbon fullerenes [26] where the endohedral specie generally does not interact strongly with the fullerene cage. However, smaller carbon fullerenes in the size range C_{20}–C_{28} have little abundance in the mass spectra. It is due to the increase in strain. However, doping with Ti, Zr, Hf, and U showed [77] abundance of predominantly doped clusters. This suggested that metal atom doping stabilizes the smaller carbon fullerenes. In recent years many theoretical studies have been carried out on such small carbon fullerenes by doping with a variety of atoms (see Zhao et al. [7]) including rare earths, and many interesting cases have been predicted.

The results on endohedral silicon cages attracted much attention, but it took some time to be realized in laboratory. Indeed, neutral Ti@Si_{16} and isoelectronic Sc@Si_{16}^- as well as V@Si_{16}^+ were found [27, 29] in highest abundances in the mass spectra with much weaker intensities of other sizes (See Figure 4.2). Further, a systematic study [78] of Ti, Zr, and Hf doped Si_n clusters showed the growth to saturate at $n = 16$, which leads to the high abundance of 16 Si atom endohedral clusters. For the case of Ti, we have shown in Figure 4.3 the BE [79] that increases continuously as the number of silicon atoms increases. These 16-silicon atom endohedral clusters have 68 valence electrons, which correspond to the electronic 2D shell closing in a jellium model. This is particularly true for the close-packed FK isomer as shown in Figure 4.4 where we have plotted [33] the gaussian broadened energy levels of FK-Ti@Si_{16} and fullerene-like Zr@Si_{16} clusters. The high symmetry of the FK isomer leads to higher degeneracy of the energy levels. The degeneracy of some states is reduced in the case of Zr@Si_{16} fullerene. Here, the *d* orbitals of the metal atom hybridize with the 2D orbitals of the Si cage. This leads to the high stability of the doped clusters [21, 80] with a gain in energy of >11 eV and a large HOMO–LUMO gap. Later experiments [28] confirmed this interpretation of the stability with 68 valence electrons.

In Figure 4.5 we show the results obtained from X-ray absorption spectroscopy on V, Ti, and Cr doped silicon cation clusters by Lau et al. [28] at the $L_{2,3}$ edges of the transition

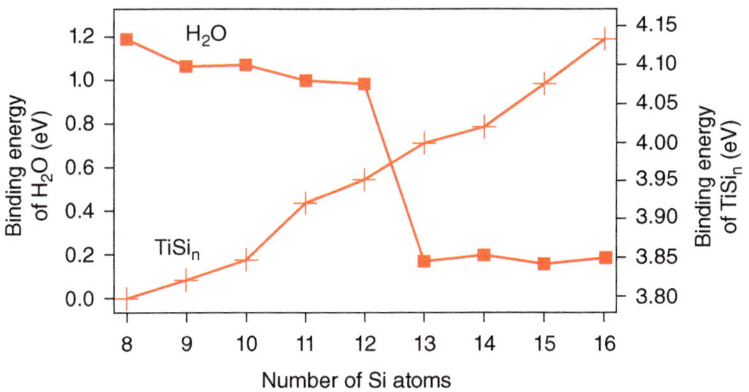

Figure 4.3 Interaction energies (left scale) between $TiSi_n$ and H_2O molecule, (n = 8–16). It drops significantly for n = 13 as Ti atom becomes inside a cage. The right scale shows the BE per atom of $TiSi_n$ clusters, which increases up to n = 16. *Source:* After Kawamura et al. [79].

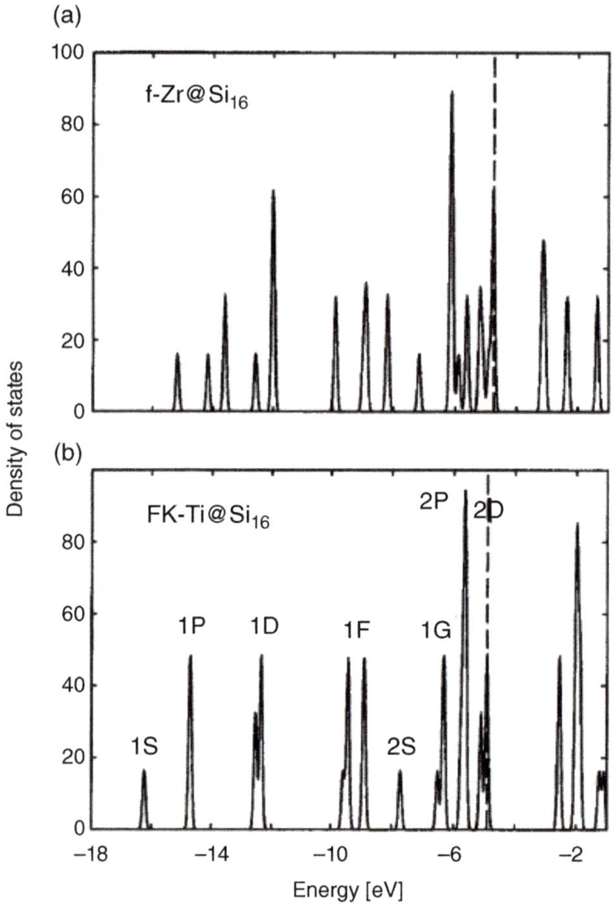

Figure 4.4 Kohn–Sham energy spectra of fullerene type f-Zr@Si_{16} and FK-Ti@Si_{16} clusters with Gaussian broadening (half width 0.05 eV) calculated with PW91 functional and planewave basis. The states are labelled with their angular momentum character as in a jellium model. For the f isomer the degeneracies are reduced due to lower symmetry. Broken line shows the HOMO. After Kumar [33]. *Source:* Copyright EDP Sciences, Springer-Verlag.

metal to probe the local electronic structure of the dopant atoms. These results show that among $[VSi_n]^+$ (n = 14–18) clusters, $[VSi_{16}]^+$ has the sharpest peak, reflecting the high degeneracy of the electronic states and high symmetry of this cluster. $[VSi_{16}]^+$ is isoelectronic with Ti@Si_{16}, which has a tetrahedrally symmetric FK polyhedron structure. $TiSi_{16}{}^+$ and $CrSi_{16}{}^+$ clusters also show similar spectral features as those of $[VSi_{16}]^+$, suggesting nearly identical local electronic states at the transition metal atoms and a similar geometric environment. Calculations [81, 82] on $CrSi_{16}$ clusters, however, showed a capped Cr@Si_{15} cage to have the lowest energy. Accordingly, the symmetry reduces and the spectral lines become broader. The cation of Ti@Si_{16} also shows Jahn-Teller distortion. Further experiments using direct and resonant core-level photoionization spectroscopy [83] with valence band photoionization curves gave the HOMO–LUMO gaps of 0.8 ± 0.2 eV for $[VSi_{15}]^+$, 2.1 ± 0.2 eV for $[VSi_{16}]^+$, and 0.5 ± 0.2 eV for $[VSi_{17}]^+$. A large HOMO–LUMO gap for

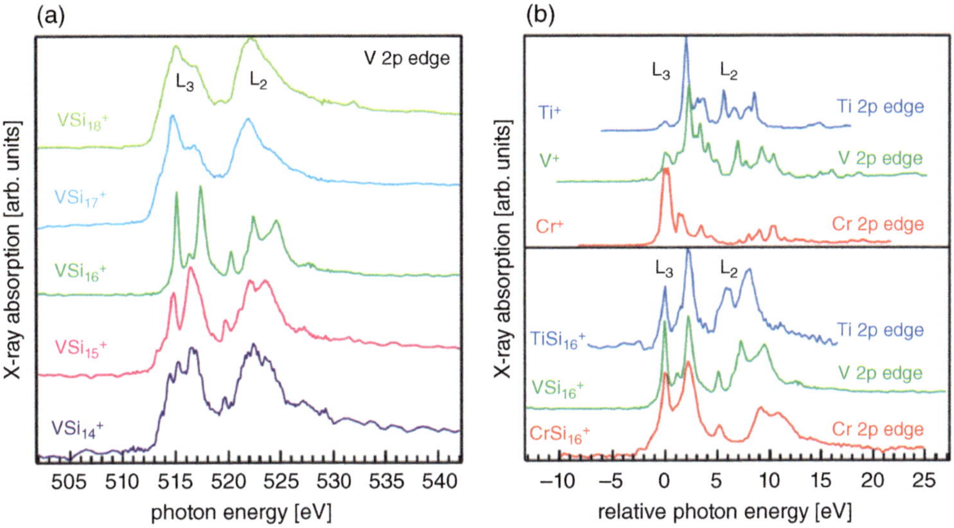

Figure 4.5 (a) $L_{2,3}$ X-ray absorption spectra of size-selected $[VSi_n]^+$ clusters (n = 14–18). The $[VSi_{16}]^+$ spectrum is characterized by the sharpest lines, indicating a highly symmetric structure with high degeneracy of the electronic levels. The double-peak structure can be linked to the electronic density of states (DOS) in tetrahedral symmetry. (b) Transition metal $L_{2,3}$ X-ray absorption spectra of $[TiSi_{16}]^+$, $[VSi_{16}]^+$, and $[CrSi_{16}]^+$ clusters in the lower panel and of bare Ti^+, V^+, and Cr^+ ions in the upper panel. The spectra are aligned at the first peak position. While bare ions show differing spectra, the relative excitation energy in doped silicon clusters is nearly identical. This surprising fact indicates almost identical local electronic structure at the dopant atom site. Relative positions of the L_2 edge shift because of decreasing $2p$ spin-orbit splitting from Cr to Ti. In the titanium spectrum, the 5 eV feature of the L_3 line is partially masked by overlapping L_2 intensity. *Source:* After Lau et al. [28].

$[VSi_{16}]^+$ supports the FK polyhedral structure for the isoelectronic $Ti@Si_{16}$. Also, a high temperature MD simulation [84] on $Ti@Si_{16}$ cluster suggested the Ti atom to remain at the centre with Si atoms around it. Further studies on the growth behaviour surprisingly showed [78] $Ti@Si_{12}$ to have a large electron affinity of 3.66 eV in a hexagonal prism structure (Figure 4.6). This is similar to a Cl atom while $Ti@Si_{16}$ has the largest ionization potential of 7.514 eV. The second order derivative of energy is also the highest for $Ti@Si_{12}$ anion in the hexagonal prism structure. One can, therefore, expect a hexagonal prism structure to be very favourable for the doping of group 5 metal atom such as $Ta@Si_{12}$ cluster as it was indeed found by Uchida et al. [85].

Among many other such clusters, $Ta@Si_{16}$ has attracted much attention (see Zhao et al. [7]) due to the almost exclusive abundance of this cluster similar to $Ti@Si_{16}$ as seen in Figure 4.2. Both Ti and Ta doped Si_{16} clusters have been produced in macroscopic quantity [35]. Our ab initio PAW pseudopotential calculations in VASP with PBE0 hybrid exchange-correlation functional predict the fullerene structure of $Ta@Si_{16}$ to be 0.423 eV lower in energy than the FK structure and the BE is 4.178 eV/atom. It has a magnetic moment of 1 μ_B due to an odd number of electrons and small distortion in the fullerene structure as shown later in Figure 4.22a. However, calculations using Perdew, Burke, and Ernzerhof (PBE) [86] exchange-correlation functional in VASP gave the FK isomer to be

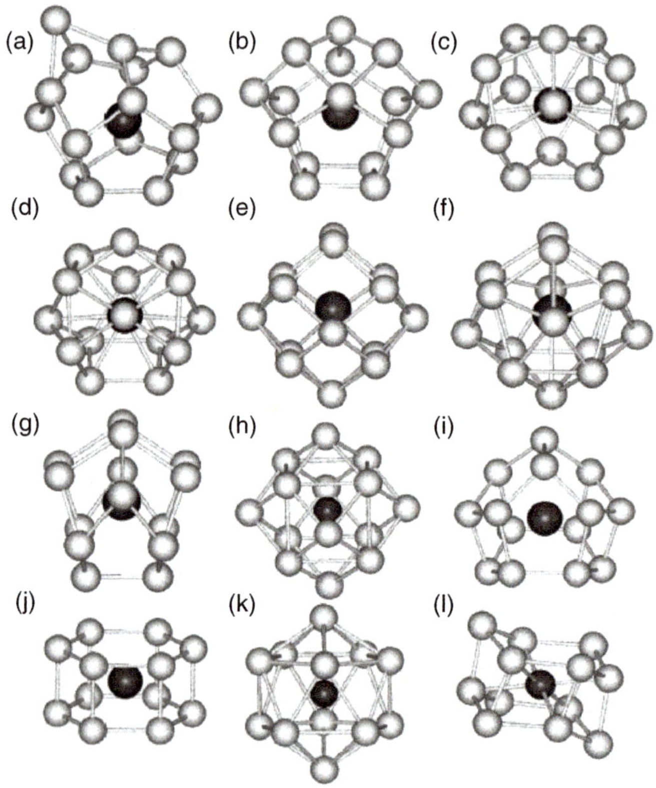

Figure 4.6 (a) M@Si_{16} (M = Cr, Mo, W) capped fullerene structure, (b) Mo@Si_{15} fullerene-like structure, (c) M@Si_{15} FK-type structure obtained from a cubic isomer, (d) another FK-type structure for M@Si_{15}, (e) cubic structure for M@Si_{15}, (f) capped decahedral structure of M@Si_{15}, (g) a fullerene-like structure of M@Si_{14}, (h) a cubic structure of M@Si_{14}, (i) another fullerene-like structure of M@Si_{14}, (j) M@Si_{12}, (k) icosahedral M@Ge_{12}, and (l) Be@Si_{12}. *Source:* After Kumar [33].

0.082 eV lower in energy than the fullerene isomer. Earlier studies [23] on Hf@Si_{16} also found the fullerene and FK as well as distorted FK isomers to lie very close in energy. However, using structure prediction method and incorporating dispersion correction in density function theory (DFT) with DFT-D2 method [87] and PBE functional, Sen and Sen [88] obtained the fullerene isomer to be most favourable for this cluster. We also performed calculations using the Gaussian program [89] and 6–311++g(d,p) basis set for Si, Lanl2dz basis set for Ta and PBE0 hybrid exchange-correlation functional. It is found that the fullerene isomer is 0.931 eV lower in energy than the FK isomer. The fullerene isomer has also been reported in experiments [35, 90, 91]. The Ta@Si_{16} cluster has one electron more than the electronic shell closing and, therefore, it behaves like an alkali atom. The structure deviates slightly from a perfect D_{4d} symmetry, and there is a small electric dipole moment of 0.67 D. The adiabatic ionization potential has been calculated to be 5.926 eV using the Gaussian program. The cation of Ta@Si_{16}, however, has a perfect D_{4d} symmetric structure as shown in Figure 4.1a. The Si–Si bond lengths between atoms on the two squares is 2.287 Å, between squares and pentagons is 2.262 Å, and between atoms on the middle ring is 2.345 Å. This isomer is 0.586 eV lower than the cation of the FK isomer. The HOMO–LUMO gap is 2.686 eV, which is similar to the value (2.629 eV) for the fullerene isomer of Hf@Si_{16}. No imaginary frequency was found, suggesting that the cation cluster is dynamically stable.

After the findings of fullerene and FK structures, a variety of other dopants have been studied in silicon cages. Ab initio calculations on Cr, Mo, and W atom encapsulation showed [81] that the M@Si_{16} cage shrinks to a M@Si_{15} (M = Cr, Mo, and W) endohedral cage with a Si atom capping it (Figure 4.6a). The latter is also the structure of these M@Si_{16} clusters. This result showed that a 15-atom Si cage was optimal for these metal atoms, and a few isomers that lie close in energy are shown in Figure 4.6b–f. Here, (b) shows the fullerene isomer obtained by removing the capping atom in (a); (c) is an FK-type isomer obtained from relaxation of 14-atom face capped cube on which one more Si atom was added on the face. The mean coordination of Si atoms on the cage is 4. Here, (d) is also an FK-type isomer, but position of one Si atom was changed; (e) is also a cubic-type isomer; and (f) is a capped decahedral isomer obtained from VASP calculation and with PW91 functional [92]. For Cr, (d) and (e) are nearly degenerate, but (e) has higher HOMO–LUMO gap of 1.537 eV. For Mo and W doping, isomer (d) is slightly more favourable than isomer (e) and the HOMO–LUMO gap was calculated to be 1.659 and 1.747 eV, respectively. These two values are even larger than that obtained for Zr@Si_{16} at the same level of theory suggesting their strong stability. These results explained the observations by Beck [71, 72]. It is interesting to note that Atobe et al. [93] obtained the strongest peak of anion of Nb@Si_{15} (isoelectronic with Mo@Si_{15}) in the mass spectrum while for the cation, Nb@Si_{16} (isoelectronic with Zr@Si_{16}) is strongly abundant (Figure 4.2) suggesting the strong dependence of the cluster abundance on the number of valence electrons and electronic shell closing. Zhao et al. [7] have recently reported results using PBE0 exchange-correlation functional and found the same isomers to be the lowest in energy for Mo and W cases with the HOMO–LUMO gaps of 2.963 and 3.032 eV, respectively, and with the embedding energy >12 eV. Here the embedding energy is the difference between the sum of energies of the cage without the endohedral atom plus the isolated endohedral atom and the energy of the doped cage. They also obtained two other isomers lying close in energy and for Cr another isomer was found to lie slightly lower in energy. Isomer (f) was found to be the lowest in energy for Ti@Si_{15} while isomer (e) lies close in energy [33]. Kawamura et al. [82] have studied the growth behaviour of Cr doped Si clusters and shown that Cr@Si_{12} has the hexagonal prism structure (Figure 4.6j) as also obtained by Khanna et al. [94]. Hiura et al. [74] also obtained a prism structure for W@Si_{12}, but the growth does not stop here. Accordingly, the nucleation conditions in the experiment of Hiura et al. [74] were such that W@Si_{12} cluster was obtained without any hydrogen. Kumar and Kawazoe [95] also studied these systems and interaction with hydrogen. Besides Cr@Si_{12}, they obtained tubular structure of W_2@Si_{18} with three hexagons of silicon in prism structure and W atoms in between. It is also interesting to note that Uchida et al. [85] produced Ta@Si_{12} hexagonal prism cluster and deposited them on a Si(111)-(7×7) substrate while Tsunoyama et al. [35] produced Ta@Si_{16} in large quantities, again supporting the role of nucleation conditions. Kawamura et al. [82] have found a strong magic behaviour of Cr@Si_{12}, Cr@Si_{14}, and Cr@Si_{15} neutral clusters from the second order derivative of energy. M@Si_{14} clusters with M = Cr and Mo have an *arachno* deltahedron structure [7, 82], and it satisfies the Wade-Mingos rule of 62 valence electrons giving rise to a large HOMO–LUMO gap of about 3 eV using PBE0 functional. Also, a large HOMO–LUMO gap (2.939 eV) was obtained [82] for Cr@Si_{12} using B3PW91 hybrid functional [96] in the Gaussian code, but the BE was found to be the highest for Cr@Si_{15}. The ionization potential of Cr@Si_{12} is also the highest (7.516 eV). The properties were found to depend significantly on the charge state of the clusters. For anion, Cr@Si_{12} is strongly magic and interestingly has the largest HOMO–LUMO gap of 2.673 eV. For cation, Cr@Si_{13} with 2.494 eV HOMO–LUMO gap and Cr@Si_{15} with 1.235 eV HOMO–LUMO gap become strongly magic. The hexagonal prism structure, which is a three-connected polyhedron (Figure 4.6j), is also favourable for M@Si_{12} with M = Ti, V, Mn, Zn, Nb, Mo, Ru, Re, and Os. According to the Wade-Mingos rule, it requires 60 valence electrons as in the case of Zn@Si_{12}, but other clusters are electron deficient [7, 97]. This also supports further growth of these clusters. M@Si_{15} clusters with M = Ru and Os have 68 valence electrons and are interesting because of electronic shell closing. These clusters were shown [33] to have a large HOMO–LUMO gap of about 1.5 eV in an FK isomer (Figure 4.6d) using PW91 exchange-correlation functional. The embedding energy as well as the binding energies are large (~13 and ~4.2 eV/atom, respectively).

Compared to endohedral Mo and W atoms, 14-atom silicon cage becomes ideal for Fe, Ru, and Os atoms, due to their smaller size. For Fe, a body centered cubic (BCC)-like structure (Figure 4.6h) becomes the lowest in energy, but for Os and Ru, a fullerene-like structure (Figure 4.6i) competes [23]. For Os doping, a D_{3h} fullerene-like cage with six pentagons and three rhombi becomes nearly degenerate with BCC-like structure [23, 34]. These results led to the question: which are the smallest and the largest cages that may be stabilized with endohedral doping of one atom? Kumar and Kawazoe [55] showed that a simple cube of Ge or Sn is the smallest cage, which can possibly be stabilized with a Be atom. Uta and King [98] also found a perfect cube for Be@Ge_8 as the smallest cluster, which can have an endohedral atom. This has 34 valence electrons that corresponds to 1F electronic shell closing in a jellium model. But, in experiments an eight-atom endohedral cluster is not observed. Rather, a 9-atom cluster derived from a tricapped prism (or a capped tetragonal antiprism) and a 10-atom bicapped tetragonal antiprism D_{4d} cluster (Figure 4.7a) are energetically more favourable and have the lowest energy [55, 98] for Be@X_{10} (X = Si, Ge, and Sn), Ni@Ge_{10}, Pt@Sn_{10}, Pt@Pb_{10}^{2-}, Al@Pb_{10}^{+} [7, 39, 99, 100]. A few other possibilities [99] are also shown in Figure 4.7, which include a threefold symmetric fullerene-like structure with three pentagons for Pt@Si_{10} (Figure 4.7b), a tetra-capped trigonal prism for Pt@Pb_{10} (Figure 4.7c), and a capped pentagon-rhombus for Ni@Si_{10} and Co@Si_{10} (Figure 4.7d). Another possibility [7] is a pentagonal prism for M@Si_{10}, M = Mn and Fe. An interesting case is that of the perfect icosahedral structure (Figure 4.6k) that has the highest symmetry and is obtained [55, 101] for Be@X_{12} (X = Ge and Sn), Al^{+}@Pb_{12} [100, 102], and Pt^{2-}@Pb_{12} [39]. All these clusters have 50 valence electrons excluding the *d* electrons of Pt and satisfy the Wade-Mingos rule. For the largest size, early attempt [103] to stabilize a Si_{20} fullerene by endohedrally doping with a Ba, Sr, Ca, Zr, and Pb atom led to distorted cages. However, later it was found that a dodecahedral cage (Figure 4.7e) can be stabilized by rare earths and Th was predicted [104] to be the only candidate to stabilize Th@Si_{20} with icosahedral symmetry, closed electronic 1F sub-shell, and the highest embedding energy of about 14 eV. It is because, among the rare earths, Th can exist in 4+ valence state. Also, anions of Y@Si_{20} and La@Si_{20} have been shown to be stable with zero spin while the

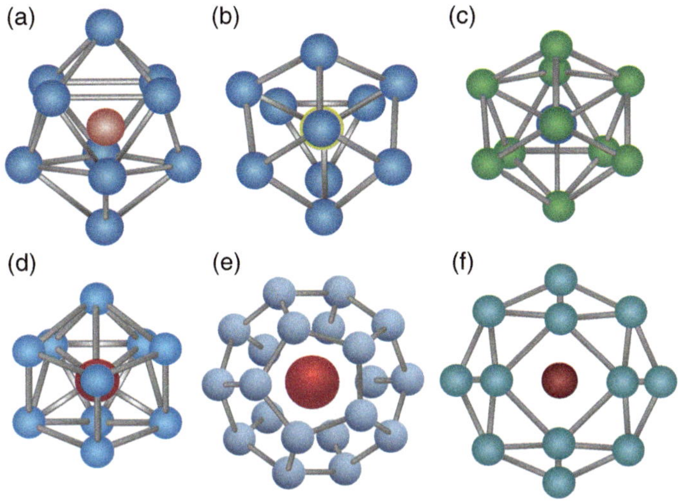

Figure 4.7 (a) Bicapped tetragonal antiprism for $Be@X_{10}$ (X = Si, Ge, Sn), $Ni@Ge_{10}$, $Pt@Sn_{10}$, $Pt@Pb_{10}^{2-}$, $Al@Pb_{10}^{+}$, (b) a fullerene-like isomer of $Pt@Si_{10}$, (c) relaxed tetra-capped trigonal prism, (d) a capped pentagon-rhombus for $Ni@Si_{10}$, (e) $Th@Si_{20}$ icosahedral fullerene, and (f) $Th@Ge_{20}$. *Source:* Adopted from Refs. [34, 99, 104, 105].

neutral clusters have a large electron affinity similar to that of a Cl atom [106]. On the other hand, for $Th@Ge_{20}$, a closed packed FK-like structure (Figure 4.7f) was obtained [105]. Besides these Dognon et al. [49] suggested a family of actinide atom doped 32 valence electron $[M@Si_{20}]_q$ clusters. In the fullerene structure of Si_{20} with each Si atom threefold coordinated with neighbouring Si atoms, the bonding is close to sp^3 type. We can consider 20 electrons in lobes pointing outwards of the cage. With 12 electrons on the central atom including charge q makes it a 32-valence electron cluster. Dognon et al. [107] have studied doping with U^{6-}, Np^{5-}, Pu^{4-}, Am^{3-}, and Cm^{2-}. The above discussed results demonstrate fullerene-like and FK polyhedral growth behaviours to be among the preferred structures, but polyhedral cages with mixed behaviours are also obtained for different cases [7]. Another detailed study [108] on anions of Y doped silicon clusters showed two growth behaviours: (i) linked clusters in which two sub-clusters are linked by the metal atom and (ii) endohedrally doped cages.

4.4 Endohedral Superatoms of Ge, Sn, and Pb

Following the extensive work on endohedral silicon clusters, many studies have been done on endohedral clusters of isoelectronic Ge and Sn. As the atomic sizes of Ge and Sn are slightly bigger than a Si atom, the growth behaviour in these cases is different. Besides the cases of $Th@Ge_{20}$ and $Zr@Ge_{16}$ discussed above that form different structures from the respective silicon cages, Ge_{12} and Sn_{12} can form [101] icosahedral cages (Figure 4.6k) around a metal atom such as a Zn atom. This is in contrast to Si for which an icosahedron is not favoured. Moreover, an empty icosahedral cage is favoured [109] for Sn_{12}^{2-} and also [110] for Pb_{12}^{2-}. This differs from the behaviour for Si and Ge cases. Atobe et al. [93]

have studied M = Sc, Ti, V, Y, Zr, Nb, Lu, Hf, and Ta doped Ge clusters from photoelectron and mass abundance spectroscopies and obtained prominent peaks for M@Ge_{16} neutral (M = Ti, Zr, and Hf) and charged clusters (cations of V, Nb, and Ta, and anions of Sc, Y, Lu) with 68 valence electrons. Interestingly, the abundances of YGe_{16}^- and $HfGe_{16}^-$ are much stronger compared with the corresponding Si_{16} cages while the abundance of $ScGe_{16}^-$ is significantly weaker than that of $ScSi_{16}^-$. The anionic MGe_{16} clusters for M = Ti, Zr, and Hf show a split off HOMO suggesting a sizeable HOMO–LUMO gap for the neutral clusters as shown in Figure 4.8. Interestingly, the HOMO–LUMO gap for all three cases of M atoms, namely, Ti, Zr, and Hf in Ge_{16} is quite similar (1.79, 1.955, and 1.979 eV, respectively, using PW91) as in all the cases the FK structure is favoured [32, 111]. This is also different from the case of silicon. The measured [93] HOMO and HOMO-1 gaps in these cases are 1.77, 1.90, and 1.83 eV, respectively, and these agree with the theoretical findings. No split-off state is seen for the MGe_{16} anions doped with M = Sc, Y, Lu, and Tb. Due to the closed electronic shell in these cases, the photoelectron spectra show two prominent peaks. On the other hand, for the MGe_{16} anions (M = V, Nb, and Ta), the two-peak feature is weakened presumably due to distortions in the structure as these systems do not correspond to electron shell closing.

In order to confirm if the metal atom was doped endohedrally, Koyasu et al. [29] and Atobe et al. [93] studied water adsorption on doped Si clusters and doped Ge, and Sn

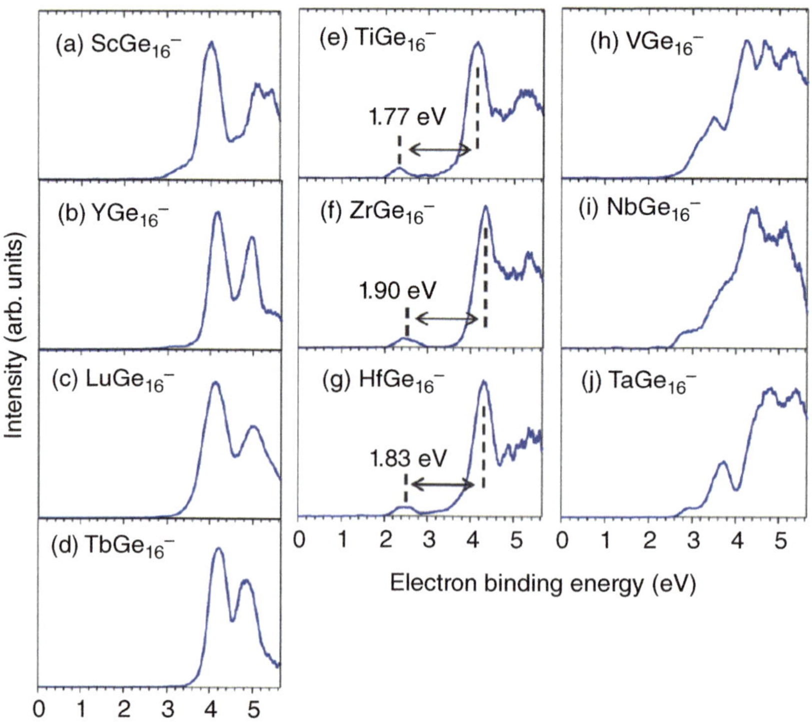

Figure 4.8 Photoelectron spectra of M@Ge_{16} (M = Sc, Y, Lu, Tb, Ti, Zr, Hf, V, Nb, and Ta) anions at 213 nm. For Ti, Zr, and Hf, a large gap of 1.77–1.9 eV between HOMO and HOMO-1 is observed. *Source:* After Atobe et al. [93].

clusters, respectively. They found that beyond a certain size, the water molecules were not adsorbed on the cluster, indicating that the metal atom is inside the cage and is not able to interact with water molecules as shown in Figure 4.9. It is seen that this size generally changes as the metal atom changes and becomes bigger as the radius of the metal atom increases. Furthermore, this size is always smaller for Ge clusters compared with silicon clusters. One can see that for the doping of a Ti atom in Si clusters, the metal atom goes inside the cage at size 13 and it becomes 12 for Ge. On the other hand, the cage size for endohedral Y in Si and Ge clusters, where no adsorption of water molecules was observed,

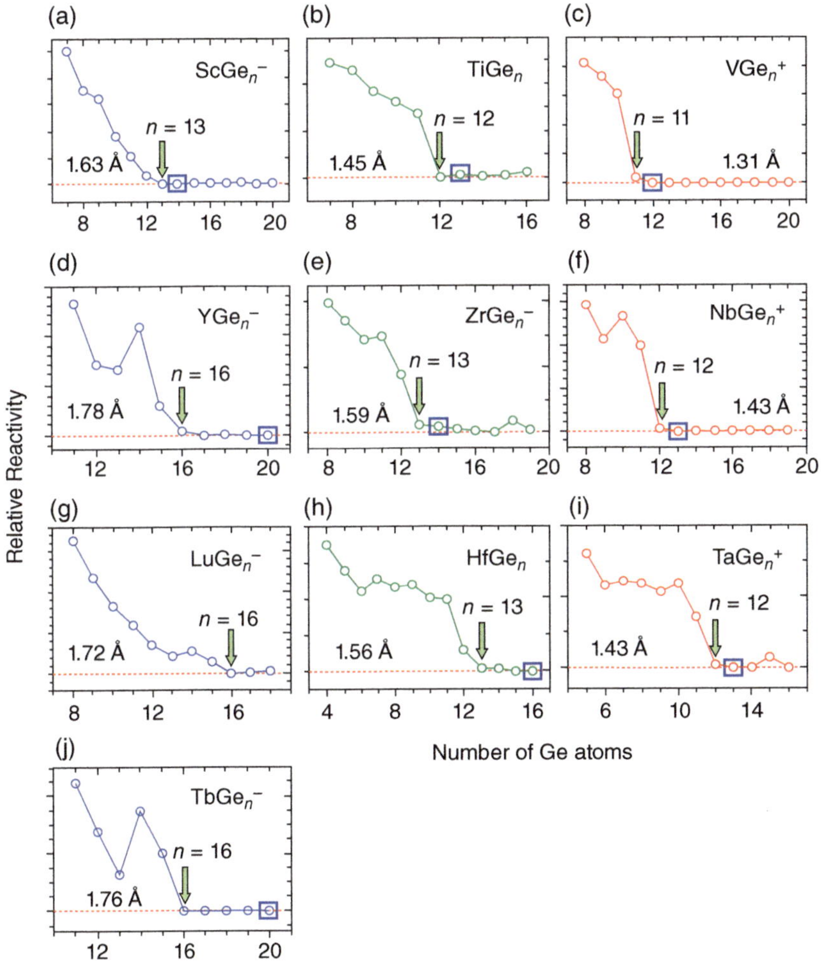

Figure 4.9 Relative reactivity of anionic/neutral/cationic $MGe_n^{-/0/+}$ clusters towards H_2O vapor. Vertical arrows show the threshold size of $MGe_n^{-/0/+}$ where the relative reactivity is lost. The metallic bond radius of the metal atom is also given to indicate the sizes of the atoms that affect the cage formation. The threshold sizes of metal–atom doped Si_n clusters are shown by a solid square for comparison. The threshold size of $MGe_n^{-/0/+}$ is always smaller than that for $MSi_n^{-/0/+}$. The change in the threshold size can be reasonably explained by metal encapsulation in a Ge cage. *Source:* After Atobe et al. [93].

is 20 and 16, respectively. Note that a $Y@Si_{20}$ fullerene was predicted for the anion cluster [106, 108]. Kawamura et al. [79] studied interaction of a water molecule with Ti doped Si_n clusters and found that the Ti atom enters the silicon cage at $n = 13$ and the BE of water molecule adsorption drops significantly as shown in Figure 4.3, in excellent agreement with experiments.

For Sn cages, there are fewer studies. In the case of $M@Sn_{16}$ anions, the results of photoelectron spectra [93] show sizeable HOMO and HOMO-1 gap suggesting a large HOMO–LUMO gap for the neutral clusters (Figure 4.10). This is larger (1.64 eV) for Zr and Hf doped clusters and slightly smaller for Ti (1.47 eV) doped cluster, but much smaller HOMO–LUMO gaps are for 15- and 17-Sn atom endohedral clusters. This suggests high symmetry structures for 16-Sn atom cages. We have performed ab initio calculations for $M@Sn_{16}$ (M = Ti, Zr, and Hf) by considering a few structures, and the optimized structures show that the FK isomer (Figure 4.1b) has the lowest energy. For the neutral Ti, Zr, and Hf doped clusters, the HOMO–LUMO gaps are, respectively, 2.50 eV (2.29 eV), 2.682 eV

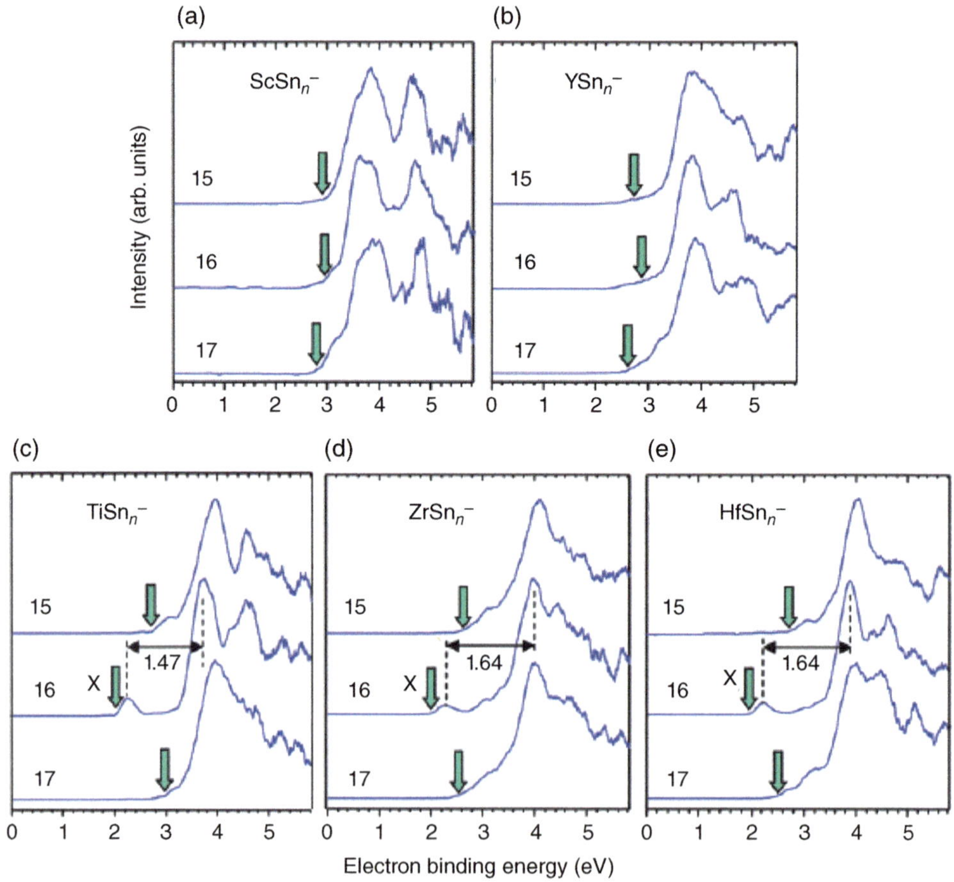

Figure 4.10 Photoelectron spectra of (a) $ScSn_n$, (b) YSn_n, (c) $TiSn_n$, (d) $ZrSn_n$, and (e) $HfSn_n$ ($n = 15$–17) anion clusters at 213 nm. For Ti, Zr, and Hf, a large gap between HOMO and HOMO-1 is observed for $n = 16$. *Source:* After Atobe et al. [93].

(2.494 eV), and 2.698 eV (2.503 eV) using PBE0 in VASP (Gaussian program with SDD basis set, which uses Stuttgart/Dresden effective core potentials [ECPs])). The deformed FK isomer also lies close in energy. The calculated trend of the HOMO–LUMO gap is in very good agreement with the experimental data. The measured gap from the anions is smaller due to the deformation of the cluster because of the extra charge that leads to Jahn–Teller distortions in the structure.

For the case of Pb clusters, there are also fewer studies (see Zhao et al. [7]) and the most prominent clusters are $[M@Pb_{12}]_q$ and then $[M@Pb_{10}]_q$ where q is the charge on the clusters. Unlike the case of Si and Ge clusters, Furuse et al. [31] reported absence of high abundance of $M@Pb_{16}$ clusters for M = Ti, Zr, and Hf. It has been shown that the nucleus independent chemical shift (NICS) value at the centre of $[Pb_{16}]^{4-}$ cage is low and positive compared with large negative values for $[Si_{16}]^{4-}$, $[Ge_{16}]^{4-}$, and $[Sn_{16}]^{4-}$ cages indicating antiaromatic character for the Pb_{16} cage. An interesting case has been the $Pu@Pb_{12}$ cluster, which has 32 electrons and is found [49, 107] to be stable.

4.5 Magnetic Superatoms

In 2003, working on the endohedral doping of group 14 clusters, it was found by Kumar and Kawazoe [55] that $Mn@Sn_{12}$ and $Mn@Ge_{12}$ clusters favoured a perfect icosahedral structure (Figure 4.6k) with $5\,\mu_B$ magnetic moments. The magnetic moments are predominantly found on the Mn atom suggesting that the hybridized *d*-type orbitals were half-filled according to the Hund's rule. Thus, similar to atoms, this superatom was called a magnetic superatom. Earlier Hund's rule was also found [54] in a perfect icosahedron of $Cu@Al_{12}$ cluster (Figure 4.6k) with the 2P electronic shell half-filled and $3\,\mu_B$ magnetic moments even though in the bulk neither Cu nor Al is magnetic. Following these results, several studies have been performed and subsequent experiments [112] showed high abundance of $MnSn_{12}$ and $MnGe_{12}$ clusters. Stern Gerlach experiments [113, 114] on $MnSn_{12}$ clusters further confirmed high magnetic moments on this cluster as predicted. Similarly, high magnetic moments have been obtained [115, 116] by endohedral doping of Fe and Cr atoms in Sn_{12} and also for the Pb_{12} cage [117]. Besides group 14 elements, magnetic superatoms have been also found for endohedral clusters of other elements such as doped clusters of sodium. In particular, $5\,\mu_B$ magnetic moments were obtained [118] in VNa_8 cluster, which arise from the half-filling of the 1D shell. Also, some doped clusters of e.g. Cu, Ag, Au, and Al have been found to have significant magnetic moments as we shall discuss in the following. More details can be found in Chapter 5.

4.6 Endohedral Clusters of Group 11 Elements

Besides group 14 elements, endohedral cages of group 11 elements (coinage metals) and in particular that of Au have been extensively studied due to the unusual and interesting catalytic behaviour of small gold clusters. Small gold clusters are planar, and there is a report of the formation of Au_{16}^- cage [119] as well as a Au_{32} triacontahedron empty cage fullerene [120]. The latter is shown in Figure 4.11d, and a Au_{16}^{2-} cage with 18 valence electrons

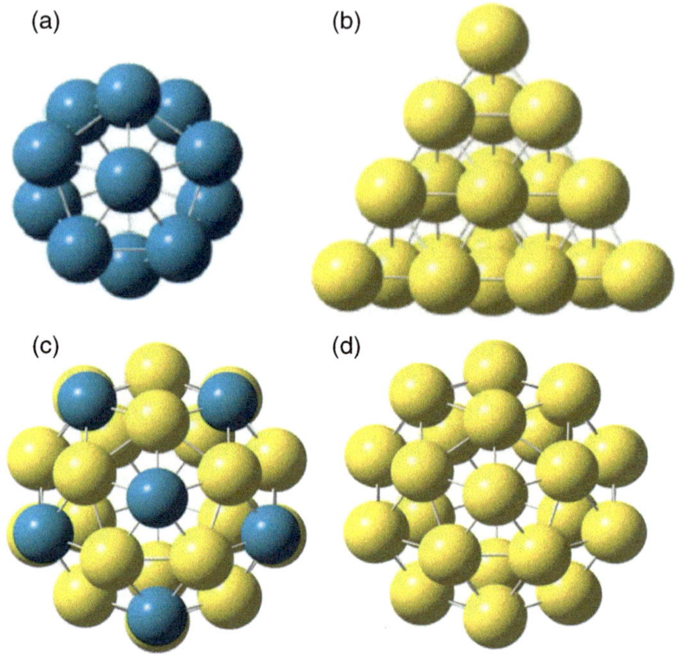

Figure 4.11 Atomic structures of (a) Al_{13}^- and (b) Au_{20}. (c) and (d) show two icosahedrally symmetric $Al@Al_{12}Au_{20}^-$ and Au_{32}, respectively. The yellow and blue balls show Au and Al atoms, respectively. *Source:* After Kumar [121].

(without considering *d* electrons on Au atoms) is shown in Figure 4.12. Also, Au_{20} is a magic cluster and has a tetrahedron structure (Figure 4.11b). These are different structural properties compared to those of Cu and Ag clusters [122] due to the importance of relativistic effects in Au. Accordingly, Au clusters are particularly interesting for endohedral doping.

Doping of small gold clusters with a suitable metal atom has been shown [123, 124] to convert a planar Au_{12} cluster into a cage cluster with effectively 18 valence electrons and high stability. One of the prominent clusters is a perfect icosahedron (Figure 4.12a) of $W@Au_{12}$ in which 12 valence electrons from Au atoms (leaving aside the 5*d* electrons of Au atoms) and 6 valence electrons on W atom occupy the 1S, 1P, and 1D orbitals in a jellium model [7, 124]. The partially occupied 1D orbitals of the icosahedral Au_{12} cage hybridize strongly with the 5*d* orbitals of W leading to a large HOMO–LUMO gap of 3.290 eV within PBE0 and high stability of this cluster. The stability of $W@Au_{12}$ is a classic example of the 18-electron rule. A similar behaviour has been obtained for $Mo@Au_{12}$ with the HOMO–LUMO gap of 3.129 eV. For these clusters, the icosahedral isomer lies lower in energy compared with cuboctahedral or decahedral isomers (Figure 4.12b, c), but for $Mn@Au_{12}$, cuboctahedral isomer is more favourable with $5\,\mu_B$ magnetic moments [125]. The icosahedral structure is also found to be favoured for the anions of V, Nb, and Ta doped Au_{12} clusters (effectively 18-electron systems) as well as for $Au@Au_{12}^{5+}$ that has eight valence electrons besides the *d* electrons on Au atoms. The latter also corresponds to electronic shell closing. Besides these clusters, 14-, 15-, and 16-atom cage clusters have been obtained [126, 127] by the doping of tetravalent atoms such as Ti, Zr, and Hf, trivalent atoms such as Sc, Y, and La,

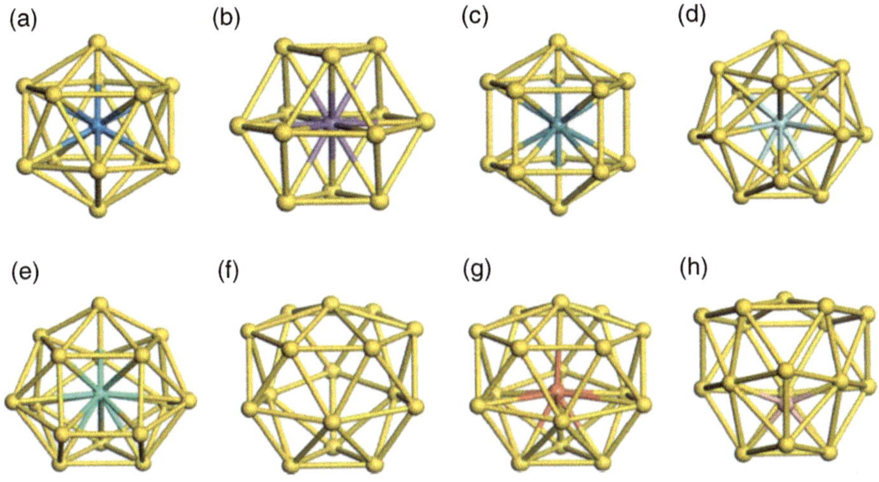

Figure 4.12 Atomic structures of (a) $W@Au_{12}$ (I_h symmetry) (also for $Mo@Au_{12}$, $V@Au_{12}^-$, $Nb@Au_{12}^-$, $Ta@Au_{12}^-$, and $Au@Au_{12}^{5+}$), (b) $Mn@Au_{12}$ (O_h symmetry), (c) $Mo@Au_{12}$ (D_{5h} symmetry), (d) $Zr@Au_{14}$ (D_{2d} symmetry) (also for $Ti@Au_{14}$ and $Hf@Au_{14}$), (e) $Gd@Au_{15}$ (C_s symmetry), (f) Au_{16}^{2-} (T_d symmetry), (g) $[Cu@Au_{16}]^-$ (C_s symmetry) (also for $Sc@Au_{16}^+$, $Y@Au_{16}^+$, and $Eu@Au_{16}$) clusters, and (h) $B@Au_{16}^-$ (C_{3v} symmetry) obtained from PBE0/SDD calculations. *Source:* After Zhao et al. [7].

and divalent atoms such as Ca. Also charged clusters (cations and anions) have been formed by endohedral doping of other suitable transition metal atoms. Figure 4.12 shows optimized structures of many endohedrally doped gold clusters [7]. These results were obtained by using SDD basis set and PBE0 hybrid functional in Gaussian program. Gao et al. [128] have studied $M@Au_{14}$ clusters and obtained large HOMO–LUMO gaps of 2.33 and 2.05 eV for M = Zr and Hf, respectively, using BP86/SDD + Au(2*f*) level of theory. Among all the Ti doped gold clusters, $Ti@Au_{14}$ is very stable and is a magic cluster. The Sc doped Au_{14} cluster has a large electron affinity (4.13 eV), which is much larger than the value for Al_{13} or Cl atom. Accordingly, this cluster is a superhalogen. Gao et al. [126] have also studied $M@Au_{15}$ clusters with M = Sc and Y and obtained their high stability. Doping of a divalent atom in Au_{16} leads to FK-type (Figure 4.1b) stable structures [126]. Similar is the case for $Cu@Au_{16}^-$ anion and $M@Au_{16}^+$ (M = Sc and Y) cations, all these clusters satisfying the 18-electron rule (Figure 4.12g). Bouwen et al. [129] and Neukermans et al. [127] have studied cations of gold clusters doped with different metal atoms (Y, Sc, Ti, V, Cr, Mn, Fe, Co, and Ni) and their photofragmentation. They reported high stability of cations of Y and Sc doped clusters, which satisfy the 18-valence electron rule. Doping of boron also leads to an FK cage, but the B atom is displaced from the centre [7]. Similar to gold, endohedral clusters of Cu and Ag have been studied [7, 130]. In particular, high abundance and stability of cations of $Sc@Ag_{16}$ and $Sc@Cu_{16}$ corresponding to 18 valence electrons (leaving aside the *d* electrons on Cu and Ag) have been obtained [131] in FK structure (Figure 4.1b). Therefore, again the endohedral doping leads to novel structures.

Yadav and Kumar [56] have shown that $Gd@Au_{15}$ (Figure 4.12e) is a magic magnetic superatom. Here, three valence electrons of Gd occupy the 1D electronic shell while the remaining seven *f* electrons remain unpaired. They hybridize with the 1F orbitals of Au_{15} giving rise to high magnetic moments with half-filled 1F-type orbitals on this cluster.

Endohedral doping of Gd leads to a large gain of about 9 eV. Doping of Mn (Eu) in Au_{16} cage leads to a magnetic superatom with 5 (7) μ_B magnetic moments [132, 133]. Further, the doping of Au_{32} empty cage has a behaviour similar to carbon fullerenes as the endohedral atom is generally not strongly bonded or is not used to stabilize the cage as the Au_{32} cage is stable by itself [121, 134, 135]. Kumar [121] stabilized a dodecahedral gold cluster with aluminium. Considering coating of 20 gold atoms on 20 faces of an Al_{13} icosahedral cluster (Figure 4.11a), the optimization led to the 12 Al atoms slightly outside the Au_{20} cage to be more favourable as shown in Figure 4.11c. The central Al atom in this triacontahedron cage interacts weakly with the cage as the distance between the central Al and cage Al (Au) atoms is 4.16 Å (3.92 Å) compared with the centre to vertex Al–Al bond length of 2.66 Å in Al_{13}^- icosahedral cluster. The BE of $Al@Al_{12}Au_{20}$ is 3.007 eV/atom using generalized gradient approximation (GGA) compared with 2.685, 2.337, and 2.457 eV/atom for Al_{13}, Au_{20}, and Au_{32}, respectively. Therefore, $Al@Al_{12}Au_{20}$ is highly stable and energetically very favourable compared with elemental clusters of Al and Au. The heat of formation, $\Delta E = E(Al@Al_{12}Au_{20}) - E(Al_{13}) - E(Au_{20})$, of the compound fullerene from the constituent Al_{13} and Au_{20} clusters is –0.55 eV/atom. This is quite high and suggests strong stability of this compound fullerene. Here, $E(X)$ is the total energy of the species X. The electron affinity of $Al@Al_{12}Au_{20}$ was calculated to be 3.326 eV, which is comparable to 3.777 eV for Al_{13}. Therefore, high abundance of the anion $Al@Al_{12}Au_{20}^-$ fullerene is very likely. Removal of the central Al atom leaves the cage stable after reoptimization. Charging the cage with two electrons makes it a highly symmetric icosahedral fullerene cage of $Al_{12}Au_{20}^{2-}$ with 58 valence electrons (besides the d electrons on Au atoms), which corresponds to the 1G electronic shell closing in a jellium model. Further replacement of the central Al atom with an Au atom led to a 58-valence electron magic anion cluster $Au@Al_{12}Au_{20}^-$ with a large HOMO–LUMO gap of about 0.99 eV within PBE. The Au atom inside the cage drifts towards the wall of the cage below an Al atom. The BE of the neutral $Al_{12}Au_{20}$ cage was calculated to be 3.017 eV/atom, which is more than the value (3.007 eV/atom) for $Al@Al_{12}Au_{20}$. The BEs of $Al_{12}Au_{20}^{2-}$ and $Al@Al_{12}Au_{20}^-$ are 3.193 and 3.108 eV/atom, respectively. Therefore, the empty cage is energetically slightly more favourable than the endohedral cage. We also studied some small endohedral clusters with eight valence electrons e.g. $B@Ag_6^+$ and $B@Li_6^+$ with an octahedral structure. In these cases, the Ag_6 and Li_6 octahedra have 1P partially occupied orbitals, which hybridize strongly with the p orbitals of B giving rise to large HOMO–LUMO gap of 3.636 and 2.789 eV, respectively, within PBE0. With eight electrons (leaving the d electrons on Ag atoms) these clusters are highly stable due to electronic shell closure. Interestingly, $Cu@Li_7$, in spite of it having eight valence electrons (leaving aside the 3d electrons on copper), does not have a large HOMO–LUMO gap. This is due to its pentagonal bipyramid structure where p orbitals of Cu are unoccupied and, therefore, do not hybridize strongly with the 1P orbitals of Li_7 cage.

4.7 Endohedral Clusters of B, Al, and Ga

Aluminium clusters have been studied for a long time. Experiments by Jarrold and Bower [136] showed low reactivity of Al_{13}^- and Al_{23}^- clusters with oxygen suggesting their magic behaviour corresponding to the closed electronic shells at 40 and 70 valence electrons, respectively, in a jellium model. In particular, high abundance was obtained for Al_{13}^- (see Chapter 1). Calculations showed [7] this cluster to have a perfect icosahedral structure with a large HOMO–LUMO gap of 2.972 eV within PBE0. As in an icosahedron, the centre to vertex distance is about 5% shorter compared with vertex to vertex; Khanna and Jena [11] suggested a neutral cluster Si@Al_{12} that is isoelectronic with Al_{13}^-, but the central Al atom is replaced with a smaller atom Si. We have calculated the electronic states of this cluster, and there is a large HOMO–LUMO gap of 3.17 eV using PBE0 hybrid exchange-correlation functional. Gong and Kumar [137] studied different endohedral dopants in icosahedral Al_{12} cage and also found B@Al_{12}^- to be strongly stable. Indeed, high abundances of Si@Al_{12} and B@Al_{12}^- clusters have been obtained in experiments by Nakajima et al. [138]. These endohedral species, although designed from consideration of electron counting and relative sizes of the dopant atom and the cage atoms, incidentally also satisfy the principle of the formation of stable endohedral clusters. The outermost orbitals of both the dopant atoms have *p* character as also the HOMO (2P type) of the Al_{12} icosahedral cage leading to strong hybridization and a large HOMO–LUMO gap. Many other cases of M@Al_{12} icosahedral clusters are possible (see Zhao et al. [7]) such as M = C, Ge, P (cation), As (cation), and Ga (anion). Interestingly, as mentioned earlier, Cu@Al_{12} is a perfect icosahedral cluster [54] with 3 μ_B magnetic moments. It satisfies the Hund's rule as the HOMO has 2P character and it is half-filled. Accordingly, it is a magnetic superatom. Interaction of this cluster with an Al atom led to another magic cluster Cu@Al_{13} with 40 valence electrons (excluding the 3*d* electrons on Cu) and a large HOMO–LUMO gap [7] of 2.817 eV using PBE0. This has also been found in high abundance in experiments [139]. Endohedral clusters of Ga have also been studied [140]. Elemental solid Ga is a molecular crystal, and the behaviour of the doped clusters is also different from those of Al in that a decahedron (similar to Figure 4.12c) becomes more favourable for Si@Ga_{12}.

In recent years, endohedral clusters of boron have attracted much attention. Elemental small boron clusters with less than 25 atoms have planar, quasi-planar, or tubular structures [141, 142], but endohedral doping leads to interesting wheel, bowl, drum, and cage structures [7]. Among the small clusters, B_9^- has a B_8 cyclic ring and one B atom at the centre of the ring (wheel structure shown in Figure 4.13a). Saha et al. [143] showed that in such clusters one can consider two electrons on each boron atom in the ring to be covalently bonded with neighbouring two boron atoms, leaving one valence electron on each B atom in the ring. The central boron atom contributes three valence electrons as do B atoms at the centres of hexagons in an alpha borophene layer [144]. This, together with an extra charge, makes this cluster an effectively 12 valence electron cluster, which is magic in a disk jellium model. It has a large HOMO–LUMO gap of 4.12 eV within B3PW91. In contrast, B_7^- and B_8^- also having boron-centred ring structures have much smaller HOMO–LUMO gaps of 2.03 and 1.41 eV, respectively. Doping of metal atoms leads [7, 145–147] to wheel-shaped boron clusters (Figure 4.13b) such as CoB_8^- (Figure 4.13b) and IrB_9 (Figure 4.13c), both with 18 valence electrons. For some metal atoms, bowl-shaped structures have been obtained [7].

Saha et al. [148] have studied the doping of 3*d*, 4*d*, and 5*d* atoms in B_{14} and B_{16}. They obtained a drum structure (Figure 4.13e) for M@B_{14} (M = Mn, Fe, Co, and Ni) but bowl-shaped structures for other metal atoms. The drum has two B_7 rings and M atom inside this drum. The Ni atom lies at the centre of the drum. It is a non-magnetic magic cluster with electronic shell closing at 24e. Co@B_{14} anion is isoelectronic with Ni@B_{14} and also has a

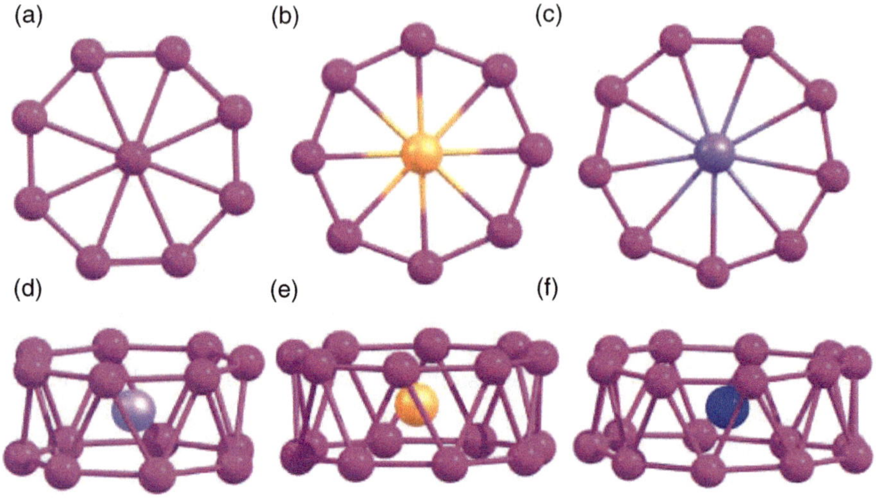

Figure 4.13 Structures of small boron clusters. (a) B_9^-, (b) Co@B_8^-, (c) Ir@B_9, (d) Ni@B_{14}, (e) Co@B_{16}^-, and (f) Ta@B_{18}^-.

large HOMO–LUMO gap of 1.81 eV compared to 1.73 eV for Ni@B_{14} within PBE. In Fe@B_{14} the Fe atom is displaced from the centre along the axis of the drum and has zero magnetic moment with the HOMO–LUMO gap of 1.30 eV. Also, M@B_{16} drum is favoured [148] for cations of M = Co, Rh, and Ir; neutrals of Fe, Ru, and Os; and anions of Mn, Tc, and Re, with two B_8 rings as shown in Figure 4.13e. All these clusters have 24 valence electrons. Popov et al. [149] performed experiments on CoB_{16} anion and showed it to have a drum structure with two B_8 rings and 2 μ_B magnetic moments. As the electronic shell closing is at 24e, two electrons are unpaired. Drum-shaped clusters have also been found in the larger size range of M@B_{18}^- with M = Ta shown in Figure 4.13f, but the structure is not ideal. Even larger clusters with Ta doping have been reported [150–152]. Further studies [152–154] on encapsulation of Cr, Mo, W, and Ta atoms have led to the finding of cage clusters with 20, 22, and 24 boron atoms as shown in Figure 4.14. M@B_{24} with M = Mo and W is a symmetric cage cluster and corresponds to a closed electronic shell with 18e. Li et al. [151] have further reported boron cages corresponding to 18 valence electron rule and with up to 28 boron atoms stabilized with a Ta endohedral atom.

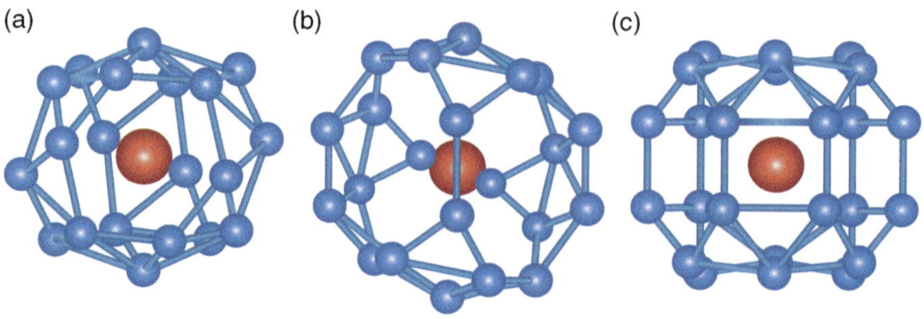

Figure 4.14 Structures of boron cages. (a) Cr@B_{20}, (b) Mo@B_{22}, and (c) Mo@B_{24}. *Source:* After Rahane et al. [153].

4.8 Hydrogenated Silicon Fullerenes

When H atoms were attached on a Cr@Si_{12} hexagonal prism cluster, it led to the development of magnetic moments on Cr as some charge on Si atoms got shared in the bonding with H atoms and effectively the Si–Si and Si–H bonding became close to the sp^3 type. This reduced the interaction of the cage with the Cr atom. These results further led to the finding [95, 155] of Si_nH_n (n = 8, 10, 12, 14, 16, and 20) empty cage fullerene structures (see $Si_{16}H_{16}$ and $Si_{20}H_{20}$ in Figure 4.15). Thus, stabilization of empty cages of Si, Ge, and Sn can take place without encapsulation of a metal atom. $Si_{20}H_{20}$ was shown [155] to be the most stable cluster energetically. Beyond this size the BE decreases slightly as hexagons of silicon with sp^2-like bonding are present in larger fullerenes. Note that silicon clathrate structures [20] such as Na_8Si_{46}, $(Ba,M)_xSi_{46}$ (M = Na and K), $A_8E_8Si_{38}$ (A = Na, K, Rb, Cs; E = Al, Ga, In) – known as type I, and Na_xSi_{136} (x = 4–23) – known as type II – are made of connected polyhedral cages such as Si_{20} and Si_{24} that are endohedrally doped with alkali and alkaline earth atoms. These materials have attracted much attention for thermoelectric applications. Similarly, there are also clathrates of Ge and Sn.

A detailed study [155] of endohedrally doped hydrogenated fullerene structures of Si, Ge, and Sn has shown that the endohedral atom interacts weakly with the cage and this leads to atom-like magnetic moments on the slaved atom. These systems can be useful models to study local properties of clathrates. Also, a halide ion template was suggested [156] to form dodecahedral (I_h symmetry) per-hydrogenated silicon fullerenes X@$Si_{20}H_{20}^-$ (X = F, Cl, Br, and I) as shown in Figure 4.16. These results stimulated experimental realization [157] of Cl@$Si_{20}H_{20}$ species in macroscopic quantity. The halide ion fits in the cage well and leads to only a very small change in the hydrogenated silicon fullerene. For the empty cluster, Si–X (in this case X represents the cage's centre), Si–Si, and Si–H bond lengths are 3.320, 2.369, and 1.496 Å, respectively. After the doping of X^-, the Si–X distances become 3.298, 3.312, 3.320, and 3.335 Å for X = F, Cl, Br, and I, respectively. Accordingly, the dodecahedral cage shrinks slightly upon inclusion of both F^- and Cl^- ions, remains unchanged for Br^-, and expands slightly for I^-. The HOMO–LUMO gap increases by about 0.5 eV compared to the undoped value of 4.51 eV obtained from B3PW91 calculation in the Gaussian program. The interaction energy of the halide ion with the cage is 3.21, 2.65, 2.21, and 1.21 eV for F^-, Cl^-, Br^-, and I^-, respectively, and there is some charge transfer from the halide ion to the cage.

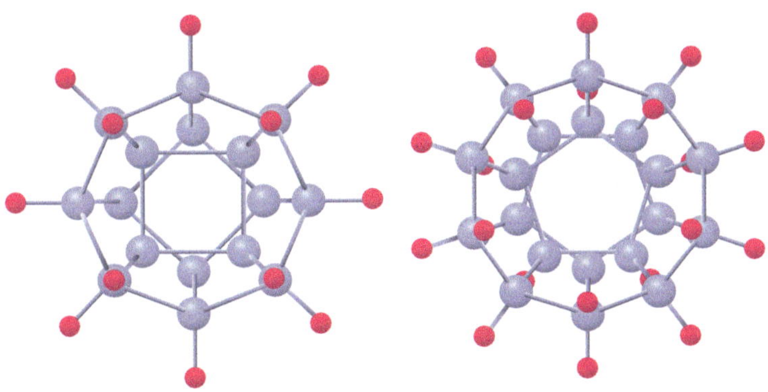

Figure 4.15 Hydrogenated silicon fullerenes $Si_{16}H_{16}$ and $Si_{20}H_{20}$. *Source:* After Kumar and Kawazoe [95].

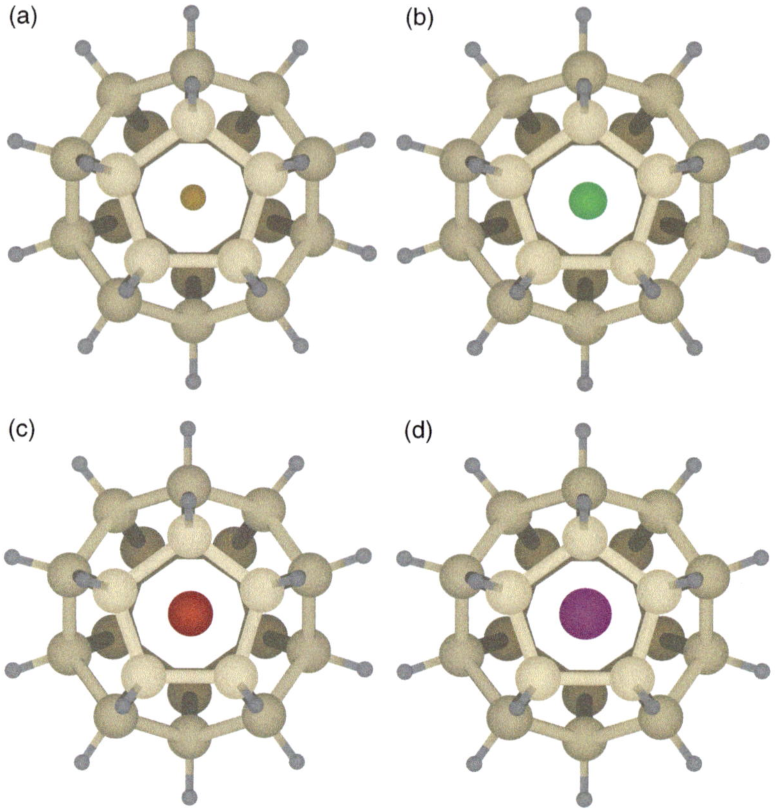

Figure 4.16 Halide ion templated $Si_{20}H_{20}$ fullerenes $X^{-}@Si_{20}H_{20}$ with (a) X = F, (b) X = Cl, (c) X = Br, and (d) X = I. *Source:* After Pichierri et al. [156].

4.9 Compound Superatoms and Other Systems

Magic clusters of II–VI compound semiconductors as well as their endohedral doping (see Zhao et al. [7]) have also been studied. A striking result is the finding [158] of magic clusters $(AB)_n$ with $n = 13, 33$, and 34 where A is a group 12 atom while B is from group 16 such as CdSe, CdS, ZnS, and ZnSe [159], which are very important from the point of view of optical properties. The atomic structure of $(CdSe)_{13}$ cluster can be described as $Se@Cd_{13}Se_{12}$ where a Se atom is encapsulated within a puckered cage of $Cd_{13}Se_{12}$ as shown in Figure 4.17a. This structure was obtained by considering the empty cage of $(CdSe)_{12}$ following $(BN)_{12}$ cage [21] and stuffing it with a CdSe molecule following the idea of endohedral silicon fullerene discussed before. However, after optimization, Cd atom from the molecule becomes part of the cage. Also, $Cd_{33}Se_{33}$ and $Cd_{34}Se_{34}$ clusters have a $Cd_{28}Se_{28}$ cage similar to the $(BN)_{28}$ empty cage [21] and it is stuffed with five and six CdSe molecules, respectively. One can encapsulate a maximum of 6 CdSe molecules in the 28 cage (Figure 4.17b), and that is why no cluster beyond this size was detected [158]. These studies have led to a great number of further works by doping metal atoms as well as assemblies of magic clusters. For example, $Cd_{13}Se_{13}$ has been doped with a Mn atom to develop assemblies of

(a) (b)

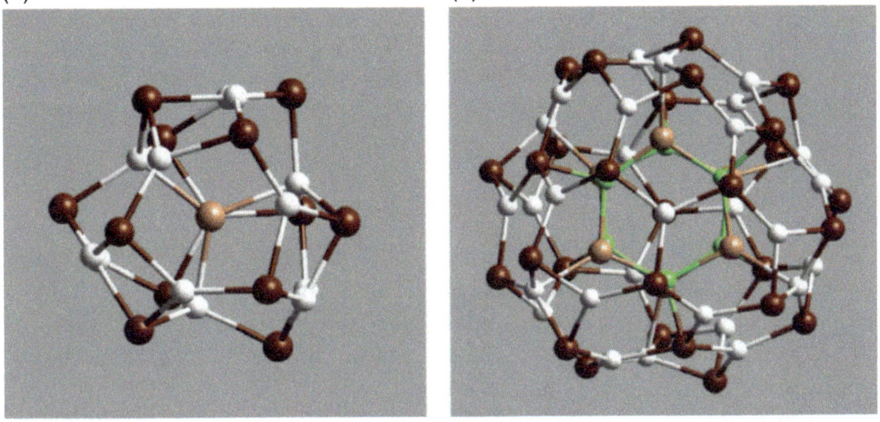

Figure 4.17 Structures of (a) $(CdSe)_{13}$ and (b) $(CdSe)_{34}$ core–cage nanoparticles viewed down a threefold symmetry axis. $(CdSe)_{13}$ has 3 four-membered and 10 six-membered rings on the cage of 12 Se (dark brown) and 13 Cd (white) ions with a Se (light brown) ion inside. $(CdSe)_{34}$ has a truncated octahedral morphology formed by a $(CdSe)_{28}$-cage (Se, dark brown; Cd, white) with 6 four-membered and 8 × 3 six-membered rings. A $(CdSe)_6$ cluster (Se, light brown; Cd, green) encapsulated inside this cage provides additional network and stability. *Source:* After Kasuya et al. [158].

magnetic superatoms [160]. We do not discuss these developments here as it has become a vast subject. However, we point to a study by Matxain et al. [161] where halide ion was endohedrally doped in a $(ZnS)_{12}$ cage. This is electronically equivalent to $(AB)_{13}$ magic clusters mentioned above. It may become possible to develop such higher symmetry species in large quantities similar to the formation of halide ion templated hydrogenated silicon fullerenes. The neutral clusters are superhalogens as they have large electron affinities, namely, 5.27 eV for $Cl@Zn_{12}S_{12}$ and 5.28 eV for $Br@Zn_{12}S_{12}$. Further, it is worth mentioning that transition metal clusters generally do not show strong size-specific magic behaviour. Many metal nanoparticles are important catalysts where it is difficult to know which size is actually the most active one. It would be interesting if such clusters could be made size selective in large quantities.

Kumar [162] showed from ab initio calculations that Nb_{12} and Ta_{12} form an empty cage icosahedral structure carrying 0 and $2\mu_B$ magnetic moments, respectively. When a Fe, Ru, and Os atom is doped inside the Ta cage, it leads to the formation of endohedral superatoms with the HOMO–LUMO gaps of 0.77, 0.75, and 0.69 eV, respectively, using PW91 exchange-correlation functional. Also, adsorption of a Ta atom on a face of $Fe@Ta_{12}$ leads to a gain of 4.6 eV in energy, which is significantly lower compared with the BE of 6.33 eV/atom for $Fe@Ta_{12}$. This shows the magic nature of these clusters. Isoelectronic charged clusters $Co@Ta_{12}^+$, $Rh@Ta_{12}^+$, and $Re@Ta_{12}^-$ were also studied. In the cases of Co and Rh doping, there is a significant HOMO–LUMO gap of 1.0 and 0.97 eV, respectively. Therefore, these icosahedrally symmetric clusters are magic and behave like a superatom. Also, Mo and W doped Nb_{14} and Ta_{14} clusters show a moderate HOMO–LUMO gap of >0.8 eV while cations of $Mn@Nb_{14}$ and $Re@Nb_{14}$ have 1.04 and 0.92 eV HOMO–LUMO gap, respectively. Often metal clusters are doped to tailor their catalytic properties. Accordingly, with suitable doping, it may become possible to have high abundances of clusters of a particular size. Further studies in this direction would be very desirable. Recently, a large heavy metal $Th@Bi_{12}^{4-}$ anionic cluster has been found [163] to have substantial π-aromaticity. It has ring current much stronger than that of benzene. It may become possible to design other metal rings with substantial π-aromaticity.

4.10 Assemblies of Superatoms

Kumar and Kawazoe [23] studied interaction between two $Zr@Si_{16}$ fullerene clusters as well as between two FK $Ti@Si_{16}$ polyhedral clusters. Due to the large HOMO–LUMO gap of $Ti@Si_{16}$, the interaction between them is weak (0.047 eV calculated within PW91) and the clusters are more than 4 Å apart. But the interaction energy is significantly larger (1.345 eV) for the dimer of the $Zr@Si_{16}$ fullerene isomer. Also, the HOMO–LUMO gap calculated with PW91 in the former decreases only slightly to 2.211 eV from 2.238 eV for the monomer, but in the dimer of the fullerene it reduces significantly to 0.673 eV from 1.58 eV for the monomer. In both cases, the structures of the individual clusters are only weakly affected. As the interaction between the FK polyhedra is very weak, we have recalculated the interaction energy as well as the HOMO–LUMO gap by incorporating van der Waals interactions and PBE0 hybrid functional in the VASP code. The calculated interaction energy is 0.130 eV. The nearest separation between the FK polyhedra is about 3.65 Å, and the HOMO–LUMO gap is 3.493 eV compared with the value of 3.824 eV for the FK polyhedron. Accordingly, the interaction remains very weak but the inter-cluster separation decreases significantly compared with the value obtained from PW91 calculation. The slight increase in the interaction energy and decrease in the separation between the clusters leave the individual FK polyhedron almost unaffected as shown in Figure 4.18. This suggests that there would be a good possibility that these superatoms can be assembled. On the other hand, the fullerenes interact at the edge of the square face as shown in Figure 4.18 and form two bonds, each with the bond length of 2.343 Å using PBE0. This is close to the value in bulk Si. The BE between the fullerenes is, however, still small (1.180 eV) but much stronger compared with the FK isomers. The HOMO–LUMO gap is reduced to 2.041 eV compared with the value of 2.768 eV for the $Zr@Si_{16}$ fullerene using PBE0. Therefore, the change in the HOMO–LUMO gap is more significant for the fullerenes compared with the FK isomer, but the gap remains sizeable. Kandalam et al. [164] have studied magnetic coupling between $Mn@Si_{12}$ magnetic superatoms.

Reis et al. [165, 166] have studied solids of the FK isomer of $M@Si_{16}$ (M = Ti, Zr, and Hf) clusters using PBE exchange-correlation functional confined to simple cubic (SC), diamond-type, BCC, face centered cubic (FCC), and hexagonal close packed (HCP) structures. They found the HCP structure to be most favourable with (a and c) lattice parameters of (8.75 and 14.36 Å), (9.05 and 14.79 Å), and (8.96 and 14.89 Å) for M = Ti, Zr, and Hf, respectively. The cohesive energy considering each cluster as a building block was only 0.20, 0.14, and 0.15 eV/cluster, respectively. The corresponding calculated band gaps are 1.3, 1.6, and 1.6 eV. The gap is reduced from the value in the cluster. Fully unconstrained quantum Langevin molecular dynamics simulations confirmed the stability of these structures at room temperature and ambient pressure. However, as we have discussed, Zr and Hf doped Si_{16} clusters favour the fullerene isomer. We have performed calculations allowing full relaxation of the lattice

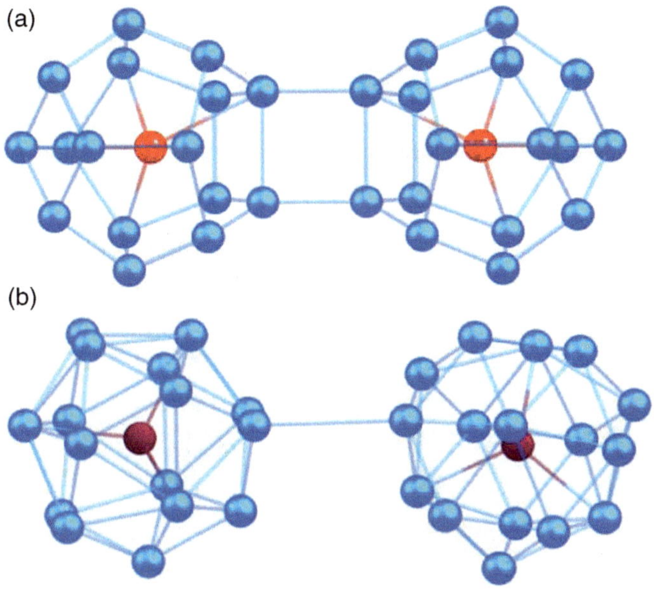

Figure 4.18 Interaction between superatoms. (a) $Zr@Si_{16}$ fullerenes and (b) FK-$Ti@Si_{16}$.

parameters and the $Ti@Si_{16}$ structure of the FK isomer using PBE functional in VASP. We find that starting with BCC, FCC, and HCP structures there are deviations from the ideal structures in all cases. The cohesive energy is calculated to be 0.64, 0.40, and 0.60 eV per cluster, respectively. The nearest neighbour separation between the cages is 2.90 Å (BCC), 2.79 Å (FCC), and 3.23 Å (HCP) while the minimum inter-cluster distance (Ti to Ti atoms) is 8.63, 8.17, and 8.63 Å, respectively. The cluster structure is little affected in BCC and HCP structures, but in the FCC case there are some distortions, although the cage remains intact. It is to be noted that PBE underestimates the band gap significantly and improved calculations would be desirable to treat weak interactions. Interestingly, when the HCP crystal of $M@Si_{16}$ clusters was doped with electron donor atoms in the interstitial space, a relatively high superconducting transition temperature of more than 80 K was predicted [166] similar to the finding of superconductivity in carbon fullerides [167]. These results suggest novel possibilities of designing superconducting and other functional materials at the nanoscale. Laurrabaquio et al. [168] and Torres et al. [169] have also performed calculations of FCC, BCC, and SC assemblies of $Ti@Si_{16}$ clusters with the fullerene isomer as the building block. Using double zeta polarized (DZP) basis set and PBE functional, they obtained BCC structure to be the most favourable with the lattice constant of 8.75 Å and cohesive energy of 5.72 eV per fullerene. However, free $Ti@Si_{16}$ cluster favours FK isomer. Accordingly, further studies would be needed to establish the correct structures of aggregates of these systems. Interestingly, further studies on NaCl- and CsCl-type bulk structures of $[Sc@Si_{16}]K$ and $[V@Si_{16}]F$ showed large cohesive energies of 5.58 and 4.51 eV per unit cell, respectively, although $[V@Si_{16}]F$ was found to be metallic.

Singh et al. [170] studied $Be@Si_{12}$ cluster that was found to have a deformed prism structure with two chair-shaped Si_6 rings stacked on top of each other as shown in Figure 4.19 (also Figure 4.61). However, when two such clusters were stacked, the chair-shaped rings transformed to flat hexagons (Figure 4.19) suggesting that the bonding in the rings transformed from sp^3 type to sp^2 type. Further continuation of such structures led to the finding of a nanotube of Si stabilized with Be atoms (Figure 4.20). The band structure of this nanotube is also shown in Figure 4.20, and this shows a metallic character. One can also add Be atoms in between the hexagonal prisms, and the nanotube remains metallic. The tubular structures of Be doped silicon were later reported from experiments [171] by depositing Be on a Si(111) surface.

Further studies [172] on doping of Mn, Fe, and Ni atoms led to the finding of magnetic nanotubular structures of silicon (see also Singh et al. [173]). These are shown in

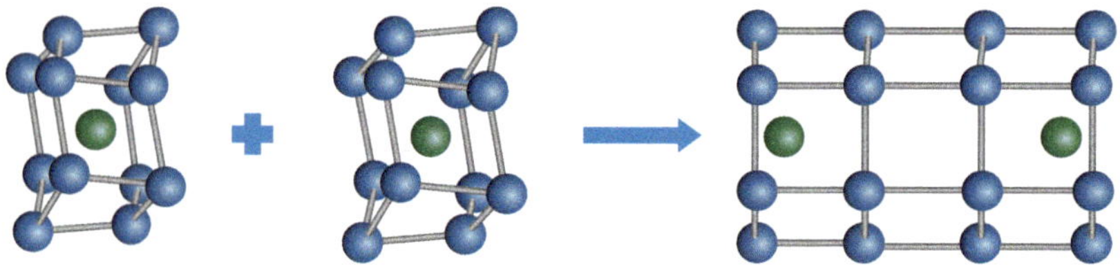

Figure 4.19 Interaction between two $Si_{12}Be$ clusters leads to the formation of nanotubular stick of silicon. *Source:* After Kumar [34].

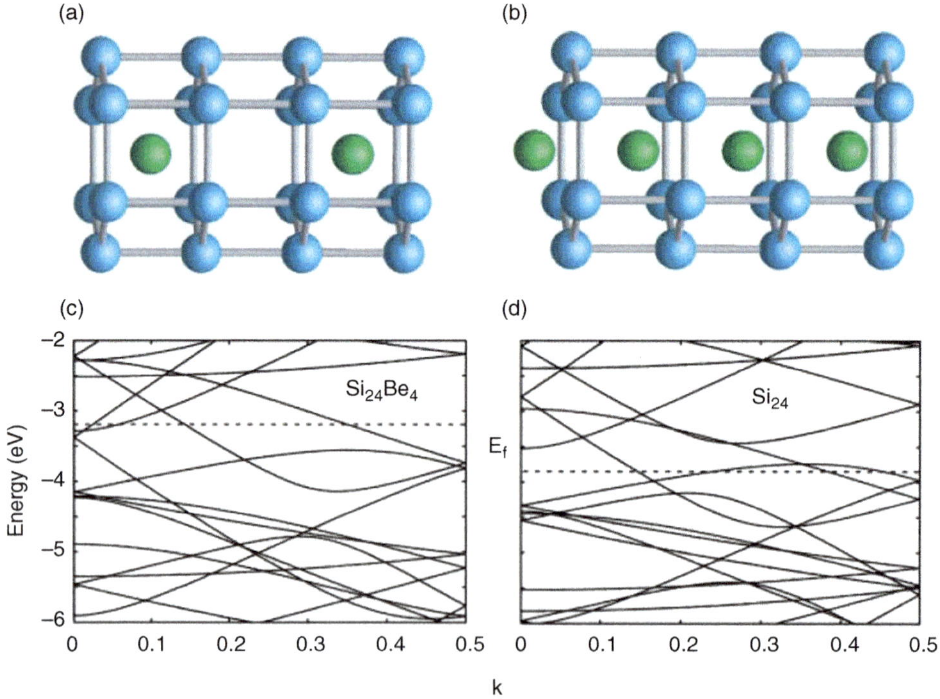

Figure 4.20 Representative units used for infinite nanotubes of Si_nBe_m. (a) $Si_{24}Be_2$ and (b) $Si_{24}Be_4$. These are metallic as can be seen from the band structure of $Si_{24}Be_4$ shown in (c). (d) Shows the band structure of Si_{24} infinite nanotube obtained by removing Be atoms in (b). This is also metallic. *Source:* After Singh et al. [170].

Figure 4.21. Fe, Co, and Mn atoms were found to drift towards one hexagon as in the middle the interaction with both the hexagons may not be optimal. In particular, Ni was found to be very close to the centre of the silicon hexagon. The Fe-doped Si nanotube was found to be ferromagnetic with $2.4\mu_B$ magnetic moments per dopant atom, which is close to the value in bulk Fe. The antiferromagnetically coupled nanotube was found to lie 0.7 eV higher in energy than the ferromagnetic state. The Mn doped hexagonal nanotube was found to favour ferromagnetically coupled pairs that interact antiferromagnetically. The Co and Ni doped nanotubes are non-magnetic. The hexagonal Si nanotube encapsulating a Cr atomic chain is ferromagnetic with the magnetic moment of only $0.34\mu_B$ per $CrSi_6$ unit. Note that a hexagonal prism of $Cr@Si_{12}$ has no magnetic moment [95]. But bonding between Si atoms of successive clusters reduces the interaction between Si and M atoms. This can lead to the development of magnetic moments in the nanotube. A similar behaviour was obtained for the hydrogenated silicon clusters as discussed above.

Nanotubes of Ge were also stabilized with metal atoms. Unlike silicon nanotubes, there is a tendency to have anti-prism stacking of Ge hexagons [173–175]. Menon et al. [176] have studied nanotubes based on pentagons that can be considered as stacking of metal-centred pentagonal prisms. Nanotubular structures with larger clusters such as those with Si_{16} or Si_{20} cages have also been studied [34, 177]. Packing of $Zr@Si_{16}$ clusters in a chain was found to lead to a semiconductor nanowire [34]. Nanoscale p–n junction such as $Sc@Si_{16}$-$V@Si_{16}$ and $Y@Si_{16}$-$Ta@Si_{16}$ obtained by replacing two Z valence atoms with $(Z-1)$ and $(Z+1)$ valence atoms in dimers of say $Ti@Si_{16}$ have been suggested [178, 179]. Also, similar to hexagonal prism structure of W_2Si_{18}, one can have $TaReSi_{18}$ that has a large

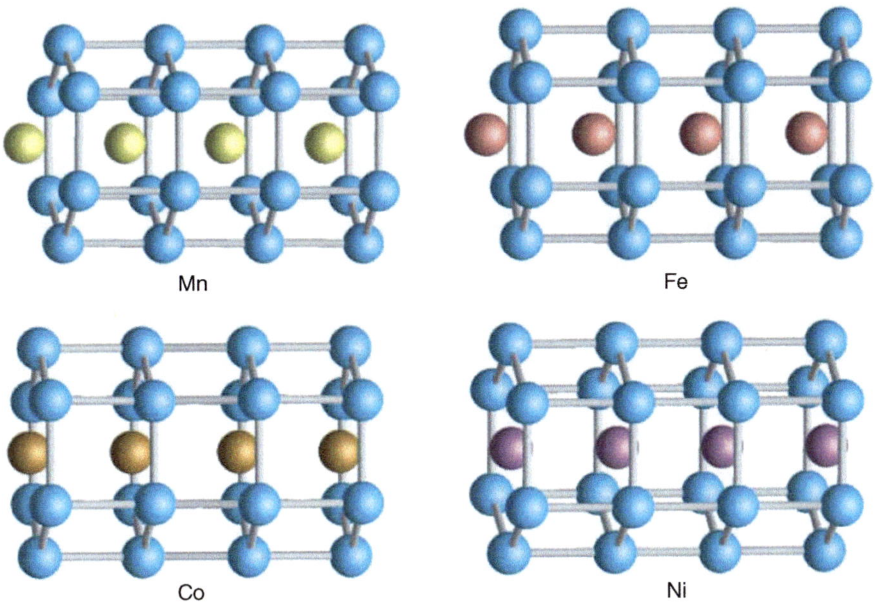

Figure 4.21 Structures of $Si_{24}M_4$ (M = Mn, Fe, Co, and Ni) infinite nanotubes. Note the shift in the positions of metal atoms. As one goes from Mn to Ni, the metal atom moves towards the centre of the silicon hexagon. *Source:* After Singh et al. [172].

HOMO–LUMO gap of 1.08 eV within GGA. Saha et al. [148] have also studied a boron nanotube by assembling the endohedrally doped disk-shaped boron clusters.

Uchida et al. [85] deposited Ta doped silicon clusters $TaSi_nH_x^+$ (n = 10–13) on a Si(111)-(7×7) surface. Among these, Ta@Si_{12} clusters had the highest fraction in all those having no H atoms. Miyazaki and Kanayama [180] also studied a 2D assembly of hexagonal Zr@Si_{12} clusters by making covalent Si—Si bonds between the clusters. The BE per $ZrSi_{12}$ unit is 6.12 eV, and this layer was predicted to be a semiconductor with an indirect band gap of 0.3 eV using PBE functional. Further 2D assemblies of V@Si_{12} hexagonal cluster were found [181] to be ferromagnetic with a large magnetic moment of 3.5 μ_B per building unit, which is predominantly localized on the V atoms. The electronic and magnetic properties of 2D assemblies of M@Si_{12} (M = 3d transition metal) have been studied by Nie et al. [182]. Ti-, V-, and Fe-doped 2D assemblies were shown to be ferromagnetic, while Mn- and Co-doped 2D structures were found to favour antiferromagnetic coupling. In the case of Cr@Si_{12} and Ni@Si_{12}, the honeycomb 2D structure is ferromagnetic. These studies could lead to new silicon-based spintronic materials.

In a series of experiments, Nakajima and coworkers have studied the formation of Ta@Si_{16}-C_{60} complexes [91] as well as deposition on different substrates [90]. Ta@Si_{16}^+ fullerenes were deposited on a C_{60}-covered highly oriented pyrolytic graphite (HOPG) substrate, C_{60}-covered Si(111)-(7×7), and Si(111)√3-Ag surface to make 2D assemblies of Ta@Si_{16} clusters as shown in Figure 4.22. The Si(111)√3-Ag surface was prepared by depositing a monolayer of Ag on Si(111)-(7×7) surface at 873 K. The interaction of C_{60} on HOPG substrate is weak, which gives it a flexibility to rotate while it can desorb as the temperature is raised. On Si(111)-(7×7) surface, however, C_{60} can be immobilized. Ta@Si_{16}^+ is an electronically closed shell cluster similar to Hf@Si_{16} and behaves like a rare gas atom. Ta_nSi_m nanoclusters with various charge states were generated in the gas phase, and Ta@Si_{16} cations were mass selected and deposited on C_{60}-terminated surfaces as mentioned above. These structures were analysed using scanning tunnelling microscopy (STM) and scanning tunnelling spectroscopy (STS) measurements under ultra-high vacuum conditions. The results are shown in Figure 4.22. Figure 4.22a and b show the wide- and molecular-scale STM images of soft-landed Ta@Si_{16} cations on C_{60}/HOPG surface.

These studies demonstrate that Ta@Si_{16} cations get densely immobilized onto these C_{60}-terminated surfaces forming a stable monolayer of Ta@Si_{16} cations. Each dot corresponds to an individual Ta@Si_{16} nanocluster. The histogram of the dot heights measured for the high-density dots in Figure 4.22f shows a peak at ~0.8 nm. This value is close to the theoretical size of isolated Ta@Si_{16} cations, which lies in the range of 0.89–0.95 nm for the C_{3v}-distorted FK and D_4d fullerene isomers. The measured heights showed that the clusters retained their fullerene cage structure. The formation of the dots with smaller heights was attributed to the diversity in the adsorption sites or fragmentation of some endohedral clusters. It was further found that Ta@Si_{16} nanoclusters changed their adsorption position from the bridge or hollow sites to atop sites when C_{60} films sparsely covered with Ta@Si_{16} nanoclusters (e.g. Figure 4.1a) were thermally annealed. The chemical states and stability of the Ta@Si_{16}/C_{60} film on HOPG were further analysed by X-ray photoelectron spectroscopy (XPS) and ultraviolet photoelectron spectroscopy (UPS), and it was revealed that each Ta@Si_{16} cluster combines with C_{60} to form a one-to-one superatomic charge transfer complex, namely, Ta@Si_{16}^+–C_{60}^-. This superatomic charge-transfer complex showed excellent

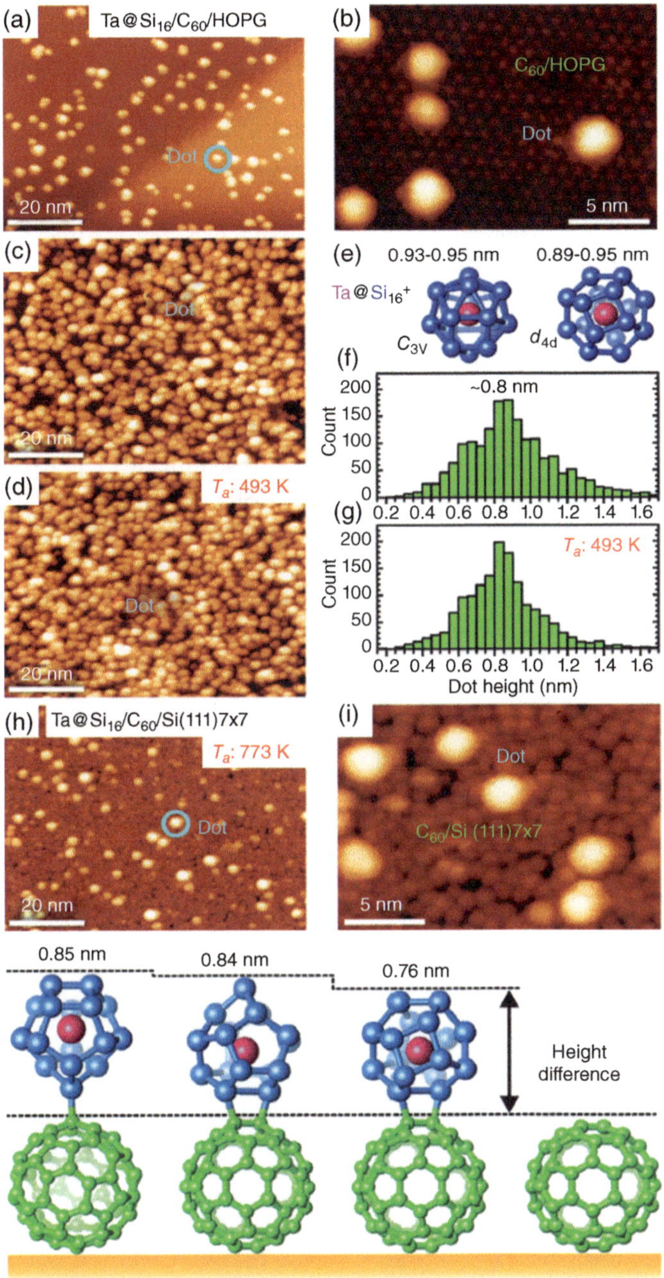

Figure 4.22 Immobilization of Ta@Si_{16} nanoclusters onto C_{60}-terminated surfaces. (a–d) STM images of C_{60}/HOPG surfaces obtained after depositing Ta@Si_{16} cations. (a) Wide- and (b) molecular-scale images obtained after the initial deposition; (c) and (d) surfaces densely covered with dots before and after annealing at 493 K, respectively. (e) Shows geometrical models and sizes of C_{3v}-deformed FK and D_{4d} fullerene isomers of Ta@Si_{16} cations; (f) and (g) histograms of the dot heights measured in (c) and (d), respectively. (h) Wide- and (i) molecular-scale STM images of the Ta@Si_{16}^{+}-deposited C_{60}/Si(111)-(7 × 7) surface obtained after annealing at 773 K. The imaging conditions (tip voltage V_{tip} and tunnelling current I_t) are −2.3 V and 2 pA for (a)–(d), 2.2 V and 5 pA for (h), and 1.8 V and 5 pA for (i). The lower panel shows models of attaching Ta@Si_{16} on C_{60} in different ways to form neutral Ta@Si_{16}–C_{60} complexes. *Source:* After Nayaka et al. [90].

thermal and chemical robustness even after heat treatment up to 720 K and oxygen exposure. These results agree with the calculations discussed below.

Our ab initio PAW pseudopotential calculations in VASP with PBE0 hybrid exchange-correlation functional show that the fullerene structure of Ta@Si_{16} is 0.423 eV lower in energy than the FK structure and its BE is 4.178 eV/atom. There is small distortion in the fullerene structure as shown in Figure 4.23a. We studied interaction of Ta@Si_{16} with a Cl atom that binds at one of the corners of the square face (Figure 4.23b) with a BE of 5.253 eV using PBE0 hybrid functional. C_{60} has a large electron affinity, and this is one of the reasons to choose C_{60} layer for the adsorption of Ta@Si_{16} in experiments. We calculated the interaction energy of a C_{60} fullerene with Ta@Si_{16} fullerene, and it shows a similar behaviour as for a Cl atom and the two are linked at vertices as shown in Figure 4.23c. This also agrees with the experimental findings. The BE calculated within PBE, however, is 0.5 eV, which is small compared with that of a Cl atom but still significant. The C–Si bond length is 1.995 Å. There is nearly one electron charge transfer to C_{60}. This again agrees with experiments. However, this charge is distributed over the C_{60} and this is the reason that the bond has become much weaker compared with the BE of a Cl atom. The metal atom (Ta) also donates about 0.8e charge to the Si cage. There is a small distortion in the Si_{16} fullerene cage. All these results of the binding of Ta@Si_{16} fullerene on C_{60} supports the experiments by Nakaya et al. [90] and Ohta et al. [91]. We have also calculated

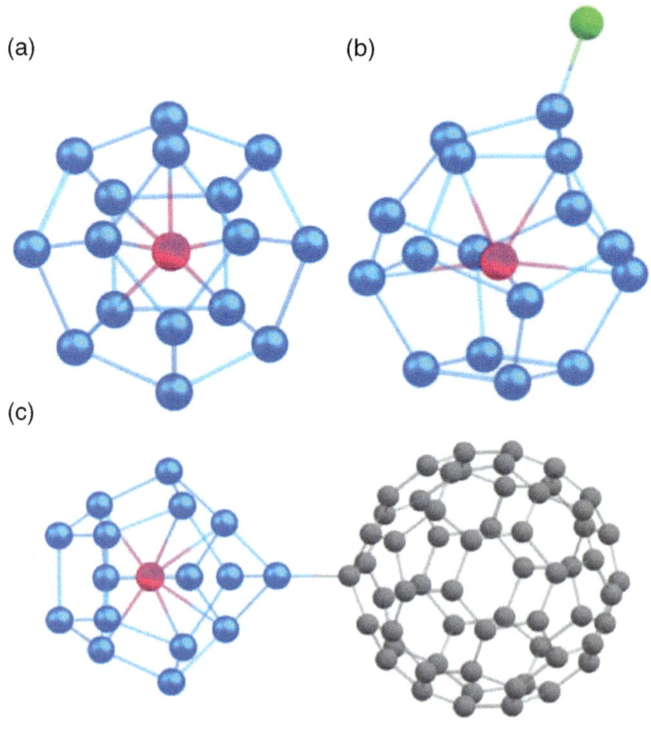

Figure 4.23 (a) A slightly distorted fullerene structure of Ta@Si_{16}, (b) Cl interaction with Ta@Si_{16}, and (c) interaction between Ta@Si_{16} and C_{60} fullerenes.

(Kumar, to be published) interaction between two $Ta@Si_{16}$ superatoms and our results show that the two interact by bonding at edges of the squares with two bonds of length ~2.505 Å. The BE within PBE0 is 1.978 eV and the HOMO–LUMO gap is 1.994 eV. The binding on vertices is 0.7 eV less favourable. Accordingly, this interaction is weaker than the typical Si–Si bond as well as the BE within a superatom. Recently, multi-layered superatom films of $Ta@Si_{16}$ clusters have also been deposited [183] on $SrTiO_3$ substrate by soft landing size-selected $Ta@Si_{16}$ clusters. Also, Liu et al. [184] have studied interaction between $Ta@Si_{16}$ clusters and also with C_{60}. The BE for the $Ta@Si_{16}$ dimer was calculated to be 2.21 eV with Ta–Ta separation of 8.49 Å and Si–Si distance of 2.51 Å while for $Ta@Si_{16}$–C_{60} the BE was calculated to be 0.42 eV. These results are in good agreement with our calculations. They have also calculated 2D assemblies of these fullerenes as well as assembly on a graphene layer. In all cases, the $Ta@Si_{16}$ cluster was found to preserve its endohedral structure. The graphene-supported $Ta@Si_{16}$ superatoms show ferromagnetic coupling.

4.11 Concluding Remarks

In summary, endohedral doping has emerged as a novel way to design clusters with significantly enhanced stability as well as size selectivity. By choosing an appropriate combination of the endohedral atom and the cage atoms, a variety of species with varying properties can be prepared in relatively large abundances. The pioneering discovery of silicon fullerenes and other polyhedral forms of silicon from calculations not only have been realized in laboratory but these findings have also led to a large number of experimental and theoretical studies on various systems. More excitingly, this progress has made it possible to synthesize some endohedral clusters such as $Ti@Si_{16}$ and $Ta@Si_{16}$ in macroscopic quantities and assemble them on substrates. This development has made it possible to realize a long-standing dream of achieving atom-precise materials with nanoclusters as the building block. We have covered many of these developments but with a focus on our work. A more complete coverage has been recently done in the review article by Zhao et al. [7], and the reader may refer to it for more details. It is also very interesting that more and more bulk materials with endohedral clusters are being discovered and even superconductivity has been found in such systems. We believe that the vast knowledge gained from studies of free endohedral clusters as well as bulk materials with endohedral clusters will stimulate further work and open new avenues for the development of novel materials for various applications.

Acknowledgements

I am very grateful to all my collaborators for discussions and fruitful cooperation that led to many new developments. I would like to thank in particular Dr. Stuti Jaiswal who contributed to the study of $Ta@Si_{16}$ clusters and assemblies. I also thankfully acknowledge the use of the high-performance computing system Magus of Shiv Nadar University where some of the calculations have been carried out.

References

1 De Heer, W.A. (1993). The physics of simple metal clusters: experimental aspects and simple models. *Rev. Mod. Phys.* 65: 611–676.

2 Kumar, V., Esfarjani, K., and Kawazoe, Y. (2002). *Clusters and Nanomaterials*, 9–88. Springer.

3 Baletto, F. and Ferrando, R. (2005). Structural properties of nanoclusters: energetic, thermodynamic, and kinetic effects. *Rev. Mod. Phys.* 77: 371–423.

4 Kumar, V. (ed.) (2008). *Nanosilicon*. Elsevier.

5 Kumar, V. (2010). Design of nanomaterials by computer simulations. In: *Handbook of Nanophysics* (ed. K. Sattler), 65–87. CRC Press.

6 Jena, P. and Sun, Q. (2018). Super atomic clusters: design rules and potential for building blocks of materials. *Chem. Rev.* 118: 5755–5870.

7 Zhao, J., Du, Q., Zhou, S., and Kumar, V. (2020). Endohedrally doped cage clusters. *Chem. Rev.* 120: 9021–9163.

8 Knight, W., Clemenger, K., de Heer, W.A. et al. (1984). Electronic shell structure and abundances of sodium clusters. *Phys. Rev. Lett.* 52: 2141–2143.

9 Kroto, H.W., Heath, J.R., O'Brien, S.C. et al. (1985). C_{60}: buckminsterfullerene. *Nature* 318: 162–163.

10 Krätschmer, W., Lamb, L.D., Fostiropoulos, K., and Huffman, D.R. (1990). Solid C_{60}: a new form of carbon. *Nature* 347: 354–358.

11 Khanna, S. and Jena, P. (1992). Assembling crystals from clusters. *Phys. Rev. Lett.* 69: 1664–1667.

12 Khanna, S. and Jena, P. (1995). Atomic clusters: building blocks for a class of solids. *Phys. Rev. B: Condens. Matter Mater Phys.* 51: 13705–13716.

13 Jena, P. (2013). Beyond the periodic table of elements: the role of superatoms. *J. Phys. Chem. Lett.* 4: 1432–1442.

14 Chevrel, R., Hirrien, M., and Sergent, M. (1986). Superconducting Chevrel phases: prospects and perspectives. *Polyhedron* 5: 87–94.

15 Simon, A. (1981). Condensed metal clusters. *Angew. Chemie Int. Ed.* 20: 1–22.

16 Fässler, T.F. (2011). *Zintl phases Principles and Recent Developments*. Springer.

17 Sevov, S.C. (2002). Intermetallic compounds. In: *Principles and Practice*, vol. 3 (eds. J.H. Westbrook and R.L. Fleischer). Wiley.

18 Sevov, S.C. and Corbett, D. (1993). Synthesis, characterization, and bonding of indium clusters. Rb2In3, a zintl phase with layers of closo-indium octahedra. *Z. Anorg. Allg. Chem.* 619: 128–132.

19 Queneau, V. and Sevov, S.C. (1997). Ge_9^{4}-: A Deltahedral Zintl Ion Now Made in the Solid-State. *Angew. Chem. Int. Ed.* 36: 1754–1756.

20 Karttunen, A.J., Fässler, T.F., Linnolahti, M., and Pakkanen, T.A. (2011). Structural principles of semiconducting group 14 Clathrate frameworks. *Inorg. Chem.* 50: 1733–1742.

21 Fowler, P.W., Heine, T., Mitchell, D. et al. (1996). Boron–nitrogen analogues of the fullerenes: the isolated-square rule. *J. Chem. Soc. Faraday Trans.* 92: 2197–2201.

22 Oku, T., Nishiwaki, A., and Narita, I. (2004). Formation and atomic structures of B_nN_n (n = 24–60) clusters studied by mass spectrometry, high-resolution electron microscopy and molecular orbital calculations. *Physica B* 351: 184–190.

23 Kumar, V. and Kawazoe, Y. (2001). Metal-encapsulated fullerenelike and cubic caged clusters of silicon. *Phys. Rev. Lett.* 87: 045503.

24 Kumar, V. and Kawazoe, Y. (2003). Erratum: metal-encapsulated fullerenelike and cubic caged clusters of silicon [Phys. Rev. Lett. 87, 045503 (2001)]. *Phys. Rev. Lett.* 91: 199901.

25 Yoo, S. and Zeng, X.C. (2005). Structures and stability of medium-sized silicon clusters. III. Reexamination of motif transition in growth pattern from Si_{15} to Si_{20}. *J. Chem. Phys.* 123: 164303.

26 Shinohara, H. and Tagmatarchis, N. (2015). *Endohedral Metallofullerenes: Fullerenes with Metal Inside*. Wiley.

27 Koyasu, K., Akutsu, M., Mitsui, M., and Nakajima, A. (2005). Selective formation of MSi_{16} (M = Sc, Ti, and V). *J. Am. Chem. Soc.* 127: 4998–4999.

28 Lau, J., Hirsch, K., Klar, P. et al. (2009). X-ray spectroscopy reveals high symmetry and electronic shell structure of transition-metal-doped silicon clusters. *Phys. Rev. A: At. Mol. Opt. Phys.* 79: 053201.

29 Koyasu, K., Atobe, J., Akutsu, M. et al. (2007). Electronic and geometric stabilities of clusters with transition metal encapsulated by silicon. *J. Phys. Chem. A* 111: 42–49.

30 Koyasu, K., Atobe, J., Furuse, S., and Nakajima, A. (2008). Anion photoelectron spectroscopy of transition metal- and lanthanide metal-silicon clusters: MSi_n^- ($n = 6$–20). *J. Chem. Phys.* 129: 214301.

31 Furuse, S., Koyasu, K., Atobe, J., and Nakajima, A. (2008). Experimental and theoretical characterization of MSi_{16}^-, MGe_{16}^-, MSn_{16}^-, and MPb_{16}^- (M = Ti, Zr, and Hf): the role of cage aromaticity. *J. Chem. Phys.* 129: 064311.

32 Kumar, V. and Kawazoe, Y. (2002). Metal-encapsulated caged clusters of germanium with large gaps and different growth behavior than silicon. *Phys. Rev. Lett.* 88: 235504.

33 Kumar, V. (2003). Novel metal-encapsulated caged clusters of silicon and germanium. *Eur. Phys. J. D* 24: 227–232.

34 Kumar, V. (2006). Alchemy at the nanoscale: magic heteroatom clusters and assemblies. *Comput. Mater. Sci.* 36: 1–11.

35. Tsunoyama, H., Akatsuka, H., Shibuta, M. et al. (2017). Development of integrated dry–wet synthesis method for metal encapsulating silicon cage superatoms of $M@Si_{16}$ (M = Ti and Ta). *J. Phys. Chem. C* 121: 20507–20516.

36 Sun, Q., Wang, Q., Jena, P. et al. Stabilization of Si_{60} cage structure. *Phys. Rev. Lett.* 90: 135503.

37 Fässler, T.F. and Hoffmann, S.D. (2004). Endohedral zintl ions: intermetalloid clusters. *Angew. Chemie Int. Ed.* 43: 6242–6247.

38 Esenturk, E.N., Fettinger, J., and Eichhorn, B. (2005). The closo-Pb_{10}^{2-} Zintl ion in the $[Ni@Pb_{10}]^{2-}$ cluster. *Chem. Commun.*: 247–249.

39 Esenturk, E.N., Fettinger, J., and Eichhorn, B. (2006). The Pb_{12}^{2-} and Pb_{10}^{2-} zintl ions and the $M@Pb_{12}^{2-}$ and $M@Pb_{10}^{2-}$ cluster series where M = Ni, Pd, *Pt. J. Am. Chem. Soc.* 128: 9178–9186.

40 Gillett-Kunnath, M.M., Paik, J.I., Jensen, S.M. et al. (2011). Metal-centered deltahedral zintl ions: synthesis of $[Ni@Sn_9]^{4-}$ by direct extraction from intermetallic precursors and of the vertex-fused dimer $[\{Ni@Sn_8(\mu\text{-}Ge)_{1/2}\}_2]^{4-}$. *Inorg. Chem.* 50: 11695–11701.

41 Yue, C.Y., Wang, M.F., Yuan, Z.D. et al. (2013). $K_{13}CoSn_{17-x}$ ($x = 0.1$): a new ternary phase containing cobalt centered $[Sn_9]$ cluster synthesized via high-temperature reaction. *Z. Anorg. Allg. Chem.* 639: 911–917.

42 Benda, C.B., Waibel, M., Köchner, T., and Fässler, T.F. (2014). Reactivity of liquid ammonia solutions of the zintl phase $K_{12}Sn_{17}$ towards mesitylcopper (I) and phosphinegold (I) chloride. *Chem. – Eur. J.* 20: 16738–16746.

43 Witzel, B.J., Klein, W., Dums, J.V. et al. (2019). Metallocages for metal anions: highly charged $[Co@Ge_9]^{5-}$ and $[Ru@Sn_9]^{6-}$ clusters featuring spherically encapsulated Co^{1-} and Ru^{2-} anions. *Angew. Chem. Int. Ed.* 58: 12908–12913.

44 Jiang, D.-e. and Dai, S. (2009). From superatomic $Au_{25}(SR)_{18}^{-}$ to superatomic M@ $Au_{24}(SR)_{18}^{q}$ core–shell clusters. *Inorg. Chem.* 48: 2720–2722.

45 Negishi, Y., Kurashige, W., Niihori, Y. et al. (2010). Isolation, structure, and stability of a dodecanethiolate-protected Pd_1Au_{24} cluster. *Phys. Chem. Chem. Phys.* 12: 6219–6225.

46 Qian, H., Jiang, D.-E., Li, G. et al. (2012). Monoplatinum doping of gold nanoclusters and catalytic application. *J. Am. Chem. Soc.* 134: 16159–16162.

47 Liu, X., Yuan, J., Yao, C. et al. (2017). Crystal and solution photoluminescence of $MAg_{24}(SR)_{18}$ (M = Ag/Pd/Pt/Au) nanoclusters and some implications for the photoluminescence mechanisms. *J. Phys. Chem. C* 121: 13848–13853.

48 Pyykkö, P. (2006). Understanding the eighteen-electron rule. *J. Organomet. Chem.* 691: 4336–4340.

49 Dognon, J.P., Clavaguéra, C., and Pyykkö, P. (2007). Towards a 32-electron principle: Pu@ Pb_{12} and related systems. *Angew. Chem. Int. Ed.* 46: 1427–1430.

50 Wade, K. (1971). The structural significance of the number of skeletal bonding electron-pairs in carboranes, the higher boranes and borane anions, and various transition-metal carbonyl cluster compounds. *J. Chem. Soc. D: Chem. Commun.*: 792–793.

51 Mingos, D.M.P. (1984). Polyhedral skeletal electron pair approach. *Acc. Chem. Res.* 17: 311-319.

52 Hirsch, A., Chen, Z., and Jiao, H. (2000). Spherical aromaticity in I_h symmetrical fullerenes: the $2(N+1)^2$ rule. *Angew. Chem. Int. Ed.* 39: 3915–3917.

53 Zhao, H., Zhou, J., and Jena, P. (2016). Stability of $B_{12}(CN)_{12}^{2-}$: implications for lithium and magnesium ion batteries. *Angew. Chemie* 55: 3704–3708.

54 Kumar, V. and Kawazoe, Y. (2001). Hund's rule in metal clusters: prediction of high magnetic moment state of $Al_{12}Cu$ from first-principles calculations. *Phys. Rev. B: Condens. Matter Mater. Phys.* 64: 115405.

55 Kumar, V. and Kawazoe, Y. (2003). Metal-doped magic clusters of Si, Ge, and Sn: the finding of a magnetic superatom. *Appl. Phys. Lett.* 83: 2677–2679.

56 Yadav, B.D. and Kumar, V. (2010). $Gd@Au_{15}$: a magic magnetic gold cluster for cancer therapy and bioimaging. *Appl. Phys. Lett.* 97: 133701.

57 Verchenko, V.Y., Zubtsovskii, A.O., Wei, Z. et al. (2019). Endohedral cluster superconductors in the Mo—Ga—Sn system explored by the joint flux technique. *Inorg. Chem.* 58: 15552–15561.

58 Srivichitranond, L.C., Seibel, E.M., Xie, W. et al. (2017). Superconductivity in a new intermetallic structure type based on endohedral $Ta@Ir_7Ge_4$ clusters. *Phys. Rev. B* 95: 174521.

59 Yvon, K. (1975). Mo_8Ga_{41}, another example of ten-coordination of the transition element. *Acta Crystallogr. B* 31: 117–120.

60 Verchenko, V.Y., Zubtsovskii, A.O., Plenkin, D.S. et al. (2020). Family of Mo_4Ga_{21}-based superconductors. *Chem. Mater.* 32: 6730–6735.

61 Verchenko, V.Y., Zubtsovskii, A.O., Tsirlin, A.A. et al. (2020). Mo_6Ga_{31} endohedral cluster superconductor. *J. Alloys Compd.* 848: 156400.

62 Shibayama, T. et al. (2007). Superconductivity in Rh_2Ga_9 and Ir_2Ga_9 without inversion symmetry. *J Phys. Soc. Jpn.* 76: 073708.

63 Grin, Y., Wedig, U., Wagner, F. et al. (1997). The analysis of 'empty space' in the $PdGa_5$ structure. *J. Alloys Compd.* 255: 203–208.

64 Xie, W., Luo, H., Phelan, B.F. et al. (2015). Endohedral gallide cluster superconductors and superconductivity in $ReGa_5$. *Proc. Natl. Acad. Sci.* 112: E7048–E7054.

65 Daghero, D., Tortello, M., Ummarino, G.A. et al. (2012). Strong-coupling d-wave superconductivity in $PuCoGa_5$ probed by point-contact spectroscopy. *Nat. Commun.* 3: 786.

66 Henning, R. and Corbett, J. (1999). Formation of isolated nickel-centered gallium clusters in $Na_{10}Ga_{10}Ni$ and a 2-D network of gallium octahedra in K_2Ga_3. *Inorg. Chem.* 38: 3883–3888.

67 Akutsu, M., Koyasu, K., Atobe, J. et al. (2006). Experimental and theoretical characterization of aluminum-based binary superatoms of $Al_{12}X$ and their cluster salts. *J. Phys. Chem. A* 110: 12073–12076.

68 Jackson, K., Kaxiras, E., and Pederson, M.R. (1994). Bonding of endohedral atoms in small carbon fullerenes. *J. Phys. Chem.* 98: 7805–7810.

69 Chen, Z., Jiao, H., Hirsch, A., and Thiel, W. (2001). The $2(N+1)^2$ rule for spherical aromaticity: further validation. *J. Mol. Model.* 7: 161–163.

70 Belomoin, G., Therrien, J., Smith, A. et al. (2002). Observation of a magic discrete family of ultrabright Si nanoparticles. *Appl. Phys. Lett.* 80: 841–843.

71 Beck, S.M. (1987). Studies of silicon cluster–metal atom compound formation in a supersonic molecular beam. *J. Chem. Phys.* 87: 4233–4234.

72 Beck, S.M. (1989). Mixed metal–silicon clusters formed by chemical reaction in a supersonic molecular beam: implications for reactions at the metal/silicon interface. *J. Chem. Phys.* 90: 6306–6312.

73 Jackson, K. and Nellermoe, B. (1996). $Zr@Si_{20}$: a strongly bound Si endohedral system. *Chem. Phys. Lett.* 254: 249–256.

74 Hiura, H., Miyazaki, T., and Kanayama, T. (2001). Formation of metal-encapsulating Si cage clusters. *Phys. Rev. Lett.* 86: 1733–1736.

75 Kresse, G. and Joubert, D. (1999). From ultrasoft pseudopotentials to the projector augmented-wave method. *Phys. Rev. B: Condens. Matter Mater. Phys.* 59: 1758–1775.

76 Blöchl, P.E. (1994). Projector augmented-wave method. *Phys. Rev.* B60: 17953.

77 Dunk, P.W., Kaiser, N.K., Mulet-Gas, M. et al. (2012). The smallest stable fullerene, $M@C_{28}$ ($M = Ti, Zr, U$): stabilization and growth from carbon vapor. *J. Am. Chem. Soc.* 134: 9380–9389.

78 Kawamura, H., Kumar, V., and Kawazoe, Y. (2005). Growth behavior of metal-doped silicon clusters Si_nM ($M = Ti, Zr, Hf$; $n = 8-16$). *Phys. Rev. B: Condens. Matter Mater. Phys.* 71: 075423.

79 Kawamura, H., Kumar, V., and Kawazoe, Y. (2004). Water adsorption on Ti-doped silicon clusters. *Phys. Rev. B: Condens. Matter Mater. Phys.* 70: 193402.

80 Kumar, V., Briere, T.M., and Kawazoe, Y. (2003). Ab initio calculations of electronic structures, polarizabilities, Raman and infrared spectra, optical gaps, and absorption spectra of $M@Si_{16}$ ($M = Ti$ and Zr) clusters. *Phys. Rev. B: Condens. Matter Mater. Phys.* 68: 155412.

81 Kumar, V. and Kawazoe, Y. (2002). Magic behavior of $Si_{15}M$ and $Si_{16}M$ (M = Cr, Mo, and W) clusters. *Phys. Rev.* B65: 073404.

82 Kawamura, H., Kumar, V., and Kawazoe, Y. (2004). Growth, magic behavior, and electronic and vibrational properties of Cr-doped Si clusters. *Phys. Rev. B: Condens. Matter Mater. Phys.* 70: 245433.

83 Lau, J., Vogel, M., Langenberg, A. et al. (2011). Communication: highest occupied molecular orbital – lowest unoccupied molecular orbital gaps of doped silicon clusters from core level spectroscopy. *J. Chem. Phys.* 134: 041102.

84 Zorriasatein, S., Joshi, K., and Kanhere, D. (2007). Dopant-induced stabilization of silicon clusters at finite temperature. *Phys. Rev. B: Condens. Matter Mater. Phys.* 75: 045117.

85 Uchida, N., Bolotov, L., Miyazaki, T., and Kanayama, T. (2003). Stability of Ta-encapsulating Si clusters on Si(111)-(7×7) surfaces. *J. Phys. D: Appl. Phys.* 36: L43–L46.

86 (a) Perdew, J.P., Burke, K., and Ernzerhof, M. (1996). Generalized gradient approximation made simple. *Phys. Rev. Lett.* 77: 3865. (b) Perdew, J.P., Burke, K., and Ernzerhof, M. (1997). Erratum to generalized gradient approximation made simple. *Phys. Rev. Lett.* 78: 1396. (E).

87 Grimme, S. (2006). Semiempirical gga-type density functional constructed with a long-range dispersion correction. *J. Comp. Chem.* 27: 1787.

88 Sen, A. and Sen, P. (2017). Properties of highly oriented pyrolytic graphite supported $TaSi_{16}$ clusters: a density functional investigation. *J. Phys. Chem. C* 121: 28490–28497.

89 Gaussian 09, Revision A.02, M. J. Frisch, G. W. Trucks et al. (2016). Gaussian, Inc. Wallingford CT.

90 Nakaya, M., Iwasa, T., Tsunoyama, H. et al. (2014). Formation of a superatom monolayer using gas-phase-synthesized Ta@Si_{16} nanocluster ions. *Nanoscale* 6: 14702–14707.

91 Ohta, T., Shibuta, M., Tsunoyama, H. et al. (2016). Charge transfer complexation of Ta-encapsulating Ta@Si_{16} superatom with C_{60}. *J. Phys. Chem. C* 120: 15265–15271.

92 Perdew, J.P. (1991). Unified theory of exchange and correlation beyond the local density approximation. In: *Electronic Structure of Solids '91* (eds. P. Ziesche and H. Eschrig), 11. Berlin: Akademie Verlag.

93 Atobe, J., Koyasu, K., Furuse, S., and Nakajima, A. (2012). Anion photoelectron spectroscopy of germanium and tin clusters containing a transition-or lanthanide-metal atom; MGe_n^- (n = 8–20) and MSn_n^- (n = 15–17) (M = Sc–V, Y–Nb, and Lu–Ta). *Phys. Chem. Chem. Phys.* 14: 9403–9410.

94 Khanna, S.N., Rao, B.K., and Jena, P. (2002). Magic numbers in metallo-inorganic clusters: chromium encapsulated in silicon cages. *Phys. Rev. Lett.* 89: 016803.

95 Kumar, V. and Kawazoe, Y. (2003). Hydrogenated silicon fullerenes: effects of H on the stability of metal-encapsulated silicon clusters. *Phys. Rev. Lett.* 90: 055502.

96 Becke, A.D. (1993). Density-functional thermochemistry. 3. The role of exact exchange. *J. Chem. Phys.* 98: 5648–5652.

97 Goicoechea, J.M. and McGrady, J.E. (2015). On the structural landscape in endohedral silicon and germanium clusters, M@Si_{12} and M@Ge_{12}. *Dalton Trans.* 44: 6755–6766.

98 Uţă, M. and King, R. (2012). Endohedral beryllium atoms in germanium clusters with eight and fewer vertices: how small can a cluster be and still encapsulate a central atom? *J. Phys. Chem. A* 116: 5227–5234.

99 Kumar, V., Singh, A.K., and Kawazoe, Y. (2004). Smallest magic caged clusters of Si, Ge, Sn, and Pb by encapsulation of transition metal atom. *Nano Lett.* 4: 677–681.

100 Neukermans, S., Janssens, E., Chen, Z. et al. (2004). Extremely stable metal-encapsulated $AlPb_{10}^+$ and $AlPb_{12}^+$ clusters: mass-spectrometric discovery and density functional theory study. *Phys. Rev. Lett.* 92: 163401.

101 Kumar, V. and Kawazoe, Y. (2002). Metal-encapsulated icosahedral superatoms of germanium and tin with large gaps: $Zn@Ge_{12}$ and $Cd@Sn_{12}$. *Appl. Phys. Lett.* 80: 859–861.

102 Bhattacharyya, S., Nguyen, T.T., De Haeck, J. et al. (2013). Mass-selected photodissociation studies of $AlPb_n^+$ clusters ($n = 7–16$): evidence for the extraordinary stability of $AlPb_{10}^+$ and $AlPb_{12}^+$. *Phys. Rev. B: Condens. Matter Mater. Phys.* 87: 054103.

103 Sun, Q., Wang, Q., Briere, T.M. et al. (2002). First-principles calculations of metal stabilized Si_{20} cages. *Phys. Rev.* B65: 235417.

104 Singh, A.K., Kumar, V., and Kawazoe, Y. (2005). Stabilizing the silicon fullerene Si_{20} by thorium encapsulation. *Phys. Rev. B: Condens. Matter Mater. Phys.* 71: 115429.

105 Singh, A.K., Kumar, V., and Kawazoe, Y. (2005). Thorium encapsulated caged clusters of germanium: $Th@Ge_n$, n = 16, 18, and 20. *J. Phys. Chem. B* 109: 15187–15189.

106 Kumar, V., Singh, A.K., and Kawazoe, Y. (2006). Charged and magnetic fullerenes of silicon by metal encapsulation: predictions from ab initio calculations. *Phys. Rev. B: Condens. Matter Mater. Phys.* 74: 125411.

107 Dognon, J.-P., Clavaguera, C., and Pyykkö, P. (2012). A new, centered 32-electron system: the predicted $[U@Si_{20}]^{6-}$-like isoelectronic series. *Chem. Sci.* 3: 2843–2848.

108 Jaiswal, S., Babar, V.P., and Kumar, V. (2013). Growth behavior, electronic structure, and vibrational properties of Si_nY anion clusters ($n = 4–20$): metal atom as linker and endohedral dopant. *Phys. Rev. B: Condens. Matter Mater. Phys.* 88: 085412.

109 Cui, L.-F., Huang, X., Wang, L.-M. et al. (2006). Sn_{12}^{2-}: stannaspherene. *J. Am. Chem. Soc.* 2006 (128): 8390–8391.

110 Cui, L.-F., Huang, X., Wang, L.-M. et al. (2006). Pb_{12}^{2-}: plumbaspherene. *J. Phys. Chem. A* 110: 10169–10172.

111 Jaiswal, S. and Kumar, V. (2016). Growth behavior and electronic structure of neutral and anion $ZrGe_n$ ($n = 1–21$) clusters. *Comput. Theor. Chem.* 1075: 87–97.

112 Neukermans, S., Wang, X., Veldeman, N. et al. (2006). Mass spectrometric stability study of binary MS_n clusters (S = Si, Ge, Sn, Pb, and M = Cr, Mn, Cu, Zn). *Int. J. Mass Spectrom.* 252: 145–150.

113 Rohrmann, U., Schäfer, S., and Schäfer, R. (2009). Size- and temperature-dependent magnetic response of molecular cage clusters: manganese doped tin clusters. *J. Phys. Chem. A* 113: 12115–12121.

114 Rohrmann, U. and Schäfer, R. (2013). Stern-Gerlach experiments on $Mn@Sn_{12}$: identification of a paramagnetic superatom and vibrationally induced spin orientation. *Phys. Rev. Lett.* 111: 133401.

115 Rohrmann, U. and Schäfer, R. (2015). Stern-Gerlach experiments on $Fe@Sn_{12}$: magnetic response of a Jahn–Teller distorted endohedrally doped molecular cage cluster. *J. Phys. Chem. C* 119: 10958–10961.

116 Chen, X., Deng, K., Liu, Y. et al. (2008). The geometric, optical, and magnetic properties of the endohedral stannaspherenes $M@Sn_{12}$ (M = Ti, V, Cr, Mn, Fe, Co, Ni). *J. Chem. Phys.* 129: 094301.

117 Chen, X., Deng, K., Liu, Y. et al. (2008). The geometric and magnetic properties of the endohedral plumbaspherene M@Pb_{12} clusters (M = Sc, Ti, V, Cr, Mn, Fe, Co, Ni). *Chem. Phys. Lett.* 462: 275–279.

118 Reveles, J.U., Clayborne, P.A., Reber, A.C. et al. (2009). Designer magnetic superatoms. *Nat. Chem.* 1: 310–315.

119 Bulusu, S., Li, X., Wang, L.-S., and Zeng, X.C. (2016). Evidence of hollow golden cages. *Proc. Natl. Acad. Sci. USA* 103: 8326–8330.

120 Gu, X., Ji, M., Wei, S.H., and Gong, X.G. (2004). Au_N clusters (N = 32, 33, 34, 35): cagelike structures of pure metal atoms. *Phys. Rev. B Condens. Matter* 70: 205401-1–205401-5.

121 Kumar, V. (2009). Coating of a layer of Au on Al_{13}: the findings of icosahedral Al@$Al_{12}Au_{20}^-$ and $Al_{12}Au_{20}^{2-}$ fullerenes using ab initio pseudopotential calculations. *Phys. Rev. B: Condens. Matter Mater. Phys.* 79: 085423.

122 Itoh, M., Kumar, V., Adschiri, T., and Kawazoe, Y. (2009). Comprehensive study of sodium, copper, and silver clusters over a wide range of sizes $2 \leq N \leq 75$. *J. Chem. Phys.* 131: 174510.

123 Pyykkö, P. and Runeberg, N. (2002). Icosahedral WAu_{12}: a predicted closed-shell species, stabilized by aurophilic attraction and relativity and in accord with the 18-electron rule. *Angew. Chem., Int. Ed.* 41: 2174–2176.

124 Li, X., Kiran, B., Li, J. et al. (2002). Experimental observation and confirmation of icosahedral W@Au_{12} and Mo@Au_{12} molecules. *Angew. Chem., Int. Ed.* 41: 4786–4789.

125 Raggi, G. and Soto, J.R. (2014). Relativistic DFT calculations of magnetic moments of pristine and thiolated Mn@Au_x (x = 6, 12). *Phys. Chem. Chem. Phys.* 16: 21506–21512.

126 Gao, Y., Bulusu, S., and Zeng, X.C. (2006). A global search of highly stable gold-covered bimetallic clusters M@Au_n (n = 8–17) endohedral gold clusters. *ChemPhysChem* 7: 2275–2278.

127 Neukermans, S., Janssens, E., Tanaka, H. et al. (2003). Element- and size-dependent electron delocalization in Au_NX^+ clusters (X = Sc, Ti, V, Cr, Mn, Fe, Co, Ni). *Phys. Rev. Lett.* 90: 033401.

128 Gao, Y., Bulusu, S., and Zeng, X.C. (2005). Gold-caged metal clusters with large HOMO–LUMO gap and high electron affinity. *J. Am. Chem. Soc.* 127: 15680–15681.

129 Bouwen, W., Vanhoutte, F., Despa, F. et al. (1999). Stability effects of $Au_nX_m^+$ (X = Cu, Al, Y, In) clusters. *Chem. Phys. Lett.* 314: 227–233.

130 Janssens, E., Neukermans, S., Nguyen, H.M.T. et al. (2005). Quenching of the magnetic moment of a transition metal dopant in silver clusters. *Phys. Rev. Lett.* 94: 113401.

131 Veldeman, N., Höltzl, T., Neukermans, S. et al. (2007). Experimental observation and computational identification of Sc@Cu_{16}^+, a stable dopant-encapsulated copper cage. *Phys. Rev. A: At. Mol. Opt. Phys.* 76: 011201.

132 Zhang, M., Zhang, H., Zhao, L. et al. (2012). Low-energy isomer identification, structural evolution, and magnetic properties in manganese-doped gold clusters $MnAu_n$ (n = 1–16). *J. Phys. Chem. A* 116: 1493–1502.

133 Shinde, P.P., Yadav, B.D., and Kumar, V. (2012). Evolution of atomic and electronic structure of magnetic Gd-doped gold clusters. *J. Mater. Sci.* 47: 7642–7652.

134 Jayasekharan, T. and Ghanty, T.K. (2010). Endohedrally doped golden fullerenes X@Au_{32} (X = Li^+, Na^+, K^+, Rb^+, Cs^+). *J. Phys. Chem. C* 114: 8787–8793.

135 Manna, D., Jayasekharan, T., and Ghanty, T.K. (2013). Structure and stability of Zn, Cd, and Hg atom doped golden fullerene (Au_{32}). *J. Phys. Chem. C* 117: 18777–18788.

136 Jarrold, M.F. and Bower, J.E. (1986). The reactions of mass selected aluminum cluster ions, Al_n^+ ($n = 4$–25), with oxygen. *J. Chem. Phys.* 85: 5373–5375.

137 Gong, X. and Kumar, V. (1993). Enhanced stability of magic clusters: a case study of icosahedral $Al_{12}X$, X = B, Al, Ga, C, Si, Ge, Ti. *As. Phys. Rev. Lett.* 70: 2078–2081.

138 Nakajima, A., Kishi, T., Sugioka, T., and Kaya, K. (1991). Electronic and geometric structures of aluminum-boron negative cluster ions ($Al_nB_m^-$). *Chem. Phys. Lett.* 187: 239–244.

139 Thomas, O.C., Zheng, W., and Bowen, K.H. Jr. (2001). Magic numbers in copper-doped aluminum cluster anions. *J. Chem. Phys.* 114: 5514–5519.

140 Henry, D.J. (2012). Structures and stability of doped gallium nanoclusters. *J. Phys. Chem. C* 116: 24814–24823.

141 Sergeeva, A.P., Popov, I.A., Piazza, Z.A. et al. (2014). Understanding boron through size-selected clusters: structure, chemical bonding, and fluxionality. *Acc. Chem. Res.* 15: 349–358.

142 Oger, E., Crawford, N.R.M., Kelting, R. et al. (2007). Boron cluster cations: transition from planar to cylindrical structures. *Angew. Chemie* 46: 8503–8506.

143 Saha, P., Rahane, A., Kumar, V., and Sukumar, N. (2016). Analysis of the electron density features of small boron clusters and the effects of doping with C, P, Al, Si, and Zn: magic B_7P and B_8Si clusters. *Phys. Scr.* 91: 053005.

144 Karthikeyan, J., Ranawat, Y.S., Murugan, P., and Kumar, V. (2018). Borophene layers on an Al (111) surface – the finding of a borophene layer with hexagonal double chains and B_9 nonagons using ab initio calculations. *Nanoscale* 10: 17198.

145 Ito, K., Pu, Z., Li, Q.-S., and Schleyer, P.V.R. (2008). Cyclic boron clusters enclosing planar hypercoordinate cobalt, iron, and nickel. *Inorg. Chem.* 47: 10906–10910.

146 Chen, T.-T., Li, W.-L., Bai, H. et al. (2019). $Re@B_8^-$ and $Re@B_9^-$: new members of the transition-metal-centered borometallic molecular wheel family. *J. Phys. Chem. A* 123: 5317–5324.

147 Galeev, T.R., Romanescu, C., Li, W.-L. et al. (2012). Observation of the highest coordination number in planar species: decacoordinated $Ta@B_{10}^-$ and $Nb@B_{10}^-$ anions. *Angew. Chem. Int. Ed.* 51: 2101–2105.

148. Saha, P., Rahane, A.B., Kumar, V., and Sukumar, N. (2017). Electronic origin of the stability of transition-metal-doped B_{14} drum-shaped boron clusters and their assembly into a nanotube. *J. Phys. Chem. C* 121: 10728–11074.

149 Popov, I.A., Jian, T., Lopez, G.V. et al. (2015). Cobalt-centred boron molecular drums with the highest coordination number in the CoB_{16}^- cluster. *Nat. Commun.* 6: 8654.

150 Li, H.-R., Liu, H., Tian, X.-X. et al. (2017). Structural transition in metal-centered boron clusters: from tubular molecular rotors $Ta@B_{21}$ and $Ta@B_{22}^+$ to cage-like endohedral metalloborospherene $Ta@B_{22}^-$. *Phys. Chem. Chem. Phys.* 19: 27025–27030.

151 Li, H.-R., Liu, H., Lu, X.-Q. et al. (2018). Cage-like $Ta@B_nq$ complexes ($n = 23$–28, $q = -1$ – $+3$) in 18-electron configurations with the highest coordination number of twenty-eight. *Nanoscale* 10: 7451–7456.

152 Li, W., Jian, T., Chen, X. et al. (2017). Observation of a metal-centered B_2-$Ta@B_{18}^-$ tubular molecular rotor and a perfect $Ta@B_{20}^-$ boron drum with the record coordination number of twenty. *Chem. Commun.* 53: 1587–1590.

153 Rahane, A.B., Saha, P., Sukumar, N., and Kumar, V. (2019). Smallest fullerene-like structures of boron with Cr, Mo, and W encapsulation. arXiv:1907.12611.

154 Lv, J., Wang, Y., Zhang, L. et al. (2015). Stabilization of fullerene-like boron cages by transition metal encapsulation. *Nanoscale* 7: 10482–10489.

155 Kumar, V. and Kawazoe, Y. (2007). Hydrogenated caged clusters of Si, Ge, and Sn and their endohedral doping with atoms: Ab initio calculations. *Phys. Rev. B: Condens. Matter Mater. Phys.* 75: 155425.

156 Pichierri, F., Kumar, V., and Kawazoe, Y. (2005). Encapsulation of halide anions in perhydrogenated silicon fullerene: $X^{-}@Si_{20}H_{20}$ (X = F, Cl, Br, I). *Chem. Phys. Lett.* 406: 341–344.

157 Tillmann, J., Wender, J.H., Bahr, U. et al. (2015). One-Step synthesis of a [20] silafullerane with an endohedral chloride ion. Angew. Chem. *Angew. Chem. Int. Ed.* 54: 5429–5433.

158 Kasuya, A., Sivamohan, R., Barnakov, Y.A. et al. (2004). Ultra-stable nanoparticles of CdSe revealed from mass spectrometry. *Nat. Mater.* 3: 99–102.

159 Nanavati, S.P., Sundararajan, V., Mahamuni, S. et al. (2009). Optical properties of zinc selenide clusters from first-principles calculations. *Phys. Rev. B* 80: 245417.

160 Yang, J., Fainblat, R., Kwon, S.G. et al. (2015). Route to the smallest doped semiconductor: Mn^{2+}-doped $(CdSe)_{13}$ clusters. *J. Am. Chem. Soc.* 137: 12776–12779.

161 Matxain, J.M., Eriksson, L.A., Formoso, E. et al. Endohedral (92007). $(X@Zn_iS_i)$ i = 4–160,± nanoclusters, X = Li, Na, K, Cl, Br. *J. Phys. Chem. C* 111: 3560–3565.

162 Kumar, V. (2013). High symmetry Nb_n and Ta_n (n = 12, 15, and 17) clusters: high magnetic moments and the finding of superatoms with doping. *Comput. Theor. Chem.* 1021: 149–154.

163 Eulenstein, A.R., Franzke, Y.J., Lichtenberger, N. et al. (2020). Substantial π-aromaticity in the anionic heavy-metal cluster $[Th@Bi_{12}]^{4-}$. *Nat. Chem.* 13: 149.

164 Kandalam, A.K., Chen, G., and Jena, P. (2008). Unique magnetic coupling between Mn doped stannaspherenes $Mn@Sn_{12}$. *Appl. Phys. Lett.* 92: 143109.

165 Reis, C., Martins, J., and Pacheco, J. (2007). Stability analysis of a bulk material built from silicon cage clusters: a first-principles approach. *Phys. Rev. B: Condens. Matter Mater. Phys.* 76: 233406.

166 Reis, C. and Pacheco, J. (2010). Bulk materials made of silicon cage clusters doped with Ti, Zr, or Hf. *J. Phys.: Condens. Matter* 22: 035501.

167 Kortan, A.R., Kopylov, N., Glarum, S. et al. (1992). Superconductivity in barium fulleride. *Nature* 360: 566–569.

168 Laurrabaquio, G.L., Torres, M.B., Fernández, E.M., and Balbás, L.C. (2010). Trends in the formation of aggregates and crystals from $M@Si_{16}$ clusters: a study from first principles calculations. *J. Math. Chem.* 48: 109–117.

169 Torres, M., Fernandez, E., and Balbás, L. (2011). Theoretical study of the structural and electronic properties of aggregates, wires, and bulk phases formed from $M@Si_{16}$ superatoms (M = Sc^{-}, Ti, V^{+}). *Int. J. Quantum Chem.* 111: 444–462.

170 Singh, A.K., Kumar, V., Briere, T.M., and Kawazoe, Y. (2002). Cluster assembled metal encapsulated thin nanotubes of silicon. *Nano Lett.* 2: 1243–1248.

171 Saranin, A.A., Zotov, A.V., Kotlyar, V.G. et al. (2004). Ordered arrays of Be-encapsulated Si nanotubes on Si (111) surface. *Nano Lett.* 4: 1469–1473.

172 Singh, A.K., Briere, T.M., Kumar, V., and Kawazoe, Y. (2003). Magnetism in transition-metal-doped silicon nanotubes. *Phys. Rev. Lett.* 91: 146802.

173 Singh, A.K., Kumar, V., and Kawazoe, Y. (2004). Metal encapsulated nanotubes of silicon and germanium. *J. Mater. Chem.* 14: 555–563.

174 Singh, A.K., Kumar, V., and Kawazoe, Y. (2005). Design of a very thin direct-band-gap semiconductor nanotube of germanium with metal encapsulation. *Phys. Rev. B: Condens. Matter Mater. Phys.* 71: 075312.

175 Singh, A.K., Kumar, V., and Kawazoe, Y. (2005). Metal encapsulated nanotubes of germanium with metal dependent electronic properties. *Eur. Phys. J. D* 34: 295–298.

176 Menon, M., Andriotis, A.N., and Froudakis, G. (2002). Structure and stability of Ni-encapsulated Si nanotube. *Nano Lett.* 2: 301–304.

177 Li, J., Wang, J., Zhao, H.-Y., and Liu, Y. (2013). Magnetic silicon nanotube: role of encapsulated europium atoms. *J. Phys. Chem.* C117: 10764–10769.

178 Kumar, V. (2006). Recent theoretical progress on electronic and structural properties of clusters: permanent electric dipoles, magnetism, novel caged structures, and their assemblies. *Comput. Mater. Sci.* 35: 375–381.

179 Iwasa, T. and Nakajima, A. (2012). Geometric, electronic, and optical properties of a superatomic heterodimer and trimer: $Sc@Si_{16}$–$V@Si_{16}$ and $Sc@Si_{16}$–$Ti@Si_{16}$–$V@Si_{16}$. *J. Phys. Chem. C* 116: 14071–14077.

180 Miyazaki, T. and Kanayama, T. (2007). Ultrathin layered semiconductor: Si-rich transition metal silicide. *Jpn. J. Appl. Phys.* 46: L28–L30.

181 Liu, Z., Wang, X., Cai, J., and Zhu, H. (2015). Room-temperature ordered spin structures in cluster-assembled single $V@Si_{12}$ sheets. *J. Phys. Chem. C* 119: 1517–1523.

182 Nie, Z., Guo, P., Zheng, J. et al. (2018). Electronic and magnetic properties of two dimensional cluster-assembled materials based on $TM@Si_{12}$ (TM = 3d transition metal) clusters. *Comput. Mater. Sci.* 146: 134–142.

183 Shibuta, M., Takano, R., and Nakajima, A. (2020). Interfacial oxidation of Ta-encapsulating Si_{16} cage superatoms ($Ta@Si_{16}$) on strontium titanate substrates. *J. Phys. Chem. C* 124: 28108.

184 Liu, J., Guo, P., Zheng, J. et al. (2020). Self-assembly of a two-dimensional sheet with $Ta@Si_{16}$ superatoms and its magnetic and photocatalytic properties. *J. Phys. Chem. C* 124: 6861–6870.

5

Magnetic Superatoms

Nicola Gaston

The MacDiarmid Institute for Advanced Materials and Nanotechnology, Department of Physics, The University of Auckland, Auckland, New Zealand

5.1 Introduction

The successful application of the superatomic concept to date has paralleled, in many ways, the anthropic principle: we can explain the existence of specific nanostructures – atomically precise clusters – through the elevated stability associated with electronic structure. The structures we see, and are able to synthesise, reflect the physical laws that govern them, just as our own existence in the universe constrains the physical laws that we observe as existing in nature. The potential for our understanding of these laws to enable the creation of new designer materials from nanostructured building blocks – one of the key promises of the field of superatom research – requires us to reach beyond the ideal cases of closed-shell, stable superatoms and press up against the limits of the model and to understand how to build reactivity, or interactivity, into the building blocks themselves. This requires us to think beyond stability alone, and one key area that has considerable evidence for eventual practical application is the study of the magnetic superatom.

The association of stability with the pairing of electrons in closed electronic shells makes the idea of a magnetic superatom a conceptual anomaly. The unpaired electrons required for magnetism come at an energetic cost. However, magnetism in the context of individual atoms is well understood to come about as a consequence of degenerate energy levels, such as found in transition metal atoms. The *d*-electrons that are the valence electrons in transition metal species prefer to align if the energies of their orbitals are truly degenerate, to avoid the energetic cost of spin-pairing. In addition, the *d*-electrons are considerably more localised than the *s*- or *p*-electrons most usually studied as the valence electrons in superatomic clusters, such as those formed from gold, aluminum, or sodium. This is even more so the case for the rare earth elements, with *f*-electrons being even more strongly localised.

The degree of localisation of an individual electron wavefunction depends on many factors. The higher angular momentum functions (*d*- or *f*-electrons) have lower principal quantum numbers relative to their energies, and thus remain more compact than *s*- or *p*-orbitals. However, the degree of localisation depends strongly on the degree of shielding

Superatoms: Principles, Synthesis and Applications, First Edition. Edited by Purusottam (Puru) Jena and Qiang Sun.
© 2022 John Wiley & Sons Ltd. Published 2022 by John Wiley & Sons Ltd.

provided by other electron shells and the occupation of a given shell; on this basis, the early transition metals have more diffuse *d*-orbitals than the late transition metals.

In this chapter, we will discuss the variety of ways in which magnetic superatoms have been proposed and demonstrated to exist and their potential for application to the synthesis of highly magnetic nanostructured materials. These ideas have considerable history, starting with the assembly of non-magnetic metal clusters [1], though the potential for magnetic species to be designed through the incorporation of transition metals was quickly realised [2, 3]. We will first touch on the first examples of magnetic superatoms; we will then discuss the ways in which the properties of magnetic superatoms are sensitive to, and able to be modulated by, their environment. Subsequently, we will consider the prospects for the properties of magnetic superatoms to be tuned synthetically – primarily through doping as a strategy for structural control. Finally, we will look at the generalisation of the superatomic model to open-shell systems, for which the stability may not be primarily controlled by superatomic shell closing, but for which the superatomic shells nonetheless provide a systematic descriptor of the electronic structure.

5.2 The Arrival of the Magnetic Superatom

It makes sense to start our introduction of magnetic superatoms with some discussion of the many studies on magnetic clusters that preceded the field. Indeed, even earlier work on transition metal complexes is relevant to many of the physical aspects of the treatment of these species. For example, the nature of the exchange coupling between two Mo atoms in a Cl-bridged dimeric complex was studied already in 1980 by Ginsberg [4]; it was concluded already then that the magnetic interaction between the metal ions was via direct metal–metal interaction, rather than super exchange mediated by the ligands.

Within this context of molecular or nanoscale magnetism, we should also mention the considerable history of molecular magnets, of interest for the design of supermagnetic materials in their own right, even when not superatomic. For example, dinuclear manganese complexes have been extensively studied in a number of contexts – tetranuclear manganese complexes being particularly relevant in photosynthesis research, as model systems for photosystem II – and in such complexes, bridging ligands between the metal centres can mediate the magnetic exchange – so called super exchange – especially in the case of oxygen or similar bridges, which when protonated can switch magnetic coupling from antiferromagnetic to ferromagnetic [5]. Such examples demonstrate the extreme sensitivity of magnetic exchange to environmental factors, and further examples of this sensitivity will be discussed later. But in the context of molecular magnet design, it is clear that the use of a new ligand to connect two metal centres can have dramatic and unforeseen consequences on the magnetic properties of the molecule; it is one of the hopes behind studying superatoms that creating magnetic clusters from a first principles understanding of stability might mitigate this, or at least enable sensible strategies to prevent it.

A first principles understanding of the stability of unpaired electrons – magnetic states – in a complex or cluster is arguably best demonstrated by the kinds of systematic studies that are the bread and butter of cluster science. A fundamental reason for the interest in manganese clusters, in addition to their relevance to photosynthesis, is the specific electron configuration of the manganese atom: with a doubly filled 4*s*-shell, and 5*d*-electrons unpaired in the 3*d*-shell, it has a particularly high magnetic moment; it also provides an exemplary illustration of the kinds of factors that influence magnetic properties for *d*-electron elements [6, 7].

The study of clusters is important in providing evidence of properties, such as magnetic behaviour, as they evolve from atoms to the bulk. Stern–Gerlach experiments on nanoscale clusters consisting of 11–99 atoms [8] suggest that in this regime neither the very small clusters nor the bulk provide good models for the dominant atomic or magnetic interactions. In particular, the magnetic moments were only around 0.3–1.4 μ_B per atom, considerably less than small clusters that will retain most *d*-electrons in an unpaired state, allowing for magnetic moments of c. 4–5 μ_B per atom. The bulk metal is antiferromagnetic in its more stable state, with a lower magnetic moment of around 1 μ_B per atom. The nanoscale clusters, however, are assumed to be coupled ferromagnetically to result in superparamagnetic properties: notably, the clusters with 13 and 19 atoms have particularly low magnetic moments, demonstrating a structural influence of the icosahedral symmetry that is possible for these specific sizes. This emergence of distinct magnetic properties in clusters that have no correspondence in the bulk metal has motivated significant further investigation.

Computational studies [9] have elucidated the ways in which both structural order and spin ordering may interact in complicated ways in these medium-sized clusters. In further theoretical work employing a Heisenberg Hamiltonian approach [10], the consequences of ferromagnetic versus antiferromagnetic interactions on neighbouring atoms, as well as on next nearest neighbours, have been studied in detail, demonstrating how sensitive the overall magnetic properties may be to small changes in interactions. In smaller clusters, calculations [11] have shown that the maximum magnetism is obtained at a size of four atoms, with each of the Mn atoms retaining all five *d*-electrons in a high spin configuration, enabling a total magnetic moment of 20 μ_B to be achieved. This high magnetic moment suggests a plethora of interesting applications – but its sensitivity to environmental factors is still evident.

What the literature to date makes clear is that the nature of this sensitivity is connected to the binding mode between Mn atoms – where the pair of 4*s* electrons are more diffuse than the localised *d*-electrons and thus provide the dominant interaction in the absence of ionisation or polarisation effects. As this interaction is a van der Waals interaction between closed-shell electron densities, it is relatively weak and enables the *d*- and *s*-electrons to operate somewhat independently. In contrast, any change in electron occupation can change the mode of binding dramatically: for example, in the Mn dimer, ionisation leads to a cation with a bond distance reduced by 10% [12].

This example of the importance of the spatial separation of the *s*- and *d*-electrons in transition metal elements is useful, especially when put alongside of the most commonly studied metals in the context of superatoms: gold. While gold is not a particularly interesting metal in the context of magnetism, this is due to the relative inertness of its 5*d*-shell, due to its closed configuration. Thus, the superatomic properties of gold clusters arise purely from the interaction of the 6*s* electrons, one per atom, which delocalise readily and result in the simple electron-counting rules that emerge from the jellium picture as applied to superatoms.

This juxtaposition of delocalised *s*-electron states that are superatomic with localised *d*-electron states that are relatively inert is the typically understood paradigm that governs the behaviour of the later transition metals. However, it is not gospel, and examples of late transition metals that break these simple rules, allowing the *d*-electrons to contribute in some sense to the superatomic states, have indeed been realised.

Nickel clusters are an excellent example, having been found to have varied and highly degenerate magnetic states in both theory and experiment [13, 14]. Analysis of their superatomic states in a computational study of their electronic states as a function of size, charge state, and multiplicity [15] has demonstrated the relevance of the superatomic stabilisation of certain clusters close to an electronic shell closing, despite the local nature of the *d*-orbitals. Optimisations of cluster structure for all possible multiplicities of clusters with 8, 9, and 10 atoms demonstrated that indeed nickel clusters behave similarly to gold, with the single valence *s*-electron contributing to superatomic states while the nine *d*-electrons per atom stay well localised on individual atoms, and therefore have the ability to result in finite magnetic moments for the whole cluster. The exact value of the magnetic moment depends on both cluster geometry and, crucially, the nature of the density functional applied, as the inclusion of exact exchange can lead to even more localised *d*-states [16]. This means that remaining discrepancies between experimentally measured and theoretical magnetic moments remain to be reconciled, as the sensitivity of magnetic properties to the inclusion of exact exchange and broken symmetry methods is well known.

However, the demonstration in this study of the ways in which the magnetic properties of superatomic clusters can be modulated by the energetic drivers of electron shell closing is useful. A cluster of 10 nickel atoms is the first size at which there is clearly an internal nickel atom around which a shell of 9 atoms creates a surface. In the cationic state (Ni_{10}^+) there are, therefore, nine *s*-electrons in principle, if one takes the d^9s^1 configuration of each nickel atom as standard, as it indeed usually is. The proximity to a superatomic shell closing, however, is a powerful motivator for configurational change and enables the internal atom to adopt a d^{10} configuration, reducing both the magnetic moment by pairing electrons within the localised *d*-shell, while reducing the *s*-electron count by one to result in an eight-electron, $1S^21P^6$ closed shell.

This realisation of a composite system from a single transition metal element echoes many of the principles that have been proposed in the literature regarding the design of magnetic superatoms. The idea that some atoms should have localised orbitals, with a separation between spin up and down levels, while others should have nearly free electron gas-like states, has led to the proposal of several classes of magnetic superatom with carefully chosen composition: VCs_8, in which the *d*-states accessible for the heavy caesium atom stabilise the magnetic moment on the central vanadium atom, $3\,\mu_B$, is a classic example [17]. Doping of manganese into a ligand-protected gold cluster, $MnAu_{24}(SH)_{18}$, retains the magnetic moment on the manganese atom of $5\,\mu_B$. While these magnetic superatoms were first proposed on the basis of simple design principles and superatomic electron-counting rules, they have led to experimental verification that these clusters exist [18]. Notably, with a single magnetic atom contributing the magnetic moment, they are somewhat less able to exhibit a wide plethora of magnetic states as is the case for pure transition metal clusters where unpaired *d*-electrons are contributed by each atom of the cluster.

Further efforts to enhance the magnetic moments of superatomic clusters by design – in particular, to look to mimic the properties of the highly magnetic rare earth elements – has been carried out with the assembly of simple boron-doped diatomic clusters, LaB and NdB, which have been determined to mimic the magnetic properties of Nd and Eu, respectively [19]. While each of these concrete discoveries adds depth to the landscape within which magnetic superatoms can be explored, it is clear that the full potential of magnetic superatoms for the design of materials requires an understanding of the variation in magnetic properties that can be obtained through environmental interactions – such as the change of site in nickel clusters enabling a change of atomic electron configuration – and an understanding of the ways in which magnetic properties can be tuned, for example, through more systematic doping of the cluster.

5.3 Tunable Superatoms

The tunability of whole classes of superatom, in particular through selective doping, has been explored in a number of systems. One of particular interest and generality is the fullerene-like cages of silicon, which have been studied as superatomic cages for transition metal atoms – including Fe, Ru, Os, Hf, and Zr – where a great advantage of such cages is that they can be systematically varied in size to accommodate guest atoms of varied sizes [20]. Such work, and its extension to cages of tin and germanium [21], was a first demonstration of the generality of the principle that doping a transition metal atom into a semiconducting cluster both preserves the magnetic moment of the guest atom ($5\,\mu_B$ for Mn, for example) and results in stable clusters of high symmetry that have high HOMO–LUMO (highest-occupied molecular orbital – lowest unoccupied molecular orbital) gaps and are therefore serious contenders for use in the synthesis of cluster-assembled materials. It should be noted that not all transition metal atoms preserve high magnetic moments in such silicon cages, however – for example, the magnetic moments of Fe atoms in such cages have been shown to be systematically quenched [22].

This generic approach has been extended to more traditional superatomic clusters with the use of more free electron metals – for example, the alkali metals – as the cage atom type. Doping of magnesium clusters with the second-row transition metals has demonstrated that the trends in magnetic moment variation are largely accounted for by the usual Hund's rule, with a maximum of five unpaired electrons reached in the middle of the series, at T_c in group 7 [23].

The possibility of instead using pure transition metal-based clusters in which the magnetic properties can be tuned through systematic changes in composition has been explored in mixed clusters of copper and nickel. For example, for a total of eight atoms, Cu_xNi_{8-x} clusters have been shown to systematically increase their magnetic moment from $0\,\mu_B$ for $x = 8$, to $8\,\mu_B$, for $x = 0$ [24]. These are all eight superatomic electron systems, with one *s*-electron delocalised over the cluster from each atom, but structural features in the intermediate regime can affect the overall magnetic properties. For example, the Cu_4Ni_4 cluster exhibits segregation of the Ni and Cu on opposite sides of the cluster, a so-called Janus effect. This results in a lower magnetic moment than expected based on an increase of $1\,\mu_B$ per Ni atom; the alignment of localised *d*-electron spin is presumably disfavoured as it could create a large magnetic dipole. Such structural details crop up in all types of cluster structures, complicating what otherwise appear to be simple rules for counting electrons.

In the search for stable clusters that might be manipulable in solution or crystalline material form, ligand-protected clusters have received considerable attention. For the most part, these clusters are described elsewhere in this book, but their ability to support magnetic properties through doping, as in the case of the $MnAu_{24}(SH)_{18}$ magnetic superatom already mentioned, deserves a little more discussion.

Doped gold clusters have been widely studied both experimentally and theoretically on account of the excellent electron delocalisation that gold enables [25] and systematic tunability of magnetic properties provided by transition metal dopants [26]. Notably, small gold clusters change shape dramatically when doped, from two-dimensional to three-dimensional. This is also true of gold clusters that are synthesised in solution with protective ligands; even pure gold clusters will change shape dramatically in the presence of even weakly bound ligands [27, 28]. The number of different classes of ligand-protected gold clusters, and their relative ease of synthesis, will continue to mean that they are a useful avenue for the introduction of magnetic properties through doping.

In discussing the tunability of electronic (and therefore magnetic) properties, substrates on which clusters may be immobilised are of similar importance to ligands in their influence on structure and electron counts. Indeed, substrates will be particularly important for the construction of nanodevices using superatomic building blocks, and also provide a first insight into the kinds of interactions that influence the self-assembly process of superatomic clusters into nanostructured materials.

Magnetic manganese clusters have been studied systematically on graphene as a function of size, demonstrating that the magnetic moments can be dramatically enhanced through this support via charge transfer processes – by as much as 186% in the case of the five-atom cluster, from 3.0 to 8.6 μ_B. Importantly, these charge transfer processes can operate in the other direction, for example, suppressing the magnetic moment of the highly ferromagnetic Mn_4 cluster by 4 μ_B per atom [29].

Similarly, iron atoms and dimers supported on polyaromatic hydrocarbons demonstrate a suppression of the magnetic moment per atom [22]. Such studies are important for the eventual assembly of materials from individual clusters as such substrates may provide a template for guided, controlled assembly of superatomic building blocks into materials. However, the modulation of the characteristics of the clusters themselves by the substrate suggests that such an approach remains a way off, for successful application to the tailored synthesis of magnetic materials.

5.4 The Delocalisation of *d*-electrons

As already mentioned, the later transition metals are well understood to have relatively strongly localised *d*-states. However, in the earlier reaches of the transition metal series, the prospects for the involvement of *d*-electrons in superatomic states are significantly more promising.

Through the design of magnetic superatoms, for example, to obtain the VCs_8 cluster already mentioned as an archetype, consisting of a magnetic *d*-electron metal atom embedded in a cluster of alkali metal clusters, there are numerous prospects for the nature of the *d*-electron states to be tuned through the choice of transition metal. A range of alkali metal coated scandium clusters have been considered, for example [30]. Scandium-doped clusters have been studied with a vast array of host species, including copper [31], silicon [32], and gold [33]; a major driver of this work has been the potential for the *d*-electrons in these very early transition metals to delocalise partially and thus hybridise with the cage molecular orbitals (or superatomic orbitals). Similar evidence has been provided for the group 3 congener, yttrium [34].

Studies of pure group 3 and 4 transition metal clusters within the context of the superatomic framework have helped to answer questions about the conditions under which delocalisation of *d*-electron density is favoured [35]. The radial extension of the *d*-orbitals in these elements with only one or two *d*-electrons is such that they can contribute into a global electronic structure. Analysis of the geometries and electronic structures of seven-atom clusters of Sc, Y, La, Ti, Zr, and Hf, which adopts the pentagonal bipyramidal structure that is a feature of icosahedral packing motifs, demonstrates that high magnetic moments can be obtained, particularly for scandium. The precise values of the magnetic moments predicted are very sensitive to the inclusion of exact exchange, which always results in an increase in magnitude; this can be understood by the tendency of exact exchange to increase the localisation of electron density.

The superatomic states obtained for the group 3 and 4 clusters are particularly instructive, with 3 and 4 valence electrons per atom available to delocalise, and thus either 21 or 28 electrons available per (neutral) cluster. The group 3 superatomic orbitals, calculated at the PBE level (without exact exchange), are shown in Figure 5.1 as a demonstration of the

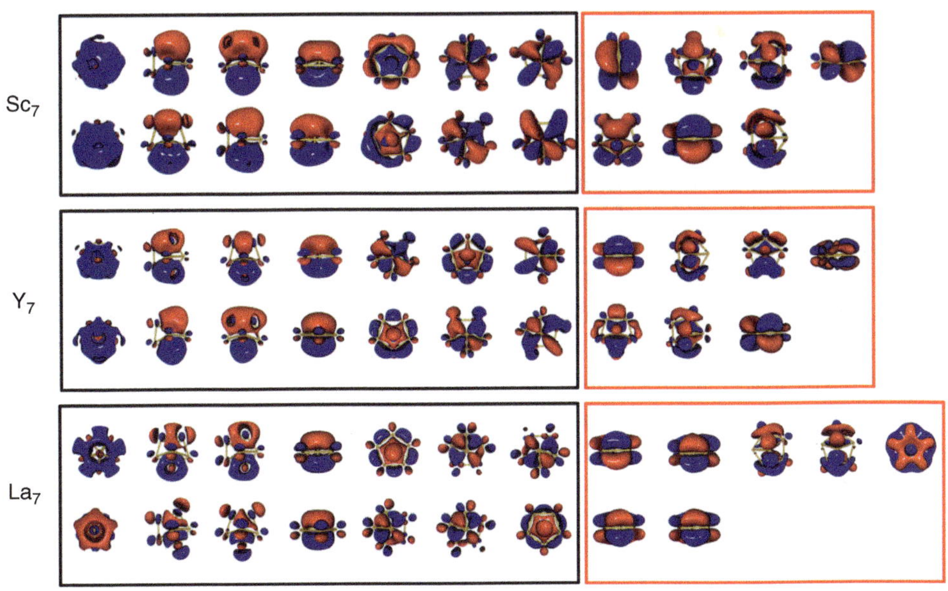

Figure 5.1 The Kohn–Sham states of the group 3 clusters of seven atoms are shown, calculated at the PBE level, with up and down spin occupied orbitals stacked vertically.

degree of delocalisation these *d*-electron-based states may have: the 14*s*-electron orbitals are outlined in each case in black and form the usual progression of $1S^21P^61D^6$ states according to the usual superatomic Aufbau rule.

The usual Aufbau rules are not sufficient for the explanation of the electronic structures of these superatoms, however. It can be seen that while there is a clear progression to a D-shell, it is not fully occupied. Before completing occupancy of the D-shell, the subsequent superatomic orbitals start the series of F-orbitals. It can be expected that this relates to the energetic stabilisation of subsets of the D- and F-shells due to the underlying symmetry of the cluster itself.

Inspection of the occupied and unoccupied superatomic 1D and 1F orbitals demonstrates that there is indeed an energetic splitting induced by the symmetry of the seven-atom cluster, with the occupied orbitals being those that enable the maximum overlap of atomic orbitals of *d*-character. This, in combination with the degeneracy of these 1D and 1F levels, means that the high magnetic moment states are strongly stabilised (and indeed, become more strongly stabilised with the inclusion of exact exchange). We therefore find ourselves with a class of magnetic superatom where the magnetic moment is not merely that of a dopant atom introduced into the cluster but emerges from the interactions of individual atomic moments. This suggests that there may be novel design principles able to be used for the further prediction of magnetic superatoms.

This idea that cluster geometry and symmetry may be able to be exploited to predict the favourable orbital overlap needed to produce a high spin cluster has been explored further [36]. The Sc_{13} cluster, which has full icosahedral symmetry, is an ideal candidate for a 39-electron superatom short of the single electron needed to close the 2P shell at 40 electrons. This is highly analogous to the case of the non-magnetic Al_{13} cluster, which has been widely studied as an archetype of trivalent metal-based superatoms. However, as in the case of scandium all the states above the 2S-orbital are formed from atomic *d*-orbitals, the symmetry of the icosahedron stabilises the F-orbitals in the xy plane preferentially, as well as splitting the G-shell; this results in a highly magnetic cluster with these states above the 2S-orbital being only singly occupied, resulting in a magnetic moment of 20 for the neutral cluster. Even the anionic cluster, with a total number of 40 electrons, retains 19 unpaired electrons, demonstrating a stark difference of this three-valence electron metal to that of its trivalent competitors in the p-block, Al and Ga.

One of the most exciting things about this demonstration of the ability of early transition metals to create such open-shell superatomic states is that this suggests alternative routes for the design of open-shell superatoms; this moves us to a mode in which the atoms able to be mimicked by designer superatoms is expanded well beyond groups of the halogens and the alkali metals, with their tendency to interact via simple electron transfer, into a regime in which more complex forms of inter-cluster bonding may be enabled between atoms. Rather than thinking of cluster-assembled materials as analogues of salts obtained through electron transfer, or weakly bound solids in which isotropic dispersion forces provide the only bonding, we might imagine the development of pseudo-covalent, directional bonds between these classes of more complex open-shell superatoms. The potential of these types of building blocks in nanostructured material design is yet to be realised.

5.5 Prospects for Nanostructured Magnetic Material Design

It has long been mooted that one of the key prospects for application of superatomic clusters is in the development of nanostructured materials where the properties of the individual building blocks are preserved. The challenge is a complicated one; the interactions between superatoms must be strong enough to produce a material in which the electronic properties of the clusters contribute to bulk properties, while yet weak enough that the clusters themselves retain some structural integrity within the final material. This is a highly non-trivial challenge, but of particular interest for magnetic interactions, which may be longer range than purely charge-based electronic interactions. They may also be able to remain decoupled, to an extent, from the bonding interactions between clusters, in analogy to the dispersion-bound manganese dimer discussed earlier [12].

One of the most fully explored systems that demonstrates the superatomic character possible in cluster-assembled materials utilises both transition metal-based clusters and fullerenes, in crystal lattices that mimic stoichiometric salts [37]. Such materials have been made with a class of transition metal-chalcogenide cluster that is ligand protected, and which has a similar size to the C_{60} fullerene. The three ligand-protected clusters, $Co_6Se_8(PEt_3)_6$, $Cr_6Te_8(PEt_3)_6$, and $Ni_9Te_6(PEt_3)_8$, form different structures, with the first two crystallising in a 1:2 stoichiometry with C_{60}, in a CdI_2-type lattice, while the latter crystallises in the NaCl structure with its 1 : 1 stoichiometry.

Significantly, the clusters have been shown to be able to interact electronically to produce a magnetically ordered phase at low temperature. Band structure calculations [38] demonstrate that while the bands formed are only weakly dispersed, and most of the energy of formation is supplied by weak van der Waals interactions between the closed-shell clusters, there is nonetheless significant spin polarisation possible. Calculations of the magnetic susceptibility, based on a semi-classical Boltzmann model that is fit to the calculated density of states, provide good agreement with experimental measurements with the assumption of a small amount of p-doping. This demonstrates that the bulk properties of these types of materials – and in particular the magnetic properties – can be well understood on the basis of interacting superatomic clusters, in analogy to atom-based clusters. This provides further evidence for the utility of the superatomic model for the description and understanding of nanostructured materials.

Beyond the small number of classes of cluster-assembled materials that clearly demonstrates that superatomic states are responsible for their formation and properties, it is worth noting that the use of the superatomic concept in the study of cluster-assembled materials is not limited to the prediction of new materials on the basis of known building blocks. One of its most important uses has instead perhaps been its ability to help us understand the essential interactions in classes of existing materials that give rise to their properties and allow us to modify them. A valuable example of this approach has been the discussion of the perovskite – a technologically useful class of material for photovoltaic solar cells – as a molecular or cluster building block-based material [39]. The potential for our models of interactions between magnetic superatoms to provide answers to existing questions about the complex structures of nanostructured magnetic materials is beyond the scope of this review of current work on magnetic superatoms but remains an exciting area of investigation.

In summary, the prediction and discovery of magnetic superatoms has been a surprising scientific journey, and a fascinating one, being based as it is on the competition and cooperation between localised and delocalised electronic states and their interactions. The superatomic concept has been placed under some severe tests as it has been extended to incorporate the complex behaviour of transition metals. However, from the results presented within this chapter it should be clear that the consequent potential for advances to be made in the discovery of nanostructured (super)-magnetic materials is hugely significant within current materials science.

References

1 Khanna, S.N. and Jena, P. (1992). *Phys. Rev. Lett.* 69: 1664.
2 Reddy, B.V., Khanna, S.N., and Jena, P. (1992). *Science* 258: 1640.
3 Khanna, S.N. and Jena, P. (1995). *Phys. Rev. B.* 51: 13705.
4 Ginsberg, A.P. (1980). Magnetic exchange in transition metal complexes. *J. Am. Chem. Soc.* 102: 111–117.
5 Pantazis, D.A., Krewald, V., Orio, M., and Neese, F. (2010). Theoretical magnetochemistry of dinuclear manganese complexes: broken symmetry density functional theory investigations on the influence of bridging motifs on structure and magnetism. *Dalton Trans.* 39: 4959–4967.
6 Liu, F., Khanna, S.N., and Jena, P. (1991). *Phys. Rev. B* 43: 8179.
7 Nayak, S.K. and Jena, P. (1998). *Chem. Phys. Lett.* 289: 473.
8 Knickelbein, M.B. (2001). Experimental observation of superparamagnetism in manganese clusters. *Phys. Rev. Lett.* 86: 5255.
9 Briere, T.M., Sluiter, M.H.F., Kumar, V., and Kawazoe, Y. (2002). Atomic structures and magnetic behaviour of Mn clusters. *Phys. Rev. B* 66: 064412.
10 López-Urias, F. and Díaz-Ortiz, A. (2003). Magnetism in Mn clusters: interplay between spin ordering and structure. *Phys. Rev. B* 68: 132405.
11 Kumar, V. and Roy, D.R. (2016). Structure, electronic, and magnetic properties of Mn_n (n=2-8) clusters: a DFT investigation. *AIP Conf. Proc.* 1728: 020245.
12 Barborini, M. (2016). Neutral, anionic, and cationic manganese dimers through density functional theory. *J. Phys. Chem. A* 10: 1716–1726.
13 Reuse, F. and Khanna, S. (1995). *Chem. Phys. Lett.* 234: 77–81.
14 Louderback, J.G. , Cox, A. J., Lising, L. J., et al. (1993). *Z. Phys. D – Atoms Mol. Clust.* 26: 301–303.
15 Gilmour, J.T.A., Hammerschmidt, L., Schacht, J., and Gaston, N. (2017). Superatomic states in nickel clusters: revising the prospects for transition metal based superatoms. *J. Chem. Phys.* 147: 154307.
16 Gilmour, J.T.A. and Gaston, N. (2020). On the influence of exact exchange on transition metal superatoms. *Phys. Chem. Chem. Phys.* 22: 772–780.
17 Reveles, J.U., Clayborne, P.A., Reber, A.C. et al. (2009). Designer magnetic superatoms. *Nature Chem.* 1: 310–315.
18 Zhang, X., Wang, Y., Wang, H. et al. (2013). On the existence of designer magnetic superatoms. *J.A.C.S.* 135: 4856–4861.
19 Cheng, S., Berkdemir, C., and Castleman, A.W. (2015). Mimicking the magnetic properties of rare earth elements using superatoms. *PNAS* 112: 4941–4945.

20 Kumar, V. and Kawazoe, Y. (2001). Metal-encapsulated and cubic caged clusters of silicon. *Phys. Rev. Lett.* 87: 045503.
21 Kumar, V. and Kawazoe, Y. (2003). Metal-doped magic clusters of Si, Ge, and Sn: the finding of a magnetic superatom. *Appl. Phys. Lett.* 83: 2677–2679.
22 Khanna, S.N., Rao, B.K., Jena, P., and Nayak, S.K. (2003). *Chem. Phys. Lett.* 373: 433.
23 Ge, G., Han, Y., Wan, J.-G. et al. (2013). First-principles prediction of magnetic superatoms in 4d-transition-metal-doped magnesium clusters. *J. Chem. Phys.* 139: 174309.
24 Gilmour, J.T.A. and Gaston, N. (2019). Tunable magnetic moments in superatomic NiCu clusters. *Electron. Struct.* 1: 035003.
25 Janssens, E., Tanaka, H., Neukermans, S. et al. (2004). Electron delocalisation in Au_NX_M (X = Sc, Ti, Cr, Fe) clusters: a density functional theory and photofragmentation study. *Phys. Rev. B* 69: 085402.
26 Zheng, B.-X., Dong, D., Ling, W., and Yang, J.-X. (2014). Density functional study on the structural, electronic, and magnetic properties of 3d transiton-metl-doped Au5 clusters. *J. Phys. Chem. A* 118: 4005–4012.
27 Mollenhauer, D. and Gaston, N. (2014). A balanced procedure for the treatment of cluster ligand interactions on gold phosphine systems in catalysis. *J. Comp. Chem.* 35: 986.
28 Mollenhauer, D. and Gaston, N. (2016). Phosphine passivated gold clusters. *Phys. Chem. Chem. Phys.* 18: 29686.
29 Liu, X. and Wang, C. (2016). Growth mode and structures of magnetic Mn clusters on graphene. *RSC Adv.* 6: 64595–64604.
30 Pradhan, K., Reveles, J.U., Sen, P., and Khanna, S.N. (2010). Enhanced magnetic moments of alkali metal coated Sc clusters: new magnetic superatoms. *J. Chem. Phys.* 132: 124302.
31 Veldeman, N., Höltzl, T., Neukermans, S. et al. (2007). Experimental observation and computational identification of Sc@Cu16+, a stable dopant encapuslated copper cage. *Phys. Rev. A* 76: 011201.
32 Xu, H.-G., Wu, M.-M., Zhang, Z.-G. et al. (2011). Structural and bonding properties of $ScSi_n^-$ (n = 2 ~ 6) clusters: photoelectron spectroscopy and density functional calculations. *Chin. Phys. B* 20: 043102.
33 Höltzl, T., Lievens, P., Veszprémi, T. et al. (2009). Comment on "tuning magnetic moments by 3d transition-metal-doped Au6 clusters". *J. Phys. Chem. C* 113: 21026–21018.
34 Gonzalez-Ramirez, H., Reveles, J.U., and Gómez-Sandoval, Z. (2013). Magnetic moments on binary yttrium-alkali superatoms. *Chem. Phys. Lett.* 583: 97–102.
35 Gilmour, J.T.A. and Gaston, N. (2019). On the involvement of d-electrons in superatomic shells: the group 3 and 4 transition metals. *Phys. Chem. Chem. Phys.* 21: 8035–8045.
36 Gilmour, J.T.A. and Gaston, N. (2020). 5-Fold symmetry in superatomic scandium clusters: exploiting favourable orbital overlap to sequester spin. *Phys. Chem. Chem. Phys.* 22: 4051–4058.
37 Roy, X., Lee, C.-H., Crowther, A.C. et al. (2013). Nanoscale atoms in solid-state chemistry. *Science* 341: 157–160.
38 Hammerschmidt, L., Schacht, J., and Gaston, N. (2016). First-principles calculations of the electronic structure and bonding in metal cluster–fullerene materials considered within the superatomic framework. *Phys. Chem. Chem. Phys.* 18: 32541–32550.
39 Fang, H. and Jena, P. (2016). *J. Phys. Chem. Lett.* 7 (8): 1596–1603.

6

Atomically Precise Synthesis of Chemically Modified Superatoms

Shinjiro Takano[1] *and Tatsuya Tsukuda*[1,2]

[1] *Department of Chemistry, School of Science, The University of Tokyo, Tokyo, Japan*
[2] *Elements Strategy Initiative for Catalysts and Batteries (ESICB), Kyoto University, Kyoto, Japan*

6.1 Introduction

6.1.1 The Concept of Superatoms

Metal clusters, which are defined as assemblies of several to hundreds of metal atoms, have attracted considerable attention as promising building components of new functional materials. For the rational development of cluster-based materials by taking advantage of their specific properties, it is important to understand several fundamental issues: the evolution of the geometrical structures as a function of the number of constituent atoms (cluster size), the origin of high stability at specific cluster sizes (magic numbers), and metal-to-nonmetal transition at a critical size [1–8]. To address these issues, the structures and physicochemical properties of naked metal clusters isolated in vacuum have been extensively studied, and it has been shown that the magic stabilities and physicochemical properties of clusters of simple metals are explained by the "jellium model." [1] The spherical jellium model assumes that the valence electrons are confined within a uniformly charged spherical volume provided by constituent metal atoms. Such confinement of the electrons leads to the quantized orbitals labeled as 1S, 1P, 1D, 2S, 1F, 2P, 1G, 2D, . . . (Figure 6.1a) [1]. The spherical jellium model predicts the closure of the electronic shells at n = 2, 8, 18, 20, 34, 40, 58, . . . for Na_n in accordance with the extraordinarily high abundance of these clusters in the mass spectrum [1, 9]. This simple model also describes the electronic structures of the clusters of coinage metals (Cu, Ag, Au) having a single valence electron in the ns orbitals (n = 4, 5, 6). The sudden drop of the electron affinities (EAs) of Ag_n and Au_n at n = 8, 20, 34, and 58 (Figure 6.1b) [2] is explained by closure of the electronic shells at these sizes. These studies clearly indicated that the stability of small Au_n and Ag_n is governed by the electronic structures, although the geometrical structures of Au_8 and Au_{20} are not spherical due to the involvement of 5d orbitals in the bonding [4, 10]. Spherical clusters of simple metals can be viewed as "superatoms" due to the formation of electronic shells, as in the case of conventional atoms.

Superatoms: Principles, Synthesis and Applications, First Edition. Edited by Purusottam (Puru) Jena and Qiang Sun.
© 2022 John Wiley & Sons Ltd. Published 2022 by John Wiley & Sons Ltd.

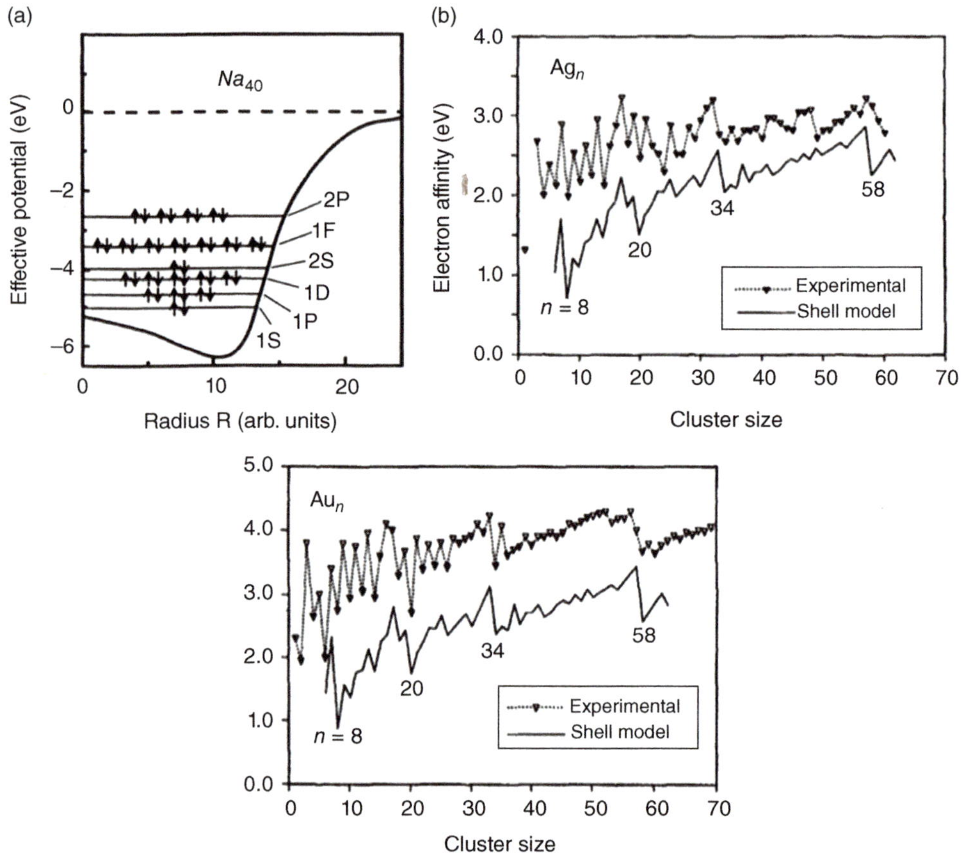

Figure 6.1 (a) Electronic shell structure of magic Na_{40} cluster. (b) EAs of Ag and Au clusters as a function of cluster size. Downward triangles and solid lines represent the experimental values and those predicted by the superatom model, respectively. *Sources:* (a) de Heer [1]. © American Physical Society. (b) Taylor et al. [2]. © American Institute of Physics.

6.1.2 Chemically Modified Au/Ag Superatoms

In order to treat Au/Ag clusters as conventional chemical compounds under ambient conditions, they must be stabilized against aggregation. One such stabilization method is to protect them by a monolayer of organic ligands. In the last two decades, an increasing number of monolayer-protected Au/Ag clusters (Au/Ag MPCs) $[(Au/Ag)_xL_y]^z$ (z = total charge) have been synthesized by using various mono-/multidentate organic ligands (L) including phosphines (R_3P/R_2P^PR_2), stibines (R_3Sb), N-heterocyclic carbenes (NHC), isocyanides (RNC), halides (X^-), cyanide (NC^-), dipyridylamides, thiolates (RS^-/S^-^S^-), selenolates (RSe^-), and alkynyls ($RC{\equiv}C^-$) [11–16]. Another advantage of protecting the Au/Ag clusters by the organic ligands is that their atomic structures including the Au/Ag cores and ligand layers can be determined by single-crystal X-ray diffractometry (SCXRD). Such structural information enables us to establish the correlation between structures and properties. Since a groundbreaking paper by Kornberg [17], the number of crystal structures of

Au/Ag MPCs has been increasing rapidly as shown in Figure 6.2 [16]. SCXRD studies have revealed that Au/Ag MPCs are composed of an Au/Ag core protected by ligand layers with a variety of motifs depending on the ligands (Table 6.1): [15] thiolates and alkynyls form unique multidentate oligomeric and network structures with Au(I) and Ag(I). The formal number of valence electrons (n^*) in $[(\mathrm{Au/Ag})_x\mathrm{L}_y]^z$ calculated according to the following scheme [18] agrees well with those predicted for naked Au/Ag superatoms:

$$n^* = Ax + By - z \tag{6.1}$$

where A and B are the number of valence electrons supplied from an individual Au/Ag atom and ligand, respectively: $A = 1$ for Au and Ag; $B = 0$ for phosphines, stibines, carbenes, and isocyanides, whereas $B = -1$ for halides, cyanides, dipyridylamides, thiolates, selenolates, and alkynyls because of their electron-withdrawing nature (Table 6.1) [15]. As shown later (Tables 6.2 and 6.3), the n^* values of Au/Ag MPCs with spherical cores agree with those predicted by the spherical jellium model, illustrating that the stability of

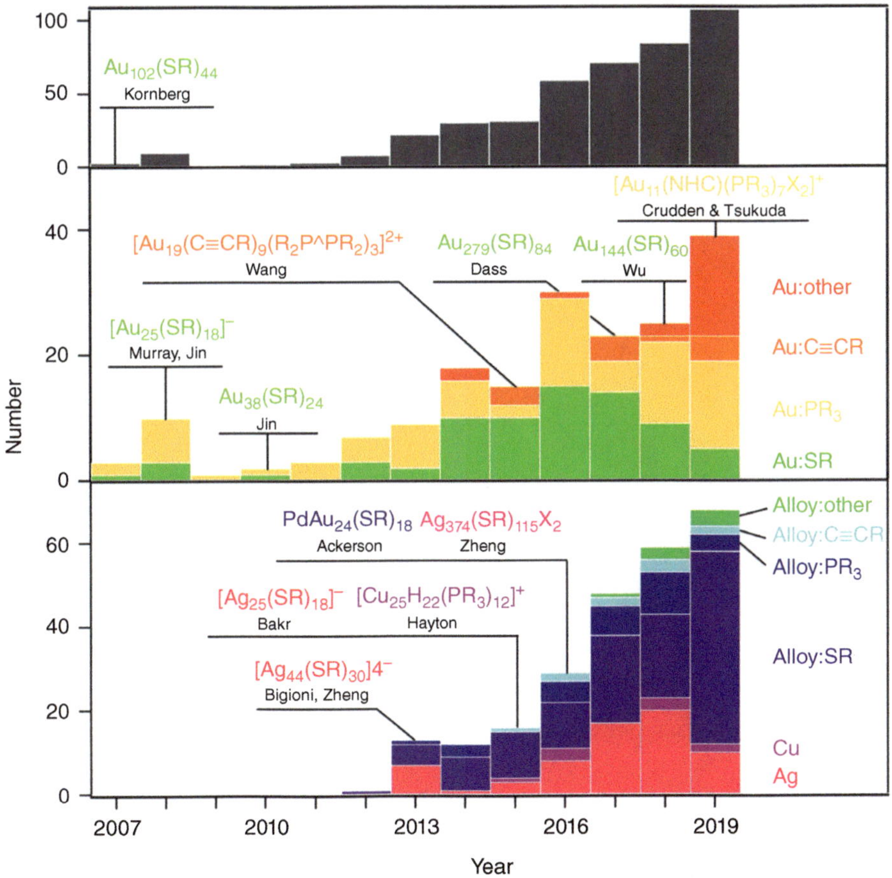

Figure 6.2 Annual numbers of crystals datasets for (a) coinage-metal MPCs, (b) Au MPCs, and (c) Ag/Cu/alloy MPCs deposited in the Cambridge Crystallographic Data Centre since 2007, with selected benchmark studies. *Source:* Hirai et al. [16]. © Royal Society of Chemistry.

Table 6.1 Interfacial structures and the *B* values (see text) of representative protective ligands.

Ligands	Interfacial structure	*B*
Phosphines		0
N-heterocyclic carbenes	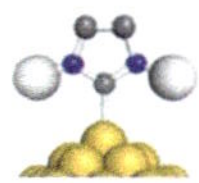	0
Halides		–1
Thiolates	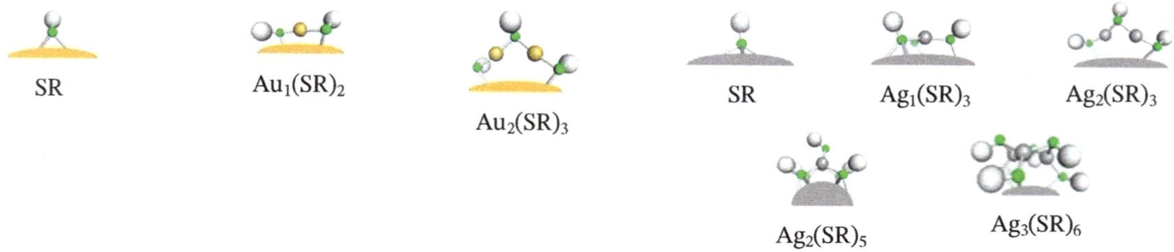 SR; $Au_1(SR)_2$; $Au_2(SR)_3$; SR; $Ag_1(SR)_3$; $Ag_2(SR)_3$; $Ag_2(SR)_5$; $Ag_3(SR)_6$	–1
Alkynyls	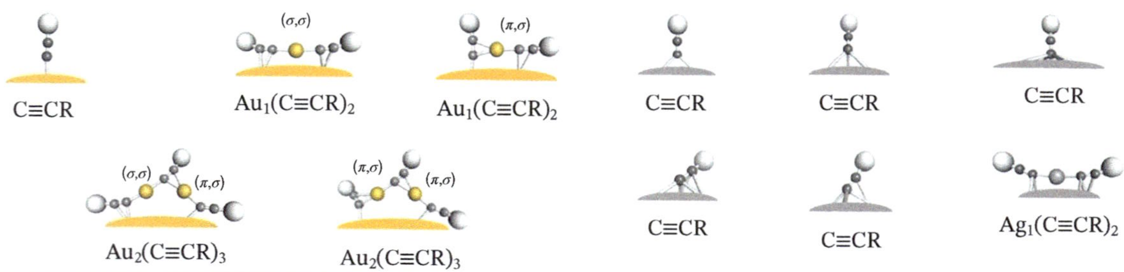 C≡CR; (σ,σ) $Au_1(C{\equiv}CR)_2$; (π,σ) $Au_1(C{\equiv}CR)_2$; C≡CR; C≡CR; C≡CR; (σ,σ) (π,σ) $Au_2(C{\equiv}CR)_3$; (π,σ) (π,σ) $Au_2(C{\equiv}CR)_3$; C≡CR; C≡CR; $Ag_1(C{\equiv}CR)_2$	–1

White balls indicate R groups in ligands. Color codes: yellow = Au; light gray = Ag; magenta = P, Sb; gray = C; blue = N; orange = X; green = S, Se.

Table 6.2 Atomic structures of representative Au MPCs.

Chemical formula	Construction scheme	Overall structure
$[Au_4(PtBu_3)_4]^{2+}$	 $Au_4(2e)$; $4{\times}PR_3$	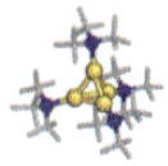
$[Au_{11}(PPh_3)_8Cl_2]^{+}$	 $Au_{11}(8e)$; $2{\times}Cl^-$; $8{\times}PR_3$	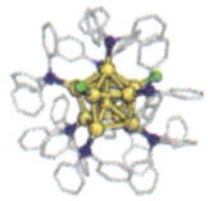
$[Au_{13}(dppe)_5Cl_2]^{3+}$	 $Au_{13}(8e)$; $2{\times}Cl^-$; $10{\times}PR_3$	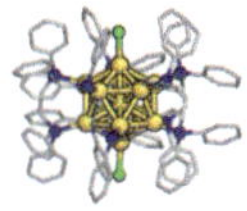
$[Au_{13}(NHC)_9Cl_3]^{2+}$	 $Au_{13}(8e)$; $3{\times}Cl^-$; $9{\times}NHC$	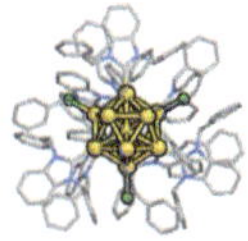
$[Au_{25}(SR)_{18}]^-$	 $Au_{13}(8e)$; $6{\times}[Au_2(SR)_3]^-$	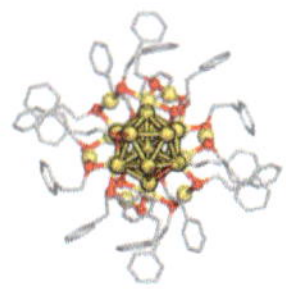
$[Au_{25}(C{\equiv}CR)_{18}]^-$	 $Au_{13}(8e)$; $6{\times}[Au_2(C{\equiv}CR)_3]^-$	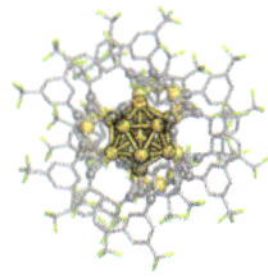
$Au_{102}(SR)_{44}$	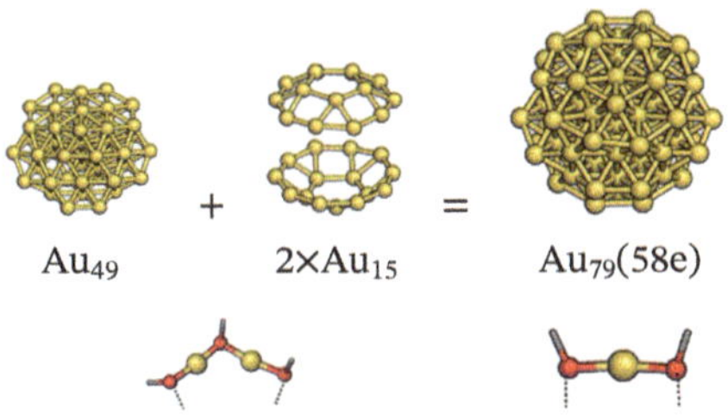 Au_{49} + $2{\times}Au_{15}$ = $Au_{79}(58e)$; $2{\times}[Au_2(SR)_3]^-$; $19{\times}[Au(SR)_2]^-$	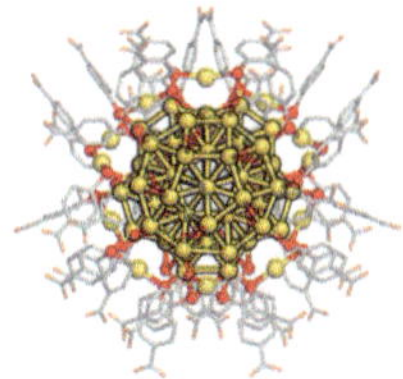
$Au_{144}(SR)_{60}$	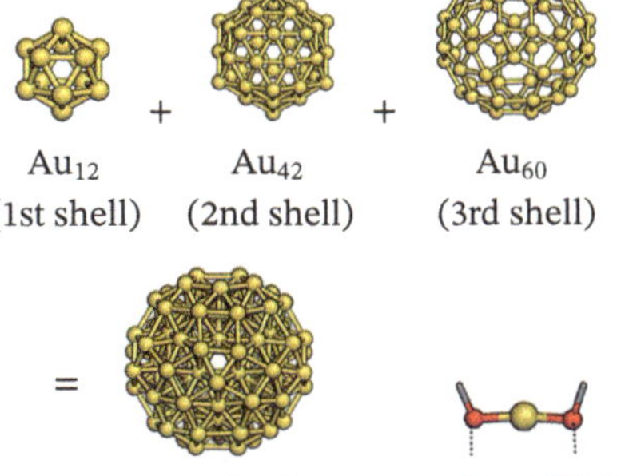 Au_{12} (1st shell) + Au_{42} (2nd shell) + Au_{60} (3rd shell) = $Au_{114}(84e)$; $30{\times}[Au(SR)_2]^-$	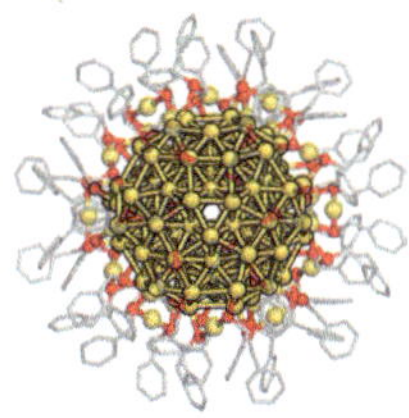

Dppe = $Ph_2PCH_2CH_2PPh_2$. Color codes: yellow = Au; dark blue = P; gray = C; light blue = N; red = S; green = Cl; light green = F; pink = O. Organic residues in the overall structure are depicted as sticks.

Table 6.3 Atomic structures of representative Ag MPCs.

Chemical formula	Construction scheme	Overall structure
$[Ag_{25}(SR)_{18}]^-$		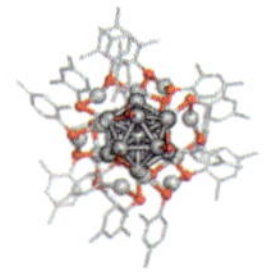
$[Ag_{29}(BDT)_{12}(PPh_3)_4]^{3-}$	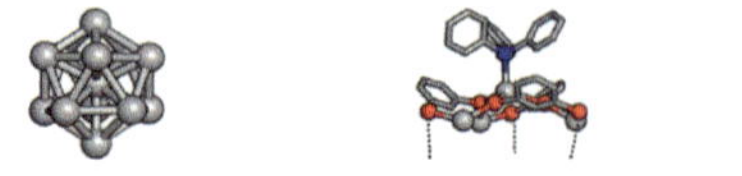	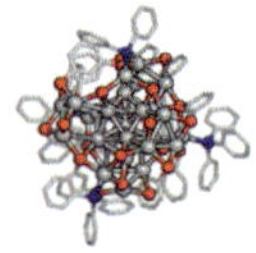
$[Au_{44}(SR)_{30}]^{4-}$	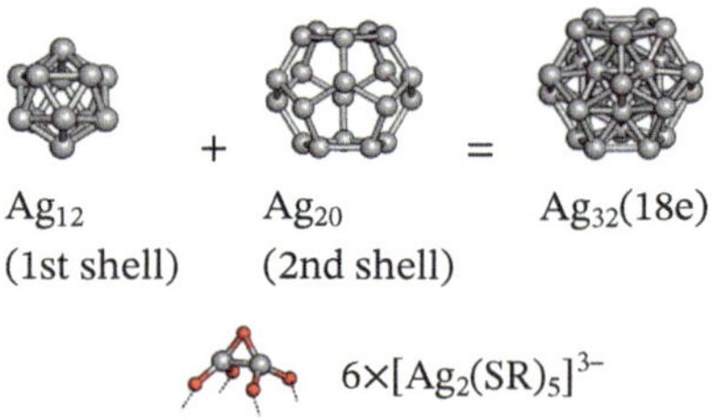	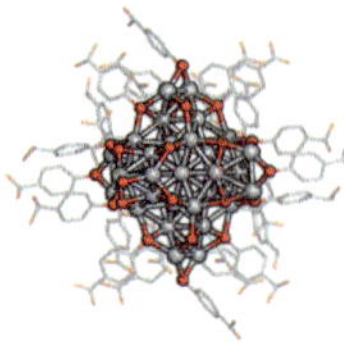
$[Ag_{112}(C{\equiv}CR)_{51}Cl_6]^{3-}$	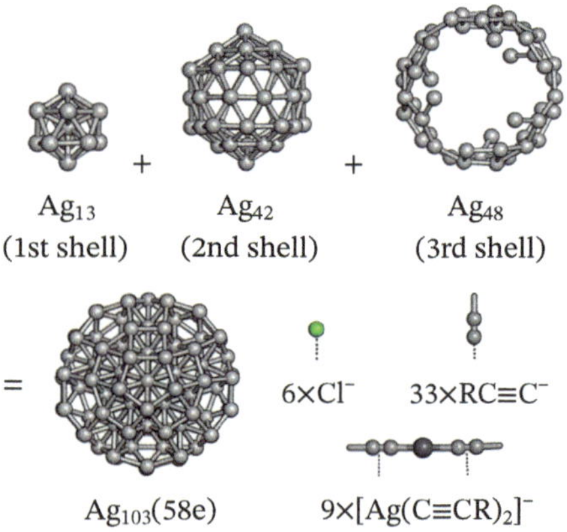	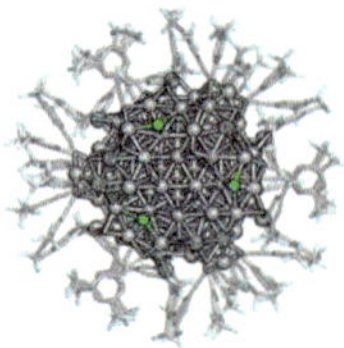
$[Ag_{137}(SR)_{64}Cl_3]^-$	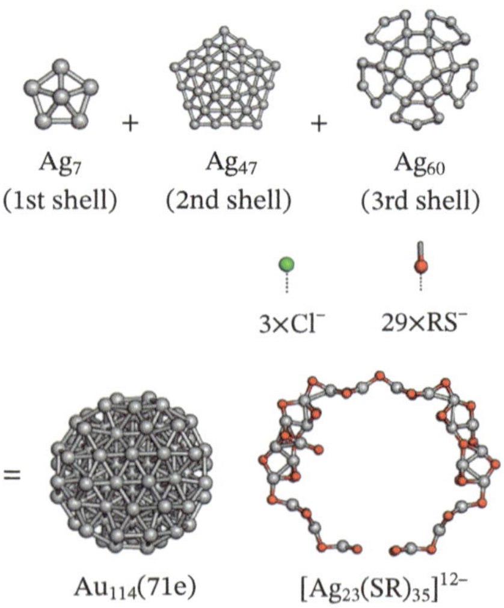	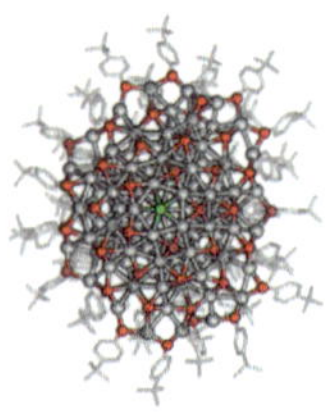

BDT = 1,3-SC_6H_4S. Color codes: silver = Ag; dark blue = P; gray = C; red = S; green = Cl; light green = F; pink = O. Organic residues in the overall structure are depicted as sticks.

chemically synthesized Au/Ag MPCs is governed by the closed electronic structures. The successful application of Eq. (6.1) for counting the number of valence electrons in the core suggests that Au/Ag MPCs can be viewed as chemically modified superatoms, in which the $\mathbf{(Au/Ag)}_X(n^*e)$ cores ($X \leq x$) are covered by the ligand layers ("divide and protect" concept) [20, 21].

The above discussion indicates that the number of valence electrons is a key descriptor of the stability and properties of the chemically modified Au/Ag superatoms as in the case of conventional atoms. Nonetheless, superatoms have additional parameters that distinguish them from conventional atoms: surface modification, composition, atomic packing, and size. The jellium potential that confines a given number of valence electrons can be deformed by changing the protecting ligands, by doping foreign elements, and by changing the size and shape [1, 15]. As a result, the physicochemical properties of superatoms can be tuned over a wider range not only by the number of valence electrons but also by these parameters. In this regard, chemically modified Au/Ag superatoms hold great promise as artificial elements at the nanoscale. In Section 6.2, we summarize how the electronic structures of $\mathbf{(Au/Ag)}_X(n^*e)$ cores are affected by structural parameters such as size, shape, and composition and rationalize their correlations within the framework of the spherical and modified jellium models to grasp the general trend of the structural effects. Section 6.3 introduces examples of synthesis methods to control the structural parameters of Au/Ag superatoms with atomic and molecular precision. The aim of this chapter is to provide a basic approach to controlling the electronic structures of chemically modified Au/Ag superatoms through atomically precise nanochemistry.

6.2 Electronic Structures of Chemically Modified Superatoms

6.2.1 Size Effects

This section introduces the geometric structures of representative Au/Ag MPCs having nearly spherical cores determined by SCXRD and then discusses the effect of size on the electronic structures. Table 6.2 shows the chemical formulas and geometric structures of magic Au MPCs having the cores of $\mathbf{Au_4}$, $\mathbf{Au_{11}}$, $\mathbf{Au_{13}}$, $\mathbf{Au_{79}}$, and $\mathbf{Au_{114}}$ [17, 22–30]. Obviously, the Au cores take nonface-centered cubic (FCC) atomic packing, such as icosahedron (Ih) and decahedron (Dh) (Table 6.2), in contrast to bulk Au. The core structures of $\mathbf{Au_4}$, $\mathbf{Au_{11}}$, and $\mathbf{Au_{13}}$ are also significantly different from the planar structures of naked Au_n clusters with comparable sizes [10]. A defective icosahedral $\mathbf{Au_{11}}$ core of $[Au_{11}(PPh_3)_8Cl_2]^+$ was also found in $Au_{11}(PPh_3)_7Cl_3$ [27] and $[Au_{11}(NHC)(PPh_3)_7Cl_2]^+$ [31]. A perfect icosahedral $\mathbf{Au_{13}}$ core is formed in $[Au_{13}(dppe)_5Cl_2]^{3+}$ (dppe = $Ph_2PCH_2CH_2PPh_2$) [26], $[Au_{13}(NHC)_9Cl_3]^{2+}$ [30], $[Au_{25}(SR)_{18}]^-$ [21, 23, 24], and $[Au_{25}(C{\equiv}CR)_{18}]^-$: [29] thiolates and alkynyls form bidentate $Au_2L_3^-$ complexes on the surface. The ubiquitous formation of $\mathbf{Au_{11}}$ and $\mathbf{Au_{13}}$, regardless of the ligands, indicates their inherent stability. The $\mathbf{Au_{79}}$ core in $Au_{102}(SR)_{44}$ consists of a Marks decahedral Au_{49} sandwiched by two bowl-shaped Au_{15} units and is protected by 2 $Au_2(SR)_3$ and 19 $Au(SR)_2$ motifs, called as "staples" [17]. The $\mathbf{Au_{114}}$ core of $Au_{144}(SR)_{60}$ is composed of three shells (a hollow Au_{12} icosahedron + an Au_{44} icosahedral shell + an icosidodecahedral Au_{60} shell) and is protected by 30 $Au(SR)_2$ staples [25, 28].

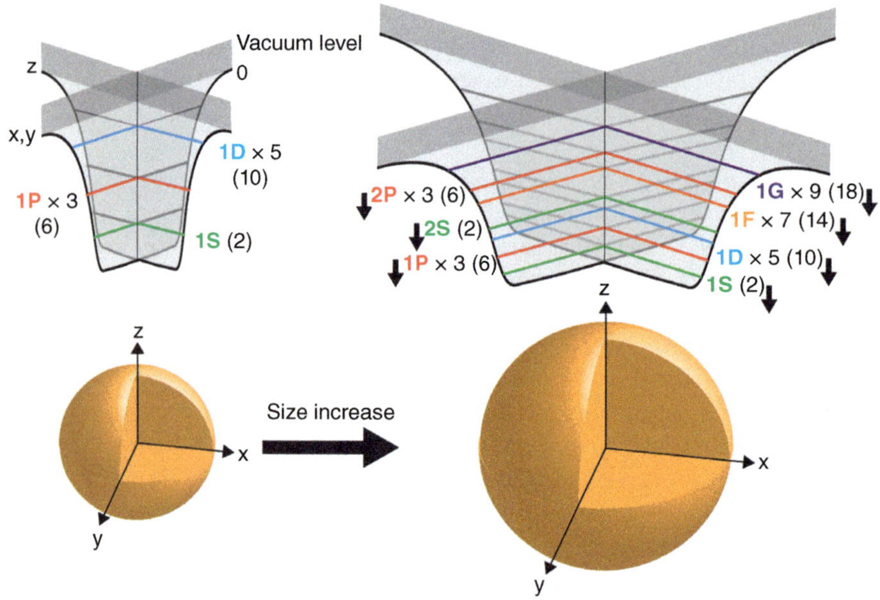

Figure 6.3 Schematic representation of size effects on spherical jellium potential. *Source:* Omoda et al. [15]. © John Wiley & Sons.

The n^* values for $\mathbf{Au_4}$(2e), $\mathbf{Au_{11}}$(8e), $\mathbf{Au_{13}}$(8e), and $\mathbf{Au_{79}}$(58e) calculated by Eq. (6.1) agree with those predicted for spherical superatoms. Theoretical calculations on $Au_{11}(PH_3)_7Cl_3$, $Au_{11}(PH_3)_7(SMe)_3$, $[Au_{13}(NHC^{Bn})_9Cl_3]^{2+}$ (NHC^{Bn} = 1,3-dibenzylbenzimida zolin-2-ylidene), $[Au_{25}(SMe)_{18}]^-$, and $Au_{102}(S\text{-}4\text{-}C_6H_4CO_2H)_{44}$ [18, 21, 30] predicted that the highest occupied molecular orbitals (HOMOs) of $\mathbf{Au_{11}}$(8e), $\mathbf{Au_{13}}$(8e), and $\mathbf{Au_{79}}$(58e) correspond to 1P, 1P, and 1G, respectively, as expected. The energy levels and shape of superatomic orbitals (SOs) of $\mathbf{Au_{13}}$(8e) are schematically shown in Figure 6.3 [15]. Gas-phase photoelectron spectroscopy on $[Au_{25}(S\text{-}nC_{12}H_{25})_{18}]^-$ has revealed that the 1P SOs are located at 2.2 eV below the vacuum level [32]. Therefore, the $\mathbf{Au_{11}}$(8e), $\mathbf{Au_{13}}$(8e), and $\mathbf{Au_{79}}$(58e) cores can be viewed as spherical superatoms with closed electronic structures. However, the n^* value for the larger core of $\mathbf{Au_{114}}$(84e) (Table 6.2) does not agree with any of the numbers predicted for the spherical superatoms. This discrepancy suggests that the stability of $\mathbf{Au_{114}}$(84e) is not governed by the electronic structure, probably because the energy gain by closure of the electronic shells in $\mathbf{Au_{114}}$ becomes trivial compared to the thermal energy. The highly symmetrical geometric structure of the $\mathbf{Au_{114}}$(84e) core suggests that the stability is governed by the closure of the geometric shells rather than the electronic shells. A transition of the origin of the magic stability from electronic shell closure to geometric shell closure was also observed in naked Na_n clusters at n ~2000 [33].

Synthesis of a series of atomically defined Au MPCs provides a novel opportunity to investigate the critical size at which metal-to-nonmetal transition occurs. Figure 6.4a shows the size-dependent optical spectra of the filmed samples of a series of $Au_x(SR)_y$ [34].

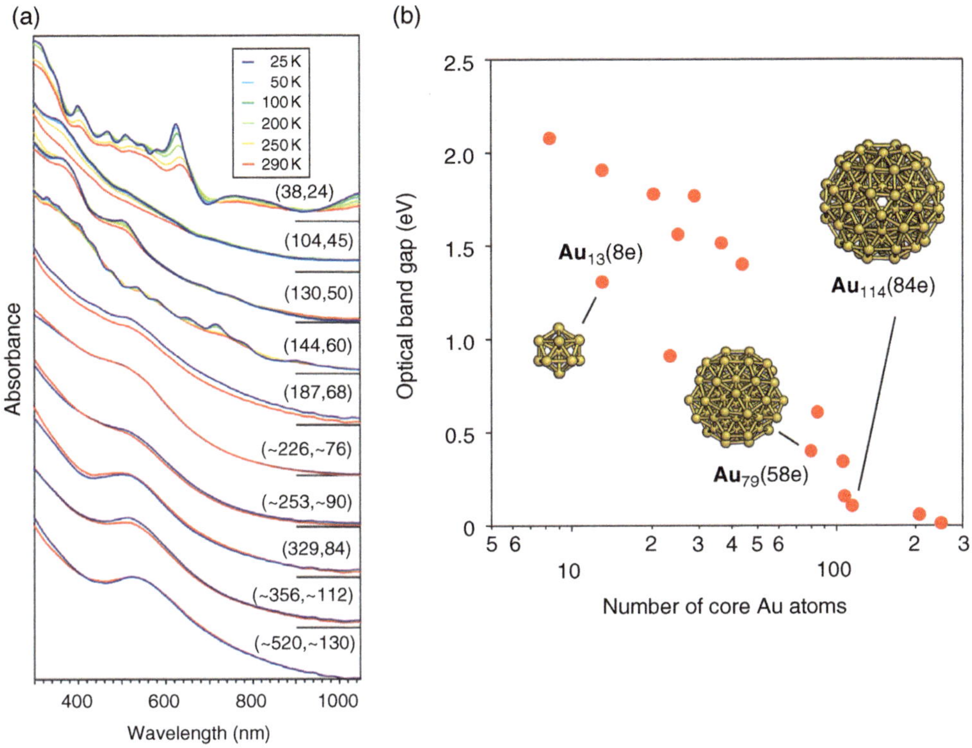

Figure 6.4 Size-dependent (a) optical absorption profiles and (b) optical band gaps of Au MPCs. The numbers in the parentheses (x, y) in panel (a) correspond to the compositions of $Au_x(SR)_y$. *Sources:* (a) Negishi et al. [34].© American Chemical Society. (b) Zhou et al. [35]. © American Chemical Society.

The $Au_x(SR)_y$ clusters smaller than $Au_{144}(SR)_{60}$ exhibit structured profiles, whereas those larger than $Au_{187}(SR)_{68}$ exhibit a localized surface plasmon resonance (LSPR) band at ~520 nm. At lower temperature, the humps and peaks of the clusters smaller than $Au_{144}(SR)_{60}$ become more prominent, in contrast the profiles for those larger than $Au_{187}(SR)_{68}$ remain unchanged [34]. These results indicate that the $Au_x(SR)_y$ clusters smaller than $Au_{144}(SR)_{60}$ are molecular-like in terms of electronic structure, while those larger than $Au_{187}(SR)_{68}$ are metallic. Optical HOMO–LUMO (HL) gaps of $Au_x(SR)_y$ estimated from the spectral onsets are plotted as a function of core size in Figure 6.4b [35]. The optical HL gap energy (ΔE_o) decreases in the order of $\mathbf{Au_4}$(2e), $\mathbf{Au_{11}}$(8e), $\mathbf{Au_{13}}$(8e), and $\mathbf{Au_{79}}$(58e) and becomes nominally zero at $\mathbf{Au_{107}}$(81e): reopening of the HL gap, that was reported for the magic Mg_n^- clusters (n = 20, 35) [36], was not observed in larger Au cores. This suggests that $\mathbf{Au_{107}}$(81e) corresponds to the smallest particles with metallic properties. However, the critical size for the nonmetal-to-metal transition depends on the evaluation criteria and the ligands used. Negishi reported that the transition takes place between $\mathbf{Au_{114}}$(84e) and $\mathbf{Au_{153}}$(119e) based on the appearance of an LSPR band [34]. Jin reported that a transition occurs between

$\mathbf{Au_{206}}$(166e) and $\mathbf{Au_{249}}$(195e) based on the lifetime of the photoexcited states [35]. The HL gap determined by electrochemical measurement (ΔE_e) and density functional theory (DFT) calculation also shows similar size-dependent behavior [37]. The reduction of the HL gap with size is understandable in the framework of the spherical jellium model. An increase in the size leads to an increase in the volume of the potential while retaining the depth (Figure 6.3). Such modulation of the potential results in the stabilization of the SOs and in the reduction of the HL gap.

Recently, an increasing number of atomically precise Ag MPCs have been reported (Figure 6.2) [11–16]. Table 6.3 shows the crystal structures of representative Ag MPCs having nearly spherical cores [19, 38–43]. The $[Ag_{25}(SR)_{18}]^-$ cluster has a similar motif with $[Au_{25}(SR)_{18}]^-$ and is composed of an icosahedral $\mathbf{Ag_{13}}$ core and six units of $Ag_2(SR)_3$ staples [42]. $[Ag_{29}(BDT)_{12}(PPh_3)_4]^{3-}$ (BDT = 1,3-SC_6H_4S) has an icosahedral $\mathbf{Ag_{13}}$ core capped by four tridentate units of $Ag_4(BDT)_3(PPh_3)$ [41]. $[Ag_{44}(SR)_{30}]^{4-}$ consists of an $\mathbf{Ag_{32}}$ core and six tetradentate $Ag_2(SR)_5$ units on the surface: [39, 40] the $\mathbf{Ag_{32}}$ core is composed of a hollow icosahedral Ag_{12} core and a dodecahedral Ag_{20} shell. $[Ag_{112}(C{\equiv}CPh\text{-}3,5\text{-}(CF_3)_2)_{51}Cl_6]^{3-}$ and $[Ag_{137}(TBBT)_{64}Cl_3]^-$ (TBBT = 4-$(CH_3)_3CC_6H_4S$) have an imperfect anti-Mackay icosahedral $\mathbf{Ag_{103}}$ core and a decahedral $\mathbf{Ag_{114}}$ core, respectively [19, 43]. The n^* values calculated for the $\mathbf{Ag_{13}}$, $\mathbf{Ag_{32}}$, and $\mathbf{Ag_{103}}$ cores (Table 6.3) are 8, 18, and 58, respectively. The high stability of the $\mathbf{Ag_{13}}$(8e), $\mathbf{Ag_{32}}$(18e), and $\mathbf{Ag_{103}}$(58e) cores can be explained by the closed electron configuration: $(1S)^2(1P)^6$, $(1S)^2(1P)^6(1D)^{10}$, and $(1S)^2(1P)^6(1D)^{10}(2S)^2(1F)^{14}(2P)^6(1G)^{18}$, respectively. In contrast, the n^* value for the $\mathbf{Ag_{114}}$ core is 71 and does not agree with those predicted for the spherical jellium model. This estimate suggests that the stabilities of the thiolate-protected $\mathbf{Ag_n}$ superatoms with $n > {\sim}100$ are not governed by the closure of the electronic structures. The appearance of the LSPR bands at ~450 mm in the optical spectra of $\mathbf{Ag_{114}}$(71e) and larger MPCs [19] suggests that metal-to-nonmetal transition takes place at n ~100 although the critical size may depend on the nature of the ligands.

A unique feature of Ag MPCs as compared to Au MPCs is that the protecting layers contain higher populations of electronegative ligands (such as mono-thiolates, di-thiolates, and alkynyl) that form 2D network and 3D caged structures with Ag(I) ions. As a result, the total charge of Ag MPCs is more negative than that of Au MPCs. For example, $[Ag_{29}(BDT)_{12}(PPh_3)_4]^{3-}$ is formally composed of $\mathbf{Ag_{13}}(8e)^{5+}$ and four units of $[Ag_4(BDT)_3(PPh_3)]^{2-}$. This increase of negative charge on the ligand layers destabilizes the HOMO levels. The electron-binding energy of $\mathbf{Ag_{13}}$(8e) of $[Ag_{29}(BDT)_{12}(PPh_3)_4]^{3-}$ was determined to be 1.0 eV [44], as much as 1 eV smaller than that of $[Ag_{25}(SR)_{18}]^-$ (2.02 ± 0.01 eV [45]). More surprisingly, photoelectron spectroscopy (PES) on $[Ag_{44}(SC_6H_3F_2)_{30}]^{4-}$ revealed that adiabatic electron affinity (AEA) of $[Ag_{44}(SC_6H_3F_2)_{30}]^{3-}$ is negative (−0.55 ± 0.05 eV) [46]. These results indicate that the 1P SOs of the $\mathbf{Ag_{13}}$ core of $[Ag_{29}(BDT)_{12}(PPh_3)_4]^{3-}$ and $[Ag_{44}(SC_6H_3F_2)_{30}]^{4-}$ are significantly upshifted by the presence of anionic ligands and suggest that they are unstable against oxidation in solution under ambient light exposure. However, the presence of a repulsive coulomb barrier (RCB) with respect to the electron detachment (Figure 6.5) may contribute to the stability against the electron loss [47–49]. The heights of RCB of $[Ag_{29}(BDT)_{12}(PPh_3)_4]^{3-}$ and $[Ag_{44}(SC_6H_3F_2)_{30}]^{4-}$ in the gas phase were estimated by PES to be 1.7 ± 0.1 eV and 2.7 eV [44, 46], although the height may be modified by solvation.

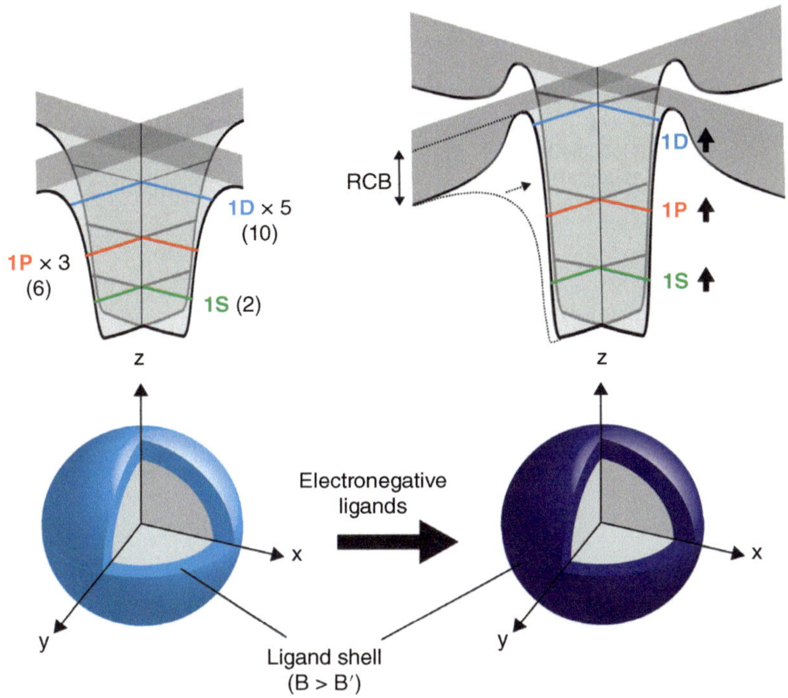

Figure 6.5 Schematic representation of the effects of anionic charge on the ligands on the spherical jellium potential of the $\mathbf{Ag_{13}}$(8e) superatomic core.

6.2.2 Composition Effects

Doping of a foreign element is a promising approach to tune the physicochemical and redox properties of Au/Ag superatoms [13–16]. In order to simplify the discussion, we focus on doping to Ih $\mathbf{Au_{13}}$(8e) and $\mathbf{Ag_{13}}$(8e) superatoms. Table 6.4 lists the crystal structures of selected examples [50–52]. The preferential dopant sites in the $\mathbf{Au_{13}/Ag_{13}}$ superatoms depend on the group of elements: Group 9 (Ir) and 10 (Pd, Pt) elements exclusively occupy the central position to form $\mathbf{Ir/Pd/Pt@Au_{12}}$ and $\mathbf{Pd/Pt@Ag_{12}}$ [50–53], whereas group 11 (Ag) and 12 (Ag, Cd, Hg) dopants occupy the surface position to form $\mathbf{Au@Ag_{\mathit{x}}Au_{12-\mathit{x}}}$ and $\mathbf{Au@Cd/HgAu_{11}}$ [52, 54].

The effect of doping on the electronic structures can be explained by considering the deformation of the jellium potential. The number of electrons provided by a single atom (A value in Eq. 6.1) to the $\mathbf{Au_{13}}$ superatomic core is dependent on the dopant elements: $A = -1$, 0, +1, and +2 for group 9, 10, 11, and 12 element, respectively. According to the two-step spherical jellium model (Figure 6.6) [55–58], the replacement of Au by a group 9 or 10 atom reduced the effective positive charge of the core, leading to the upshift of the SOs. This simple prediction on the energy shift of the SOs has been verified by differential pulse voltammetry (DPV) [53] and gas-phase PES [59]. DPV measurements revealed that the 1P SOs of $\mathbf{Au_{13}}$(8e) of $[Au_{13}(dppe)_5Cl_2]^{3+}$ are destabilized by ~0.4 eV upon doping a single Pt or Pd atom and further by ~0.5 eV by doping an Ir atom [53]. The preference of the six-electron states in $\mathbf{Pd@Au_{12}}$ of $PdAu_{24}(SR)_{18}$ and $\mathbf{Pt@Au_{12}}$ of $PtAu_{24}(SR)_{18}$ is ascribed to

Table 6.4 Atomic structures of representative MAu_{12} and MAg_{12} superatoms.

Chemical formula	Construction scheme		Overall structure
$MAu_{24}(SR)_{18}$ M = Pd, Pt	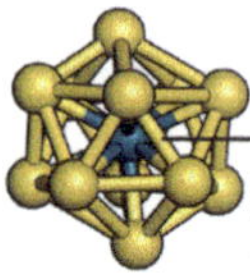 $M@Au_{12}(6e)$	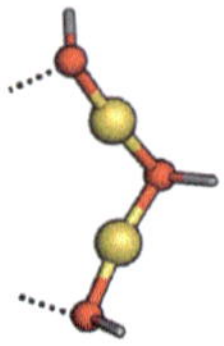 $6\times[Au_2(SR)_3]^-$	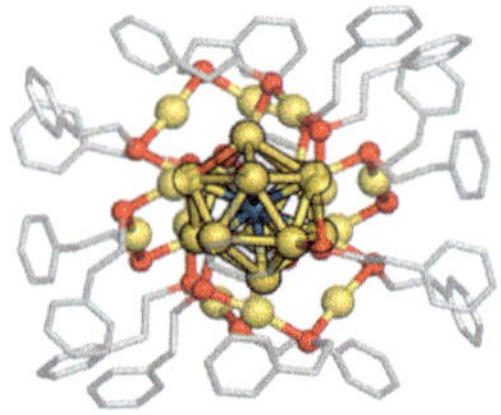
$MAu_{24}(SR)_{18}$ M = Cd, Hg	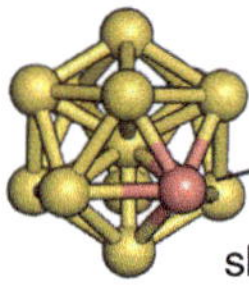 $Au@MAu_{11}(8e)$	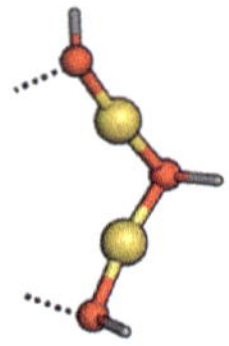 $6\times[Au_2(SR)_3]^-$	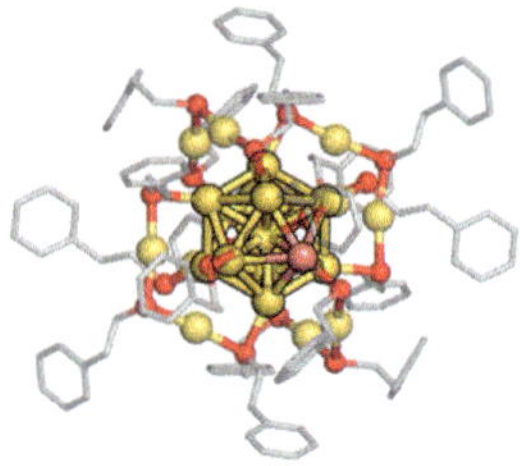
$[MAg_{24}(SR)_{18}]^{2-}$ M = Pd, Pt	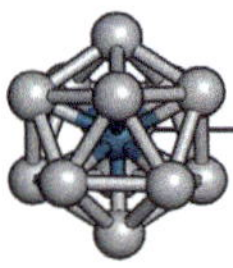 $M@Ag_{12}(8e)$	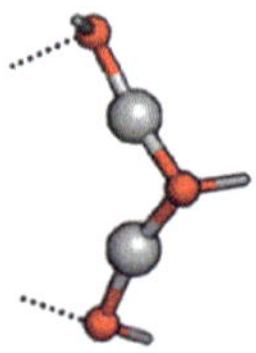 $6\times[Ag_2(SR)_3]^-$	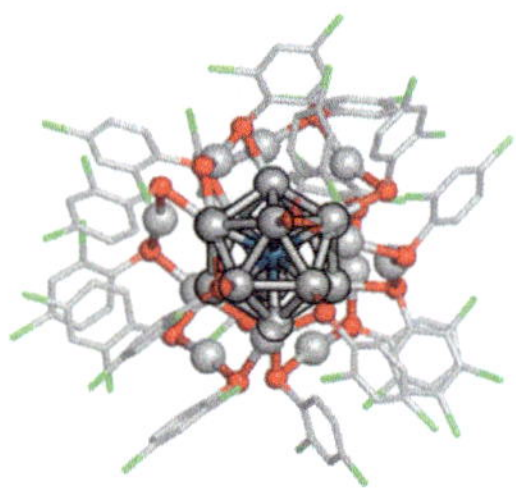

Color codes: yellow = Au; silver = Ag; blue = Pd, Pt; pink = Cd, Hg; red = S; green = Cl. Organic residues in the overall structure are depicted as sticks.

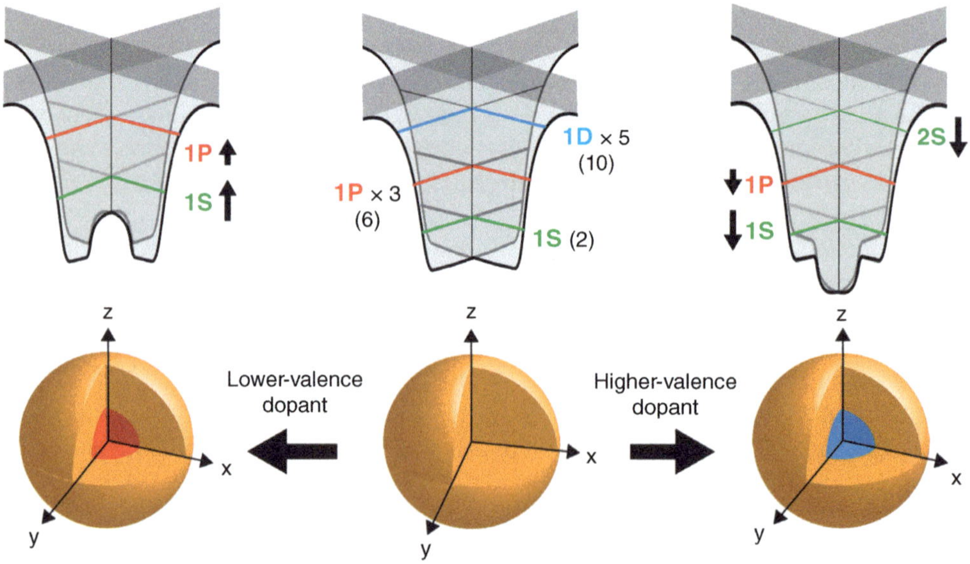

Figure 6.6 Schematic representation of heterometal doping to Au superatom. *Source:* Omoda et al. [15]. © John Wiley & Sons.

the upshift of the 1P SOs (Figure 6.6, left) [60]. Spontaneous and efficient electron transfer (ET) from **Pt@Au$_{12}$**(8e) of $[PtAu_{24}(SR)_{18}]^{2-}$ to **Pt@Au$_{12}$**(6e) of $[PtAu_{24}(SR)_{18}]^0$ to form open-shell **Pt@Au$_{12}$**(7e) also suggests the instability of **Pt@Au$_{12}$**(8e) due to the upshift of the 1P SOs [61]. PES on $[Ag_{25}(2,4\text{-DMBT})_{18}]^-$ (2,4-DMBT = 2,4-$(CH_3)_2C_6H_3S$) and $[MAg_{24}(2,4\text{-DMBT})_{18}]^{2-}$ (M = Pt, Pd) revealed that the electron-binding energy of the **Ag$_{13}$** core is significantly reduced by replacing the central Ag^+ atom with a Pt or Pd atom [59]. The higher-energy shift of the 1P SOs was ascribed to the weaker binding of valence electrons in $M@(Ag^+)_{12}$ (M = Pd, Pt) compared to that in $Ag^+@(Ag^+)_{12}$ due to the reduction in formal charge of the core potential (Figure 6.6, left) and the upward shift of the apparent vacuum level by the presence of a RCB for $M@(Ag^+)_{12}$ (Figure 6.5).

In contrast, the SOs are energetically stabilized by doping Cd or Hg due to the increase in the effective positive charge of the core although the simple model in Figure 6.6 is not directly applicable because the Cd or Hg dopant is located on the superatom surface [52]. DPV results showed that the 1P SOs of **Au@ZAu$_{11}$**(8e) of $[Au_{24}Z(SR)_{18}]^0$ (Z = Cd or Hg) were stabilized by ~0.5 eV as compared to those of **Au$_{13}$**(8e) [52]. The **Au@ZAu$_{11}$**(8e) superatom is polarized due to the deformation of the 1S and 1P SOs by the surface doping. These results provide a simple guiding principle for the engineering of SOs.

A recent theoretical study proposed that an H atom behaves as an Au atom in $Au_{25}(SR)_{18}$ and $PtAu_{24}(SR)_{18}$ and that the H atom contributes its 1s electron to the superatomic electron count [62] while H behaves as an electron-withdrawing ligand in Ag clusters [63–65]. An analogy between Au and H has also been proposed for bare Au clusters: experimental and theoretical studies demonstrated that HAu_n and Au_{n+1} are electronically and geometrically equivalent [66]. Tsukuda reported mass spectrometric detection of H-doped superatoms **HAu$_9$**(8e) and **HPdAu$_8$**(8e) by the reaction of $[Au_9(PPh_3)_8]^{3+}$ and $[PdAu_8(PPh_3)_8]^{2+}$, respectively, with $NaBH_4$ [67–69]. DFT studies on the model cluster $[HAu_9(PMe_3)_8]^{2+}$ predicted that the H^- is adsorbed onto the central unligated Au atom of the **Au$_9$**(6e) core of $[Au_9(PMe_3)_8]^{3+}$ to form **HAu$_9$**(8e) (Figure 6.7a) [67]. This adsorption of H^- induces spherization of the **HAu$_9$**(8e) core and results in degeneration of three 1P SOs (Figure 6.7a). The hydrogen in the **HPdAu$_8$**(8e) core of $[HPdAu_8(PPh_3)_8]^{2+}$ exhibits similar features (Figure 6.7b) except for the possibility of another isomer in which the H atom occupies the lateral Pd–Au bond [68]. The $[Au_{20}(PPh_3)_{12}H_3]^{3+}$ cluster has an **Au$_{20}$H$_3$**(20e) core in which three H atoms are bonded to the bridging sites between the Au_9 and Au_{11} units [70].

6.2.3 Shape Effects

One of the unique features of superatoms is that their electronic structures are strongly coupled with their geometric structures (shape and atomic packing). We first discuss the correlation between electronic structures of two types of nonspherical superatom: oblate and prolate cores. The **Au$_7$**, **Au$_8$**, and **Au$_9$** cores of $[Au_7(PPh_3)_7]^+$, $[Au_8(PPh_3)_7]^{2+}$, $[Au_8(PPh_3)_8]^{2+}$, $[Au_8(BINAP)_3(PPh_3)_2]^{2+}$ (BINAP = 2,2′-bis(diphenylphosphino)-1,1′-binaphthyl), and $[Au_9(PR_3)_8]^{3+}$ are viewed as oblate superatoms (Table 6.5) [71–74]. The n^* values of all of these cores are 6, which do not meet the shell-closing numbers of spherical superatoms. The electronic stability of **Au$_7$**(6e), **Au$_8$**(6e), and **Au$_9$**(6e) cores can be explained by distortion of the spherical jellium potential into an oblate. When the spherical jellium is

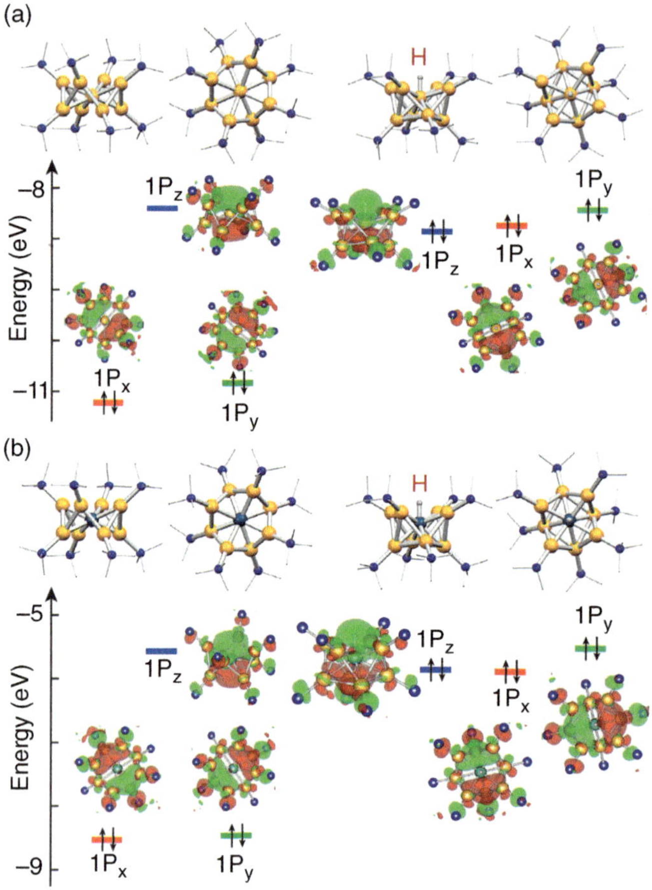

Figure 6.7 Optimized structures and the corresponding energy diagrams of (a) $[Au_9(PMe_3)_8]^{3+}$ and $[HAu_9(PMe_3)_8]^{2+}$ and (b) $[PdAu_8(PMe_3)_8]^{2+}$ and $[HPdAu_8(PMe_3)_8]^{+}$. Color codes: yellow = Au; blue = P; dark green = Pd; white = H. Methyl groups are depicted as sticks. Kohn-Sham orbitals are depicted with isodensity values at 0.025e levels. *Source:* Takano et al. [69]. © American Chemical Society.

compressed along the z-axis, the $1P_z$ SO is destabilized more than the other degenerate $1P_x$ and $1P_y$ SOs. Consequently, three 1P orbitals originally degenerated in a spherical potential split into two subsets $1P_{x,y}$ and $1P_z$ (Figure 6.8). As a result, the electronic shell of an oblate superatom is closed with six electrons. The icosahedral $\mathbf{Au_{13}}$(8e) core of $[Au_{25}(SR)_{18}]^-$ becomes distorted to an oblate shape upon oxidation due to the Jahn-Teller effect [75]. In contrast, the $\mathbf{Au_6}$, $\mathbf{Au_7}$, $\mathbf{Au_{20}}$, and $\mathbf{Au_{23}}$ cores of $[Au_6(dppp)_4]^{2+}$ (dppp = $Ph_2PCH_2CH_2CH_2PPh_2$) [76], $Au_{22}(ETP)_{18}$ (ETP = C_4SH_3-C≡C) [77], $[Au_{20}(PPhPy_2)_{10}Cl_4]^{2+}$ [78], and $Au_{38}(PET)_{24}$ (PET = $PhCH_2CH_2S$) [79] (Table 6.6) can be viewed as prolate superatoms. The calculated n^* values for $\mathbf{Au_6}$ and $\mathbf{Au_7}$ are 4, whereas those for $\mathbf{Au_{20}}$ and $\mathbf{Au_{23}}$ are 14. Both of these n^* values do not agree with the shell-closing numbers of spherical superatoms. The electronic shell closing at n^* = 4 and 14 can be explained by

Table 6.5 Atomic structures of Au MPCs with oblate superatomic cores.

Chemical formula	Core structure	Overall structure
$[Au_7(PPh_3)_7]^+$	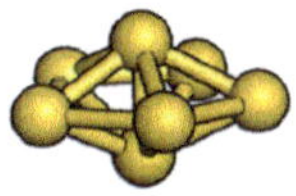 $Au_7(6e)$	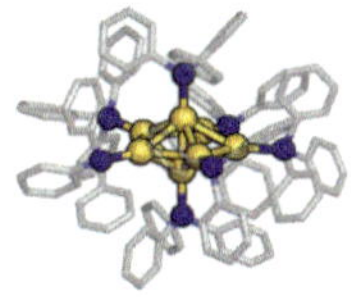
$[Au_8(PPh_3)_7]^{2+}$	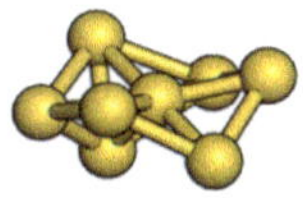 $Au_8(6e)$	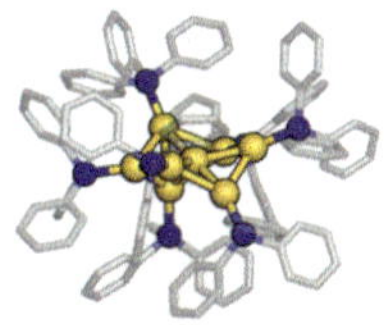
$[Au_8(PPh_3)_8]^{2+}$	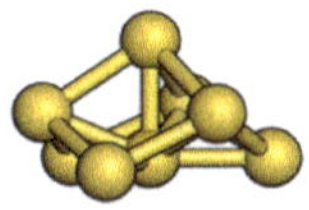 $Au_8(6e)$	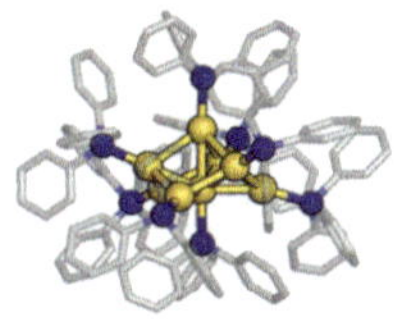
$[Au_8(R\text{-BINAP})_3(PPh_3)_2]^{2+}$	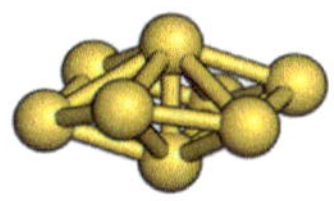 $Au_8(6e)$	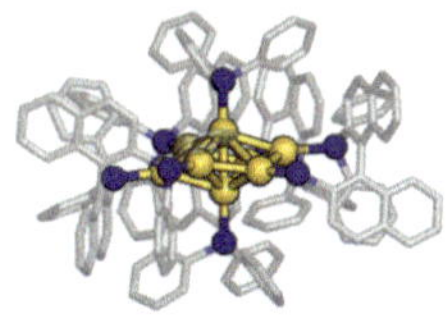
$[Au_9(PPh_3)_8]^{3+}$	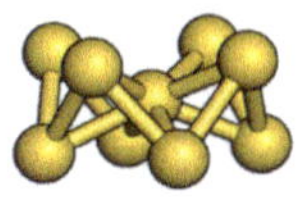 $Au_9(6e)$	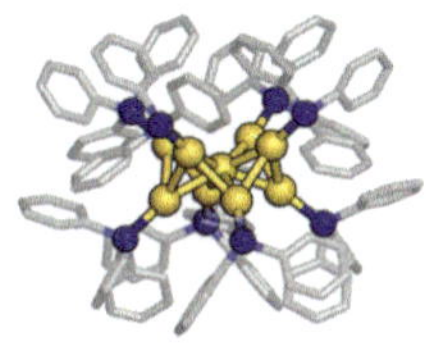

BINAP = 2,2'-Bis(diphenylphosphino)-1,1'-binaphthyl. Color codes: yellow = Au; blue = P; gray = C. Organic residues in the overall structure are depicted as sticks.

distortion of the spherical jellium potential into a prolate. When the spherical jellium is elongated along the z-axis, the 1P and 1D SOs distributed along the z-axis are stabilized, leading to the split into subgroups as shown in Figure 6.8. Consequently, the $\mathbf{Au_6}$ and $\mathbf{Au_7}$ cores form the closed electron configuration $(1S)^2(1P_z)^2$ with 4 electrons, while the $\mathbf{Au_{20}}$ and $\mathbf{Au_{23}}$ cores take the closed electron configuration $(1S)^2(1P_z)^2(1P_{x,y})^4(1D_{z2})^2(1D_{xz,yz})^4$ with 14 electrons.

Interestingly, the electronic structures of some prolate superatoms can be understood alternatively by viewing them as quasi-molecules composed of spherical superatoms (superatomic molecules) [80–82]. Bi-icosahedral $\mathbf{Au_{23}}$(14e) [79] and $\mathbf{Au_{25}}$(14e) [83] cores of $Au_{38}(SR)_{24}$ and

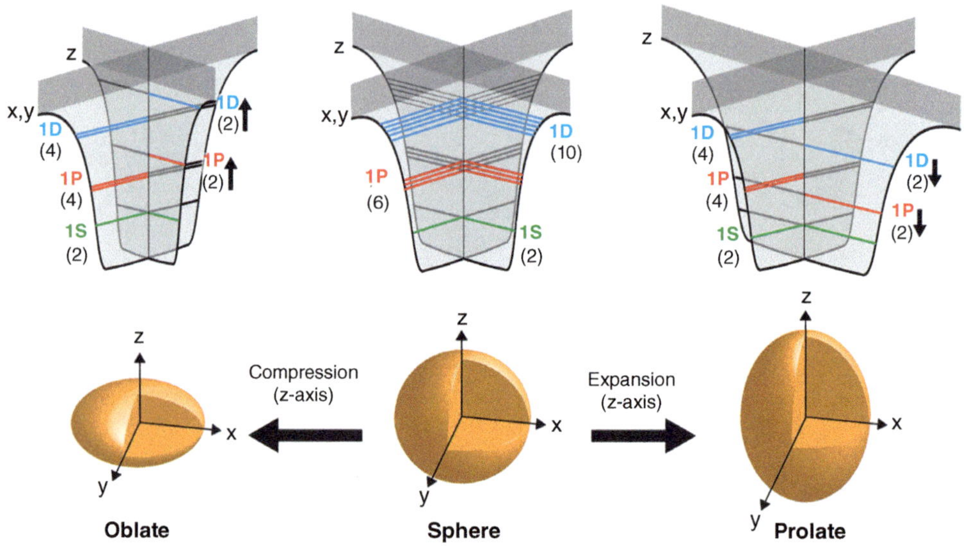

Figure 6.8 Schematic representation of jellium potentials before and after deformation into oblate and prolate. The numbers in parentheses represent the maximum numbers of valence electrons for the individual SOs. *Source:* Omoda et al. [15]. © John Wiley & Sons.

$Au_{25}(SDpp)_{11}$ (SDpp = 2,6-$Ph_2C_6H_3S$) correspond to dimers of open-shell $\mathbf{Au_{13}}$(7e) superatoms linked by sharing a facet $(Au^+)_3$ or a vertex atom Au^+, respectively (Figure 6.9):

$$2\mathbf{Au_{13}}(7e)-(Au^+)_3=\mathbf{Au_{23}}(\mathbf{14}e) \tag{6.2}$$

$$2\mathbf{Au_{13}}(7e)-Au^+=\mathbf{Au_{25}}(\mathbf{14}e) \tag{6.3}$$

According to the "super valence bond theory" developed by Yang and Chen, the electron configuration of $\mathbf{Au_{23}}$(14e) is expressed as $(1\Sigma)^2(1\Sigma^*)^2(1\Pi)^4(2\Sigma)^2(1\Pi^*)^4$ [80]. The 2Σ and 1Π orbitals are constructed by bonding interaction between 1P SOs of $\mathbf{Au_{13}}$(7e), whereas the $1\Pi^*$ orbital is constructed by antibonding interaction between 1P SOs (Figure 6.10a). The shape and energy ordering of these Kohn-Sham orbitals of $\mathbf{Au_{23}}$(14e) correspond well to those of the SOs predicted for a prolate superatom (Figure 6.8) [84]. The electron configuration of bi-icosahedral $\mathbf{Au_{23}}$(14e) is similar to that of an F_2 molecule and the bond order between the two $\mathbf{Au_{13}}$ units is calculated to be one, the same as in the case of F_2. Similarly, the prolate $\mathbf{Au_{20}}$(14e) superatom of $[Au_{20}(PPhPy_2)_{10}Cl_4]^{2+}$ can be viewed as a dimer of $\mathbf{Au_{11}}$(7e) superatoms linked by sharing an edge $(Au^+)_2$ (Figure 6.9): [82]

$$2\mathbf{Au_{11}}(7e)-(Au^+)_2=\mathbf{Au_{20}}(\mathbf{14}e) \tag{6.4}$$

The electron configuration is similar to that of bi-icosahedral $\mathbf{Au_{23}}$(14e) (Figure 6.10b).

In contrast, some prolate superatoms can be viewed as quasi-molecules of closed-shell superatoms. The $\mathbf{Au_6}$(4e) core in Table 6.6 corresponds to a dimer of a planar $\mathbf{Au_3}$(2e),

Table 6.6 Atomic structures of Au MPCs with prolate superatomic cores.

Chemical formula	Core structure	Overall structure
$[Au_6(dppp)_4]^{2+}$	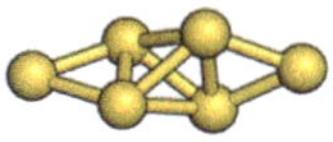 $Au_6(4e)$	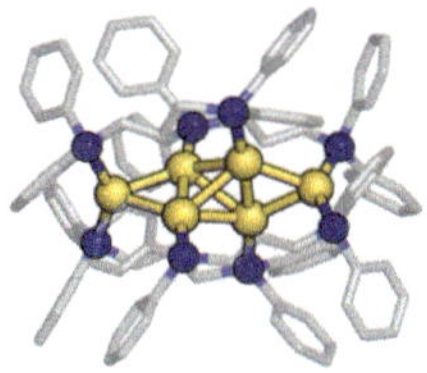
$Au_{22}(C{\equiv}CR)_{18}$	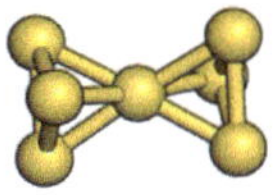 $Au_7(4e)$	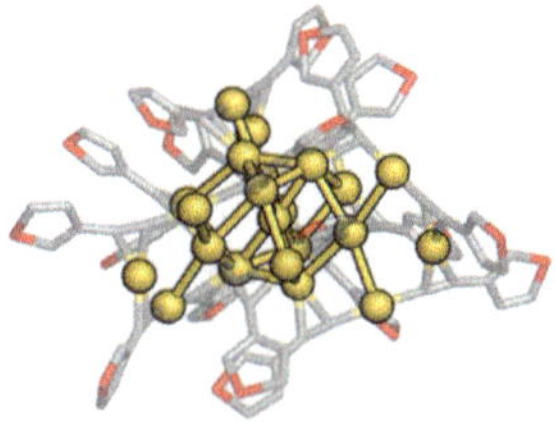
$[Au_{20}(PPhPy_2)_{10}Cl_4]^{2+}$	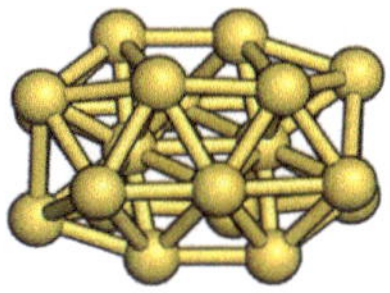 $Au_{20}(14e)$	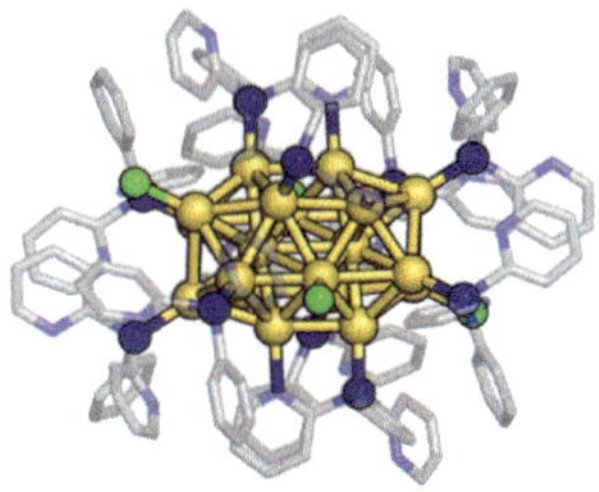
$Au_{38}(SR)_{24}$	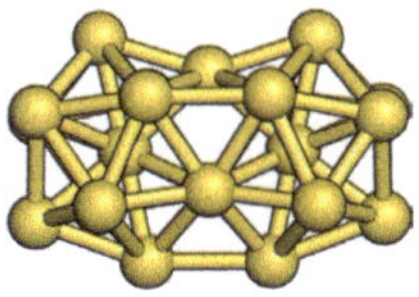 $Au_{23}(14e)$	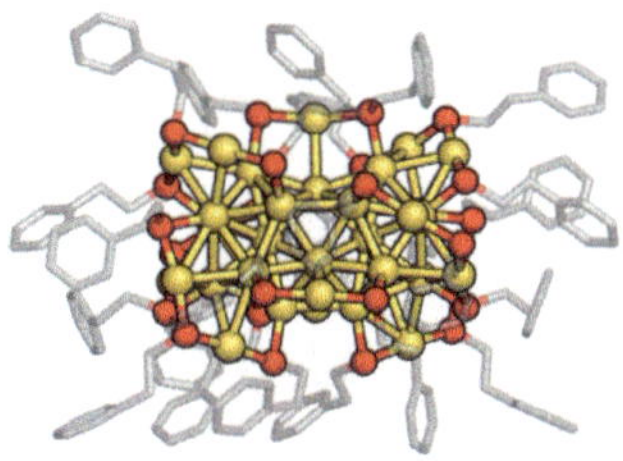

dppp = $Ph_2PCH_2CH_2CH_2PPh_2$. Color codes: yellow = Au; blue = P; green = Cl; red = S; light blue = N; gray = C. Organic residues in the overall structure are depicted as sticks.

whereas the $\mathbf{Au_7}$(4e) core is a dimer of a tetrahedral $\mathbf{Au_4}$(2e) connected by sharing a vertex atom Au^+:

$$2\,\mathbf{Au_3}(2e) = \mathbf{Au_6}(4e) \tag{6.5}$$

$$2\,\mathbf{Au_4}(2e) - Au^+ = \mathbf{Au_7}(4e) \tag{6.6}$$

Bi-, tri-, and penta-icosahedral cores $\mathbf{Au_{25}}$(16e), $\mathbf{Au_{37}}$(24e), and $\mathbf{Au_{60}}$(40e) are constructed in $[Au_{25}(SR)_5(PPh_3)_{10}Cl_2]^{2+}$ [85], $[Au_{37}(SR)_{10}(PPh_3)_{10}Cl_2]^+$ [86], and

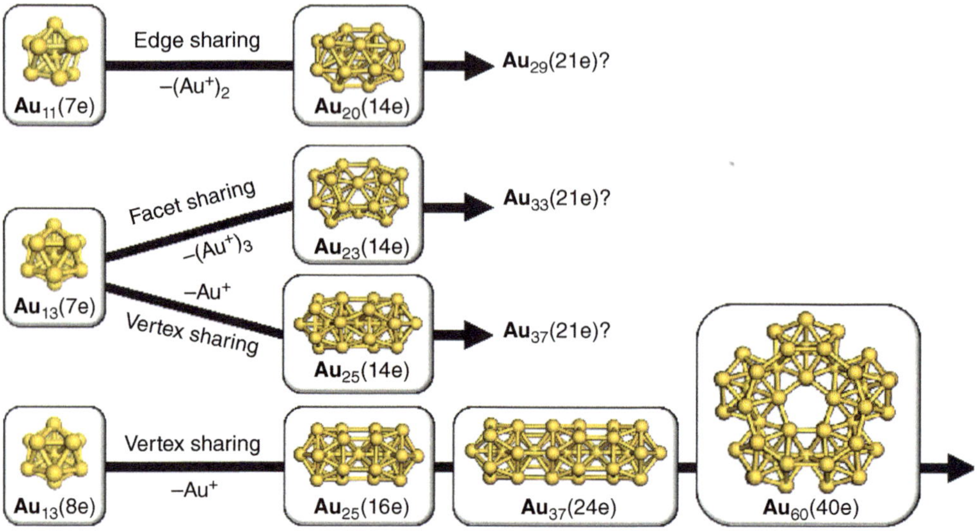

Figure 6.9 Series of superatomic molecules of $\mathbf{Au_{11}}$(7e) and $\mathbf{Au_{13}}$(7/8e) with different bonding schemes.

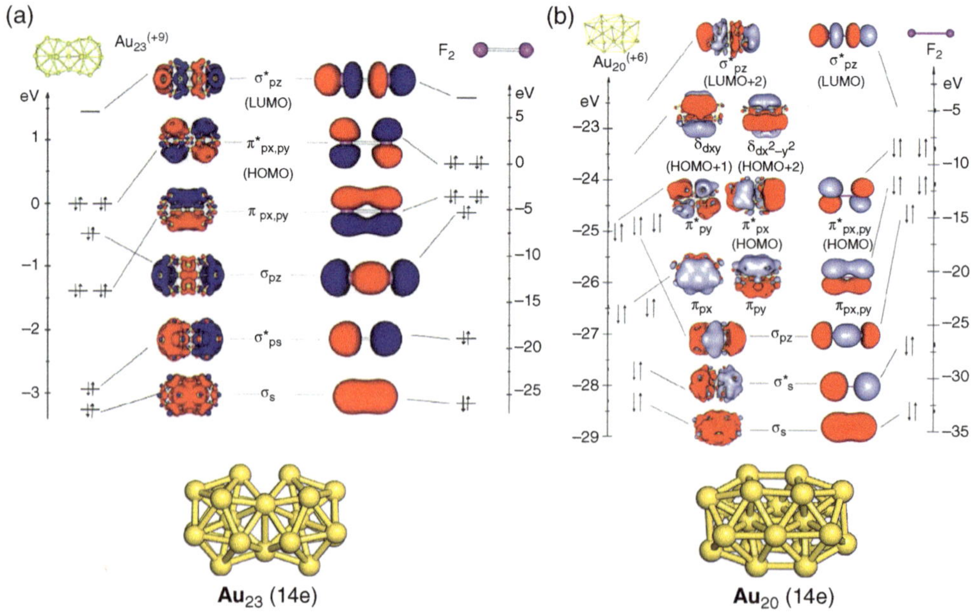

Figure 6.10 Energy diagram of the SOs for (a) $\mathbf{Au_{23}}$(14e) and (b) $\mathbf{Au_{20}}$(14e). *Sources:* (a) Cheng et al. [80]. © Royal Society of Chemistry, 2013. (b) Yuan et al. [82]. © American Chemical Society.

$[Au_{60}(SeR)_{15}(PPh_3)_{10}Se_2]^+$ [87] (Figure 6.9), respectively, by sharing vertex atom(s) of the closed-shell $\mathbf{Au_{13}}$(8e):

$$2\mathbf{Au_{13}}(8e) - Au^+ = \mathbf{Au_{25}}(\mathbf{16}e) \quad (6.7)$$

$$3\mathbf{Au_{13}}(8e) - 2Au^+ = \mathbf{Au_{37}}(\mathbf{24}e) \quad (6.8)$$

$$5\mathbf{Au_{13}}(8e) - 5Au^+ = \mathbf{Au_{60}}(\mathbf{40}e) \quad (6.9)$$

The bonding interaction between the $\mathbf{Au_{13}}$ moieties in this series is relatively weak, as expected from the fact that the formal bond order is zero. Recently, Zheng and Gao proposed in the "grand unified model" that all the clusters can be viewed as the assemblies of small superatomic components such as triangular $\mathbf{Au_3}$(2e), tetrahedral $\mathbf{Au_4}$(2e), and icosahedral $\mathbf{Au_{13}}$(8e) [88]. These examples strongly suggest that a unique class of quasi-molecules can be developed by bonding various superatomic units via various bonding schemes.

A mechanical stress imposed on the superatomic core due to multidentate ligation and/or steric repulsion between the adjacent ligands induces a slight dislocation of the constituent atoms from the equilibrium positions in a less- or non-perturbed state. For example, enantiomeric pairs of superatoms can be produced by using intrinsically chiral ligands, such as $\mathbf{Au_{11}}$(8e) and $\mathbf{Au_8}$(6e) cores of $[Au_{11}(R/S\text{-DIOP})_4Cl_2]^+$ (DIOP = *O*-isopropylidene-2,3-dihydroxy-1,4-bis(diphenylphosphino)butane) and $[Au_8(R/S\text{-BINAP})_3(PPh_3)_2]^{2+}$, respectively (Figure 6.11a and b) [74]. It was proposed that circular dichroism (CD) in the visible region originates from a chiral core structure quantified by the Hausdorff chirality

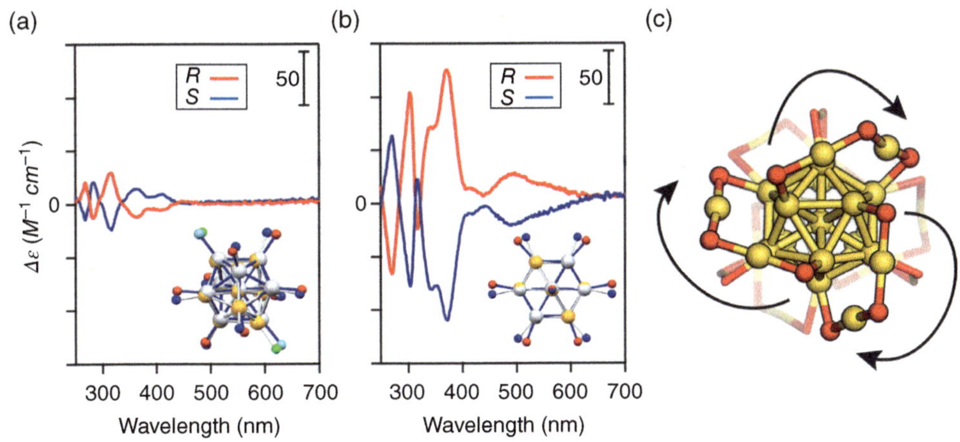

Figure 6.11 CD spectra of (a) $[Au_{11}(R/S\text{-DIOP})_4Cl_2]^+$ and (b) $[Au_8(R/S\text{-BINAP})_3(PPh_3)_2]^{2+}$. Insets show superimposed structures of each enantiomeric pair. (c) The structure of $Au_{38}(SR)_{24}$ showing a chiral configuration of the $Au_2(SR)_3$ units. Color codes: yellow = Au of *R* isomer; silver = Au of *S* isomer; blue = P of *R* isomer; red = P of *S* isomer; green = Cl of *R* isomer; light blue = Cl of *S* isomer; white bonds = *R* isomer; blue bonds = *S* isomer. *Source:* Takano and Tsukuda [74]. © American Chemical Society.

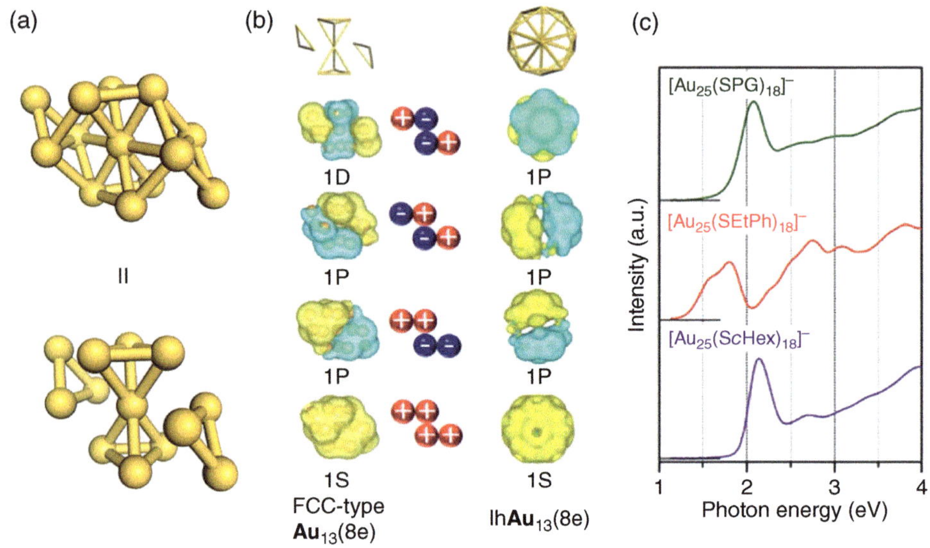

Figure 6.12 (a) Atomic structure of anisotropic $\mathbf{Au_{13}}$(8e) observed in $[Au_{23}(ScHex)_{16}]^-$, $Au_{24}(SAdm)_{16}$, and $[Au_{25}(SPG)_{18}]^-$. The decomposed structure is shown below. (b) Electron density diagram of occupied extended Hückel molecular orbitals of FCC $\mathbf{Au_{13}}$(8e) (left) and Ih $\mathbf{Au_{13}}$(8e) (right). (c) UV-vis optical absorption spectra of $[Au_{25}(SPG)_{18}]^-$, $[Au_{25}(SEtPh)_{18}]^-$, and $[Au_{23}(ScHex)_{16}]^-$. *Sources:* (b and c) Omoda et al. [91] and Xiong et al. [92]. © American Chemical Society.

measure [89, 90] and the location of chiral binaphthyl groups having a diffuse π electron in close proximity to the Au core. Chiral superatomic cores can also be produced by using achiral ligands by forming a multidentate, chiral network on the surface of superatoms as in the case of the $\mathbf{Au_{23}}$(14e) core of $Au_{38}(SR)_{24}$ (Figure 6.11c) [79]. When the degree of distortion exceeds a critical level, the superatom isomerizes to distinctly different atomic packings. For example, $\mathbf{Au_{13}}$(8e) with an FCC-based structure was formed in $[Au_{23}(ScHex)_{16}]^-$ (ScHex = ScC_6H_{11}) and $[Au_{25}(SPG)_{18}]^-$ (H-SPG = N-(2-mercaptopropionyl)glycine) (Figure 6.12a) [91, 93] probably due to the steric repulsion between the thiolate ligands having a bulky α-carbon. Theoretical calculation showed that the electron configurations of FCC $\mathbf{Au_{13}}$(8e) and Ih $\mathbf{Au_{13}}$(8e) are expressed as $(1S)^2(1P)^4(1D)^2$ and $(1S)^2(1P)^6$, respectively (Figure 6.12b) [92]. Owing to the different electron configuration, the optical absorption spectra of the Ih and the FCC $\mathbf{Au_{13}}$(8e) differ significantly (Figure 6.12c) [91]. These examples of structural isomers show that the atomic arrangement and electronic structure of superatoms can be tuned by chemical modification.

6.3 Atomically Precise Synthesis of Chemically Modified Superatoms

6.3.1 Size Control

Size-controlled synthesis is one of the most important tasks in the study of superatoms since every single atom in a superatom affects its properties. A fundamental step of the synthesis is the formation of a metallic core by reducing metal ions in the presence of protecting ligands (or metal complexes). According to the La-Mer mechanism [94], formation of the cluster is initiated by nucleation followed by growth of the nuclei. In order to obtain small, monodisperse clusters, all the ions in solution should be reduced instantaneously by using a strong reducing agent such as $NaBH_4$. However, it is generally difficult to control kinetically the cluster size with atomic resolution because of the stochastic nature of the nucleation and growth process. This section introduces several strategies to overcome this inherent problem.

6.3.1.1 Top-down Approach: Size Focusing

Since Au/Ag MPCs obtained by reduction are polydisperse in terms of cluster size, a post-synthetic treatment is necessary to obtain atomically pure MPCs. The most effective strategy is a top-down method called "size focusing." [11, 12] This method allows us to reduce the population of less-stable species present in the as-prepared statistical mixture and, in the ideal case, to convert them to chemically robust clusters. In practice, chemical etching by free ligands added in excess has been the most widely used approach. Figure 6.13 schematically shows size focusing by chemical etching of polydisperse mixtures. This approach utilizes the essential difference in chemical robustness between the kinetically trapped species, as clearly demonstrated by glutathione (SG)-protected Au clusters $Au_x(SG)_y$ with $(x, y) = (10, 10), (15, 13), (18, 14), (22, 16), (25, 18), (29, 20), (33, 22)$, and $(39, 24)$ [95]. $Au_x(SG)_y$ clusters with $x > 25$ were converted to $Au_{25}(SG)_{18}$ in the presence of an excess amount of glutathione (GSH), whereas $Au_x(SG)_y$ clusters with $x < 25$ were oxidized to Au(I)SG complexes (Figure 6.14a) [95]. The size-focusing process can be monitored by mass spectrometry and optical spectroscopy. matrix assisted laser desorption/ionization (MALDI) mass spectrometry has demonstrated that a polydisperse mixture of $Au_x(SR)_y$ was focused into $Au_{38}(SR)_{24}$ by the reaction with an excess amount of RSH (Figure 6.14b) [96]. Conversion of polydisperse Au:dppe clusters into $[Au_{13}(dppe)_5Cl_2]^{3+}$ by hydrochloric acid (HCl) was monitored by optical spectroscopy (Figure 6.14c) [26]. The size-focusing treatment affords a mixture of a limited number of chemically and thermodynamically robust clusters. Surprisingly, the compositions of Au/Ag MPCs are atomically defined at this stage, reflecting the composition-specific stability. Thus, the last step to obtain the atomically pure form

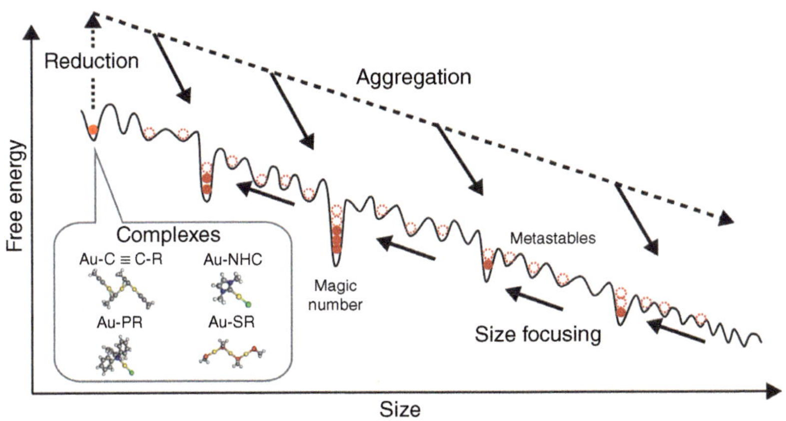

Figure 6.13 Schematic illustration of the preparation and size-focusing strategy of chemically modified superatoms.

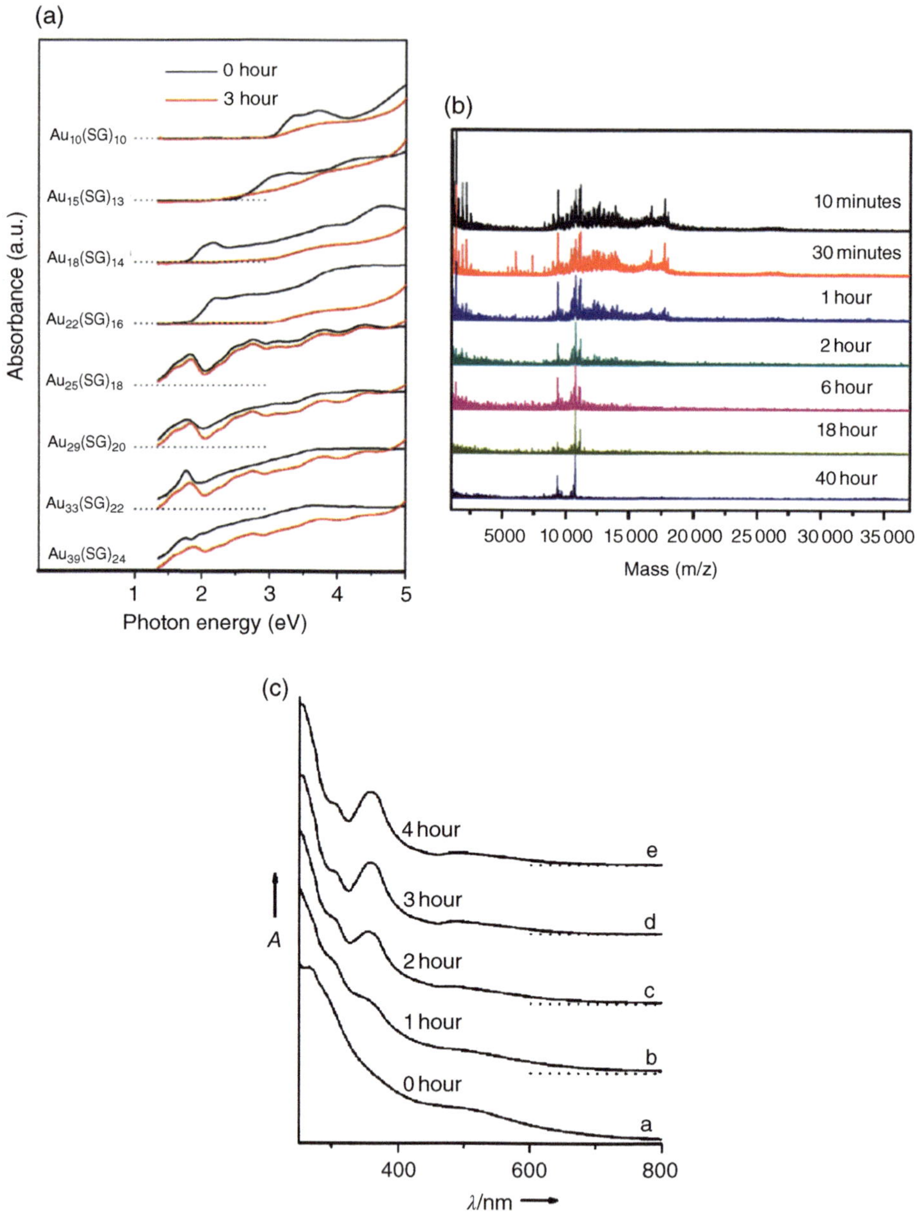

Figure 6.14 (a) Optical spectra of size-selected $Au_x(SG)_y$ before (0 hour) and after (3 hours) etching reactions with GSH. (b) MALDI mass spectra of $Au_x(PET)_y$ products as a function of reaction time with PET. (c) Time course of optical absorption spectra of polydisperse Au:dppe clusters by HCl treatment. *Sources:* (a) Shichibu et al. [95]. © John Wiley & Sons. (b) Qian et al. [96]. © American Chemical Society. (c) Shichibu and Konishi [26]. © John Wiley & Sons.

of clusters is fractionation. Fractionation methods include fractional precipitation/extraction with a specific solvent, chromatography, and crystallization.

Fractional precipitation/extraction: One of the MPCs in the mixture can be isolated by fractional precipitation or fractional extraction based on the difference in the solubility between the desired and other MPCs to a specific solvent. This method was successfully used to isolate water-soluble $Au_{102}(pMBA)_{44}$ (pMBA = S-$pC_6H_4CO_2H$) [97]. Fractional precipitation/extraction is a powerful technique if the crude mixture contains a limited number of MPCs and the solubility of the target cluster is known. However, a major drawback is poor resolving power, especially between the MPCs with similar compositions, in addition to the necessity of finding a suitable solvent system for separation by trial and error.

Chromatography: Size exclusion chromatography (SEC) can separate MPCs based on their hydrodynamic diameters (D_{hyd}). The MPCs having a larger D_{hyd} elute faster than those with smaller D_{hyd}. The resolving power of SEC can be improved by a recycling mode, in which the MPCs are injected into the column successively. This approach has been successfully used to isolate $Au_{38}(SR)_{24}$ and $Au_{55}(SR)_{31}$ and to purify $Au_{279}(TBBT)_{84}$ (Figure 6.15) [98, 99]. Normal-phase chromatography (NPC) using a silica or alumina column can isolate neutral MPCs from ionic MPCs by taking advantage of their significantly different polarities. Preparative thin-layer chromatography (PTLC) has a higher resolving power than NPC and has been successfully used to separate two isomers of $Au_{38}(SR)_{24}$ and those of $Au_{144}(SR)_{60}$ and $Au_{25}(SR)_{18}$ [100, 101]. Reverse-phase chromatography (RPC) allows MPCs to be fractionated by utilizing the weak interaction between the ligand layers of MPCs and organic residues on the beads packed in the column. The interaction strength can be tuned by changing the polarity of a mobile phase or by adding an ion-pair reagent such as alkylammonium salt. A typical example is the isolation of a series of $Au_x(SR)_y$ (Figure 6.16a and b) [34]. The reverse-phase high performance liquid chromatography (HPLC) technique with an ion-pair reagent can be applied to hydrophilic MPCs. The ion-pair reagent is always required for the separation of hydrophilic MPCs because the polar ligand moiety can weakly interact with the beads. The most straightforward separation method is hydrophilic interaction liquid chromatography (HILIC). The $Au_x(SG)_y$ clusters can be separated with atomic precision by using an HILIC column (Figure 6.16a–c) [102]. High-resolution separation by HPLC coupled with MS is a powerful means of optimizing the synthesis conditions of the desired MPCs [103], although the method is not suitable for large-scale preparation due to the limit on the amount of samples that can be loaded into an analytical HPLC column. A racemic mixture of MPCs can be separated by HPLC. Figure 6.17 shows an example of separation of enantiomers of $Au_{38}(PET)_{24}$. This cluster possesses a screw axis along the long axis by the ligation of $Au_2(SR)_3$ oligomers (Figure 6.11c) [104].

Polyacrylamide gel electrophoresis (PAGE): Early work by Whetten demonstrated that fractionation of hydrophilic MPCs is possible using PAGE [105, 106]. The key to achieving atomically precise fractionation is to use a gel that is much denser than those used in bioscience. PAGE was successfully used to fractionate $Au_x(SG)_y$ [107], $Au_x(SPG)_y$ [91], $Au_x(m\text{MBA})_y$ (mMBA = S-$mC_6H_4CO_2H$) [108], and $Ag_x(SG)_y$ [109] (Figure 6.18). electrospray ionization (ESI)–MS of the MPCs contained in individual bands revealed that the composition was atomically defined and MPCs with a difference of a few atoms can be clearly separated.

6.3.1.2 Bottom-up Approach: Size Convergence

The yield of magic MPCs can be enhanced or maximized by the bottom-up approach: kinetically controlled growth or sterically limited growth. A successful example is the kinetically controlled synthesis of water-soluble $Au_{25}(pMBA)_{18}$ clusters [110]. The reduction rate of Au ions was reduced by using a stoichiometric amount of reductant in a basic

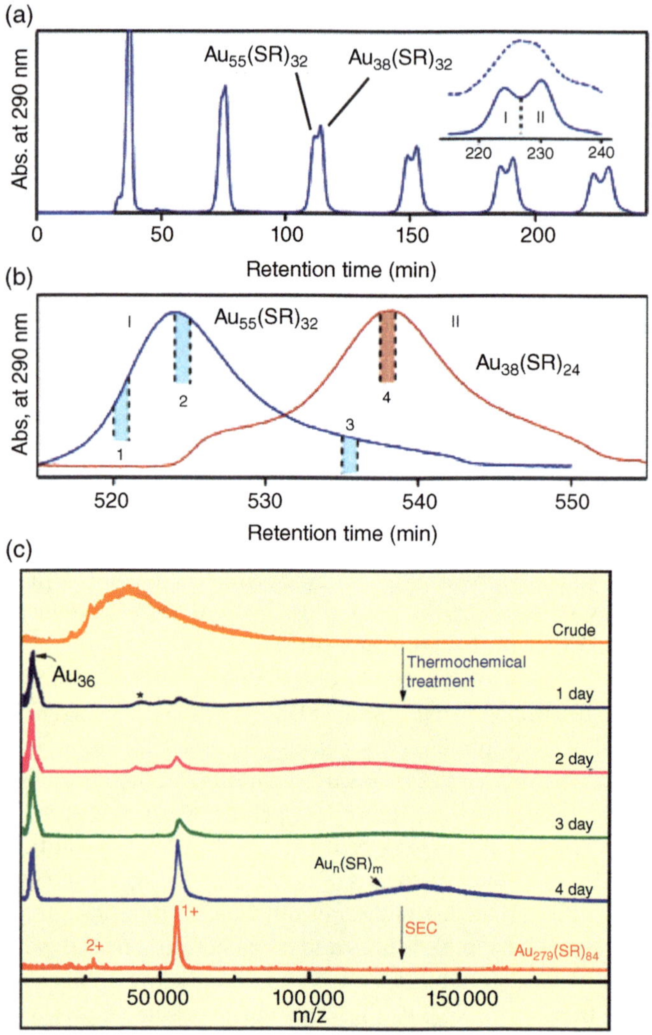

Figure 6.15 (a, b) SEC separation of $Au_{55}(SC_{18}H_{37})_{32}$ and $Au_{38}(SC_{18}H_{37})_{24}$ in a recycling mode. (c) SEC isolation of $Au_{279}(TBBT)_{84}$. *Sources:* (a) Tsunoyama et al. [98]. © American Chemical Society. (b) Sakthivel et al. [99]. © American Chemical Society.

media. *In situ* mass analysis of the slow production process revealed that a step-by-step growth of small clusters formed at the first stage into $Au_{25}(pMBA)_{18}$ (Figure 6.19a). Another example is the hydride-mediated, selective transformation of $[Au_9(PPh_3)_8]^{3+}$ to $[Au_{11}(PPh_3)_8Cl_2]^+$. *In situ* ESI–MS elucidated that $[Au_9(PPh_3)_8]^{3+}$ undergoes sequential growth after addition of hydride (H^-) (Figure 6.19b):

$$\left[HAu_9(PPh_3)_8\right]^{2+} + AuClPPh_3 \rightarrow \left[Au_{10}(PPh_3)_8Cl\right]^+ + PPh_3 + H^+ \tag{6.10}$$

$$\left[Au_{10}(PPh_3)_8Cl\right]^+ + AuClPPh_3 \rightarrow \left[Au_{11}(PPh_3)_8Cl_2\right]^+ + PPh_3 \tag{6.11}$$

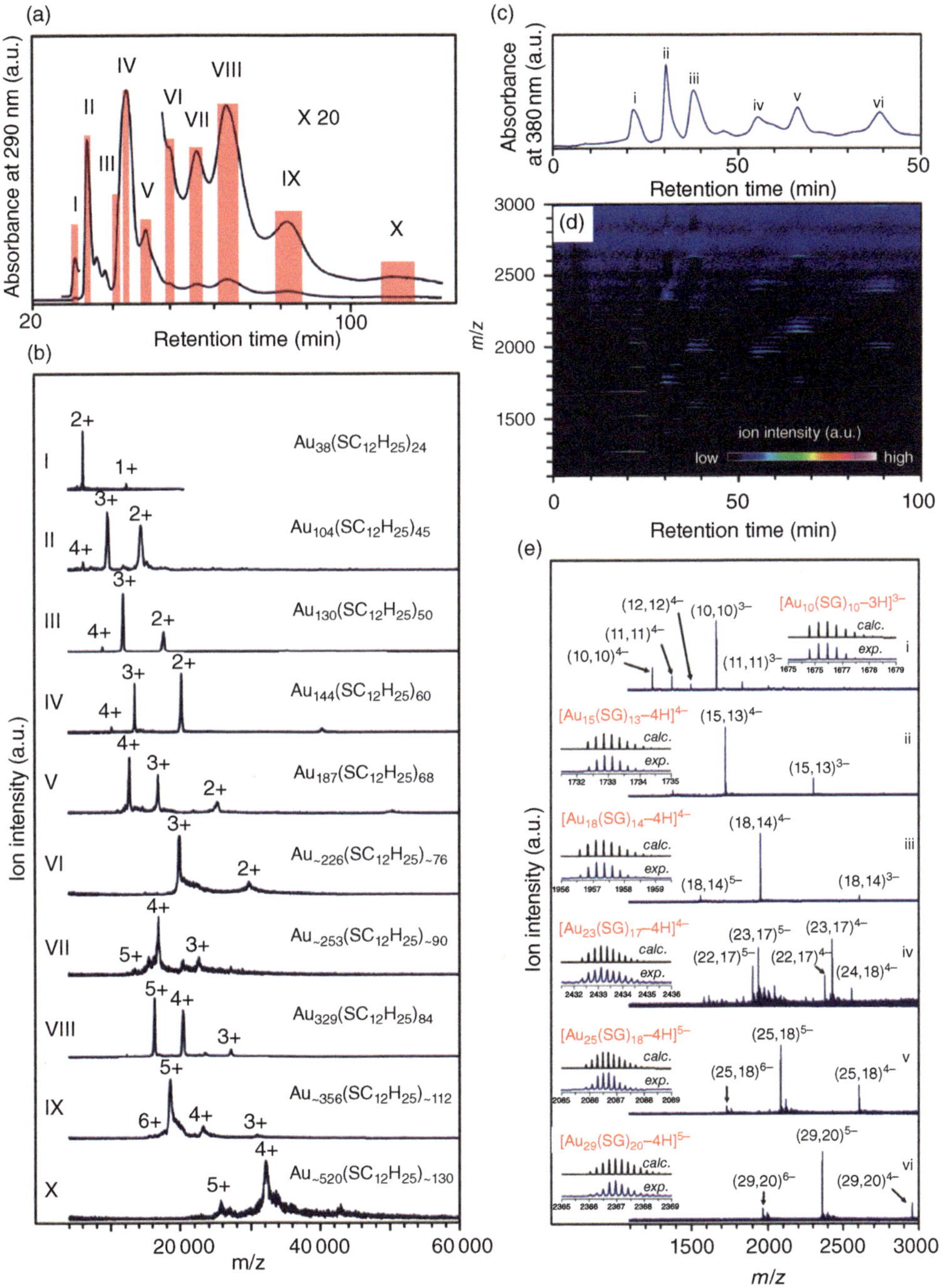

Figure 6.16 (a, b) Reverse-phase HPLC separation of a series of $Au_x(SR)_y$ clusters. (c–e) HPLC separation of $Au_x(SG)_y$. *Sources:* (a and b) Negishi et al. [34]. © American Chemical Society. (c–e) Niihori et al. [102]. © Royal Society of Chemistry.

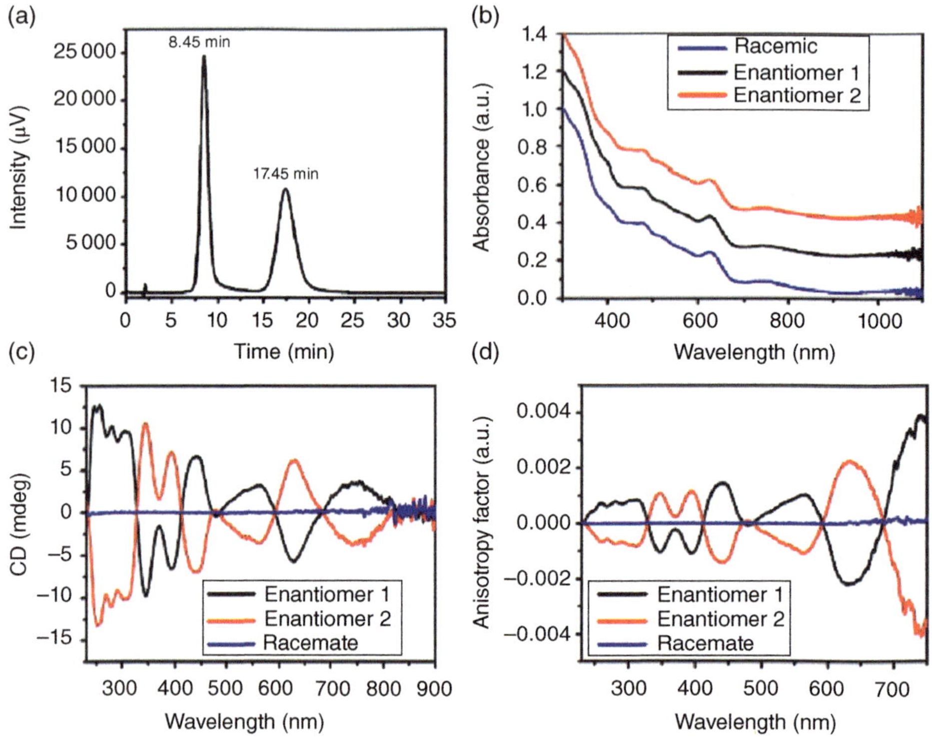

Figure 6.17 Separation of enantiomers of $Au_{38}(PET)_{24}$. (a) Chiral HPLC chromatogram of a racemic mixture of $Au_{38}(PET)_{24}$. (b) Optical absorption spectra, (c) CD spectra, and (d) anisotropy factor ($\Delta\varepsilon/\varepsilon$) of racemic and separated enantiomers, respectively. *Source:* Dolamic et al. [104]. © Springer Nature.

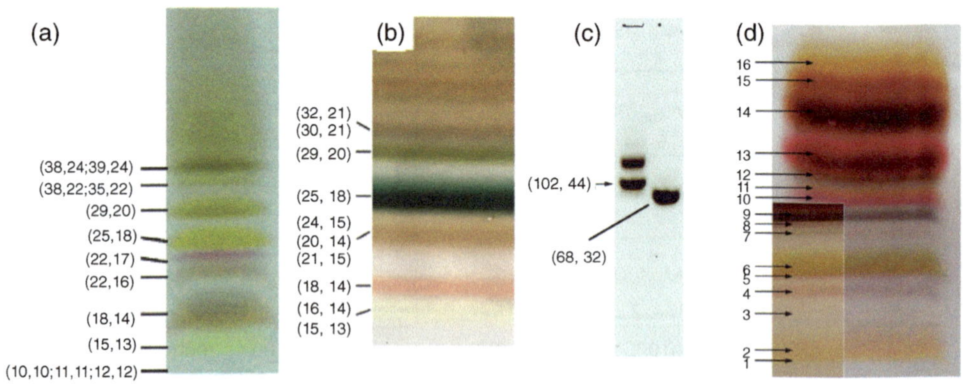

Figure 6.18 PAGE separation of (a) $Au_x(SG)_y$, (b) $Au_x(SPG)_y$, (c) $Au_x(mMBA)_y$, and (d) $Ag_x(SG)_y$. The compositions are given by (x, y). *Sources:* (a) Reprinted with permission Negishi et al. [107]. © American Chemical Society. (b) Reprinted with permission Omoda et al. [91]. © American Chemical Society. (c) Reprinted with permission Azubel et al. [108]. © American Association for the Advancement of Science-AAAS. (d) Reprinted with permission Kumar et al. [109]. © American Chemical Society.

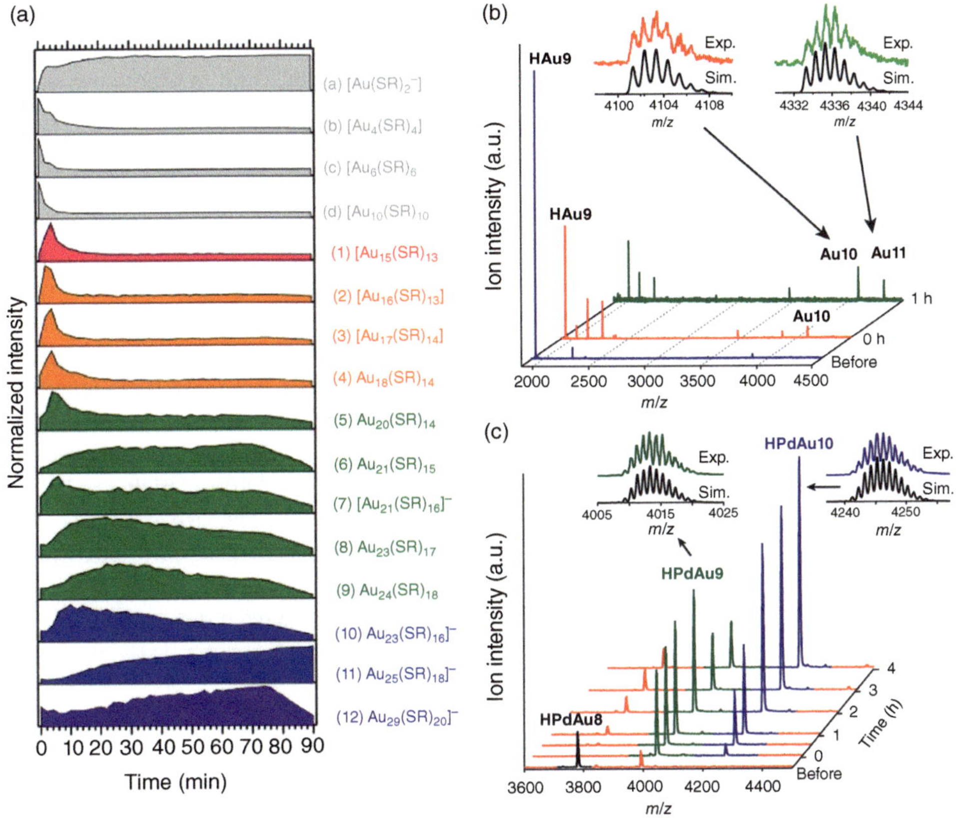

Figure 6.19 (a) Time course of the abundance of each reaction intermediate in the synthesis of $Au_{25}(pMBA)_{18}$. (b, c) *In situ* ESI mass spectra during the growth reactions of hydride-doped $\mathbf{HAu_9}$ and $\mathbf{HPdAu_8}$ superatoms. *Sources:* (a) Chen et al. [110]. © American Chemical Society. (b, c) Takano et al. [69]. © American Chemical Society.

$[HAu_9(PPh_3)_8]^{2+}$ exhibits high nucleophilicity toward Au(I)ClPPh$_3$ complex because the occupied 1P$_z$ SO is located at a high energy and is sterically exposed (Figure 6.7a) [67–69]. The hydrogen in $[HAu_9(PPh_3)_8]^{2+}$ was released in the form of a proton, leaving $[Au_{10}(PPh_3)_8Cl]^+$, upon reaction with the first Au(I)ClPPh$_3$. Termination of the growth at $[Au_{11}(PPh_3)_8Cl_2]^+$ is probably associated with the steric hinderance caused by bulky PPh$_3$ ligands. In contrast, $[PdAu_8(PPh_3)_8]^{2+}$ underwent the following growth process upon hydride addition (Figure 6.19c):

$$\left[HPdAu_8\left(PPh_3\right)_8\right]^+ + AuClPPh_3 \rightarrow \left[HPdAu_9\left(PPh_3\right)_8Cl\right]^+ + PPh_3 + H^+ \qquad (6.12)$$

$$\left[HPdAu_9\left(PPh_3\right)_8Cl\right]^+ + AuClPPh_3 \rightarrow \left[HPdAu_{10}\left(PPh_3\right)_8Cl_2\right]^+ + PPh_3 \qquad (6.13)$$

In contrast to Eq. (6.10), the hydrogen in $[HPdAu_8(PPh_3)_8]^+$ was retained during the growth, forming an unprecedented $[HPdAu_{10}(PPh_3)_8Cl_2]^+$. The hydride-mediated growth method is a promising approach to scaling up the yields of MPCs and to producing novel MPCs that cannot be produced by the conventional reduction reaction.

6.3.1.3 Template Method

The cluster size can be controlled by limiting the number of metal precursor ions using a template. Dendrimers having a given number of metal coordinating sites such as amine and imine are promising candidates for templates of superatoms. The initial step is to condense a given number of precursor metal ions to the coordination sites within the dendrimer. Then, reduction of these ions results in the formation of atomically precise superatoms within a cavity of the dendrimer. Recently, one of the most famous magic clusters, $\mathbf{Al_{13}}^-$(40e) [6, 7], with an icosahedral structure and closed electronic shell was synthesized within the phenylazomethine dendrimer (Figure 6.20a) [111]. The size of the $\mathbf{Al_{13}}^-$ superatoms was investigated by mass spectrometry (Figure 6.20b) and high-resolution transmission electron microscopy (HRTEM) (Figure 6.20c and d). The size and properties of the superatoms can be controlled by using structurally designed dendrimers. Major remaining challenges are determination of the atomic structures of such fluxional species by SCXRD and scaling up the yields of the superatoms using special dendrimers whose synthesis is not trivial.

6.3.1.4 Kinetic Control

Linear polymers such as polyvinyl pyrrolidone (PVP) are suitable stabilizers of Au superatoms for their catalytic application because the superatoms are stabilized against aggregation while exposing the surface [112, 113]. However, Au clusters stabilized by PVP cannot be fractionated by the methods introduced in Section 6.3.1.1 and cannot be crystallized for SCXRD. Therefore, cluster size must be controlled kinetically. One of the reasons for the heterogeneity of the resulting Au clusters is the inhomogeneity in the mixing of two solutions of the Au ions and reducing agent. To overcome this problem, two solutions containing the Au ions and $NaBH_4$ were mixed homogeneously by employing a microfluidic device: individual solutions continuously supplied by syringe pumps were laminated into substreams with a thickness of ~70 mm and fused together by using a SIMM-V2 (Figure 6.21a) [114]. Size distributions of Au:PVP prepared at 0° C by the microfluidic and conventional batch mixing methods were compared by MALDI mass spectrometry (Figure 6.21b) [115]. Au:PVP prepared by the microfluidic method contains a few magic clusters such as $\mathbf{Au_{35\pm1}}$, $\mathbf{Au_{43}}$, and $\mathbf{Au_{58}}$ in contrast to a continuous distribution of Au:PVP prepared by the conventional batch mixing. By immersing the micromixer in a heat bath (40° C), the size distribution is shifted to $\mathbf{Au_{\sim70}}$, $\mathbf{Au_{\sim107}}$, $\mathbf{Au_{\sim130}}$, and $\mathbf{Au_{\sim150}}$. The high abundance of some of the clusters such as $\mathbf{Au_{34}}$ and $\mathbf{Au_{58}}$ can be explained by the electronic shell closure [2] although the magic behaviors of other clusters are not yet understood. It may be possible to synthesize atomically precise Au superatoms in the future by optimizing the conditions of the microfluidic method.

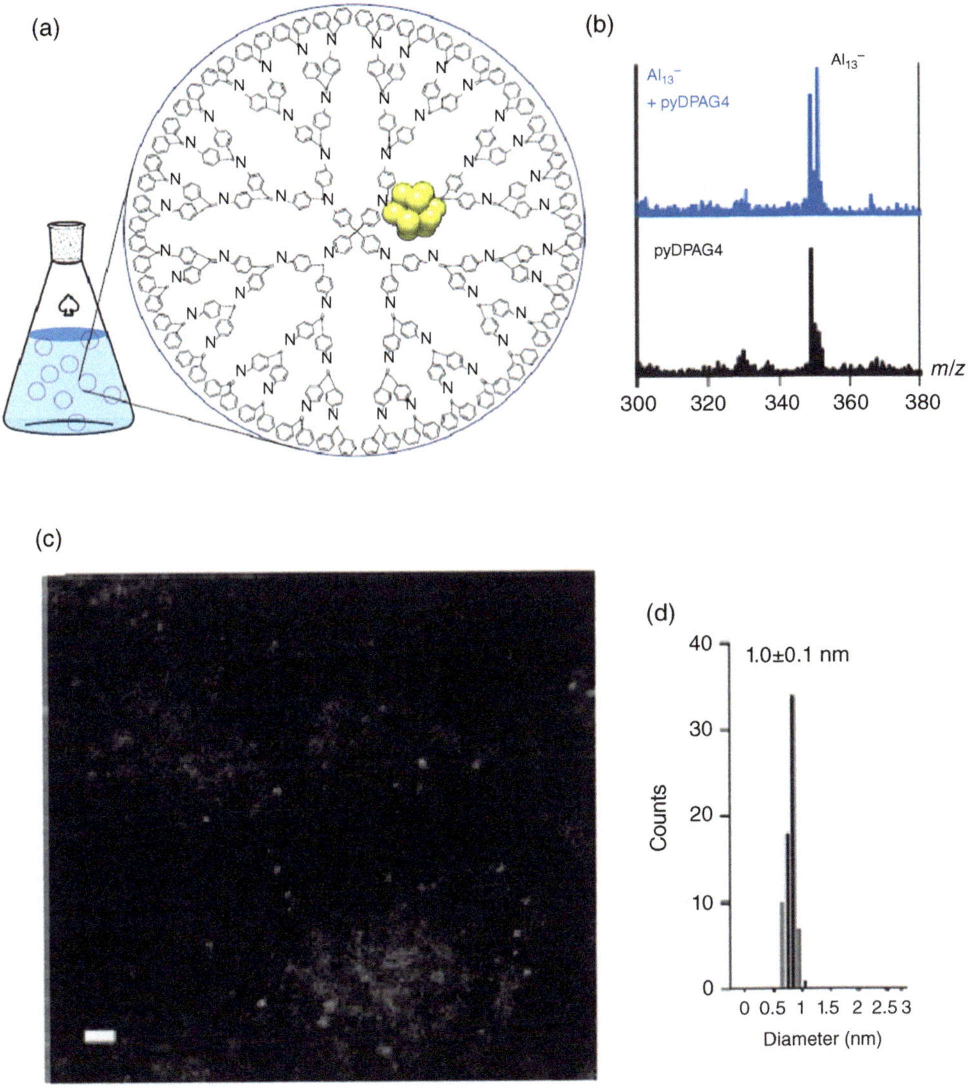

Figure 6.20 (a) A schematic image, (b) mass analysis, (c) HRTEM image, and (d) size distribution of the $\mathbf{Al_{13}}$ superatom within a phenylazomethine dendrimer. *Source:* Kambe et al. [111].
© Springer Nature.

6.3.2 Composition Control

6.3.2.1 Co-reduction Method

The easiest way of doping other transition metals to the Au/Ag superatoms is co-reduction of the precursor metal ions in the presence of ligands [14]. For efficient mixing, the reduction rates of the precursor metal ions should be comparable. Doped superatoms such as $\mathbf{Ir/Pd/Pt@Au_{12}}$ and $\mathbf{Pd/Pt@Ag_{12}}$ (Table 6.4) have been synthesized by this approach, but

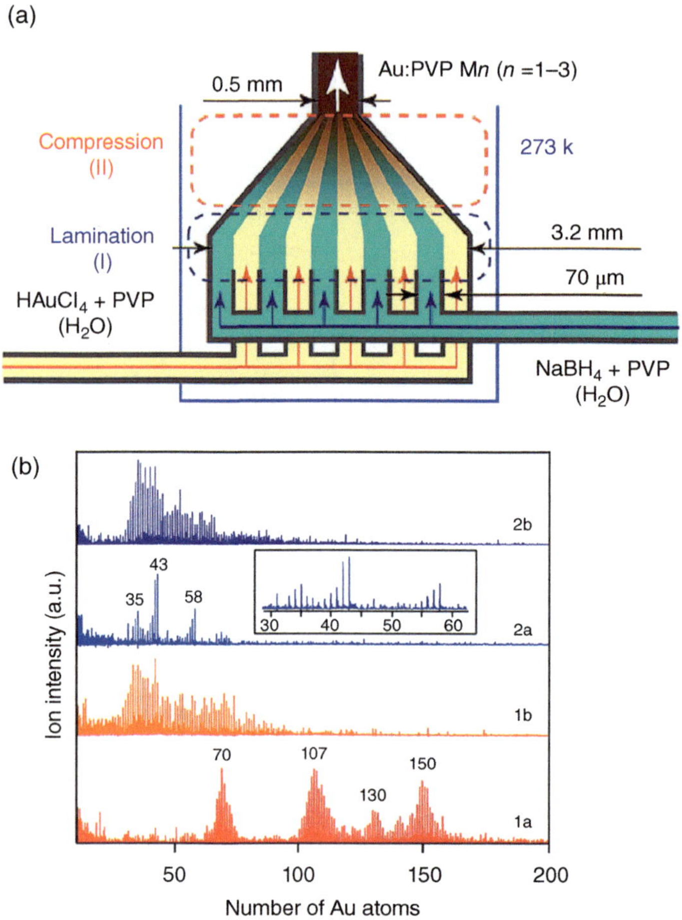

Figure 6.21 (a) Schematic illustration of the synthesis of Au:PVP using a microfluidic device. (b) Negative-mode MALDI mass spectra of Au:PVP prepared by batch mixing at 0° C (**2b**) and at 40° C (**1b**), and the micromixing at 0° C (**2a**) and at 40° C (**1a**), respectively. *Sources:* (a) Tsunoyama et al. [114]. © American Chemical Society. (b) Tsunoyama and Tsukuda [115]. © American Chemical Society.

the yields of atomically precise doped superatoms are generally very low. This is due to the stochastic nature of the aggregation of the metal atoms. Even when the dopants have preferential sites, the products have distributions in not only the size but also the mixing mode. As a result, the final yield of the desired products is low even though they can be isolated in the atomically pure form out of the mixtures. Two methods to improve the yields are explained below.

6.3.2.2 Antigalvanic Method

Metal exchange of preformed Au/Ag MPCs is a promising method to produce multimetallic superatoms efficiently and selectively [116, 117]. The leading approach is the galvanic reaction (GR) that has been widely used to prepare alloy nanoparticles and bulk alloy.

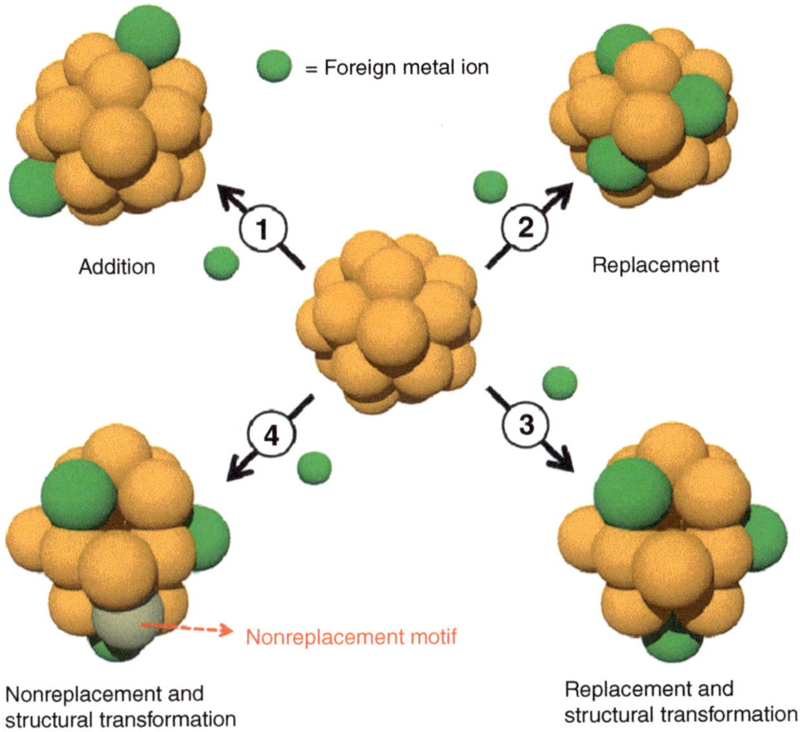

Figure 6.22 Schematic illustration of the four modes of AGRs. *Source:* Gan et al. [116]. © American Chemical Society.

The driving force for the GR is a difference in the electrochemical potential between the two metals: a noble metal cation is spontaneously reduced by a less-noble metal in solution. For example, Toshima successfully doped Au atoms onto the corners of cuboctahedral $Pd_{\sim 147}$: PVP and proposed that Au dopants act as active sites for aerobic oxidation reaction [118].

Murray and Wu found that this classical rule fails at ultrasmall sizes: Murray [119] introduced Ag atoms into $[Au_{25}(PET)_{18}]^-$ by the reaction with Ag^+, and Wu [116] revealed that Ag^+ or Cu^+ ions also react with $[Au_{25}(PET)_{18}]^0$ and other Au/Ag MPCs. Wu's group further demonstrated that this antigalvanic reaction (AGR) is a general process in ultrasmall metal clusters and is associated with the increase in reducibility of smaller metal clusters. So far, four modes of alloying processes have been achieved by AGR (Figure 6.22): (i) addition, (ii) replacement, (iii) replacement and structural transformation, and (iv) nonreplacement and structural transformation. Recently, Pt and Cd co-doped **Pt@CdAu$_{11}$**(8e) superatoms have been successfully synthesized by the AGR between Cd ion with the **Pt@Au$_{12}$**(8e) superatoms preformed by the co-reduction method [120]. The AGR provides another way of synthesizing atomically precise multimetallic clusters that are otherwise difficult to obtain.

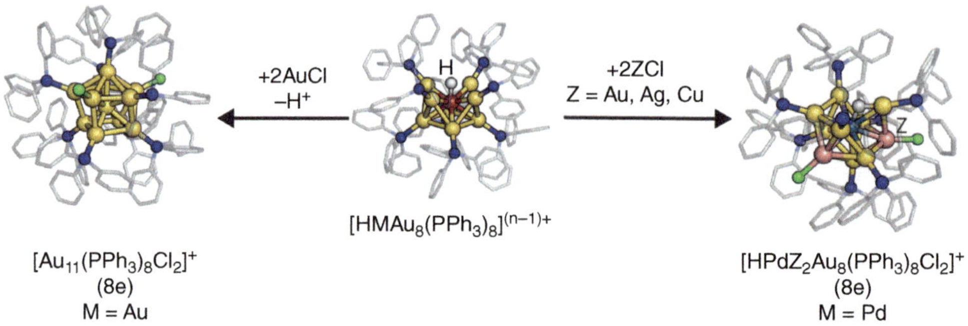

Figure 6.23 Formation of hydrogen-containing Au-based superatom and its transformation. Color codes: yellow = Au; blue = P; dark red = Au, Pd; dark green = Pd; green = Cl; white = H; pink = Au, Ag, Cu; gray = C. Methyl groups are depicted as sticks. Organic residues are depicted as sticks.

6.3.2.3 Hydride-Mediated Transformation

New types of hydrogen-doped superatomic $\mathbf{HPdAu_8Z_2}$ cores (Z = Cu, Ag, Au) have been produced by the sequential addition of two Z(I)Cl units to $\mathbf{HPdAu_8}$ (Figure 6.7) according to the following addition reactions: [121]

$$\left[\mathrm{HPdAu_8(PPh_3)_8}\right]^+ + \mathrm{Z(I)ClPPh_3} \rightarrow \left[\mathrm{HPdAu_8Z(PPh_3)_8Cl}\right]^+ + \mathrm{PPh_3} \tag{6.14}$$

$$\left[\mathrm{HPdAu_8Z(PPh_3)_8Cl}\right]^+ + \mathrm{Z(I)ClPPh_3} \rightarrow \left[\mathrm{HPdAu_{10}Z_2(PPh_3)_8Cl_2}\right]^+ + \mathrm{PPh_3} \tag{6.15}$$

The most striking point is the retention of the H atom during the growth process. SCXRD determined that the metal atoms in $\mathbf{HPdAu_8Z_2}$ exhibit atomic packing similar to that of $\mathbf{Au_{11}}$(8e). The Pd atom is located at the center, and two Z atoms occupy the equivalent sites on the surface and are coordinated by the Cl ligands regardless of Z (Figure 6.23). Although the H atom was not identified by SCXRD due to the small electron density, DFT calculations and nuclear magnetic resonance (NMR) spectroscopy suggest that the H atom is located at one of the four bridging sites of the radial Pd–Au bonds. The transformation from $\mathbf{HPdAu_8}$ to $\mathbf{HPdAu_8Z_2}$ proceeds via electrophilic attack of Z(I)Cl to the occupied $1P_z$ SO of $\mathbf{HPdAu_8}$ because of its higher accessibility along the z-axis through an open space of the ligand layer compared to the $1P_x$ and $1P_y$ SOs (Figure 6.7). These hydride-mediated transformations are selective, efficient, and scalable and will be developed as a novel bottom-up atomically precise method of synthesizing chemically modified multimetallic superatoms.

6.3.3 Shape Control

The examples in Figure 6.9 imply that a class of quasi-molecules can be developed by bonding superatomic units through various bonding schemes. However, their synthesis relies on serendipity and the scope of the structures obtained by the conventional chemical reduction methods is limited: hetero-superatomic molecules in particular are rarely obtained. Therefore, there is strong demand for targeted and efficient synthesis methods to exploit

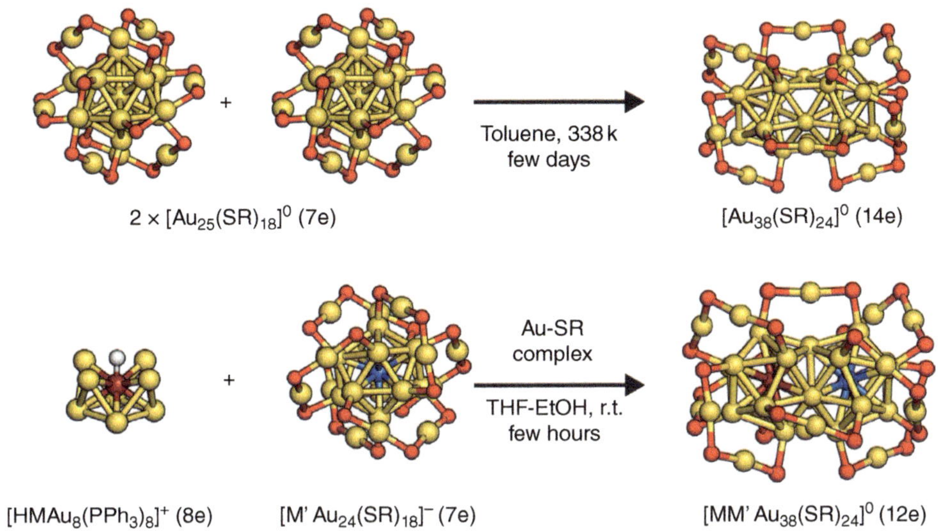

Figure 6.24 Examples of fusion reactions of preformed MPCs. Color codes: yellow = Au; red = S; dark red = Pd, Pt; sky blue = Pd, Pt; white = H. Organic residues are omitted for clarity. *Sources:* Modified from Dainese et al. [122] and Ito et al. [123]. © John Wiley & Sons.

the world of new molecular systems at nanoscale. This section introduces an emerging trend in the synthesis of superatomic molecules.

Maran reported the formation of $Au_{38}(SR)_{24}$ by dimerization of $[Au_{25}(SR)_{18}]^0$ (R = C_3H_7, C_4H_9, C_8H_{17}, C_2H_2Ph) (Figure 6.24) [122]. In this reaction, two $\mathbf{Au_{13}}$(7e) superatomic cores are fused into a bi-icosahedral $\mathbf{Au_{23}}$(14e):

$$\mathbf{Au_{13}}(7e) + \mathbf{Au_{13}}(7e) \rightarrow \mathbf{Au_{23}}(14e) + 3\,Au^{+} \tag{6.16}$$

The faster reaction with smaller thiolates suggests that the overlap of SOs is crucial. Recently, Takano and Tsukuda synthesized **MM'**$\mathbf{Au_{21}}$(12e) (M, M' = Pd, Pt) by reacting $[HMAu_8(PPh_3)_8]^+$ with $[M'Au_{24}(PET)_{18}]^-$ (Figure 6.24) [123]. SCXRD revealed that **MM'**$\mathbf{Au_{21}}$(12e) corresponds to homodimer and heterodimer of **M@**$\mathbf{Au_{12}}$(6e) and **M'@**$\mathbf{Au_{12}}$(6e). This is the first example of targeted synthesis of heterodimers of superatoms via fusion reactions of the corresponding precursors. It would be interesting to study the scope of hydride-mediated fusion reactions to create larger and hetero superatomic molecules in the future.

Anisotropic growth may be another possible route to synthesizing superatomic molecules. Tsukuda has proposed that gold ultrathin nanorods (diameters of <2 nm and aspect ratios of <30) are generated by the oriented attachment mechanism in the slow reduction of Au(I) ions by tri-isopropylsilane in the presence of oleylamine (OA) [124]. Slow reduction of Au(I)-meba polymers (meba = 4-(2-mercaptoethyl)benzoic acid) under basic conditions (pH > 12) yielded green-colored products with the formula of $Au_{76}(meba)_{44}$ [125]. Powder XRD studies suggested that the Au core has a one-dimensional FCC structure that is elongated along the {100} direction; a penta-cuboctahedral Au_{49} core constructed from the Au_{13} cuboctahedrons by sharing their (100) facets has been proposed. These results suggest that superatomic

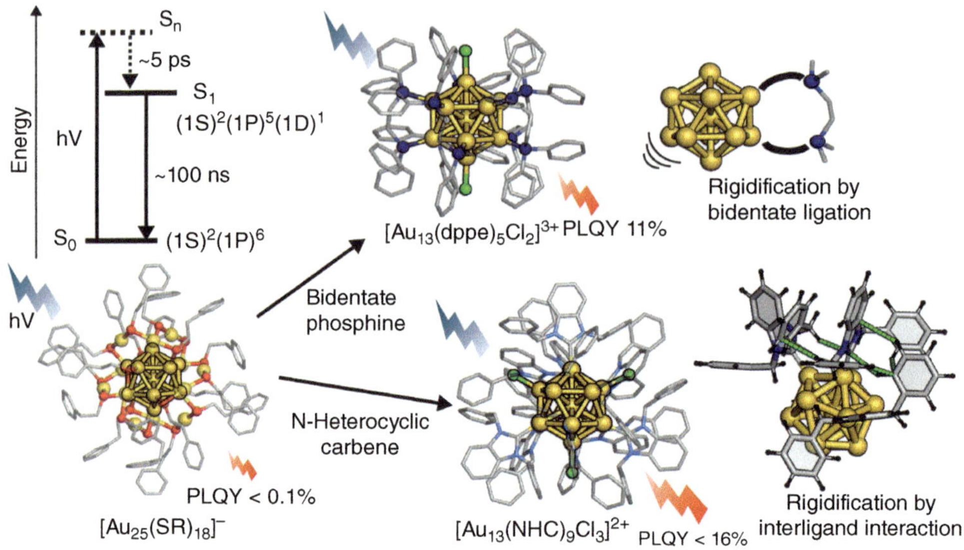

Figure 6.25 Luminescent properties in chemically modified Au superatoms depending on the surface-protecting motif. Color codes: yellow = Au; blue = P; red = S, gray = C; light blue = N; green = Cl. Organic residues are depicted as sticks except for the schematic illustration of the multipoint interactions in $[Au_{13}(NHC)_9Cl_3]^{2+}$.

oligomers or polymers may be produced by anisotropic nucleation and growth of Au(0) atoms formed by the slow reduction of 1-D Au(I) polymers formed via aurophilic interaction.

6.3.4 Surface Control

Ligands do not just protect the superatoms from aggregation but also affect the stabilities and fundamental properties of the superatoms. Modification with designed ligands will be necessary to develop functional materials based on superatoms. For example, the photoluminescence quantum yield (PLQY) of the $\mathbf{Au_{13}}$(8e) superatom is enhanced from <0.5% [120] to 11 and 16% by replacing $Au_2(SR)_3$ layers with bidentate phosphines [126, 127] or monodentate NHC, [30] respectively (Figure 6.25). This is probably due to the rigidification of the $\mathbf{Au_{13}}$(8e) superatom by bidentate fixation and by CH–π and π–π interactions between adjacent ligands. This fixation suppresses the nonradiative relaxation of the photoexcited states via suppression of soft lateral motion of the core surface. Methods of controlling the surface modification are introduced below.

6.3.4.1 Ligand Exchange

The exchange of ligands on preformed Au/Ag superatoms with other ligands having higher affinity is a direct way of tuning their properties by surface modification. For example, the thermal stabilities of $\mathbf{Au_{11}}$(8e) and $\mathbf{Au_{13}}$(8e) are enhanced by replacing phosphine with NHC [31] and thiolate with selenolate [128], respectively. However, the superatomic cores

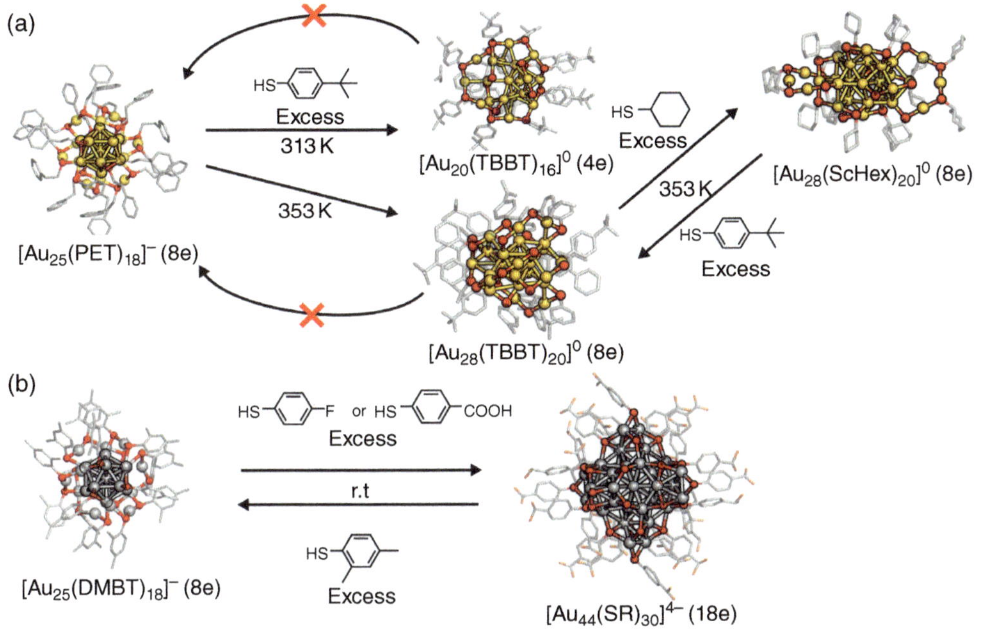

Figure 6.26 Examples of LEIST and reversible LEIST of (a) Au MPC and (b) Ag MPC. TBBT = $(CH_3)_3CC_6H_4S$; *Sc*Hex = *c*-$C_6H_{11}S$; DMBT = 2,4-$(CH_3)_2C_6H_3S$. Color codes: yellow = Au; gray = silver; red = S. Organic moieties of the ligands are depicted as sticks. *Sources:* (a and b) Modified from Chen et al. [129] and Bootharaju et al. [130]. © John Wiley & Sons.

are not always retained during the ligand exchange but are transformed to superatomic cores with different sizes and structures. For example, the $\mathbf{Au_{13}}$(8e) core of $[Au_{25}(PET)_{18}]^-$ was transformed by the reaction with TBBT-H (TBBT-H = 4-$(CH_3)_3CC_6H_4SH$) to the $\mathbf{Au_7}$(4e) core in $[Au_{20}(TBBT)_{16}]^0$ or the $\mathbf{Au_{14}}$(8e) core in $[Au_{28}(TBBT)_{20}]^0$, depending on the reaction temperature (Figure 6.26a) [131, 132]. Such ligand-exchange-induced size/structure transformation (LEIST) [133] has also been observed for the $\mathbf{Au_{23}}$(14e) core of $Au_{38}(PET)_{24}$ [134, 135] and the $\mathbf{Au_{114}}$(84e) core of $Au_{144}(PET)_{60}$ [136, 137]. Although the detailed mechanism remains unclear, complex reactions between the MPCs induced by destabilization due to steric hinderance between adjacent ligands are involved in the LEIST [138]. Reversible LEIST was observed: the $\mathbf{Au_{13}}$(8e) core of $Au_{28}(TBBT)_{20}$ was transformed to another $\mathbf{Au_{13}}$(8e) core in $Au_{28}(ScHex)_{20}$ (Figure 6.26a) [129]. Bakr demonstrated by ESI–MS analysis that conversion of the $\mathbf{Ag_{13}}$(8e) core of $[Ag_{25}(2,4\text{-DMBT})_{18}]^-$ to the $\mathbf{Ag_{32}}$(18e) core of $[Ag_{44}(4\text{-FTP})_{30}]^{4-}$ (4-FTP = 4-FC_6H_4S) proceeded via dimerization, while the reverse conversion proceeded by dissociative disproportionation to form Ag_{25} and smaller Ag(I)-SR fragments (Figure 6.26b) [130]. These examples illustrate that LEIST allows us to create new superatoms that cannot be synthesized by the direct reduction method. However, it is still difficult to predict what type of transformation will be induced by ligand exchange.

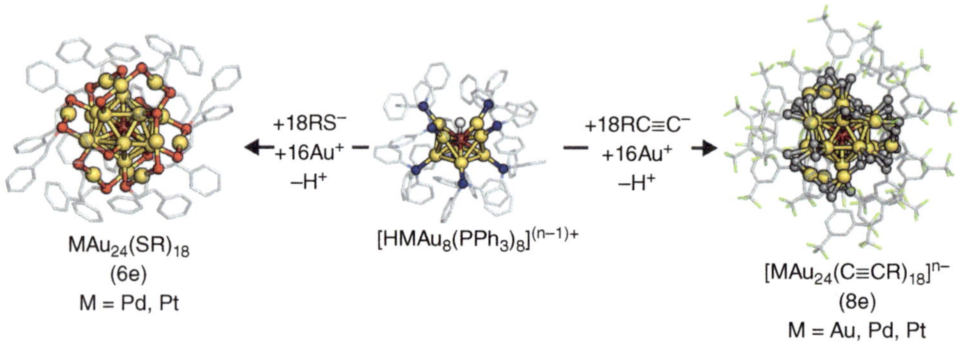

Figure 6.27 Hydride-mediated surface transformation reactions. Color codes: yellow = Au; dark red = M (Au, Pd, Pt); blue = P; red = S; gray = C; light green = F. Organic residues are depicted as sticks.

6.3.4.2 Hydrogen-Mediated Transformation

The phosphine ligands of the **$HMAu_8$** (M = Pd, Pt) superatoms can be replaced with thiolates or alkynyls by reacting with the corresponding Au(I) complexes [139]. The **$HMAu_8$** (M = Pd, Pt) superatom undergoes nucleophilic attack on the Au(I)-SR complex, selectively yielding $MAu_{24}(SR)_{18}$ with an icosahedral **$M@Au_{12}$**(6e) superatomic core (Figure 6.27). The production of $MAu_{24}(SR)_{18}$ via this transformation reaction is efficient and selective for primary thiolates: the yield of the final product is >50% with 200 mg scale. The selectivity of $MAu_{24}(SR)_{18}$ production is much higher than that by the co-reduction of thiolate complexes of Au(I) and Pd(II) ions. An analysis of the by-products suggests that the transformation reaction proceeds as follows:

$$\left[HMAu_8\left(PPh_3\right)_8\right]^+ + 24AuClSMe_2 + 18RSH \rightarrow MAu_{24}\left(SR\right)_{18} + 8AuClPPh_3 + 24SMe_2 + 16HCl + H_2 + H^+ \tag{6.17}$$

The PPh_3 ligands that are released are captured in the form of $AuClPPh_3$ and hinder the back-attack to the $MAu_{24}(SR)_{18}$ product. The putative intermediate $[MAu_{24}(SR)_{18}]^{2-}$ having the **$M@Au_{12}$**(8e) core might be oxidized under ambient conditions due to its low oxidation potential as revealed by electrochemistry (Figure 6.6). Nonetheless, this method has a limitation in the scope: the yield is reduced for secondary thiolates and becomes negligible for tertiary thiolates probably due to the larger steric demand. Unprecedented clusters $[MAu_{24}(C{\equiv}CR)_{18}]^{2-}$ having an icosahedral **$M@Au_{12}$**(8e) superatom were obtained by the reaction of $[HMAu_8(PPh_3)_8]^{(n-1)-}$ and Au-C≡CR complexes (Figure 6.27). The formation of an eight-electron superatom in $[MAu_{24}(C{\equiv}CR)_{18}]^{2-}$ can be ascribed to the upshift of the oxidation potential of the alkynyl-protected superatoms by ~1.2 eV.

6.4 Summary

This chapter examined chemically synthesized Au- and Ag-based superatoms protected by a monolayer of organic ligands, including thiols, phosphines, and terminal alkynes. SCXRD revealed the formation of ligand-specific interfacial structures on the metallic cores.

Phosphines, carbenes, and halides are directly coordinated to the metallic core through the P–M, C–M, and X–M bonds (M = Au/Ag). In contrast, thiolates and alkynyls not only directly coordinate to the core surface by the S–M and C–M bonds (M = Au/Ag) but they also form multidentate 2D or 3D oligomers containing Au(I)/Ag(I), as schematically presented in Table 6.1. These ligand layers not only protect the metal cores from aggregation but also contribute to their electronic structures. A simple electron-counting scheme (Eq. 6.1) shows that the numbers of the valence electrons in Au/Ag MPCs with spherical metallic cores agree with those of naked spherical jellium clusters. The successful application of Eq. (6.1) for counting the number of valence electrons in Au/Ag MPCs suggests that they can be viewed as Au/Ag superatomic cores and ligand layers in the framework of a "divide and protect" concept.

Section 6.2 summarized the impact of structural factors, such as size, composition, and shapes, on the electronic structures of the superatomic cores. The size-dependent evolution of the HL gap of spherical superatoms, doping effects on redox properties of spherical superatoms, and electronic shell closures of nonspherical superatoms were qualitatively explained by considering how the volume, shape, and depth of the jellium potential are affected by these structural factors. The reduction of the HL gap with size is ascribed to the increase of the volume of the jellium potential and the occupation of SOs in high-energy and dense regions (Figure 6.3). The effects of doping on the electronic structures of superatoms can be qualitatively predicted by a two-step spherical jellium model, which is based on the idea that the host and dopant atoms provide different but uniform background potential. When the dopant atom has higher valency than the host atoms, the SOs are stabilized because the effective potential becomes more attractive, and vice versa (Figure 6.6). The electronic shell-closing behaviors of prolate and oblate superatoms were explained by the degeneration of SOs induced by deformation of the jellium potentials (Figure 6.8). Some prolate superatoms can be viewed as quasi-molecules made of spherical superatoms. Their bonding scheme can be understood by supervalence bond theories. Although such structural simplification blurs the detailed structural information obtained by SCXRD, these qualitative pictures provide a rational design principle for the development of new functionalized superatoms.

Section 6.3 focused on the experimental methods for precisely controlling the structural parameters of chemically modified Au/Ag superatoms. Conventional methods are based on kinetically controlled reduction of organometallic complexes, followed by size focusing by chemical etching and fractionation. Atomic-level monodispersity of the products is confirmed by mass spectrometry. Although these methods have successfully yielded atomically precise pure and alloyed MPCs, there is much room for improving the selectivity and scale of synthesis. Section 6.3 also introduced several transformation reactions recently developed for targeted, large-scale synthesis. The methods include AGRs for controlled doping and LEISTs. Hydride-mediated transformation is a promising, bottom-up approach to controlling the core size, composition, surface modification, and fusion of superatoms.

References

1 de Heer, W.A. (1993). *Rev. Mod. Phys. 65*: 611–676.

2 Taylor, K.J., Pettiette-Hall, C.L., Cheshonvsky, O., and Smally, R.E. (1992). *J. Chem. Phys. 96*: 3319–3329.

3 Khanna, S.N. and Jena, P. (1995). *Phys. Rev. B 51*: 13705–13716.
4 Häkkinen, H. (2008). *Chem. Soc. Rev. 37*: 1847–1859.
5 Jena, P. (2013). *J. Phys. Chem. Lett. 4*: 1432–1442.
6 Luo, Z. and Castleman, A.W. Jr. (2014). *Acc. Chem. Res. 47*: 2931–2940.
7 Luo, Z., Castleman, A.W. Jr., and Khanna, S.N. (2016). *Chem. Rev. 116*: 14456–14492.
8 Jena, P. and Sun, Q. (2018). *Chem. Rev. 118*: 5755–5870.
9 Knight, W.D., Clemenger, K., de Heer, W.A. et al. (1984). *Phys. Rev. Lett. 52*: 2141–2143.
10 Wang, L.-M. and Wang, L.-S. (2012). *Nanoscale 4*: 4038–4053.
11 Jin, R., Zeng, Z., Zhou, M., and Chen, Y. (2016). *Chem. Rev. 116*: 10346–10413.
12 Chakraborty, L. and Pradeep, T. (2017). *Chem. Rev. 117*: 8208–8271.
13 Kang, X., Li, Y., Zhu, M., and Jin, R. (2020). *Chem. Soc. Rev. 49*: 6443–6514.
14 Negishi, Y., Kawawaki, T., Imai, Y. et al. (2020).*Chem. Eur. J. 26*: 16150–16193.
15 Omoda, T., Takano, S., and Tsukuda, T. (2021). *Small 17*: 2001439.
16 Hirai, H., Ito, S., Takano, S. et al. (2020). *Chem. Sci. 11*: 12233–12248.
17 Jadzinsky, P.D., Calero, G., Ackerson, C.J. et al. (2007). *Science 318*: 430–433.
18 Walter, M., Akola, J., Lopez-Acevedo, O. et al. (2008). *Proc. Natl. Acad. Sci. U. S. A. 105*: 9157.
19 Yang, H., Wang, Y., Chen, X. et al. (2016). *Nat. Commun. 7*: 12809.
20 Häkkinen, H., Walter, M., and Grönbeck, H. (2006). *J. Phys. Chem. B 110*: 9927–9931.
21 Akola, J., Walter, M., Whetten, R.L. et al. (2008). *J. Am. Chem. Soc. 130*: 3756–3757.
22 Zeller, E., Beruda, H., and Schmidbaur, H. (1993). *Inorg. Chem. 32*: 3203–3204.
23 Heaven, M.W., Dass, A., White, P.S. et al. (2008). *J. Am. Chem. Soc. 130*: 3754–3755.
24 Zhu, M., Aikens, C.M., Hollander, F.J. et al. (2008). *J. Am. Chem. Soc. 130*: 5883–5885.
25 Lopez-Acevedo, O., Akola, J., Whetten, R.L. et al. (2009). *J. Phys. Chem. C 113*: 5035–5038.
26 Shichibu, Y. and Konishi, K. (2010). *Small 6*: 1216–1220.
27 McKenzie, L.C., Zaikova, T.O., and Hutchison, J.E. (2014). *J. Am. Chem. Soc. 136*: 13426–13435.
28 Yan, N., Xia, N., Liao, L. et al. (2018). *Sci. Adv. 4*: eaat7259.
29 Li, J.-J., Guan, Z.-J., Hu, F., and Wang, Q.-M. (2019). *Angew. Chem. Int. Ed. 58*: 1083–1087.
30 Narouz, M.R., Takano, S., Lummis, P.A. et al. (2019). *J. Am. Chem. Soc. 141*: 14997–15002.
31 Narouz, M.R., Osten, K.M., Unsworth, P.J. et al. (2019). *Nat. Chem. 11*: 419–425.
32 Hirata, K., Yamashita, K., Muramatsu, S. et al. (2017). *Nanoscale 9*: 13409–13412.
33 Martin, T.P., Bergmann, T., Göhlich, H., and Lange, T. (1991). *J. Phys. Chem. 95*: 6421–6429.
34 Negishi, Y., Nakazaki, T., Malola, S. et al. (2015). *J. Am. Chem. Soc. 137*: 1206–1212.
35 Zhou, M., Higaki, T., Li, Y. et al. (2019). *J. Am. Chem. Soc. 141*: 19754–19764.
36 Thomas, O.C., Zheng, S., Xu, S., and Bowen, K.H. Jr. (2002). *Phys. Rev. Lett. 89*: 213403.
37 Dass, A., Sakthivel, N.A., Jupally, V.R. et al. (2020). *ACS Energy Lett. 5*: 207–214.
38 Bakr, O.M., Amendola, V., Aikens, C.M. et al. (2009). *Angew. Chem. Int. Ed. 48*: 5921–5926.
39 Desireddy, A., Conn, B.E., Guo, J. et al. (2013). *Nature 501*: 399–402.
40 Yang, H., Wang, Y., Huang, H. et al. (2013). *Nat. Commun. 4*: 2422.
41 AbdulHalim, L.G., Bootharaju, M.S., Tang, Q. et al. (2015). *J. Am. Chem. Soc. 137*: 11970–11975.
42 Joshi, C.P., Bootharaju, M.S., Alhilaly, M.J., and Bakr, O.M. (2015). *J. Am. Chem. Soc. 137*: 11578–11581.
43 Hu, F., Li, J.-J., Guan, Z.-J. et al. (2020). *Angew. Chem. Int. Ed. 59*: 5312–5315.

44 Veenstra, A.P., Monzel, L., Baksi, A. et al. (2020). *J. Phys. Chem. Lett. 11*: 2675–2681.
45 Hirata, K., Kim, K., Nakamura, K. et al. (2019). *J. Phys. Chem. C 123*: 13174–13179.
46 Tasaka, Y., Nakamura, K., Malola, S. et al. (2020). *J. Phys. Chem. Lett. 11*: 3069–3074.
47 Scheller, M.K., Compton, R.N., and Cederbaum, L.S. (1995). *Science 270*: 1160–1166.
48 Wang, X.-B., Ding, C.-F., and Wang, L.-S. (1998). *Phys. Rev. Lett. 81*: 3351–3354.
49 Wang, X.-B. and Wang, L.-S. (1999). *Nature 400*: 245–248.
50 Yan, J., Su, H., Yang, H. et al. (2015). *J. Am. Chem. Soc. 137*: 11880–11883.
51 Tofanelli, M.A., Ni, T.W., Philllips, B.D., and Ackerson, C.J. (2016). *Inorg. Chem. 55*: 999–1001.
52 Fei, W., Antonello, S., Dainese, T. et al. (2019). *J. Am. Chem. Soc. 141*: 16033–16045.
53 Hirai, H., Takano, S., Nakamura, T., and Tsukuda, T. (2020). *Inorg. Chem. 59*: 17889–17895.
54 Kumara, C., Aikens, C.M., and Dass, A. (2014). *J. Phys. Chem. Lett. 5*: 461–466.
55 Zhang, S.B., Cohen, M.L., and Chou, M.Y. (1987). *Phys. Rev. B 36*: 3455–3458.
56 Baladrón, C. and Alonso, J.A. (1988). *Phys. B 154*: 73–81.
57 Yannouleas, C., Jena, P., and Khanna, S.N. (1992). *Phys. Rev. B 46*: 9751–9760.
58 Janssens, E., Neukermans, S., and Lievens, P. (2004). *Curr. Opin. Solid State Mater. Sci. 8*: 185–193.
59 Kim, K., Hirata, K., Nakamura, K. et al. (2019). *Angew. Chem. Int. Ed. 58*: 11637–11641.
60 Kwak, K., Tang, Q., Kim, M. et al. (2015). *J. Am. Chem. Soc. 137*: 10833–10840.
61 Suyama, M., Takano, S., Nakamura, T., and Tsukuda, T. (2019). *J. Am. Chem. Soc. 141*: 14048–14051.
62 Hu, G., Tang, Q., Lee, D. et al. (2017). *Chem. Mater. 29*: 4840–4847.
63 Bootharaju, M.S., Dey, R., Gevers, L.E. et al. (2016). *J. Am. Chem. Soc. 138*: 13770.
64 Yuan, X., Sun, C., Li, X. et al. (2019). *J. Am. Chem. Soc. 141*: 11905–11911.
65 Sun, C., Teo, B.K., Deng, C. et al. (2021). *Coord. Chem. Rev. 427*: 213576.
66 Buckart, S., Ganteför, G., Kim, Y.D., and Jena, P. (2003). *J. Am. Chem. Soc. 125*: 14205–14209.
67 Takano, S., Hirai, H., Muramatsu, S., and Tsukuda, T. (2018). *J. Am. Chem. Soc. 140*: 8380–8383.
68 Takano, S., Hirai, H., Muramatsu, S., and Tsukuda, T. (2018). *J. Am. Chem. Soc. 140*: 12314–12317.
69 Takano, S., Hasegawa, S., Suyama, M., and Tsukuda, T. (2018). *Acc. Chem. Res. 51*: 3074–3083.
70 Yuan, S.-F., Li, J.-J., Guan, Z.-J. et al. (2020). *Chem. Commun. 56*: 7037–7040.
71 van der Velden, J.W.A., Bour, J.J., Bosman, W.P. et al. (1981). *Soc. Chem. Comm.*: 1218–1219.
72 Schulz-Dobrick, M. and Jansen, M. (2006). *Eur. J. Inorg. Chem.*: 4498–4502.
73 Schulz-Dobrick, M. and Jansen, M. (2008). *Angew. Chem. Int. Ed. 47*: 2256–2259.
74 Takano, S. and Tsukuda, T. (2016). *J. Phys. Chem. Lett. 7*: 4509–4513.
75 Tofanelli, M.A., Salorinne, K., Ni, T.W. et al. (2016). *Chem. Sci. 7*: 1882–1890.
76 van der Velden, J.W.A., Bour, J.J., Steggerda, J.J. et al. (1982). *Inorg. Chem. 21*: 4321–4324.
77 Ito, S., Takano, S., and Tsukuda, T. (2019). *J. Phys. Chem. Lett. 10*: 6892–6896.
78 Wan, X.-K., Lin, Z.-W., and Wang, Q.-M. (2012). *J. Am. Chem. Soc. 134*: 14750–14752.
79 Qian, H., Eckenhoff, W.T., Zhu, Y. et al. (2010). *J. Am. Chem. Soc. 132*: 8280–8281.
80 Cheng, L., Ren, C., Zhang, X., and Yang, J. (2013). *Nanoscale 5*: 1475–1478.

81 Cheng, L. and Yang, J. (2013). *J. Chem. Phys. 138*: 141101.
82 Yuan, Y., Cheng, L., and Yang, J. (2013). *J. Phys. Chem. C 117*: 13276–13282.
83 Nishigaki, J.-i., Yamazoe, S., Kohara, S. et al. (2014). *Chem. Commun. 50*: 839–841.
84 Mingos, D.M.P. (2015). *Dalton Trans. 44*: 6680–6695.
85 Shichibu, Y., Negishi, Y., Watanabe, T. et al. (2007). *J. Phys. Chem. C 111*: 7845–7847.
86 Jin, R., Liu, C., Zhao, S. et al. (2015). *ACS Nano 9*: 8530–8536.
87 Song, Y., Fu, F., Zhang, J. et al. (2015). *Angew. Chem. Int. Ed. 54*: 8430–8434.
88 Xu, W.W., Zeng, X.C., and Gao, Y. (2018). *Acc. Chem. Res. 51*: 2739–2747.
89 Garzón, I.L., Beltrán, M.R., González, G. et al. (2003). *Eur. Phys. J. D 24*: 105–109.
90 Pelayo, J.J., Whetten, R.L., and Garzón, I.L. (2015). *J. Phys. Chem. C 119*: 28666–28678.
91 Omoda, T., Takano, S., Yamazoe, S. et al. (2018). *J. Phys. Chem. C 122*: 13199–13204.
92 Xiong, L., Yang, S., Sun, X. et al. (2018). *J. Phys. Chem. C 122*: 14898–14907.
93 Das, A., Li, T., Nobusada, K. et al. (2013). *J. Am. Chem. Soc. 135*: 18264–18267.
94 LaMer, V.K., Dinegar, R.H., and Am, J. (1950). *Chem. Soc. 72*: 4847–4854.
95 Shichibu, Y., Negishi, Y., Tsunoyama, H. et al. (2007). *Small 3*: 835–839.
96 Qian, H., Zhu, Y., and Jin, R. (2009). *ACS Nano 3*: 3795–3803.
97 Levi-Kalisman, Y., Jadzinsky, P.D., Kalisman, N. et al. (2011). *J. Am. Chem. Soc. 133*: 2976–2982.
98 Tsunoyama, H., Negishi, Y., and Tsukuda, T. (2006). *J. Am. Chem. Soc. 128*: 6036–6037.
99 Sakthivel, N.A., Theivendran, S., Ganeshraj, V. et al. (2017). *J. Am. Chem. Soc. 139*: 15450–15459.
100 Tian, S., Li, Y.-Z., Li, M.-B. et al. (2015). *Nat. Commun. 6*: 8667.
101 Ghosh, A., Hassinen, J., Pukkinen, P. et al. (2014). *Anal. Chem. 86*: 12185–12190.
102 Niihori, Y., Shima, D., Yoshida, K. et al. (2018). *Nanoscale 10*: 1641–1649.
103 Niihori, Y., Yoshida, K., Hossain, S. et al. (2019). *Bull. Chem. Soc. Jpn. 92*: 664–695.
104 Dolamic, I., Knoppe, S., Dass, A., and Bürgi, T. (2012). *Nat. Commun. 3*: 798.
105 Schaaff, T.G., Knight, G., Shafigullin, M.N. et al. (1998). *J. Phys. Chem. B 102*: 10643–10646.
106 Schaaff, T.G. and Whetten, R.L. (2000). *J. Phys. Chem. B 104*: 2630–2641.
107 Negishi, Y., Nobusada, K., and Tsukuda, T. (2005). *J. Am. Chem. Soc. 127*: 5261–5270.
108 Azubel, M., Koivisto, J., Malola, S. et al. (2014). *Science 345*: 909–912.
109 Kumar, S., Bolan, M.D., and Bigioni, T.P. (2010). *J. Am. Chem. Soc. 132*: 13141–13143.
110 Chen, T., Fung, V., Yao, Q. et al. (2018). *J. Am. Chem. Soc. 140*: 11370–11377.
111 Kambe, T., Haruta, N., Imaoka, T., and Yamamoto, K. (2017). *Nat. Commun. 8*: 2046.
112 Tsukuda, T., Tsunoyama, H., and Sakurai, H. (2011). *Chem. Asian J. 6*: 736–748.
113 Yamazoe, S., Koyasu, K., and Tsukuda, T. (2014). *Acc. Chem. Res. 47*: 816–824.
114 Tsunoyama, H., Ichikuni, N., and Tsukuda, T. (2008). *Langmuir 24*: 11327–11330.
115 Tsunoyama, H. and Tsukuda, T. (2009). *J. Am. Chem. Soc. 131*: 18216–18217.
116 Gan, Z., Xia, N., and Wu, Z. (2018). *Acc. Chem. Res. 51*: 2774–2783.
117 Wang, S., Li, Q., Kang, X., and Zhu, M. (2018). *Acc. Chem. Res. 51*: 2784–2792.
118 Zhang, H., Watanabe, T., Okumura, M. et al. (2012). *Nat. Mater. 11*: 49–52.
119 Choi, J.-P., Fields-Zinna, C.A., Stiles, R.L. et al. (2010). *J. Phys. Chem. C 114*: 15890–15896.
120 Suyama, M., Takano, S., and Tsukuda, T. (2020). *J. Phys. Chem. C 124*: 23923–23929.
121 Hirai, H., Takano, S., and Tsukuda, T. (2019). *ACS Omega 4*: 7070–7075.

122 Dainese, T., Antonello, S., Bogialli, S. et al. (2018). *ACS Nano 12*: 7057–7066.

123 Ito, E., Takano, S., Nakamura, T., and Tsukuda, T. (2021). *Angew. Chem. Int. Ed.* 60: 645–649.

124 Takahata, R. and Tsukuda, T. (2019). *Chem. Lett. 48*: 906–915.

125 Takano, S., Yamazoe, S., Koyasu, K., and Tsukuda, T. (2015). *J. Am. Chem. Soc. 137*: 7027–7030.

126 Sugiuchi, M., Shichibu, Y., Nakanishi, T. et al. (2015). *Chem. Commun. 51*: 13519–13522.

127 Gao, Z.-H., Dong, J., Zhang, Q.-F., and Wang, L.-S. (2020). *Nanoscale Adv. 2*: 4902–4907.

128 Kurashige, W., Yamaguchi, M., Nobusada, K., and Negishi, Y. (2012). *J. Phys. Chem. Lett. 3*: 2649–2652.

129 Chen, Y., Liu, C., Tang, Q. et al. (2016). *J. Am. Chem. Soc. 138*: 1482–1485.

130 Bootharaju, M.S., Joshi, C.P., Alhilaly, M.J., and Bakr, O.M. (2016). *Chem. Mater. 28*: 3292–3297.

131 Zeng, C., Li, T., Das, A. et al. (2013). *J. Am. Chem. Soc. 135*: 10011–10013.

132 Zeng, C., Liu, C., Chen, Y. et al. (2014). *J. Am. Chem. Soc. 136*: 11922–11925.

133 Zeng, C., Chen, Y., Das, A., and Jin, R. (2015). *J. Phys. Chem. Lett. 6*: 2976–2986.

134 Zeng, C., Qian, H., Li, T. et al. (2012). *Angew. Chem. Int. Ed. 51*: 13114–13118.

135 Rambukwella, M., Sementa, L., Fortunelli, A., and Dass, A. (2017). *J. Phys. Chem. C 121*: 14929–14935.

136 Zeng, C., Chen, Y., Kirschbaum, K. et al. (2015). *Sci. Adv. 1*: e1500045.

137 Nimmala, P.R., Theivendran, S., Barcaro, G. et al. (2015). *J. Phys. Chem. Lett. 6*: 2134–2139.

138 George, A., Sundar, A., Nair, A.S. et al. (2019). *J. Phys. Chem. Lett. 10*: 4571–4576.

139 Takano, S., Ito, S., and Tsukuda, T. (2019). *J. Am. Chem. Soc. 141*: 15994–16002.

7

Atomically Precise Noble Metals in the Nanoscale, Stabilized by Ligands

Hannu Häkkinen

Departments of Physics and Chemistry, Nanoscience Center, University of Jyväskylä, Jyväskylä, Finland

7.1 Introduction

The quest to develop synthesis methods for stabilizing nanoscale metals by wet chemistry is driven by two-fold motivation. First, analogous to science of gas-phase metal clusters, which has a strong research tradition spanning over several decades [1], metal clusters stabilized by ligands offer a diverse set of systems for detailed fundamental, experimental, and theoretical studies of how size, shape, and composition of the metal core affects physical properties. Compared to gas-phase metal clusters, the ligand layer adds a dual role: it determines the chemical stability of the metal core but also affects the electronic structure of the core. A tremendous added value for computational research is the fact that a large database of definite atom-scale structures, determined by single-crystal X-ray experiments, is now available. Second, wet chemistry naturally yields macroscopic amounts of the product materials, facilitating investigations of cluster materials to advance understanding of structure-property correlations.

This chapter begins by a brief introduction to the fundamental concepts on nanoscale metals, the so-called Kubo energy gap and electronic shell structure. Adding the ligands to protect the metal cluster affects not only the chemical stability but also drastically the electronic structure by changing the count of delocalized, "free," electrons in the metal core. A fundamental and transparent counting formula to account for the free electrons is discussed. This has consequences for optical properties and chemical reactivity as discussed through two selected case studies. The chapter also offers an example demonstrating that both electronic and atomic (geometric) shell closings can play a simultaneous role in stabilizing two very different-size clusters in the same synthesis. Finally, two potential applications of ligand-stabilized superatoms in biology and materials science are discussed.

Superatoms: Principles, Synthesis and Applications, First Edition. Edited by Purusottam (Puru) Jena and Qiang Sun.
© 2022 John Wiley & Sons Ltd. Published 2022 by John Wiley & Sons Ltd.

7.2 Fundamentals

7.2.1 Free Electron Model and the Kubo Gap

Every undergraduate textbook of solid-state physics contains a discussion of a "perfect theorist's metal": the three-dimensional electron gas of N electrons confined in volume V, where the electrons interact only by virtue of being fermions, i.e. obeying the Pauli exclusion principle. In such system, the dispersion relation for the electron states is parabolic ($E \propto \mathbf{k}^2$) with the wave vector $\mathbf{k}$ in reciprocal space. N electrons fill up the $N/2$ available quantum states up to the Fermi energy E_F. The density of electron states $D(E)$ follows a square root as a function of the electron energy

$$D(E) = \frac{3}{2} N \frac{\sqrt{E}}{E_F^{3/2}} \tag{7.1}$$

which leads to a simple estimate of D at the Fermi energy

$$D(E_F) = \frac{3N}{2E_F} \tag{7.2}$$

In a macroscopic metal, the density of states (DOS) has a finite value at E_F. As discussed long ago by Fröhlich [2] and Kubo [3], confining electrons in a nanometer-scale volume discretizes the allowed levels and thus can create an energy gap Δ also at the Fermi level (Figure 7.1). The existence of this fundamental gap (or "band gap") changes profoundly the physical properties of the system since the external perturbations thus need to have a minimum energy to excite electrons over the bandgap. Conventionally, the magnitude of the gap has been taken as a measure to classify systems into "metallic" (with Δ approaching zero), "semiconducting," or "insulating." For practical matters, systems have been considered "metallic" when Δ is equal or less than the thermal energy at room temperature ($k_B T_R \approx 25$ meV). This criterion leads to interesting predictions for a critical size (diameter d_{cr}) where Δ exceeds $k_B T_R$ as the size of the system is decreased. From Eq. (7.2), the mean spacing of electron levels at E_F is simply

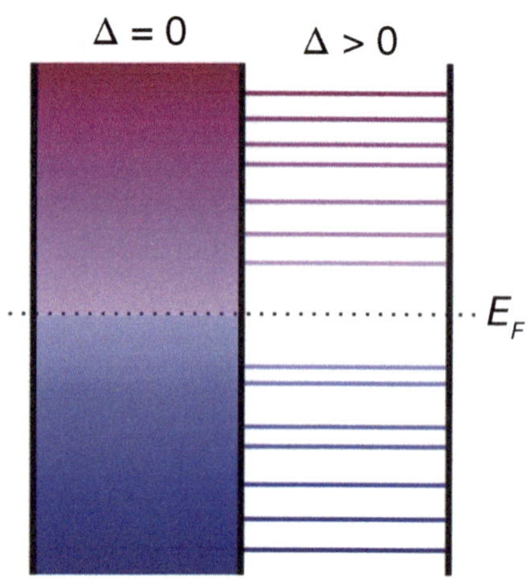

Figure 7.1 A schematic representation of the electron states in bulk metal (left) and in a nanocluster (right). The energy gap Δ is zero in the bulk metal while the gap is nonzero in the cluster. E_F denotes the Fermi energy. *Source:* Häkkinen [4]. © Taylor & Francis.

Table 7.1 Critical diameters and number of electrons for metal clusters when the "Kubo gap" equals the thermal energy. Values for the free electron density and Fermi energy are taken from Kittel [5].

	Z	n_{free} (1/nm^3)	E_F (eV)	N_{cr}	d_{cr} (nm)
Li	1	47.0	4.72	126	1.7
Na	1	26.5	3.23	86	1.8
K	1	14.0	2.12	57	2.0
Cu	1	84.5	7.00	187	1.6
Ag	1	58.5	5.48	146	1.7
Au	1	59.0	5.51	147	1.7
Mg	2	86.0	7.13	190	1.6
Ca	2	46.0	4.68	125	1.7
Zn	2	131.0	9.39	250	1.5
Al	3	180.6	11.63	310	1.5
Ga	3	153.0	10.35	276	1.5

Source: Modified from Kittel [5]. © John Wiley & Sons.

$$\Delta = \frac{1}{D(E_F)} = \frac{2E_F}{3N} \tag{7.3}$$

Note that Eq. (7.3) is equivalent to Kubo's original result ($\Delta = 4E_F/3N_{\text{atoms}}$), which was derived by assuming a monovalent metal and ignoring the spin degeneracy of electron states [3]. Clearly, Δ approaches zero as $1/N$. Knowing the value of the Fermi energy from the free electron model for the metal in question as a material parameter, Eq. (7.3) yields predictions for the critical size, both in terms of diameter (d_{cr}) and the number of electrons ($N_{cr} = 2\,E_F/3\Delta$), when $\Delta \approx k_B T_R$ and quantum size effects will emerge. Table 7.1 shows these estimates [4]. It is interesting to note that while the values of N_{cr} vary for more than sixfold, the varying valence electron densities equalize the differences and the d_{cr} values are consistently within 1.6–2 nm for different metals.

In addition to creating a fundamental bandgap at the Fermi energy, decreasing the size of metal clusters into nanometer-scale has another fundamental effect for single-electron levels, as will be discussed next.

7.2.2 Electron Shell Structure

Shell structure, i.e. bunching of single-electron levels into groups of (nearly) degenerate energy, is a fundamental property of fermions confined in small systems. It is observed for nucleons in atomic nuclei, for electrons in ordinary atoms, in metal nanoclusters and in semiconductor quantum dots, and arises from the symmetry properties of the Schrödinger equation governing the solutions to their quantum states. Historically, the concept of fermionic shell structure and symmetry principle in quantum mechanics were tied together in the Nobel Prize of Physics 1963. Half of the prize was given to Eugene Paul Wigner "for his contribution to the theory of the atomic nucleus and the elementary particles, particularly through the discovery and application of fundamental symmetry principles" and the other half was shared by Maria Goeppert Mayer and J. Hans D. Jensen "for their discoveries concerning nuclear shell structure." [6] In fact, theories first worked out for nuclear shell structure in 1950s greatly fertilized the field of metal cluster physics in 1980s and 1990s.

The basics of the shell structure can be grasped by considering quantum mechanics of noninteracting fermions in simple forms of radial potentials, solved routinely as an exercise in undergraduate physics courses. Figure 7.2 shows the degeneracies ($2(2l+1)$) of solutions to the radial Schrödinger equation for the 3D infinite spherical well with a "square box shape" organized to the increasing order in energy according to the angular momentum (l) subshells [7]. Note that it is customary to label each set of subshells by a "principal" quantum number starting always from $n=1$ (1s, 1p, 1d, . . .; 2s, 2p, 2d, . . .). This notation differs from electron shells in hydrogen-like atoms where additional dependencies between quantum numbers arise due to the $1/r$ Coulomb potential between electrons and the point-like nucleus. Figure 7.2 clearly shows that angular momentum subshells at higher energies start to group together; this is seen already in the set of 2d–1h–3s subshells that accommodate electrons between the major shell-closing ("magic") numbers of 58 and 92. Furthermore, details of the radial potential affect sensitively this grouping. For instance, the infinite harmonic well ($V \propto r^2$) has high degenerate levels at strong magic numbers of 40, 70, 112, 168, . . ., but when the shape of the potential is modified by increasing anharmonicity, finally reaching a shape of the square well, many of the subshells split and group together in a new way. This gives rise to the often-quoted series of magic numbers 2, 8, 18,

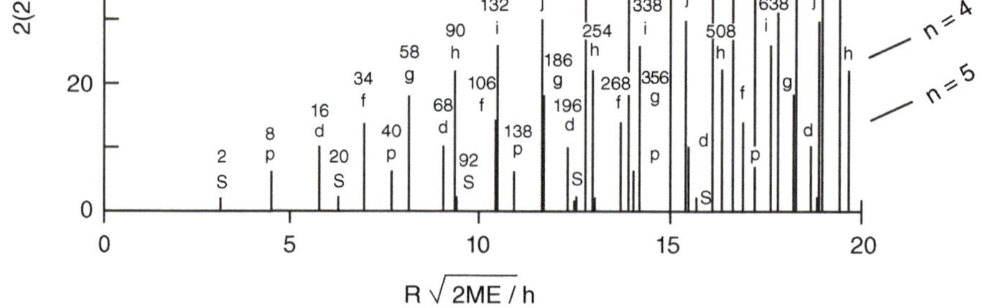

Figure 7.2 Shell structure of noninteracting fermions in an infinite spherical well. Solutions to the radial Schrödinger equation are grouped into subshells according to the radial quantum number (l) with degeneracy of $2(2l+1)$. The cumulative number of fermions needed to fill all levels up to each subshell is shown. The shells are ordered according to an increasing momentum (x-axis). *Source:* Martin et al. [7]. © American Chemical Society.

20, 34, 40, 58, ... For a more detailed discussion, see Chapters 1 and 6 of this book and the classic reviews of deHeer [8] and Brack [9].

The simplest theoretical model to account for electron–electron interactions in metal clusters is the so-called jellium model [9, 10]. Figure 7.3 shows the optimal shapes of electron clusters from 2 to 8 electrons, calculated from the "ultimate jellium model," which takes unbiased background deformations into account [11]. It is important to realize that the obtained shapes are free from any shape constraints, i.e. all possible multipolar deformations can occur. The evolution of the shape is easily understood by considering shapes of the single-particle s, p_x, p_y, and p_z orbitals. Three- and four-electron clusters have one and two electrons in one of the p orbitals – say p_x, which makes the electron density axially symmetric and prolate. The initial symmetry degeneration of p orbitals is broken by the Jahn–Teller deformation, which favors shape deformation rather than a symmetric shape with paramagnetic spin configuration (such as occupation of spin triplet $1s^2$ $1p_x^{2/3}p_y^{2/3}p_z^{2/3}$ for a spherical four-electron cluster). The fifth and sixth electrons go to the p_y orbital, making the six-electron cluster axially symmetric oblate. Quasi-2D models for circular, hexagonal, and triangular shapes predict six as a shell-closing number. Clusters with seven and eight electrons fill the remaining p_z orbital, which makes the eight-electron cluster a magic, spherical $1s^2$ $1p^6$ octet.

Atomistic density functional theory (DFT) calculations for sodium clusters have predicted minimum-energy structures that rather well reflect the shapes of the electron density. As a

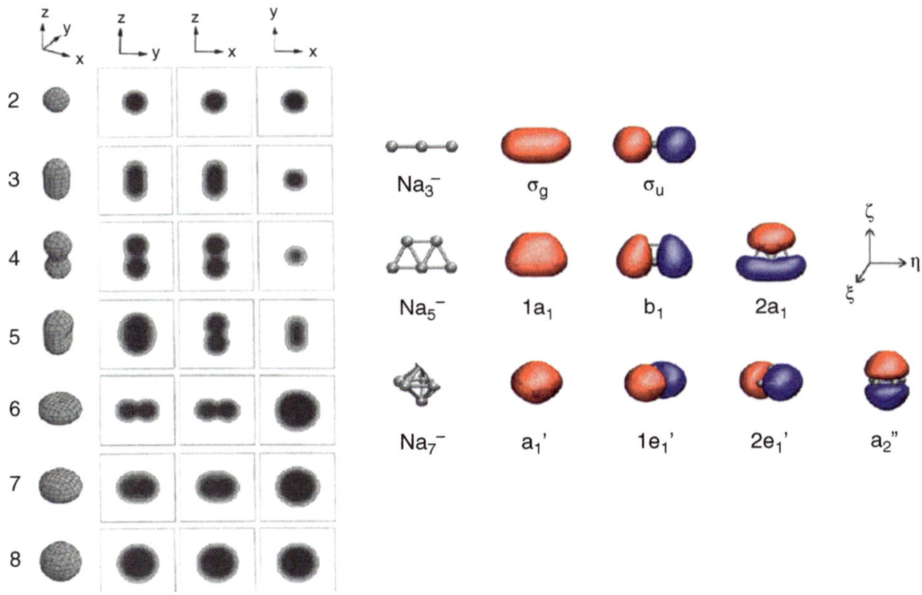

Figure 7.3 Left: Ground-state densities of electron cluster from two to eight electrons, obtained from the "ultimate" jellium model. The 3D constant density surface is plotted at the value of 38% of the bulk density. *Source:* Koskinen et al. [11]. © Springer Nature. Right: orbitals for the valence electrons of Na_3^-, Na_5^-, and Na_7^- from atomistic DFT calculations. Angular-dependent PE spectroscopy of these clusters can be interpreted by using this orbital scheme. *Source:* Bartels et al. [12]. © American Chemical Society.

recent example, photoelectron (PE) imaging spectroscopy of sodium cluster anions Na_3^- (4e), Na_5^- (6e), and Na_7^- (8e) was interpreted [12] in terms of atomic structures and electron orbitals as shown in Figure 7.3.

7.2.3 Ligand-Stabilized Metal Clusters as Superatoms

In recent years, ever-increasing numbers of chemically stabilized ("passivated") metal clusters have been synthesized, and successful structural characterizations have opened a fascinating view on how metal atoms organize in the nanometer scale [13–17]. The large database of solved structures has given a treasure trove for theoretical analysis. It has been confirmed numerous times that electron shell structure does play a role and is a useful concept for understanding electronic and optical properties of chemically stabilized metal clusters (the so-called monolayer protected clusters or MPCs). This has led to the use of the concept of "superatom" in the field of MPCs as well, just as it has been used for decades in the field of gas-phase metal clusters [14].

Generally, the composition of the ligand-stabilized clusters, that have *N* metal atoms (*M*), *S* charge-donating ligands (*L*), *Y* charge-withdrawing ligands (*X*, such as halides, thiolates, alkynyls), and an overall charge *z*, can be written as $[L_S\ M_N\ X_Y]^z$. In addition, the ligand layer may include "neutral" ligands that do not affect the free electron count (such as phosphines). The count of free electrons is defined by the simple formula

$$n_e = N\nu_M - Y\nu_X + S\nu_L - z \tag{7.4}$$

where ν_M, ν_X, and ν_L are the valence numbers of the metal and the ligands. It is clear that this formula may be generalized to intermetallic clusters as well, including all the metal types and their valences.

As an example, the electronic structure of three different systems, consisting of different metals (Au, Al, Ga) and different organic ligand layers, was compared by DFT calculations [18]. Figure 7.4 shows their compositions and atomic structures as determined from single-crystal X-ray diffraction: $Au_{102}(SC_6H_5COOH)_{44}$ (Ref. [19]), $Al_{50}(C_5(CH_3)_5)_{12}$ (Ref. [20]), and $Ga_{23}(N(Si(CH_3)_3)_2)_{11}$ (Ref. [21]). All the ligands in these systems are one-electron-withdrawing ligands, so Eq. (7.4) yields the free electron counts of 58 (102 − 44), 138 (3 × 50 − 12), and 58 (3 × 23 − 11) for the gold, aluminum, and gallium compounds, respectively. The symmetries of highest occupied molecular orbital (HOMO) and lowest unoccupied molecular orbital (LUMO) orbitals were analyzed by means of spherical harmonics, and the results are shown in Figure 7.5. It is seen that for both the 58 electron systems (Au_{102} and Ga_{23}), the symmetry changes from 1G to 1H over the HOMO–LUMO gaps (with a strong mixing of 2D for the gallium cluster). Hence, for the 138 electron Al_{50} cluster, the symmetry changes from (predominantly) 1I to 2G. These results are in agreement with the shell model in an approximately spherical system.

7.2.3.1 Case Study: The $(Ag_{44}(SR)_{30})^{4-}$ Superatom

In 2013, two independent papers [22, 23] reported simultaneously the synthesis, atomic structure, and optical properties of a 44-atom silver cluster, stabilized by 30 thiolates. This was a remarkable breakthrough as silver was thought to be a problematic metal for achieving large nanoclusters that would be ambient stable for extended periods of time.

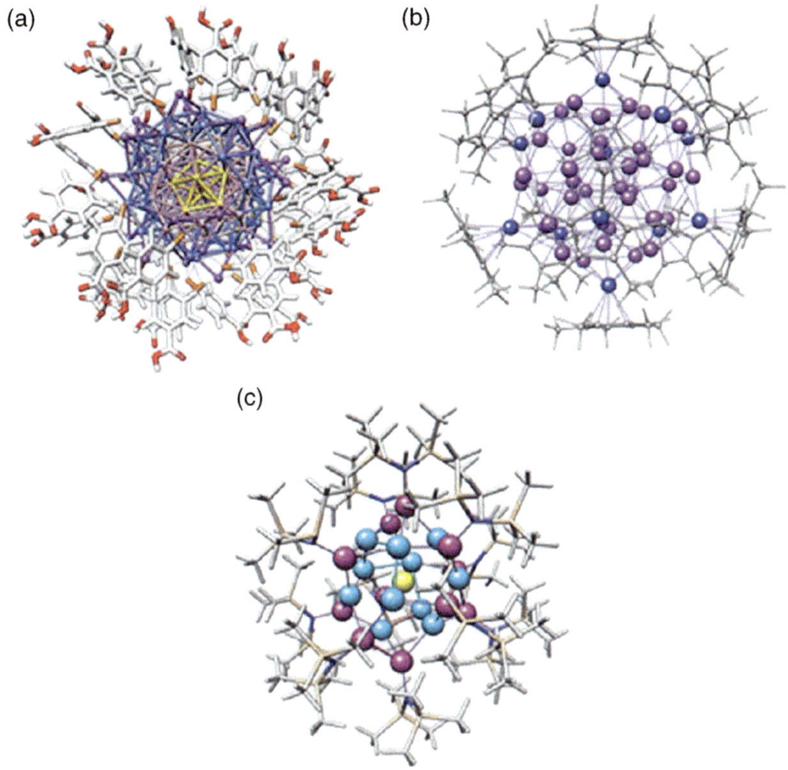

Figure 7.4 Visualization of ligand-stabilized metal nanoclusters composed of three different metals: (a) protected gold cluster (58e) $Au_{102}(SC_6H_5COOH)_{44}$; (b) protected aluminum cluster (138e) $Al_{50}(C_5(CH_3)_5)_{12}$; (c) protected gallium cluster (58e) $Ga_{23}(N(Si(CH_3)_3)_2)_{11}$. The colored spheres denote metal atoms in various atom shells of the core, and the organic layer is shown by sticks. A few ligands have been removed in the front for a better view of the core. *Source:* Lopez-Acevedo et al. [18]. © American Physical Society.

Both papers also reported scale-up of the synthesis to macroscopic quantities (10–100 g) of the cluster material, in striking difference to thiolate-stabilized gold clusters where scale-up of the synthesis remains a considerable challenge. Interestingly, the crystal unit cell contained four countercations per silver cluster, giving the cluster's chemical composition as $(Ag_{44}(SR)_{30})^{4-}$ indicating an unusually high negative charge. Equation (7.4) then yields 18 free electrons in the system, suggesting a superatom electronic configuration $1S^2$ $1P^6$ $1D^{10}$. Figure 7.6 shows the atomic structure of $Ag_{44}(SR)_{30}{}^{4-}$ where SR = $SPhF_2$. The metal core is a two-shell Keplerate Ag_{12}(icosahedron)@Ag_{20}(dodecahedron) structure, covered by six 3D $Ag_2(SR)_5$ units on its surface. It is remarkable to find out that the metal core has one-atom vacancy at the center. This atomic structure coincides with the 18-electron shell structure in a natural way, since the 2S electronic state is pushed up in the energy due to the missing atom at the center, stabilizing a significant HOMO–LUMO energy gap (DFT-PBE (Perdew-Burke-Ernzerhof) value of 0.8 eV) at 18 electrons instead of 20 electrons.

Due to the high negative charge of $Ag_{44}(SR)_{30}{}^{4-}$, anion PE spectroscopy is a suitable experimental technique to study the electronic structure of this superatom in the gas phase.

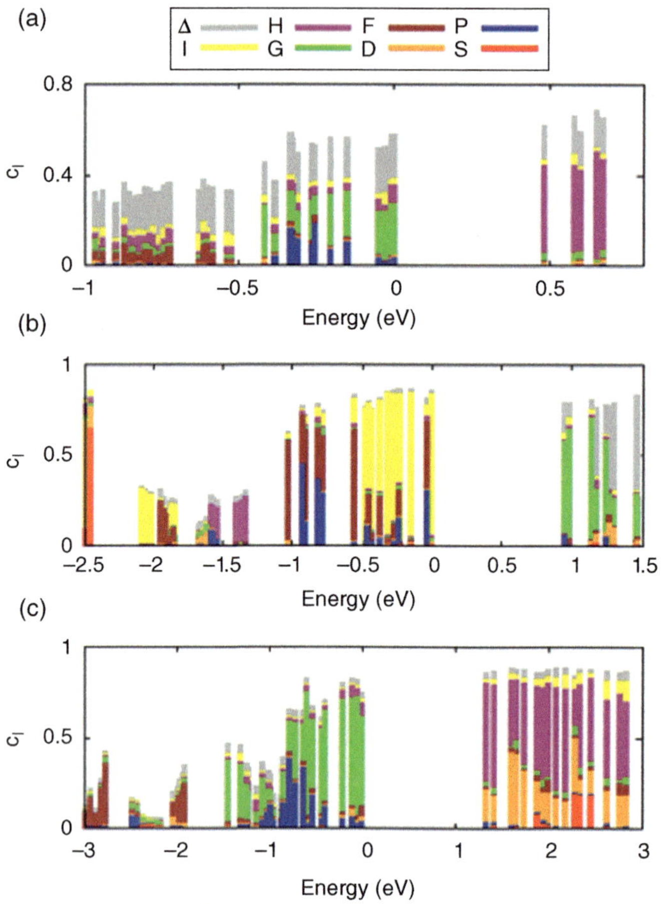

Figure 7.5 Superatom analysis: angular momentum weights for Kohn–Sham orbitals for the systems shown in Figure 7.4. The HOMO state is at zero energy. (a) protected gold cluster, (b) protected aluminum cluster, (c) protected gallium cluster. *Source:* Lopez-Acevedo et al. [18]. © American Physical Society.

Tsukuda's group measured recently the PE spectra of $Ag_{44}(SPhF_2)_{30}{}^{4-}$ clusters by using 3.49 and 4.66 eV photons [24]. The results revealed the existence of a repulsive Coulomb barrier (RCB) in the electron detachment process $Ag_{44}(SPhF_2)_{30}{}^{4-} \rightarrow Ag_{44}(SPhF_2)_{30}{}^{3-}$, as expected for a multicharged anion, but surprisingly also a negative adiabatic electron affinity (AEA). This means that the total energy of the resulting anion $Ag_{44}(SPhF_2)_{30}{}^{3-}$ is lower than that of the parent anion after relaxation. This clearly stresses the importance of the RCB to stabilize the parent cluster in 4-charge and 18-electron superatom configuration.

Theoretical modeling of the measured PE spectra was based on DFT calculations of electron detachment energies from 138 topmost single-electron states below HOMO, spanning a range of electron-binding energies up to 3.5 eV. In addition, a simple "potential box" model with a tunneling barrier was built, to account for electron tunneling through the RCB. A convoluted spectrum of these individual energies, taking into account the tunneling model, was compared to the experimental data, explaining the origin of peaks

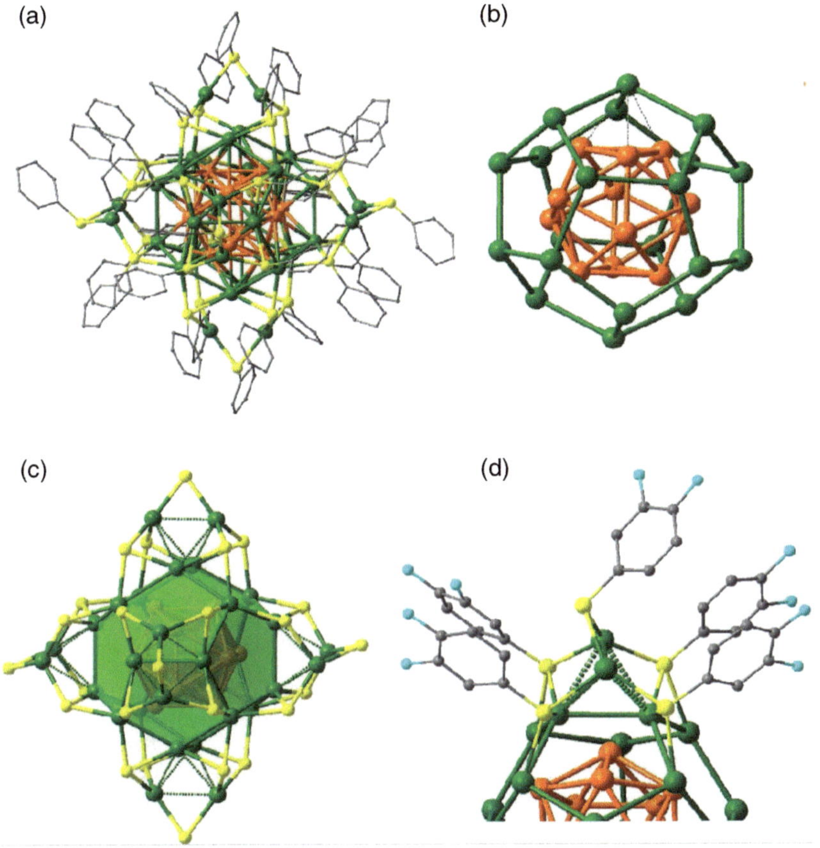

Figure 7.6 The atomic structure of $Ag_{44}(SPhF_2)_{30}^{4-}$. (a) Full structure; (b) metal core; (c) core + ligand layer; (d) zoom-up of one $Ag_2(SR)_5$ ligand unit. *Source:* Yang et al. [22]. © Springer Nature.

"A" and "B" in terms of the superatom and ligand states (Figure 7.7). Furthermore, both vertical detachment energy (VDE, electron removal without structural relaxation) and AEA were computed. The computed AEA was indeed negative confirming the experimental observation. This is one of the few successful cases so far where the electronic structure of a superatom synthesized by wet chemistry has been investigated also in gas phase.

7.2.4 Transition from Electronic to Atomic Shells

As Table 7.1 indicates, the simple free electron model predicts that for many metal nanoparticles, the Kubo gap closes for a particle size (diameter) of 1.5–2 nm. For monovalent group 11 metals Cu (. . .4*s*), Ag(. . .5*s*), and Au(. . .6*s*), the critical size is around 150–190 free electrons. As discussed in Chapter 6, signatures for a transition from "molecular" to "metallic" behavior in ligand-stabilized nanoclusters are seen exactly in this size range. This range is extensively studied for thiolate-stabilized gold clusters. A notable cluster composition in this size range is $Au_{144}(SR)_{60}$, which was first experimentally isolated and

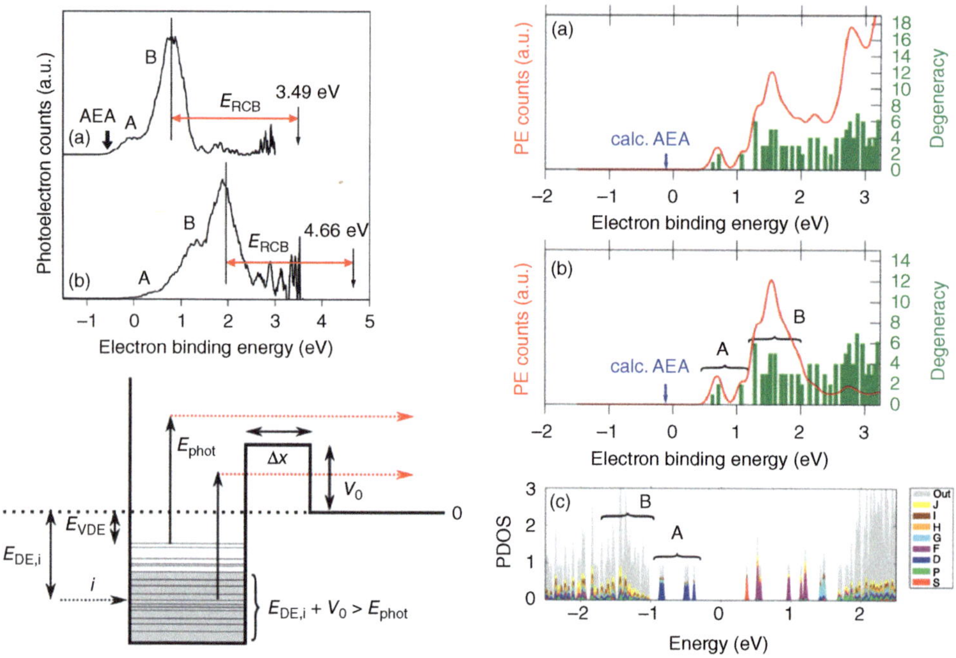

Figure 7.7 (top left): measured PE spectra of $Ag_{44}(SPhF_2)_{30}{}^{4-}$ using two different photon energies; (bottom left): (right): convoluted theoretical spectrum of electron detachment energies without (a) and with (b) the RCB tunneling model, (c) projection of frontier orbitals to spherical harmonics to reveal the superatomic electron states below and above the HOMO–LUMO energy (centered at zero). Detachment from states "A" and "B" assigned to corresponding features in the measured PE spectra. *Source:* Tasaka et al. [24]. © American Chemical Society.

long referred to as a "29 kDa compound," [25] later to be characterized from precision mass spectrometry as $Au_{144}(SR)_{60}$ (ref. [26]). The first theoretical model for its atomic structure and electronic shell structure was presented also in 2009 [27], to be proven essentially correct almost 10 years later by a single-crystal X-ray structure of the compound $Au_{144}(SCH_2Ph)_{60}$ (ref. [28]). The model (Figure 7.8) suggested and the experiment confirmed a gold core Au_{114} (Au_{12}@Au_{42}@Au_{60}) protected by 60 RS-Au-SR units. This high (icosahedral) symmetry of the core supports also a rather well-defined electronic shell structure, but the HOMO–LUMO gap at 84 free electrons is "accidentally" vanishingly small (Figure 7.9), supporting a "metallic" behavior seen in the computed charging energy curve and early evidence from electrochemistry for this cluster [29, 30]. However, remnants of the shell structure are seen in both experimental and computed optical absorption spectra [31–33]. In general, the transition from "molecular" to "metallic" regime for ligand-stabilized metal clusters is not sharp (see also discussion in Chapter 6). It depends on the particular metal and on the chemical character of the stabilizing ligand. For instance, a *p*-MBA-protected (para mercapto benzoic acid) gold cluster $Au_{210-230}(p\text{-MBA})_{70-80}$ (ref. [34]) with about 140–150 free electrons shows a weak plasmonic absorption band usually assigned as a sign of "metallic" behavior, but a slightly larger thiolate-stabilized gold cluster $Au_{246}(SR)_{80}$ (ref. [35]) with 166 free electrons still shows a nonplasmonic absorption due to

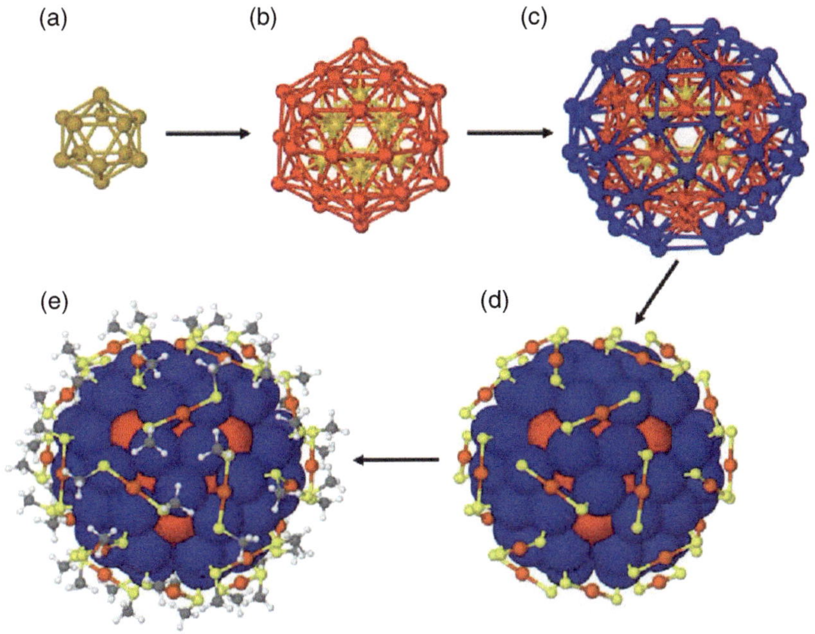

Figure 7.8 Predicted structure of the $Au_{144}(SR)_{60}$ cluster. (a)–(c) Show the concentric 12-atom (hollow), 42-atom, and 60-atom Au shells of the 114-atom gold core, respectively. (d) Shows the S-Au-S arrangement of the 30 RS-Au-SR units covering the surface (blue) of the 114-atom gold core, and (e) shows all the atoms. Color code in (d), (e): S: yellow, Au(I): orange, C: gray, H: white. *Source:* Modified from Lopez-Acevedo et al. [27]. © American Chemical Society.

electronic shell structure and a finite HOMO–LUMO energy gap (calculated as 0.2 eV) supported by high symmetry of the core [36].

Traditionally, the relative importance of the electronic shell structure (electronic effects, HOMO–LUMO energy gap) and atomic shell structure (atom packing effects) for stabilizing extra stable or "magic" cluster compositions has been long discussed in the science of gas-phase metal clusters. In the wet chemistry synthesis of ligand-stabilized metal clusters, the atomistic processes to reach stable cluster compositions and structures are still not fully understood (see discussion in Chapter 6). There is a recent demonstration, however, that a single synthesis can simultaneously produce (at least) two very different sizes and compositions that are apparently stabilized by different mechanisms [37]. The synthesis was done for Ag-Au intermetallic clusters stabilized by both thiolates and phosphines. Remarkably, the co-crystal (Figure 7.10) of two different clusters showed compositions of $(AuAg)_{45}(SR)_{27}(PPh_3)_6$ and $(AuAg)_{267}(SR)_{80}$, the former being an 18-electron system and the latter 187-electron system. The $(AuAg)_{267}$ cluster has a multishell icosahedral structure $Ag@M_{12}$ (Mackay)$@M_{42}$ (Mackay)$@M_{92}$ (Mackay)$@M_{120}$ (anti-Mackay) (M = Au/Ag) and the 80 thiolates are all either threefold or fourfold coordinated on top of the outermost metal shell in a slightly distorted truncated icosahedral symmetry (Figure 7.11). The smaller cluster $(AuAg)_{45}$ has an atomic hole at the center, destabilizing the 2S superatom electronic shell over the calculated HOMO–LUMO energy gap of 0.7 eV (Figure 7.12a). Notably, all the models for the larger cluster show distinct electronic shells around the

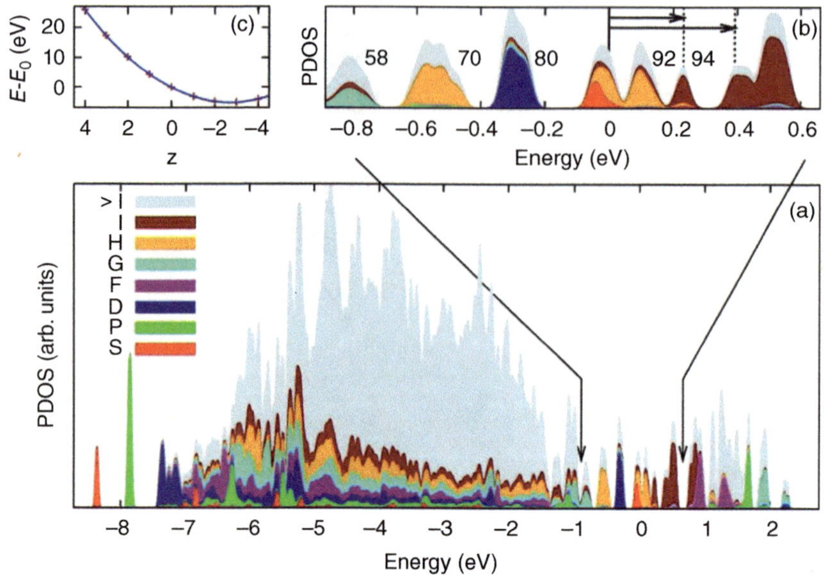

Figure 7.9 (a) Projected density of electron states (PDOS) within the Au_{114} core region of $Au_{144}(SR)_{60}$. The color-filled curves denote weights of various angular momenta up to $L = 6$. Note the manifold of states in the energy range below −7 and −1 eV to +1 eV, where each state can be described by a single major angular momentum component. These states are the Au(6s)-derived delocalized "metallic" states in the core. Gray denotes angular momenta $L > 6$. The energy range of −7 to −1 eV includes the Au(5d)-derived states. (b) Shows the region around the HOMO (zero energy), with the lowest optically allowed transitions (from H to I symmetry) marked by arrows. The numbers denote subshell closing numbers in the spherical electron gas model for metal clusters in the $1H$–$2D$–$3S$–$1I$ manifold of states. (c) Shows the calculated total energies for charge states $-4 \leqslant z \leqslant 4$ and a quadratic fit to $E(Q) - E(Q = 0)$. No apparent energy gap is seen in the charging curve, consistent with a vanishingly small HOMO–LUMO energy gap. *Source:* Lopez-Acevedo et al. [27]. © American Chemical Society.

HOMO energy, with gaps of 0.1–0.2 eV at 190 and 198 electrons (**Figure 7.12b–d**). One can conclude that electronic effects are a defining factor to stabilize the smaller cluster in the co-crystal, while atomic packing may dominate in the energetic stability of the larger one. However, taking into account that the M_{267} is essentially one anti-Mackay layer larger than the previously discussed M_{144}, one can understand that the prevailing icosahedral symmetry supports the symmetries of electronic shells as well in the larger cluster just like in M_{144}.

7.3 Applications

7.3.1 Catalysis

Equation (7.4) predicts that the free electron count n_e of ligand-stabilized metal clusters could be controlled by changing the number of electron-withdrawing or -donating ligands. In particular, partially protected clusters could be made electronically "active" toward electron transfer reaction such as dioxygen binding and reduction, by promoting one or more

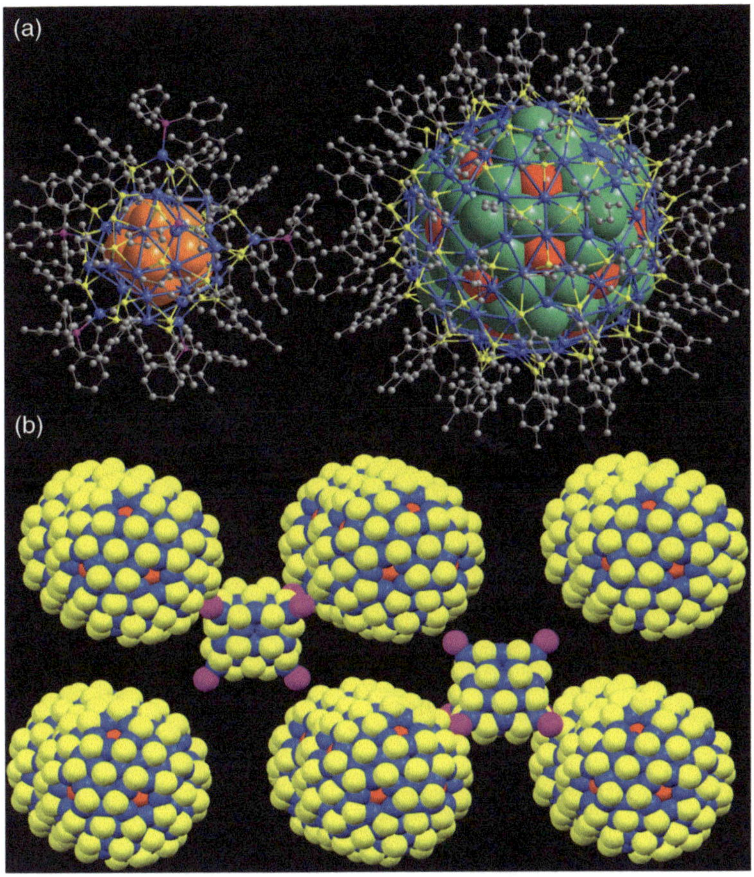

Figure 7.10 Co-crystal structure of $(AuAg)_{267}$·$(AuAg)_{45}$. (a) The total structures of the clusters; (b) unit cell of the three-dimensional structure in space-filling model. All hydrogen and carbon atoms are omitted for clarity. Color code: orange Au, sea green and red AuAg, blue Ag, yellow S, magenta P, gray C. *Source:* Yan et al. [37]. © Springer Nature.

electrons over the HOMO–LUMO gap of the protected, stable cluster. Several ligand-protected gold clusters were studied in this respect via DFT calculations [38]. A surprisingly clear correlation between the HOMO–LUMO gap of the (nonactivated) cluster and dioxygen binding to the activated cluster was found (Table 7.2 and Figure 7.13). In the limit where the HOMO–LUMO gap approaches zero, the binding of dioxygen turns endothermic. Figure 7.14 illustrates the electron transfer upon dioxygen activation. One phosphine and one chlorine ligand were removed from the fully protected, neutral $Au_{11}(PH_3)Cl_3$ cluster, which renders the partially protected $Au_{11}(PH_3)Cl_3$ to be a nine-electron system where the electron at the HOMO level is actually transferred over the HOMO–LUMO gap of the parent cluster. When O_2 and CO are coadsorbed on the available gold sites at the partially protected cluster, the HOMO electron transfers to the $2\pi^*$ orbital of O_2 and activates the O-O bond, making the cluster effectively an eight-electron system again. CO oxidation was studied as a model reaction, and it was found to proceed with low activation barriers on clusters with

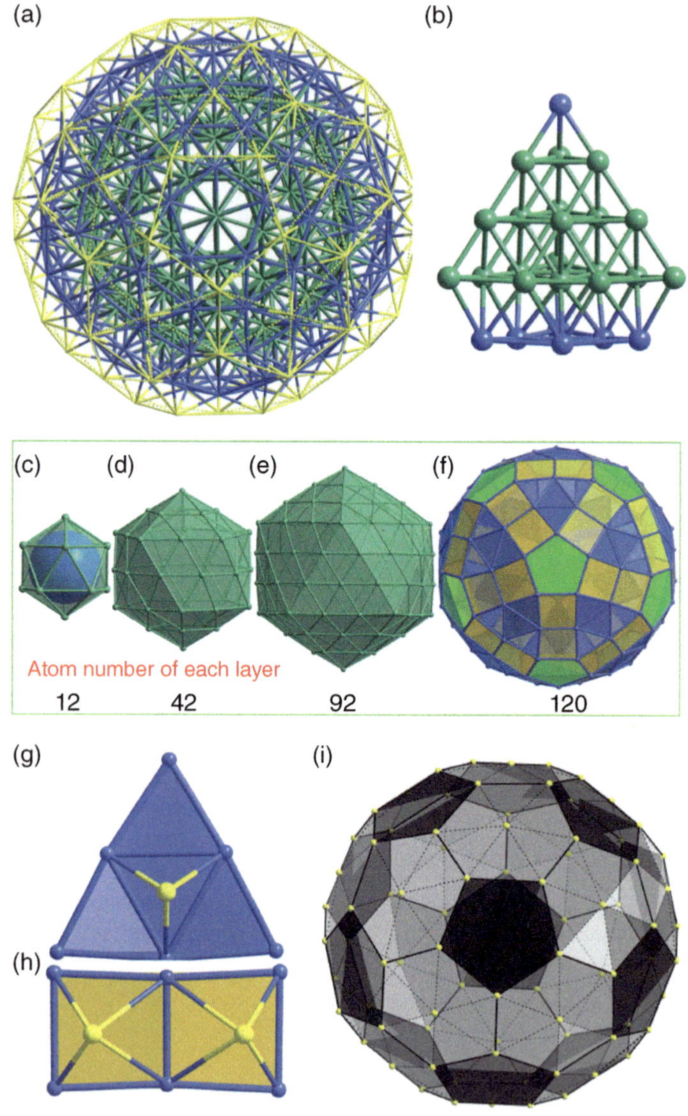

Figure 7.11 (a) Buckyball-shaped $(AuAg)_{267}$ nanoparticle. (b) Layered ABCAC packing (or ABCB if the central Ag atom is ignored) of metal atoms. (c)–(f) Anatomy of metal framework of $(AuAg)_{267}$: the first Mackay icosahedral shell with 12 M atoms (c); the second Mackay icosahedral shell with 42 M atoms (d); the third Mackay icosahedral shell with 92 M atoms (e); and the fourth anti-Mackay shell with 120 Ag atoms (f). (g) Threefold coordinated thiolate centered on six-Ag-atom triangle (highlighted in blue in panels (f) and (g). (h) Two tetradentate thiolates capped on two edge-shared squares (highlighted in yellow in panels (f) and (h)). (i) Sulfur atoms (yellow) arranged in a slightly distorted truncated icosahedron (buckyball). All hydrogen and carbon atoms are omitted for clarity. Color code: sea green and red AuAg, blue Ag, yellow S. *Source:* Yan et al. [37]. © Springer Nature.

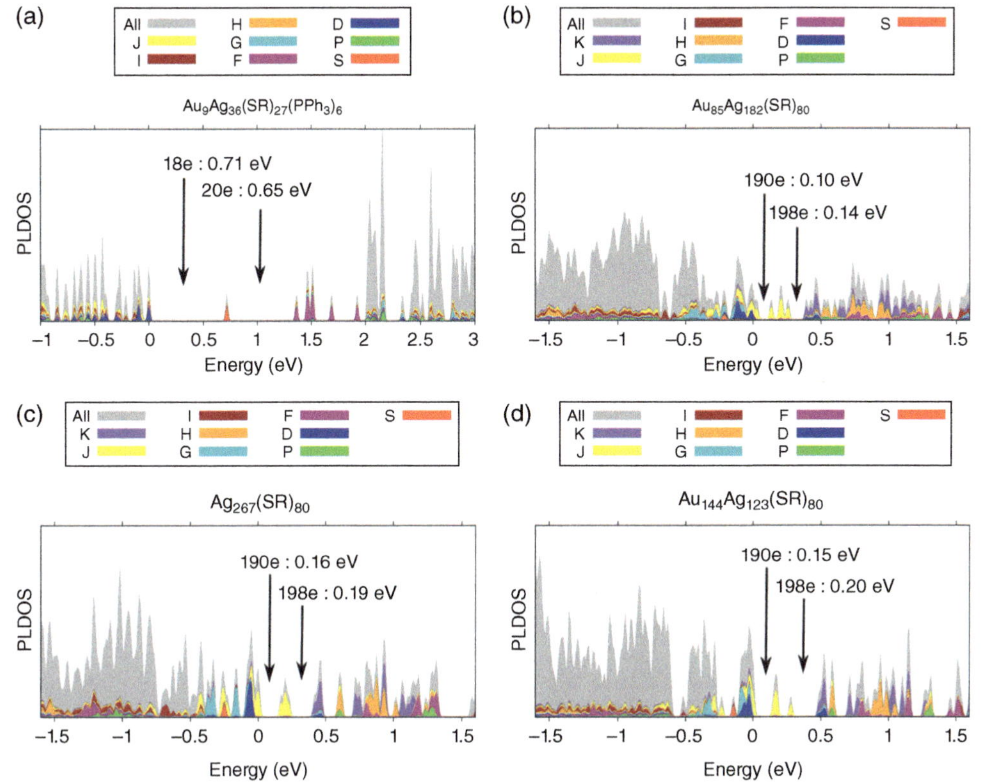

Figure 7.12 Calculated electronic structures of $(AuAg)_{267}$ and $(AuAg)_{45}$. (a) Projected local density of states (PLDOS) of the $(AuAg)_{45}$ cluster. (b)–(d) The same analysis for a hypothetical all-silver Ag_{267} and for two intermetallic clusters with compositions of $Au_{85}Ag_{182}$ and $Au_{144}Ag_{123}$, all computed in the nanoparticle observed crystal structure of $(AuAg)_{267}$. The various colors denote the weights of orbitals projected onto spherical harmonics, centered at the cluster center-of-mass. The HOMO energy is at zero. The major energy gaps around or above the HOMO energy are indicated by arrows. *Source:* Yan et al. [37]. © Springer Nature.

Table 7.2 The correlation between the electronic structure and binding characteristics of O_2 for partially protected gold clusters 1–5, with binding of dioxygen at Au(111) as reference. Negative binding energy denotes endothermic adsorption. D and D_C are the overall and metal core diameters, respectively; n_e and n'_e are the free electron counts of the fully and partially protected clusters; and HL gap is the HOMO–LUMO gap of the fully protected cluster, which is activated toward oxygen adsorption by removing part of the ligand layer. $BE(O_2)$ is the binding energy of dioxygen on the activated clusters, and d(Au-O) and d(O–O) are the binding distance of dioxygen to gold and O–O bond distance, respectively.

		D (nm)	D_C (nm)	n_e	HL gap (eV)	Ligand removed	n'_e	$BE(O_2)$ (eV)	d(Au-O) Å	d(O–O) Å
1	$Au_{11}(PH_3)_7Cl_3$	1.2	0.9	8	2.03	Cl	9	0.95	2.17	1.31
2	$Au_{25}(SR)_{18}^{-1}$	1.6	0.9	8	1.25	$Au_2(SR)_3$	9	0.72	2.24	1.31
3	$Au_{39}(PH_3)Cl_6^{-1}$	1.8	1.5	34	0.85	Cl	35	0.59	2.25	1.32
4	$Au_{102}(SR)_{44}$	2.2	1.5	58	0.53	$Au(SR)_2$	59	0.08	2.19	1.30
5	$Au_{144}(SR)_{60}$	2.4	1.7	84	0.08	$Au(SR)_2$	85	−0.15	2.22	1.29
6	Au(111) surface							−0.54	2.26	1.24

Source: Lopez-Acevedo et al. [38]. © Springer Nature.

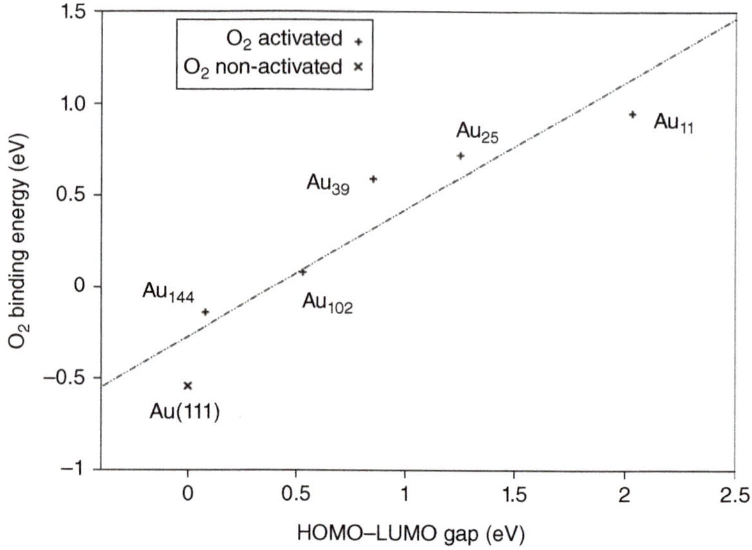

Figure 7.13 Correlation between the binding energy of O_2 to partially protected clusters 1–5, shown in Table 7.2, and the HOMO–LUMO gap of the corresponding fully protected cluster. The clusters are labeled by the gold atom count. Although all the clusters activate the O–O bond, only clusters 1–3 (derived from Au_{11}, Au_{25}, and Au_{39}) bind O_2 appreciably (by more than 0.5 eV). For comparison, the result for O_2 adsorption on the Au(111) surface is also shown. In that case the O_2 remains neutral and the O–O bond is not activated. Note that the binding of O_2 to the Au144 cluster and Au(111) surface is metastable (endothermic). *Source:* Lopez-Acevedo et al. [38]. © Springer Nature.

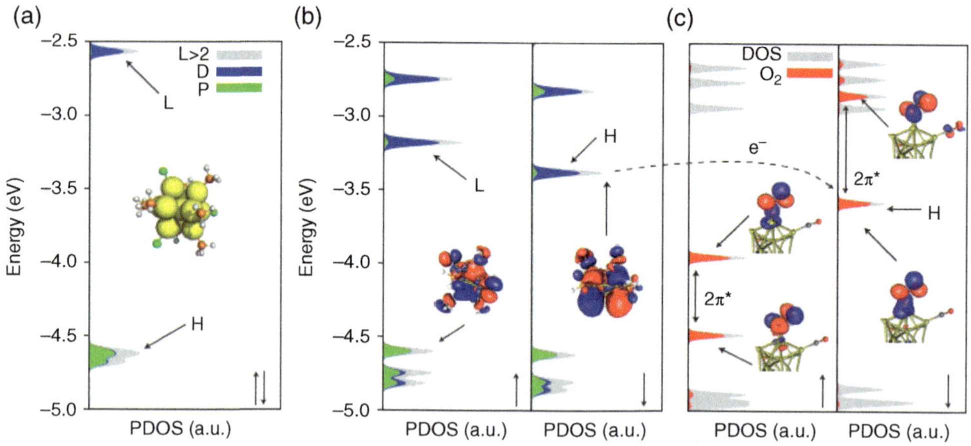

Figure 7.14 (a) HOMO and LUMO states of the fully protected $Au_{11}(PH_3)_7Cl_3$ cluster. These states are composed of Au(6s) states delocalized over the metal core. The green and blue show the major center-of-mass angular momentum character assigned to the states by projection to spherical harmonics inside the metal, indicating that the HOMO has the superatom P symmetry and LUMO has D symmetry. (b) Spin-polarized states in the partially protected $Au_{11}(PH_3)_6Cl_2$ cluster. H and L refer to HOMO and LUMO, respectively. The HOMO of the partially protected cluster has D symmetry, as does the LUMO of the fully protected cluster in (a). One electron has thus been transferred to this state over the HOMO–LUMO gap of the fully protected cluster. The P and D symmetries of the H- and L-orbitals are visualized. (c) On O_2 adsorption, the D-symmetric HOMO state in (b) is depleted completely and the electron is transferred to one of the $2\pi^*$ states of O_2, lying in the HOMO–LUMO gap of the gold cluster. This $2\pi^*$ state is empty in the neutral O_2 triplet. Au, yellow; Cl, green; P, orange; H, white; C, gray; O, red. *Source:* Lopez-Acevedo et al. [38]. © Springer Nature.

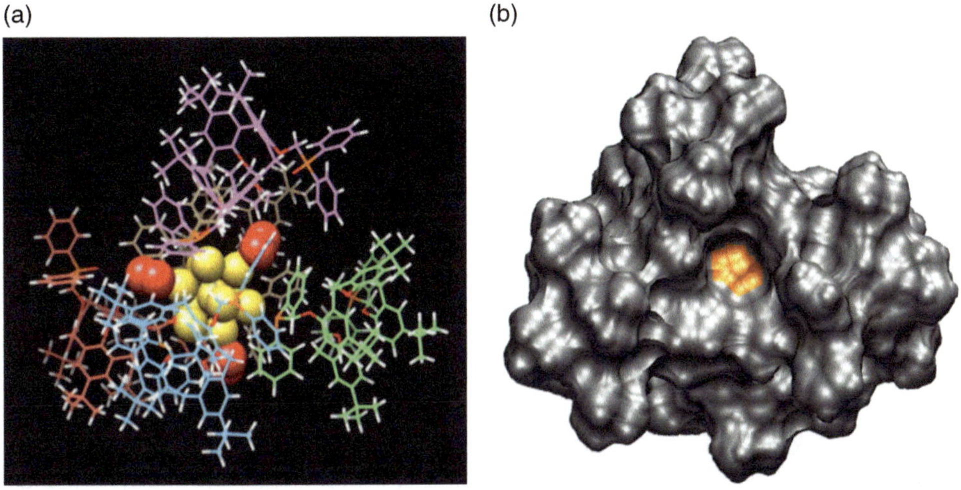

Figure 7.15 (a) The model structure of $Au_{11}L_5(O_2)_3$ complex; (b) Solvent-accessible surface of $Au_{11}L_5$, showing one of the three Au sites available for binding reactants. In (a), oxygen molecules are bound at these sites. Water molecule has been used as the probe giving the area for binding at the Au site as 1.7–1.9 Å^2. L = diphosphine calix(4)arene. *Source:* Chen and Häkkinen [40]. © American Chemical Society.

partially removed ligand layer. Mechanisms like this may be important to understand the reactivity of the ligand layer and intercluster reactions.

Experimental evidence for the viability of subnanometer Au_{11} clusters for small-molecule activation was provided by the Katz group [39]. In an interesting experiment, they synthesized gold clusters by using diphosphine-calixarene macrocycles as stabilizing ligands. Characterization by mass spectrometry suggested that the smallest stable clusters were about the size of Au_{11}. Thiol titration experiments showed that 25% of the Au atoms in the cluster are still accessible to naphthalene thiols. This corresponds to three gold sites. Chen and Häkkinen made an atomistic model using the known phosphine-stabilized Au_{11} cluster for DFT calculations (Figure 7.15). It was shown that diphosphine-calixarenes form a porous molecular layer around the cluster, exposing three gold sites for small molecule adsorption. Three O_2 could be adsorbed to these sites, and each of them was activated to a superoxo state by electron transfer from gold, by exactly the same mechanism as discussed above (Figure 7.14). The calixarene-stabilized Au_{11} serves thus as a bioinspired model allowing steric access to reactive metal centers, in analogy to reaction centers in enzymes.

7.3.2 Biological and Medical Applications

Over the last decade or so, colloidal gold nanoparticles have been exploited widely in biomedicine and bioimaging with applications in drug delivery, image-contrast agents, and photothermal therapy [41–43]. However, despite definite proofs of their efficacy in various applications, *in vivo* biocompatibility and toxicity remain in dispute. Nanoparticles' relatively large size (tens of nanometers) and eventual tendency for agglomeration to bigger assemblies hinder their metabolism leading to accumulation in some organs like liver and spleen [43]. Due to their much smaller size (1–3 nm) and molecule-level tunability of the ligand layer, ligand-stabilized gold nanoclusters offer an attractive alternative to colloidal nanoparticles for bioapplications [44]. In the following, a case study applying Au_{102}(*p*-MBA)$_{44}$ clusters to bioimaging of viruses is discussed.

7.3.2.1 Case Study: Imaging of Enteroviruses

Development of precise protocols for accurate site-specific conjugation of monodisperse inorganic nanoparticles to biological material is one of the challenges in contemporary bionanoscience and nanomedicine. As shown by Kornberg group previously, *p*-MBA-protected gold nanoparticles show affinity toward antibodies and DNA [45–47]. Recent studies have shown that Au_{102}(*p*-MBA)$_{44}$ clusters [19] can be site-specifically attached to capsids of enteroviruses, utilizing either covalent interactions with cysteine residues of the capsid [48] or noncovalent interactions with the hydrophobic pockets in the capsid [49]. For covalent linking of Au_{102}(*p*-MBA)$_{44}$ clusters to enteroviruses, the clusters were functionalized by a maleimide group that is known to react with free thiol end of a cysteine side group. The functional group was covalently attached to the cluster surface with a C_6OH arm making an ester bond to the carboxylic acid group at the cluster surface (Figure 7.16).

Two enteroviruses, echovirus one (EV1) and coxsackievirus B3 (CVB3), were selected to test the binding hypothesis. While these viruses share the overall icosahedral packing of viral capsids and are of the same size (about 35 nm in diameter), their distribution of available cysteine sites at the capsid surface is different. Conjugation reaction to these viruses was studied in the time scale of a few hours to two days, producing tens of bound clusters on the capsids (Figure 7.17). Control experiments with nonfunctionalized clusters did not show binding to viruses. Careful analysis of the transmission electron microscopy (TEM) images, center-to-center distribution of the clusters, and comparisons to distributions of available cysteine sites in the X-ray structures of native EV1 and CVB3 viruses confirmed that site-specific covalent binding to capsid cysteines was achieved. This paves way to tracking studies of viruses during their endocytic pathways and in the endosomes of the cell.

Viruses are dynamic objects, and little is known on the various factors and chemical conditions that initiate the uncoating process inside cells, that is, opening of the virus capsid and release of the genome. The so-called hydrophobic pockets, containing natural lipids, are believed to play an important role in the mechanical stability of enteroviruses. Another Au_{102}-based probe containing lipid-like molecules that mimic known antivirals was developed as well, and its conjugation to EV1 was studied by nuclear magnetic resonance (NMR) spectroscopy and TEM imaging [49]. The results showed that hydrophobic pockets could be targeted by these probes, but due to the noncovalent binding, only a fraction of the available 60 pockets can be occupied at any given time. Nevertheless, the results suggested that, like the physiological hydrophobic content, the probe may be released upon virus uncoating. Consequently, gold-containing and/or fluorescently labeled probes may be used to track and visualize enteroviruses during the early phases of infection opening new avenues to follow virus uncoating in cells.

In a follow-up work, extensive molecular dynamics simulations [50] were performed to estimate binding free energies of lipid-like drug and drug-like molecules into these pockets (Figure 7.18). The results showed that the drug-like molecules bind to the pockets slightly stronger than the natural lipids, confirming the experimental results indicating competition

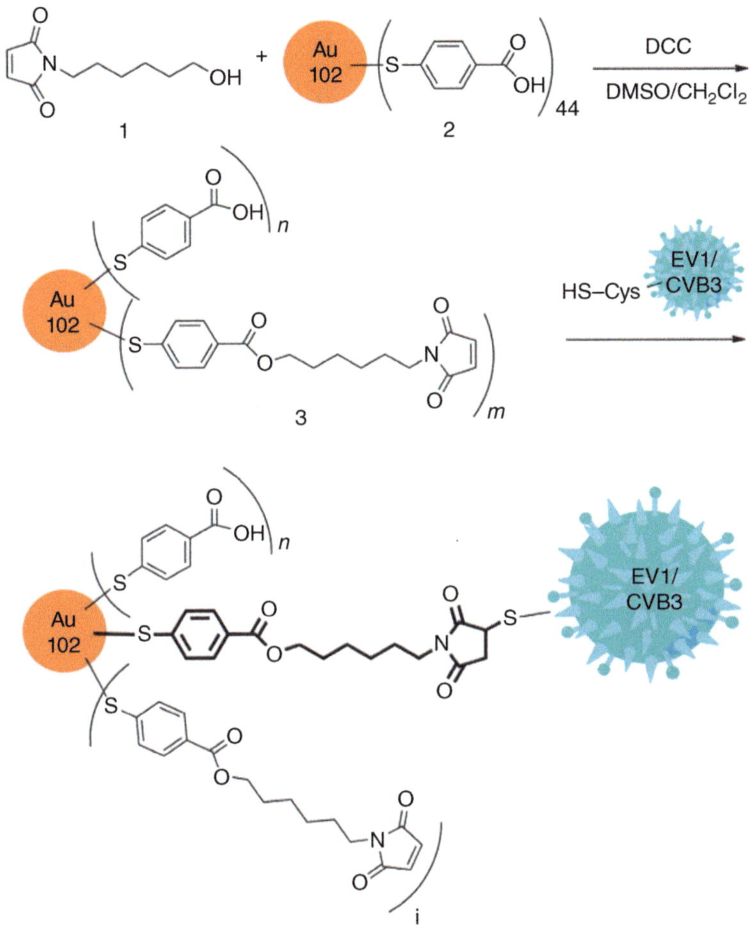

Figure 7.16 A schematic representation of functionalization of Au_{102}(*p*-MBA)$_{44}$ for covalent binding to enteroviruses. *Source:* Marjomäki et al. [48]. © National Academy of Sciences.

to these binding sites. Interestingly, the presence of Au_{102} clusters modified the binding of the drug-like molecules, making it dependent on the pH. Au_{102} induced a stronger binding at lower pH. This indicates that simulations can give valuable additional molecular-scale information that can help in the design of optimal strategies for gold–virus bioconjugation for virus detection and manipulation.

7.3.3 Self-Assembling Cluster Materials from Superatoms

Early successful realizations of nanoparticle materials included gold nanoparticle assemblies where individual particles were linked by dithiols [51], DNA [52, 53], or using viral proteins as scaffolds [54, 55]. Atomically precise ligand-stabilized metal nanoclusters can be regarded, in principle, as excellent building blocks for nanomaterials with desired, tunable properties, controlled even to a greater degree than what was possible with the

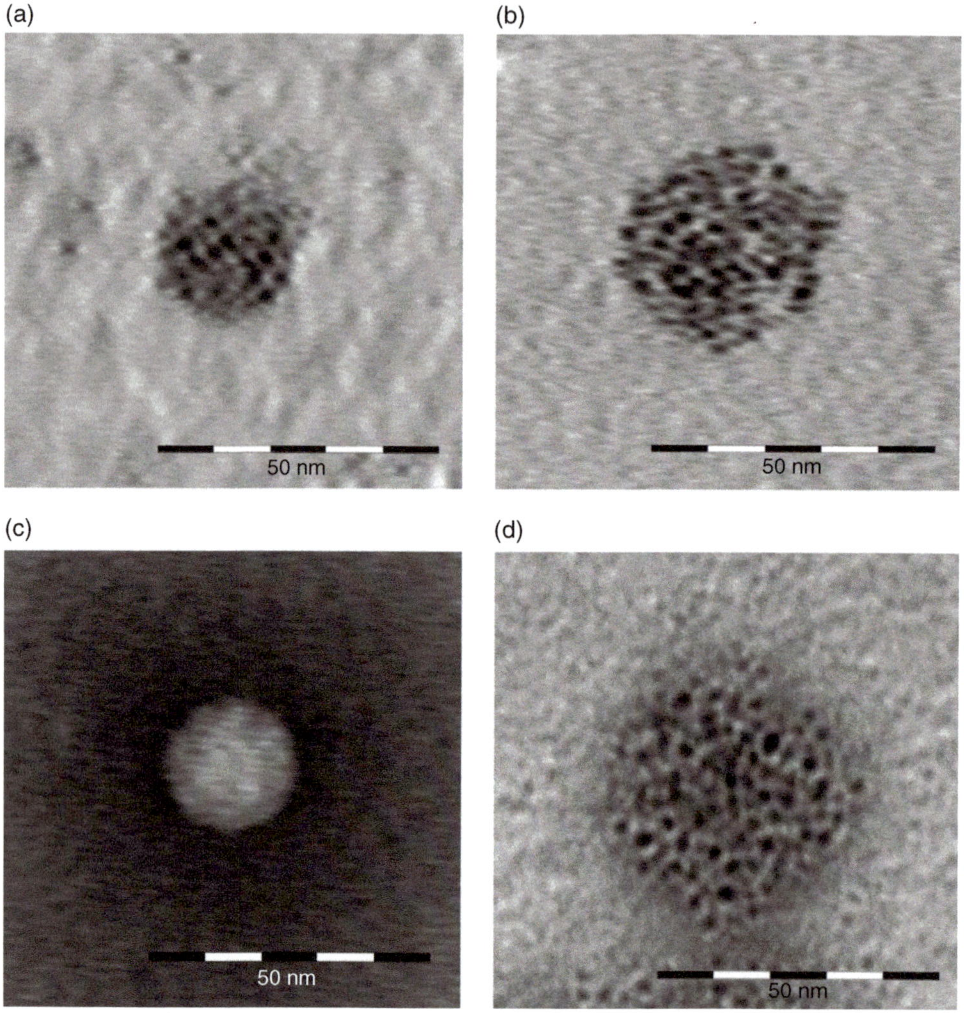

Figure 7.17 TEM images of enteroviruses. (a) CVB3 incubated with functionalized gold clusters for two days, without column purification (some unbound gold clusters are visible in the grid); (b) the same but after column purification. (c) A control sample of CVB3 incubated with nonfunctionalized $Au_{102}(p\text{-MBA})_{44}$. The dark halo around the virus is produced by metal-containing staining agent. (d) EV1 incubated with functionalized clusters for one day (staining agent has been used in the imaging). *Source:* Reproduced with permission from Marjomäki et al. [48]. © National Academy of Sciences.

early nanoparticle materials. The challenge in the bottom-up synthesis of cluster-assembled nanomaterials lies in finding the proper chemistry regarding the interaction strength, topology, and dimensionality for reaching high-quality macroscopic samples. As discussed elsewhere in this chapter and in Chapter 6, there is now an extensive literature of successful syntheses to produce a large variety of gold-, silver-, and copper-based ligand-stabilized clusters in 1–3 nm range. This has inspired efforts toward nanoscale and macroscale assemblies of these clusters [56–65] that might have a variety of applications in the fields of (bio)chemical sensing, plasmonics, and optoelectronics. The present success can be seen

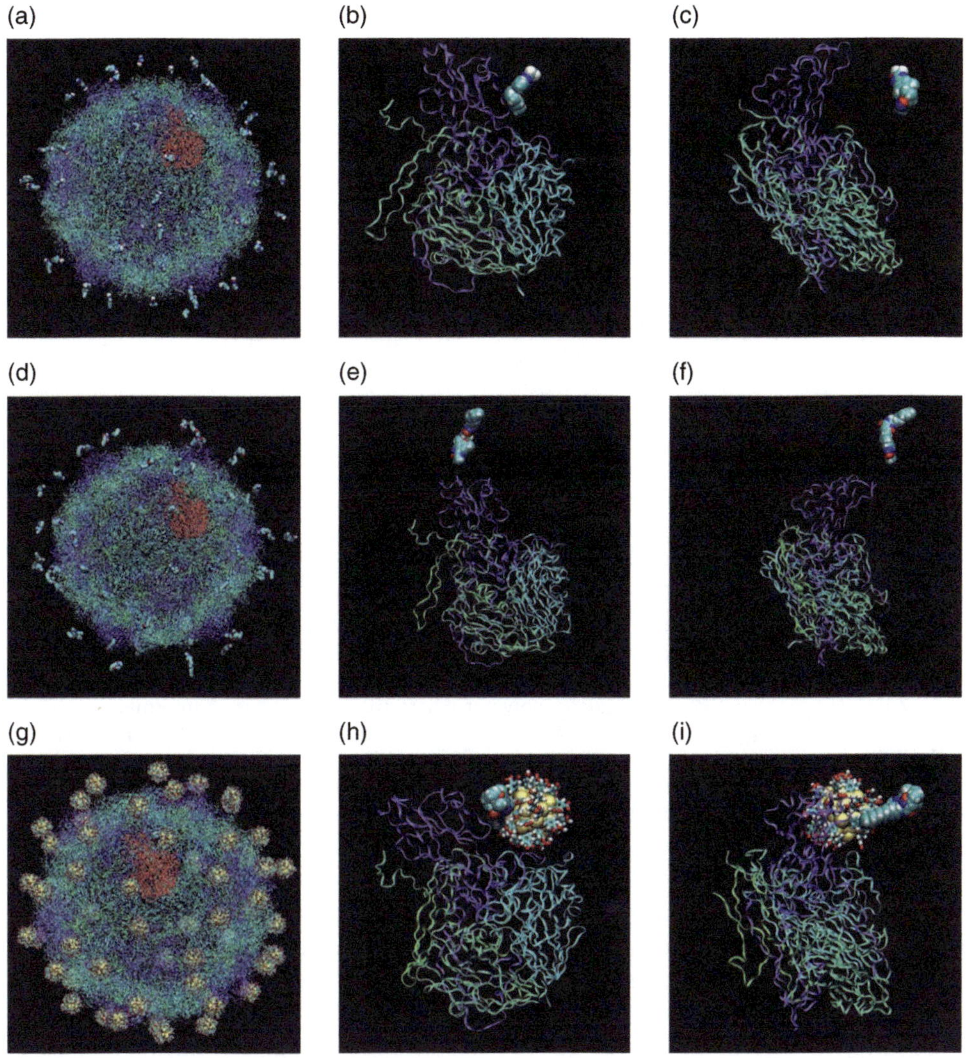

Figure 7.18 End configurations of directed molecular dynamics (MD) simulations determining binding free energies of linkers to 60 hydrophobic pockets of an EV1 virus. (a–c) Full virus and 1/60 unit close-ups with the pleconaril (drug) molecule, (d–f) with drug-like molecule ("kirtan"), and (g–i) with kirtan–Au_{102} conjugate. Solvent molecules (water and counterions) are not shown. *Source:* Pohjolainen et al. [50]. © American Chemical Society.

as the chemical realization of the early concept for potential assembly of cluster materials considering simple metal clusters as the basic building blocks [66, 67].

7.3.3.1 Case Study: Polymeric 1D Cluster Materials

Zheng's group demonstrated recently [68] a novel type of a cluster-assembled macroscopic single-crystal material where intermetallic $(AuAg)_{34}$ clusters are linked to 1D polymers (**Figure 7.19**). The metal cores of the clusters are stabilized by 1-ethynyladamantane

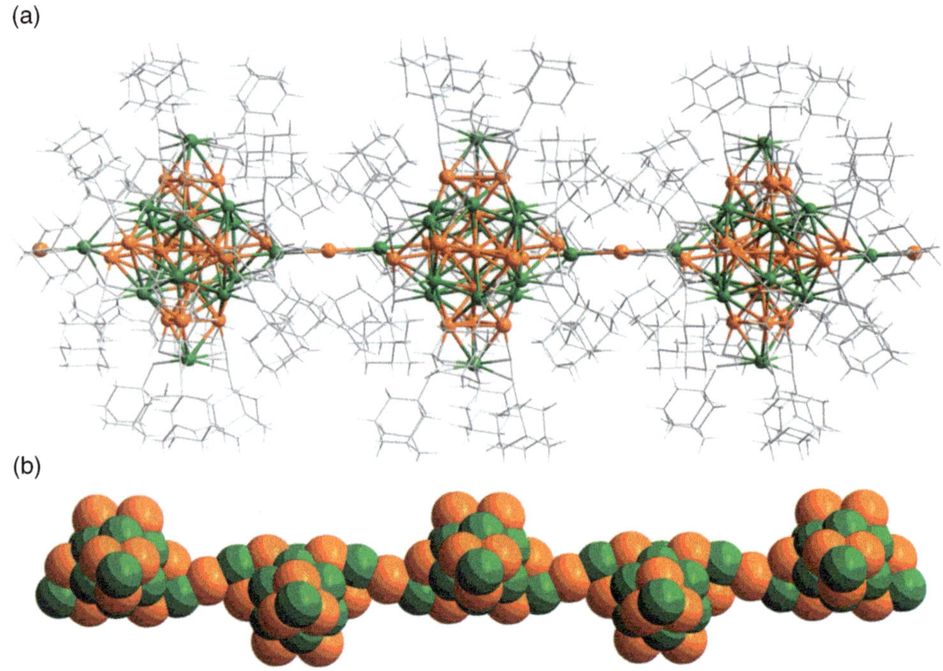

Figure 7.19 Atomic structure of the cluster polymer, where the basic unit is an intermetallic 34-atom Au(orange)/Ag(green) cluster, protected by 1-ethynyladamantane. The linking atom is Au. (a) Total structure viewed approximately orthogonal to the polymer axis (parallel direction). (b) Space-filling view of the metal atoms (viewed from the top direction in (a)). Colors: golden and green, Au/Ag; gray, C. All hydrogen atoms are omitted for clarity. *Source:* Yuan et al. [68]. © Springer Nature.

molecules, which are sterically quite bulky, isolating the different polymer chains from each other in the polymer cross directions. The electronic effect of these molecules to the metal core is similar to thiolates and other electron-withdrawing ligands. The chemical composition of the material indicated 14 free electrons per each $(AuAg)_{34}$ cluster. Interestingly, this electron count fits qualitatively with the strongly nonspherical shape of the clusters, and DFT calculations yielded a clear HOMO–LUMO energy gap of 1.3–1.5 eV for an individual cluster, depending on the details of the cluster model (occupation of Au/Ag atoms in atom sites). A periodic bulk model of the cluster polymer was also built, and DFT band-structure calculations predicted a semiconductor-type behavior with about 1.3 eV bandgap (Figure 7.20).

The observed single-crystal atomic structure and DFT calculations indicated a 1D semiconductor behavior for the cluster material. To test these predictions, directional conducting properties of the polymer single crystals were measured in a field effect transistor (FET) setup (Figure 7.21). *I–V* measurements showed a large anisotropy in the conductivity, with the conductivity along the polymer *c*-axis being about 1800-fold as compared to the cross-direction (*a*-axis in Figure 7.21). Field-effect studies indicated p-type conductivity and hole mobility of the order of $0.02\,cm^2/V/s$ and an ON/OFF ratio up to about 4000.

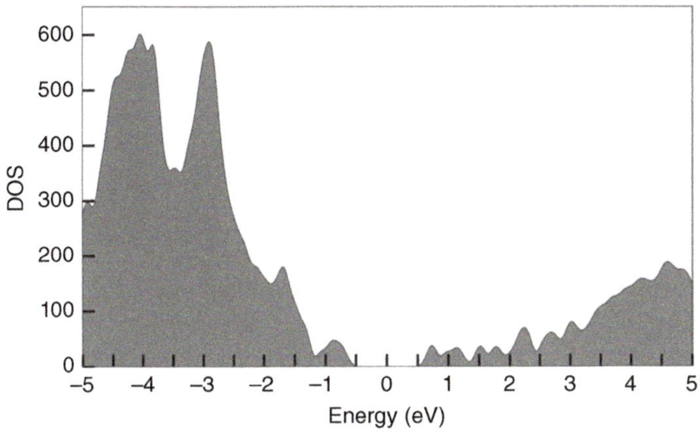

Figure 7.20 DFT-computed electronic DOS of the cluster polymer crystal. The integration over the Brillouin zone was done in a 4 × 4 × 4 Monkhorst–Pack k-point mesh. The apparent bandgap, determined as the peak separation closest to the gap, is about 1.3 eV. *Source:* Yuan et al. [68]. © Springer Nature.

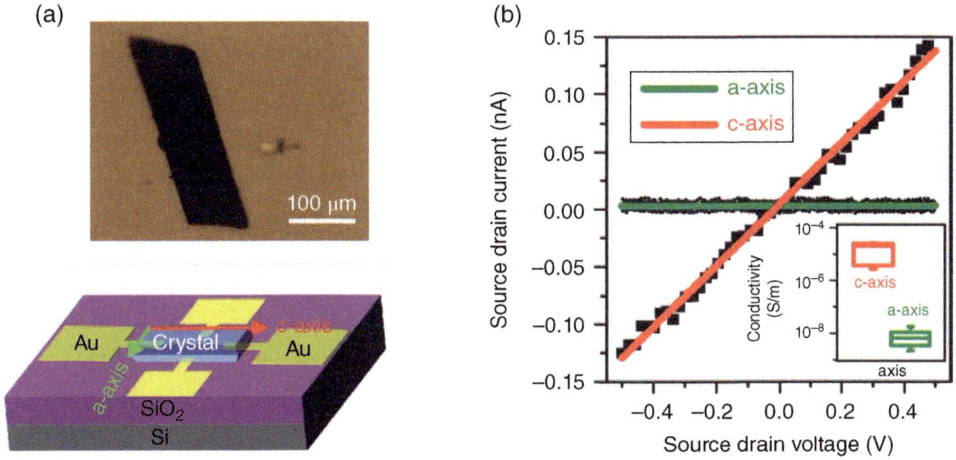

Figure 7.21 (a) A single crystal of $(AuAg)_{34}$ cluster polymers (top) and the layout of the polymer crystal FET measurement (bottom). (b) *I*–*V* plot of the polymer crystal along *a*-axis and *c*-axis, respectively, with the range of corresponding conductivity values shown in the inset. *Source:* Yuan et al. [68]. © Springer Nature.

7.4 Summary and Outlook

This chapter has discussed ligand-stabilized metal nanoclusters as a special class of superatoms. This subfield of cluster science has advanced greatly over the last decade, due to significant developments in synthesis, characterization methods, and increasing computing power available for large-scale atomistic simulations of realistic systems. Theory, simulations, and experiments have contributed hand-in-hand for many conceptual advances regarding understanding of the electronic structure, optical properties, chemical and catalytic activity, and self-assembly of these superatoms, as also discussed in this chapter through selected case studies. The great variety of factors building into synthesis of these nanomaterials allow to optimize the clusters for specific applications by changing the metal content and size of the metal core as well as solubility and chemical composition of the ligand layer. For example, the special electronic shell structure, well developed in the smallest ligand-stabilized superatoms, opens up possibilities to use the discrete quantum properties for efficient fluorescence labels in biological applications. In the area of catalysis, ligand-stabilized superatoms offer both well-defined systems to investigate fundamental factors affecting selectivity and efficiency in many important reactions and truly novel single-site catalysts at the interface between homogeneous and heterogeneous catalysis. Future efforts in theoretical understanding of these systems will require larger simulation models taking into account also the superatom–environment interactions [69]. Machine learning may offer many conceptual advances in the coming years by yielding atomistic interaction potentials for dynamical simulations that have a numerical accuracy inherited from electronic structure theory, available for use with just a tiny fraction of needed computational power as compared to DFT. In conclusion, the field holds many promises and exciting challenges to be solved for many years to come.

References

1 Jena, P. and Sun, Q. (2018). *Chem. Rev.* 118: 5755–5870.
2 Fröhlich, H. (1937). *Physica* 6: 406.
3 Kubo, R. (1962). *J. Phys. Soc. Jpn.* 17: 975.
4 Häkkinen, H. (2016). *Adv. Phys. X* 1: 467–491.
5 Kittel, C. (1996). *Introduction to Solid State Physics*, 7e. Wiley.
6 The Nobel Prize in physics 1963. http://www.nobelprize.org/nobel_prizes/physics/laureates/1963/
7 Martin, T.P., Bergmann, T., Goldich, H., and Lange, T. (1991). *J. Phys. Chem.* 95: 6421.
8 de Heer, W.A. (1993). *Rev. Mod. Phys.* 65: 611.
9 Brack, M. (1993). *Rev. Mod. Phys.* 65: 677.
10 Tran, H.T. and Perdew, J.P. (2003). *Am. J. Phys.* 71: 1048.
11 Koskinen, M., Lipas, P.O., and Manninen, M. (1995). *Z. Phys. D* 35: 285.
12 Bartels, C., Hock, C., Kuenen, R., and Issendorff, B.V. (2014). *J. Phys. Chem. A* 118: 8270.
13 Tsukuda, T. and Häkkinen, H. (eds.) (2015). *Protected Metal Clusters: From Fundamentals to Applications*. Elsevier: Amsterdam.
14 Walter, M., Akola, J., Lopez-Acevedo, O. et al. (2008). *Proc. Natl. Acad. Sci. U. S. A.* 105: 9157.
15 Häkkinen, H. (2012). *Nat. Chem.* 4: 443.
16 Fernando, A., Weerawardene, K.L.D.M., Karimova, N.V., and Aikens, C.M. (2015). *Chem. Rev.* 115: 6112.
17 Chakraborty, I. and Pradeep, T. (2017). *Chem. Rev.* 117: 8208–8271.
18 Lopez-Acevedo, O., Clayborne, P.A., and Häkkinen, H. (2011). *Phys. Rev. B* 84: 035434.
19 Jadzinsky, P.D., Calero, G., Ackerson, C.J. et al. (2007). *Science* 318: 430.
20 Vollet, J., Hartig, J.R., and Schnöckel, H. (2004). *Angew. Chem. Int. Ed.* 43: 3186.

21 Hartig, J., Stößer, A., Hauser, P., and Schnöckel, H. (2007). *Angew. Chem. Int. Ed.* 46: 1658.

22 Yang, H., Wang, Y., Huang, H. et al. (2013). *Nat. Commun.* 4: 2422.

23 Desireddy, A., Conn, B.E., Guo, J. et al. (2013). *Nature* 501: 399–402.

24 Tasaka, Y., Nakamura, K., Malola, S. et al. (2020). *J. Phys. Chem. Lett.* 11: 3069–3074.

25 Whetten, R.L., Khoury, J.T., Alvarez, M.M. et al. (1996). *Adv. Mater.* 8: 428.

26 Qian, H. and Jin, R. (2009). *Nano Lett.* 9: 4083–4087.

27 Lopez-Acevedo, O., Akola, J., Whetten, R.L. et al. (2009). *J. Phys. Chem. C* 113: 5035.

28 Yan, N., Xia, N., Liao, L. et al. (2018). *Sci. Adv.* 4: eaat7259.

29 Hicks, J.F., Miles, D.T., and Murray, R.W. (2003). *J. Am. Chem. Soc.* 124: 13322.

30 Quinn, B.M., Liljeroth, P., Ruiz, P. et al. (2003). *J. Am. Chem. Soc.* 125: 6644.

31 Negishi, Y., Nakazaki, T., Malola, S. et al. (2015). *J. Am. Chem. Soc.* 137: 1206.

32 Weissker, H.C., Escobar, H.B., Thanthirige, V.D. et al. (2014). *Nat. Commun.* 5: 3785.

33 Malola, S., Lehtovaara, L., and Häkkinen, H. (2014). *J. Phys. Chem. C* 118: 20002–20008.

34 Sokolowska, K., Malola, S., Lahtinen, M. et al. (2019). *J. Phys. Chem. C* 123: 2602–2612.

35 Zhou, M., Zeng, C.J., Song, Y.B. et al. (2017). *Angew. Chem. Int. Ed.* 56: 16257–16261.

36 Malola, S., Kaappa, S., and Häkkinen, H. (2019). *J. Phys. Chem. C* 123: 20655–20663.

37 Yan, J., Malola, S., Hu, C. et al. (2018). *Nat. Commun.* 9: 3357.

38 Lopez-Acevedo, O., Kacprzak, K.A., Akola, J., and Häkkinen, H. (2010). *Nat. Chem.* 2: 329.

39 De Silva, N., Ha, J.M., Solovyov, A. et al. (2012). *Nat. Chem.* 2: 1062.

40 Chen, X. and Häkkinen, H.J. (2013). *Am. Chem. Soc.* 135: 12944–12947.

41 Boisselier, E. and Astruc, D. (2009). *Chem. Soc. Rev.* 38: 1759–1782.

42 Yeh, Y.C., Creran, B., and Rotello, V.M. (2012). *Nanoscale* 4: 1871.

43 Adewale, O.B., Davids, H., Cairncross, L., and Roux, S. (2019). *Int. J. Toxicol.* 38: 357.

44 Matus, M.F. and Häkkinen, H. (2020). *Small*, in revision. https://doi.org/10.1002/smll.202005499.

45 Ackerson, C.J., Jadzinsky, P.D., Sexton, J.Z. et al. (2010). *Bioconjug. Chem.* 21: 214–218.

46 Ackerson, C., Sykes, M., and Kornberg, R.D. (2005). *Proc. Natl. Acad. Sci. U. S. A.* 102: 13383–13385.

47 Ackerson, C., Jadzinsky, P., Jensen, G., and Kornberg, R.D. (2006). *J. Am. Chem. Soc.* 128: 2635–2640.

48 Marjomäki, V., Lahtinen, T., Martikainen, M. et al. (2014). *Proc. Natl. Acad. Sci. U. S. A.* 111: 1277.

49 Martikainen, M., Salorinne, K., Lahtinen, T. et al. (2015). *Nanoscale* 7: 17457.

50 Pohjolainen, E., Malola, S., Groenhof, G., and Häkkinen, H. (2017). *Bioconjug. Chem.* 28: 2327–2339.

51 Brust, M., Schriffin, D.J., Bethell, D., and Kiely, C.J. (1995). *Adv. Mater.* 7: 795–797.

52 Mirkin, C.A., Letsinger, R.L., Mucic, R.C., and Storhoff, J.J. (1996). *Nature* 382: 607–609.

53 Alivisatos, A.P., Johnsson, K.P., Peng, X. et al. (1996). *Nature* 382: 609–611.

54 Blum, A.S., Soto, C.M., Wilson, C.D. et al. (2004). *Nano Lett.* 4: 867–870.

55 Zahr, O.K. and Blum, A.S. (2012). *Nano Lett.* 12: 629–633.

56 Lahtinen, T., Malola, S., Lahtinen, M. et al. (2016). *Nanoscale* 8: 18665–18674.

57 Nonappa, N., Lahtinen, T., Haataja, J.S. et al. (2017). *Angew. Chem. Int. Ed.* 55: 16035–16038.

58 De Nardi, M., Antonello, S., Jiang, D. et al. (2014). *ACS Nano* 8: 8505–8512.

59 Li, Q., Russel, J.C., Luo, T.Y. et al. (2018). *Nat. Commun.* 9: 3871.

60 Wang, Z.Y., Wang, M.Q., Li, Y.L. et al. (2018). *Am. Chem. Soc.* 140: 1069–1076.
61 Alhilaly, M.J., Huang, R.W., Naphade, R. et al. (2019). *Am. Chem. Soc.* 141: 9585–9592.
62 Yuan, P., Chen, R., Zhang, X. et al. (2019). *Angew. Chem. Int. Ed.* 58: 835–839.
63 Wu, Z., Liu, J., Gao, Y. et al. (2015). *Am. Chem. Soc.* 137: 12906–12913.
64 Liu, Y., Yao, D., and Zhang, H. (2018). *ACS Appl. Mater. Interfaces* 10: 12071–12080.
65 Wen, Z.R., Guan, Z.J., Zhang, Y. et al. (2019). *Chem. Commun.* 55: 12992–12995.
66 Khanna, S.N. and Jena, P. (1992). *Phys. Rev. Lett.* 69: 1664.
67 Khanna, S.N. and Jena, P. (1995). *Phys. Rev. B* 51: 13705.
68 Yuan, P., Zhang, R., Selenius, E. et al. (2020). *Nature Comm.* 11: 2229.
69 Malola, S. and Häkkinen, H. (2020). *Nature Comm.* 12: 2197.

8

Superatoms as Building Blocks of 2D Materials

Zhifeng Liu

School of Physical Science and Technology, Inner Mongolia University, Hohhot, Inner Mongolia, China

8.1 Introduction

When a layered material is thinned to its physical limits, it exhibits novel properties different from its bulk counterpart. The prominent example is graphene, a single layer of sp^2-hybridized carbon atoms arranged in a two-dimensional (2D) honeycomb lattice, which was first mechanically exfoliated from graphite by Novoselov et al. in 2004 [1]. Owing to the long-range π-conjugation, graphene holds extraordinary mechanical, thermal, and electronic properties (e.g. high Young's modulus, excellent thermal conductivity, and giant carrier mobility from the Dirac state) [2–6], which have led to extensive interest from both fundamental science to industry. The remarkable properties enable graphene to have potential applications in a wide range of areas, such as lubricant [7], high-speed electronic and optical devices [8, 9], transistors [10], energy generation and storage [11], photocatalysts [12], and biological/chemical sensors [13–15].

Following the discovery of graphene, numerous 2D materials have been explored in the past few years [16–31], such as graphene-allotropes (e.g. penta-graphene [31], graphdiyne, and graphyne [30]), elemental 2D-Xenes (e.g. silicene [29, 32], germanene [33, 34] and stanene [35], phosphorene [36], arsenene [37], antimonene [37], bismuthine [38], borophene [39, 40], and other metallenes [20]), transition metal dichalcogenides (TMDCs) [41, 42], metal nitrides/carbides (MXenes) [26, 43], transition metal halides (TMHs) [44, 45], transition metal oxides [46], perovskite-type oxides [47], and so on. As with graphene, these 2D materials also hold rich properties, broadening possible application areas. For instance, unlike the zero-gap graphene, the isoelectronic h-BN monolayer [48–51] consisting of alternating boron and nitride atoms arranged in a honeycomb lattice is a good electrical insulator with a large bandgap of 5.97 eV. Because of this, it was proved to be an ideal substrate for graphene electronics [52]. In addition to graphene, the elemental 2D-Xenes of group IVA (i.e. silicene [29, 32], germanene [33, 34], and stanene [35]) were also confirmed to be Dirac semimetals without considering spin-orbital coupling (SOC). However, when the SOC effect is turn on, the Dirac cone opens a sizable bandgap (1.55 meV for silicene [53],

Superatoms: Principles, Synthesis and Applications, First Edition. Edited by Purusottam (Puru) Jena and Qiang Sun.
© 2022 John Wiley & Sons Ltd. Published 2022 by John Wiley & Sons Ltd.

23.9 meV for germanene [53], and 73.5 meV for stanene [54]) due to their buckled honeycomb lattice and strong SOC. This makes the experimental detection of quantum spin Hall effect possible. As the neighbor of carbon in periodic table, boron atoms can also form stable monolayer on an appropriate substrate, which is referred to as borophene [39, 40]. Owing to the multicenter bonding characteristics, borophene exhibits metallicity. Moreover, the 2D monolayers of the group-VA elements (i.e. phosphorene [36], arsenene [37], antimonene [37], and bismuthine [38]) are confirmed as semiconductors with significant fundamental bandgaps, thus making them promising candidates for future electronic devices [21].

Structurally, nearly all these 2D materials to date are atomic crystals, where atoms are the building blocks. Generally, in these materials each atom has its own intrinsic characteristics that determine whether it can bond with other atoms and what crystal structures they will form. However, the atoms themselves cannot be altered. Thus, the crystal structures of atomic materials are relatively simple. In this context, there is a growing interest in designing 2D materials with complex, hierarchical, or tunable structures [55–57]. Cluster-assembled or superatomic materials [58–60] with stable clusters (superatoms [61–64]), instead of atoms, as the building blocks, are attractive targets for designing the complex 2D materials with tailored properties due to the following two reasons: (i) in the traditional periodic table, there are 82 stable and nonradioactive atomic building blocks that can be used to assemble crystalline materials [65]. In contrast, the number of potential superatomic building blocks is infinite. As is known, clusters are the ultimate nanoparticles [64] where their atomic composition is known exactly and the quantum confinement plays a leading role because the electron wavelength is comparable to the particle size. Thus, the properties of clusters are sensitive to the addition or removal of a single atom, even an electron. Moreover, the electron energy levels in cluster are much more complex compared with that of an atom, which are governed not only by the size, composition, and charged state of cluster but also by its structural symmetry. These facts make clusters provide unlimited possibilities for searching for new superatomic building blocks with various structure and unique properties. (ii) Cluster-assembled materials have hierarchical structure at two levels: internal structure of building blocks determined by atomic interactions in a relatively small length scale, and periodic lattice pattern formed by inter-cluster interaction in a larger length scale. The coupling between these interactions in different levels, not existing atomic crystals, dominates the collective properties of cluster-assembled materials. As a consequence, compared with atomic crystals there is more flexibility for tuning the materials properties and more opportunities for designing new functional materials and discovering novel phenomena.

Inspired by the successful synthesis of C_{60} solid [66, 67] and the concept of "assembling crystals from clusters" [58, 59], considerable research has been carried out on the structure, stability, properties, experimental synthesis, and possible applications of cluster-assembled materials. Although some excellent reviews [63–65, 68–78] have been devoted to discuss various aspects of cluster-assembled materials, most of them focus on assembling three-dimensional (3D) solid materials. In this chapter, we will mainly discuss the 2D superatomic materials. As the research in this direction is still at an embryonic stage, there lacks a general theory to cover most of the reported works. In this regard, it is hard to sum up the principles of assembling superatoms, theoretical model for inter-cluster interaction, and the rules for tuning their properties. Our objective is to review current advances in both experiment and theory for the 2D superatomic materials, and highlight their particular collective properties, hoping to provide useful reference for researchers to enter this field and to know what has been achieved.

8.2 Fullerene-assembled 2D Materials

8.2.1 C_{60}-assembled Monolayer

In 1985, a remarkably stable cluster, C_{60} (buckminsterfullerene), was produced during the vaporization of graphite by laser irradiation [79], which has a truncated icosahedron structure containing 12 pentagons and 20 hexagons. Owing to the ideal spherical hollow structure with icosahedral symmetry, C_{60} exhibits a series of atom-like unoccupied molecular orbitals, namely, superatomic molecular orbitals (SAMOs) or "superatom states" formed by a long-range polarization interaction, which has been confirmed by low-temperature scanning tunneling microscopy (STM) experiment combined with density functional theory (DFT) calculation [80, 81]. Moreover, the remarkable stability of C_{60} makes it serve as atom-like building blocks of various fullerites [66] and fullerides [82], with applications in different fields, such as molecular electronics, organic photovoltaics, cosmetics, and healthcare. Among abundant C_{60}-based assemblies, 2D C_{60} monolayers including the freestanding ones and that on different substrates have attracted significant attention. This is due to (i) superconductivity in the C_{60} monolayer with a T_C as high as the one in 3D bulk systems [82, 83] and (ii) the possibility of controlling, at a fine scale, the charge state, orientation, and inter-C_{60} distance. This makes C_{60}-based 2D materials extremely attractive for a wide range of applications [84–92].

The structural and cohesive properties of a freestanding monolayer of C_{60} (ML-C_{60}) have been explored by means of total energy DFT calculations [93]. As shown in Figure 8.1a, the cohesive energy is zero when the C_{60}-C_{60} distance is 13.0 Å, which indicates that each building block in ML-C_{60} is isolated without any interaction between them. Furthermore, as the C_{60}-C_{60} distance decreases one can see that the cohesive energy curve has two

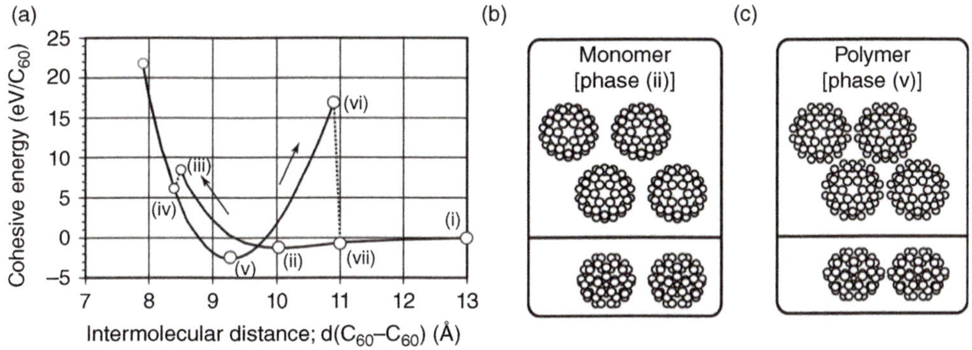

Figure 8.1 (a) Cohesive energy per C_{60} as a function of C_{60}-C_{60} inter-cluster distance. Arrows indicate the direction of the compression/tension stroke. The optimized atomic arrangements of (b) monomer phase (ii) and (c) polymer phase (v) for ML-C_{60}. *Source:* Nakamura et al. [93]. © American Physical Society.

minima (ii and v in Figure 8.1a), corresponding to two different phases of ML-C_{60}: the monomer phase assembled by van der Waals (vdW) interaction (Figure 8.1b) and the polymer-I by covalent binding (Figure 8.1c), indicating that ML-C_{60} is a bistable system [93].

8.2.1.1 Freestanding vdW C_{60} Monolayer

For vdW monomer phase (ii), the minimum in energy is shallow at a C_{60}-C_{60} distance of 10.05 Å and the calculated cohesive energy of 1.31 eV [93], slightly smaller than that of 3D face-centered-cubic (fcc) C_{60} (1.6 eV [94]) and much smaller than the typical C-C bond energy (more than 3 eV). To examine whether the freestanding infinite vdW C_{60} monolayer is stable, Reddy et al. [95] carried out molecular dynamics (MD) simulations for a hexagonally arranged C_{60} monolayer. Their results show that vdW interactions are strong enough to enforce a stable in-plane monolayer structure with the same C_{60} orientation at 0 K. As temperature increases, each C_{60} cluster in the hexagonal lattice structure (Figure 8.2a) undergoes random thermal fluctuations with both translational and rotational degrees of freedom. When temperature is less than 600 K, the average out-plane distance of all the C_{60}

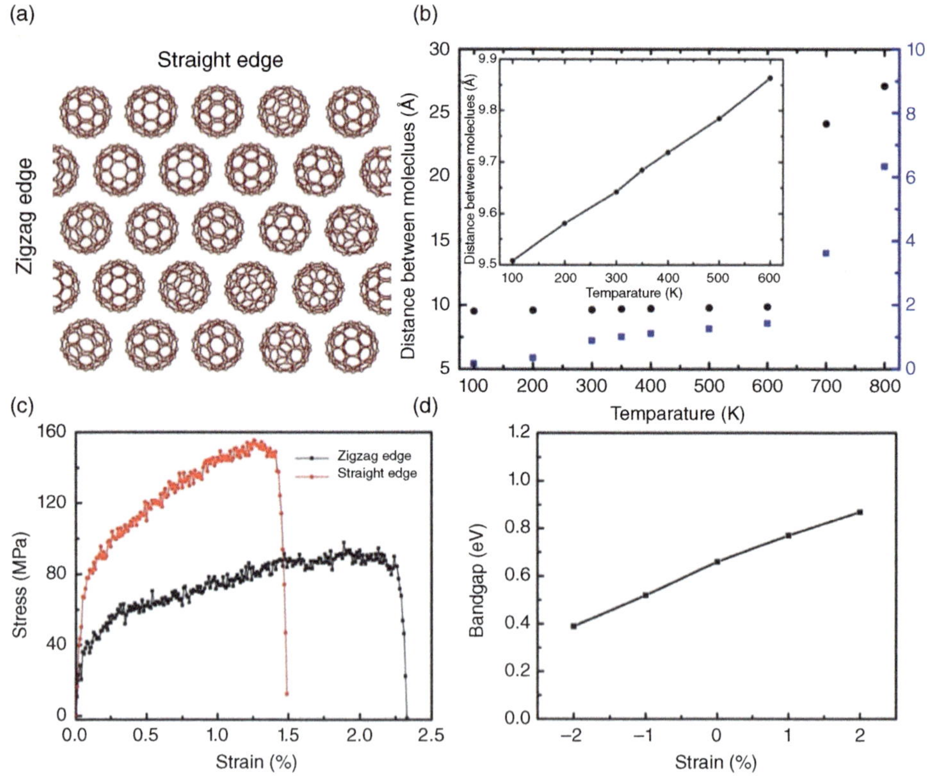

Figure 8.2 (a) The energetically optimized hexagonal structure of C_{60} monolayer after random perturbations. (b) The variation of the average C_{60}-C_{60} distance (black) and the average out-of-plane distance (blue) of the C_{60} clusters from its equilibrium as a function of temperature. (c) The stress-strain curves of the vdW C_{60} monolayer under uniaxial tensile loading along the zigzag and straight edge directions. (d) The variation of calculated bandgap with biaxial strain for the C_{60} monolayer. *Source:* Reddy et al. [95]. © Springer Nature.

clusters increases slowly (Figure 8.2b). For temperatures larger than this, the distance increases sharply, implying the occurrence of structural collapse in the vdW C_{60} monolayer. Such degradation temperature is about a half of that in 3D C_{60} solid crystal, which is experimentally measured to be 970~1300 K [96, 97]. Additionally, the variation of the average C_{60}-C_{60} distance is also examined as the temperature increases, from which the coefficient of thermal expansion of vdW C_{60} monolayer is calculated to be $8 \times 10^{-5}\ K^{-1}$ [95].

Figure 8.2c shows the stress-strain curves of vdW C_{60} monolayer under uniaxial loading along the zigzag and straight edge directions. One can find that the stress rises sharply in the beginning and then increases slowly in a relatively large range of strain. Such transition can be attributed to the rotation of C_{60} clusters, which causes the slope to change from the sharp to slow increase portion of the stress-strain curve. The ultimate tensile stress is calculated to be 90 MPa at 2.3% and 155 MPa at 1.5% in the zigzag and straight edge directions, respectively. On the basis of these, the elastic modulus is estimated to be 55 and 100 GPa for the zigzag and the straight edge directions, respectively [95].

The DFT calculations show that the vdW C_{60} monolayer with an inter-cluster distance of 9.5 Å is a semiconductor with a bandgap of 0.66 eV [95]. When the distance is 10.05 Å, the bandgap is calculated to be 1.01 eV [93], which is in good agreement with the experimental value (~1 eV) obtained from a scanning tunneling spectroscopy (STS) measurement of ML-C_{60} on the Si(111) $\sqrt{3} \times \sqrt{3}R30°$-Ag surface with weak bond between ML-C_{60} and the substrate [85]. When ~ 2% biaxial strain is applied on vdW ML-C_{60}, the bandgap can be tuned from 0.4 to 0.8 eV (Figure 8.2d), which is consistent with the result that the energy gap will decrease by application of external pressure [93].

Since the isolated C_{60} holds atom-like diffuse orbitals, namely, SAMOs [80], one may naturally want to know about the hybridized states of SAMOs for different C_{60} assembled materials. Recently, a series of delocalized SAMOs were identified in a 2D C_{60} monolayer on a highly oriented pyrolytic graphite (HOPG) utilizing resonant angle-solved two-photon photoemission spectroscopy (AR-2PPE). At an energy higher than the lowest unoccupied molecular orbital (LUMO), four spectral peaks are observed, which are denoted by SA1-SA4 along with the energy order in Figure 8.3a. To characterize these SAs obtained by AR-2PPE, Shibuta et al. [98] performed DFT calculations of the electronic band structure (Figure 8.3b) of a freestanding 2D vdW ML-C_{60} with C_{60}-C_{60} distance set to 10 Å. At the Γ point, one can see that there are several highly dispersive nearly free electron (NFE) bands (**154**, **156**, and **158**) at approximately $E_F + 6$ eV. As shown in Figure 8.3c–e, the orbital distributions of these bands extend outside the C_{60} hollow core, essentially distributing between C_{60} clusters. This indicates that the highly dispersive NFE bands are formed by C_{60} diffuse SAMOs through strong inter-cluster hybridization. Moreover, by comparing the calculated negative effective masses (Table 8.1) of these bands with that of experimentally observed SAs, one can conclude that the SA1, SA2, and SA3 should be mainly dominated by the calculated bands of **154**, **156,** and **158,** respectively. Thus, the DFT calculations of NFE bands of the freestanding 2D vdW ML-C_{60} have quantitatively reproduced the experimental observations on the ML-C_{60}/HOPG system.

8.2.1.2 Freestanding Covalent Polymerized C_{60} Monolayer

For the fcc C_{60} films, Rao et al. [99] revealed that exposure to visible or ultraviolet light can cause the polymerization of solid C_{60} films by reversible [2+2] cycloaddition reactions between two adjacent C_{60} clusters. Under high pressure and high temperature, various

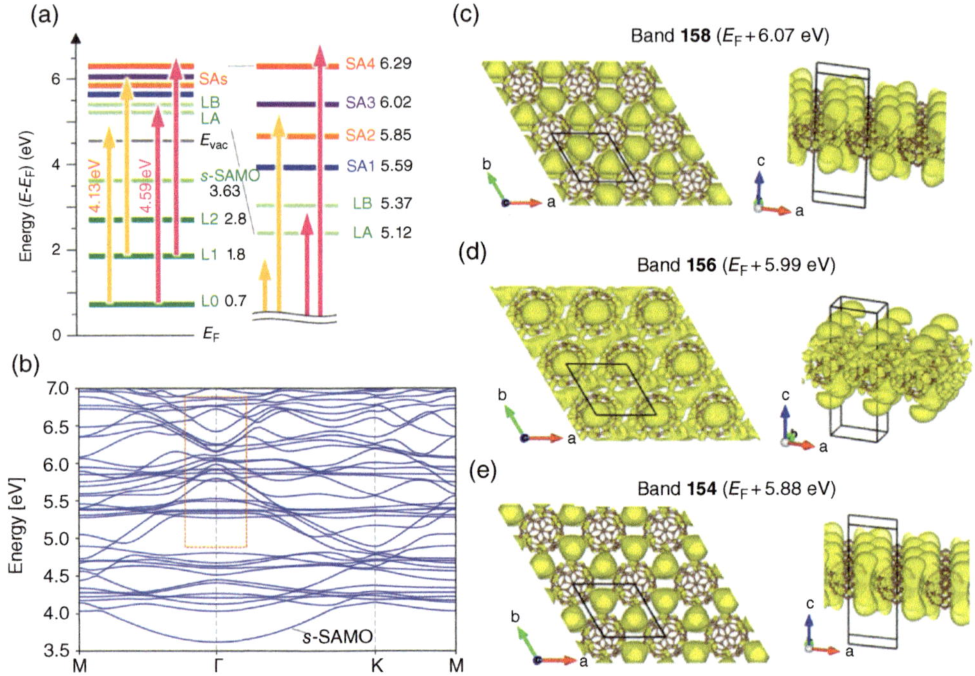

Figure 8.3 (a) The AR-2PPE obtained energy diagram of the 2D ML-C_{60}, in which the LUMO, LUMO+1, and s-SAMO are denoted by L_0, L_1 and L_2, respectively. The vertical arrows indicate the resonance excitations of SAs from L_0 and L_1. (b) The calculated electronic band structures of 2D vdW ML-C_{60} with 10 Å C_{60}-C_{60} distance. Energy is aligned with respect to the band 135 (s-SAMO, i.e. E_F + 3.63 eV). The calculated orbital distributions for the (c) SA3: band 158, (d) SA2: band 156, and (e) SA1: band 154 of the 2D vdW ML-C_{60}. *Source:* Shibuta et al. [98]. © American Chemical Society.

Table 8.1 Experimental energies (E_0) at the Γ point and effective electron masses (m_{eff}) of unoccupied bands SA1, SA2, and SA3 (or 154, 156, and 158) for the studied 2D ML-C_{60}. The corresponding DFT calculated results (E_{0cal} and $m_{eff,cal}$) of freestanding 2D ML-C_{60} are also presented.

	E_0 (eV)	E_{0cal} (eV)	m_{eff} (m_e)	$m_{eff,cal}$ (m_e)
SA3	6.02	6.00 (158)	−0.69	−0.06
SA2	5.85	5.92 (156)	−0.17	−0.10
SA1	5.59	5.81 (154)	−0.11	−0.14

Source: Shibuta et al. [98]. © American Chemical Society.

bulk polymerized C_{60} phases can also be obtained, such as rhombohedral, orthorhombic, and tetragonal phases, possessing different properties [100–106]. Among them, the rhombohedral phase is the dominant one, which is made of layers of hexagonal close-packed (HCP) C_{60} clusters. The measured nearest distance between C_{60} clusters in different layers is ~9.8 Å, suggesting the domination of weak vdW interaction between C_{60} layers [100, 101]. However, the in-layer C_{60}-C_{60} distance is only 9.2 Å due to the formation of C–C covalent double bonds by [2+2] cycloaddition reactions [107]. Such layered structural features brought attention to the properties of a 2D single layer of the new C_{60} phases by neglecting interlayer interaction [93, 107–109].

Based on the synthesized rhombohedral C_{60} phases [100], Xu et al. [108] first studied the structure, stability, and conducting properties of 2D rhombohedral ML-C_{60}. In this 2D layer, each individual C_{60} is connected to six neighbors by [2+2] cycloaddition of "66" bonds (C–C double bond shared by two six-membered rings), as shown in Figure 8.4a. The C_{60}-C_{60} equilibrium distance is calculated to be 9.17 Å from the tight-binding model [108], which is slightly larger than that (9.03 Å [93]) obtained from the DFT calculation. The cohesive energy of the hexagonal ML-C_{60} with respect to the free C_{60} is estimated to be 2.1 eV/C_{60} (2.66 eV/C_{60} in Ref. [93]), much larger than the mentioned vdW ML-C_{60} (1.3 eV/C_{60} [93]) due to the strong interaction between C_{60} fullerenes with a covalent C-C bond length of 1.64 Å [108]. From the obtained elastic constants c_{11} (2.1 Mbar) and c_{12} (0.2 Mbar), the elastic modulus $M = (c_{11} + c_{12})/2$ of the hexagonal ML-C_{60} is estimated to be 1.1 Mbar [108], which is much larger than the bulk modulus of vdW fcc C_{60} (0.14 Mbar [67]). This result is also consistent with the formation of covalent bonds between C_{60} clusters in the 2D hexagonal layer. Moreover, similar calculations have also been carried out for the 2D tetragonal ML-C_{60} (see Figure 8.4b and Table 8.2).

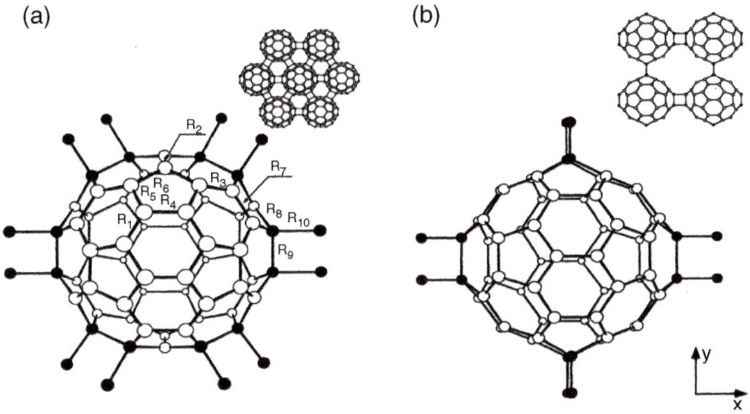

Figure 8.4 (a) The building block C_{60} in the single hexagonal C_{60} layer (seven C_{60} fullerenes are shown in the inset) and (b) tetragonal C_{60} layer (four C_{60} fullerenes are shown in the inset). The dark atoms are those involved in inter-cluster bonding. *Source:* Xu and Scuseria [108]. © American Physical Society.

Table 8.2 Binding energies E with respect to C_{60}, inter-cluster center–center distances d, intermolecular bond lengths R of [2+2] cycloaddition, and electronic bandgaps E_g for two different C_{60} layers.

	E (eV)	d (Å)	R (Å)	E_g (eV)
Hexagonal C_{60} layer	2.1	9.17, 9.03	1.64	1.0
Tetragonal C_{60} layer	0.9	9.06(x), 9.13(y)	1.64(x), 1.64(x)	1.2

Source: Xu and Scuseria [108]. © American Physical Society.

The analysis of electronic properties shows that the 2D hexagonal ML-C_{60} linking by 66/66 bands is a semiconductor with a bandgap of 1.0 eV [108]. However, Okotub et al. [109] revealed that the rotation of C_{60} can dramatically change the energy band dispersion of the 2D hexagonal ML-C_{60}. With different linking bonds, three polymer structures were proposed, including polymer-I (66/66), II (65/56), and III (65/66), as shown in Figure 8.5a, c, and e, respectively. From the band structure of polymer-I (Figure 8.5b), one can see that it is an indirect semiconductor with a bandgap of 0.81 eV [109], which is consistent with the result in Ref. [108] (1.0 eV) based on the density of state calculations. When the C_{60} clusters are connected by 65/56 bonds (shared by a hexagon and a pentagon), the conduction and valence bands meet at K point, indicating metallic properties of polymer-II (Figure 8.5d). Compared to that of polymer-I, the electronic band structure of polymer-II is characterized by increasing the width of the lowest branch of the conduction band and shifting the top of the valence band toward higher energy. This is because of an enhancement of the

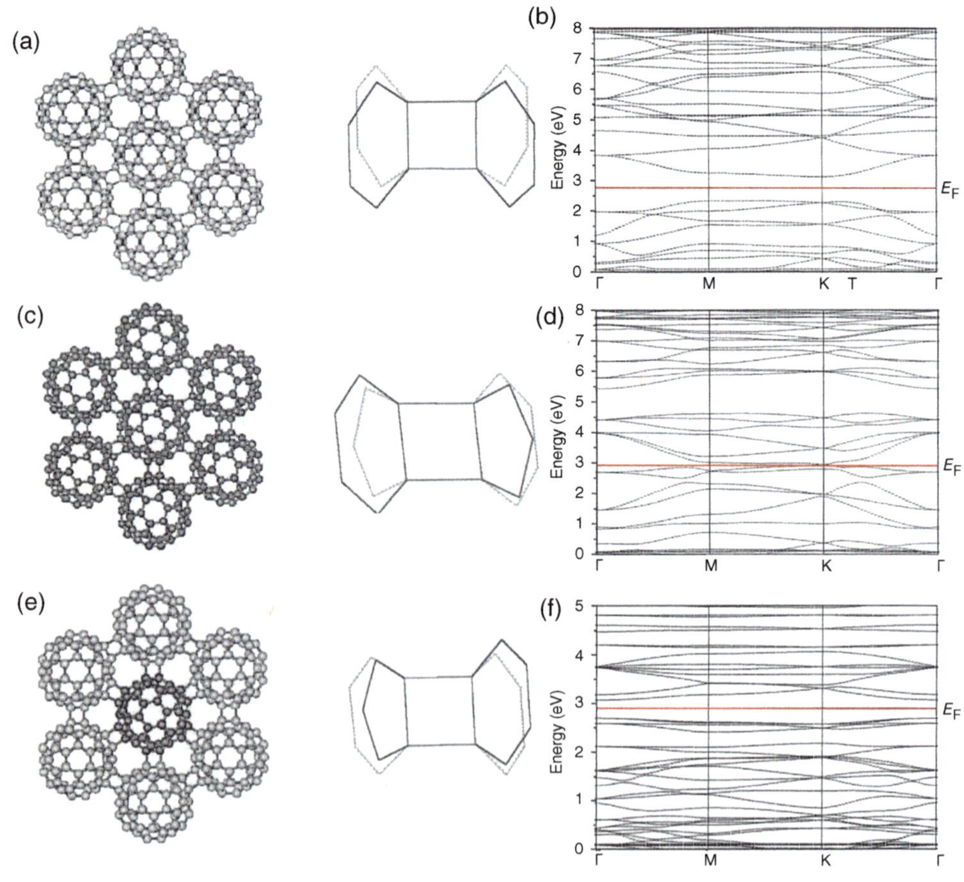

Figure 8.5 The structural models of three 2D C_{60} polymers with different inter-cluster bonds: (a) polymer-I with 66/66 bonding, (b) polymer-II with 65/56 bonding, and (c) polymer-III with 65/66 bonding. Changes in the orientation of the fullerenes are shown by different levels of shading. The corresponding electronic structure of (d) polymer-I, (e) polymer-II, and (f) polymer-III. *Source:* Okotrub et al. [109]. © AIP Publishing.

inter-cluster overlap of π-type orbitals. In polymer-III with 65/66 bonds, the C_{60} clusters hold the cage orientations in both polymer-I and -II (Figure 8.5e). The band structure (Figure 8.5f) shows that it is a direct semiconductor with a bandgap of 0.38 eV, about half of that in the indirect bandgap polymer-I. These results reflect together that the C_{60}-C_{60} interactions play a primary role in determining the electronic properties of the hexagonal polymers.

8.2.2 C_n ($n = 20, 26, 32, 36$)-assembled Monolayers

Similar to C_{60}, small fullerenes (C_n, $n < 60$) have also been extensively studied due to their remarkable reactivity, which allows the design of novel complexes having fascinating physical properties [110]. In experiment, small fullerenes can be observed in soot and synthesized through appropriate chemical methods [111–114]. Interestingly, these small fullerenes hold one or more fused pentagons that contain bond angles close to that of sp^3 C atoms, implying that these C atoms possess a dangling bond rather than conventional π states. Because of this, small fullerenes intrinsically tend to form polymerized or clathrate structure. Similar to C_{60} assembled materials, polymeric small fullerenes are also expected to have unusual electronic structures that are different from that of conventional carbon allotropes [115, 116]. Based on first-principles calculations, some interesting 2D materials assembled by small fullerenes (e.g. C_{20} [117], C_{26} [118, 119], C_{32} [120], and C_{36} [121]) have been proposed, which are designed by sharing carbon rings or covalent interaction.

As the smallest member of fullerene family, C_{20} cage has unique atomic configuration, which is made up of 12 pentagons with high curvature and strong strain. Although it is not the ground state structure of C_{20} and violates the so-called isolated pentagon rule, it is dynamically stable and has already been successfully synthesized [111]. Encouraged by this, the studies of using C_{20} cage cluster as a building block to design new carbon crystals [115, 122–125] have become one of the topics in the research area of carbon allotropes. On the basis of advances of 1D and 3D C_{20}-assembled structure, Shen et al. [117] proposed the first 2D C_{20}-based sheet, named Hexa-C_{20}. By sharing face, the C_{20} cage clusters assemble into a periodic sheet with a 34-atom unit cell with $P6/mmm$ symmetry (see Figure 8.6a–c). Their calculations proved that such 2D structure is not only dynamically stable but also thermally stable at temperatures as high as 1500 K. The band structure (Figure 8.6d) shows that Hexa-C_{20} is a semiconductor with a quasi-direct bandgap of 3.28 eV. Remarkably, by solving the linearized phonon Boltzmann transport equation, the authors confirmed that the assembled Hexa-C_{20} displays an ultrahigh intrinsic lattice thermal conductivity κ_{lat} of 1132 W/m/K at 300 K (Figure 8.6e), which is much larger than that of many carbon materials, such as penta-graphene [126, 127] and graphyne [128], indicating its great potential in thermal nanodevices. The analysis of phonons uncovers that three-phonon scattering is predominant in thermal transport [117].

Compared with C_{20} cage [111], the larger C_{26} fullerene has three more hexagonal faces beside 12 pentagons (see Figure 8.7a). Owing to this structural feature, it can form a 2D hexagonal covalent network by sharing the three hexagonal faces with adjacent ones, leading to a graphene-like structure consisting of C_{26} polyhedrons (Figure 8.7b). Based on first-principles geometry optimization, Maruyama et al. [118, 119] revealed that the C_{26} polymer has two different metastable structures [C_{26}(I) and C_{26}(II) phases] with total energies larger

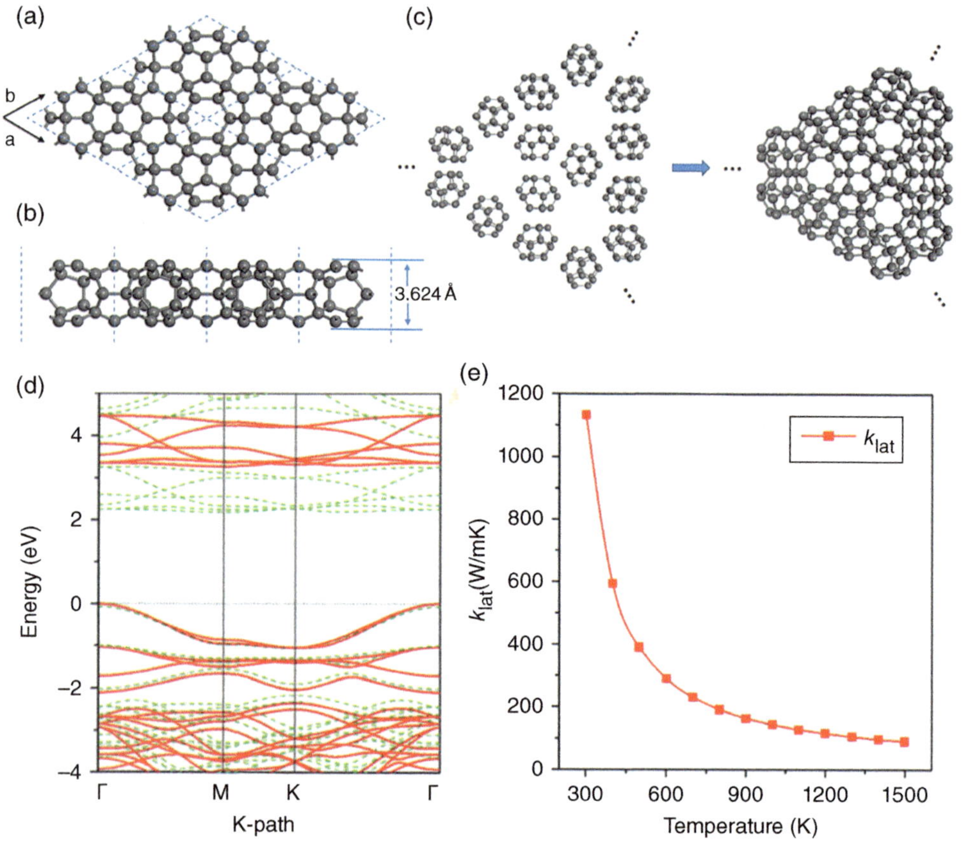

Figure 8.6 The optimized atomic structure of Hexa-C_{20} from (a) top and (b) side views. (c) The assembling process from the C_{20} cage to the Hexa-C_{20} sheet by face sharing. (d) Electronic band structure of Hexa-C_{20}; (e) variation of κ_{lat} with temperature in Hexa-C_{20}. *Source:* Shen et al. [117]. © Royal Society of Chemistry.

than that of diamond by about 0.6 eV/atom. Figure 8.7c shows the optimized structure of the stable phase C_{26}(I). Although the main skeleton of C_{26} cage is preserved, local distortion can be found. Specifically, octagonal rings are formed at the fused region of adjacent C_{26} polyhedrons due to the dissociation of the bond shared by two pentagons and a hexagon [118]. Therefore, the resultant 2D polymeric phase is composed of distorted C_{26} polyhedrons. Interestingly, the formation of fused octagons can result in outstanding electronic properties. As is illustrated in Figure 8.7d, one can find that there exist linear dispersion bands, which cross each other slightly above the Fermi level. Since the existence of Dirac massless fermions, such assembled sheet is anticipated to hold high carrier mobility, with a promise for further high-speed electronics.

As for the C_{32} fullerene, the isomer with D_{3h} symmetry consists of 12 pentagons and 6 hexagons [129]. Based on such structural feature, it can form a honeycomb network by constructing three four-membered rings with three adjacent fullerenes [120] (see Figure 8.8a). In this regard, the 2D polymeric C_{32} can be considered as a C_{32}-graphene. Unlike the above metallic C_{26} monolayer, DFT band structure calculations show that

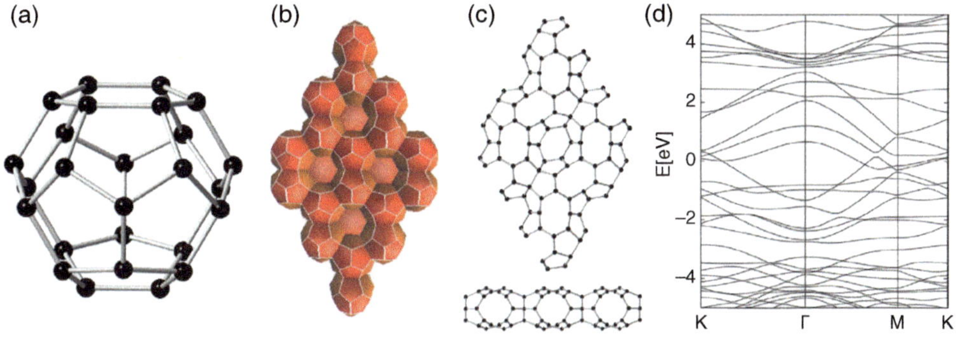

Figure 8.7 (a) The optimized atomic structure of C_{26} cluster. (b) Initial structure of C_{26} assembled 2D sheet. (c) Top and side views of the optimized structures of C_{26}(I) sheet. (d) Electronic band structure of the polymeric C_{26}(I) sheet. *Source:* Maruyama and Okada [118]. © The Physical Society of Japan.

C_{32}-graphene is a semiconductor with a direct bandgap of 1.5 eV at the Γ point (Figure 8.8c). Around the fundamental gap, the dispersion of electron states is relatively small with a width of about 0.5 eV due to the distribution of π electrons separated by the fourfold coordinated C atoms connecting adjacent C_{32}. When the pores of C_{32}-graphene are filled by the larger C_{36} fullerene (Figure 8.8b), the electronic structure changes from semiconductor to metal with localized states crossing the Fermi level (Figure 8.8d). Up to now, few works involve the assembly of mixed clusters with different sizes. The C_{36}-filled C_{32} sheet provides a good prototype for further studies to explore new 2D superatomic materials with different types of clusters as building blocks.

The research interest in C_{36} fullerene comes from the arc production [116] of polycrystalline carbon solid consisting of C_{36} cluster with D_{6h} [130] symmetry (Figure 8.9a). When the D_{6h}-C_{36} are connected with each other by two covalent bonds, a 2D trigonal lattice can be formed (Figure 8.9b), in which every C_{36} polyhedron is surrounded by six similar building

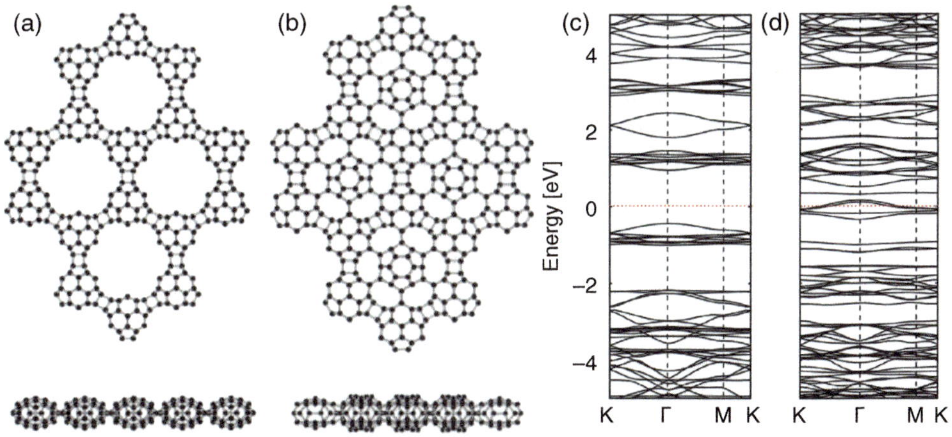

Figure 8.8 The optimized structures of (a) C_{32}-graphene and (b) C_{36}-filled C_{32}-graphene. The electronic structures of (c) C_{32}-graphene and (d) C_{36}-filled C_{32}-graphene. *Source:* Maruyama et al. [120]. © The Japan Society of Applied Physics.

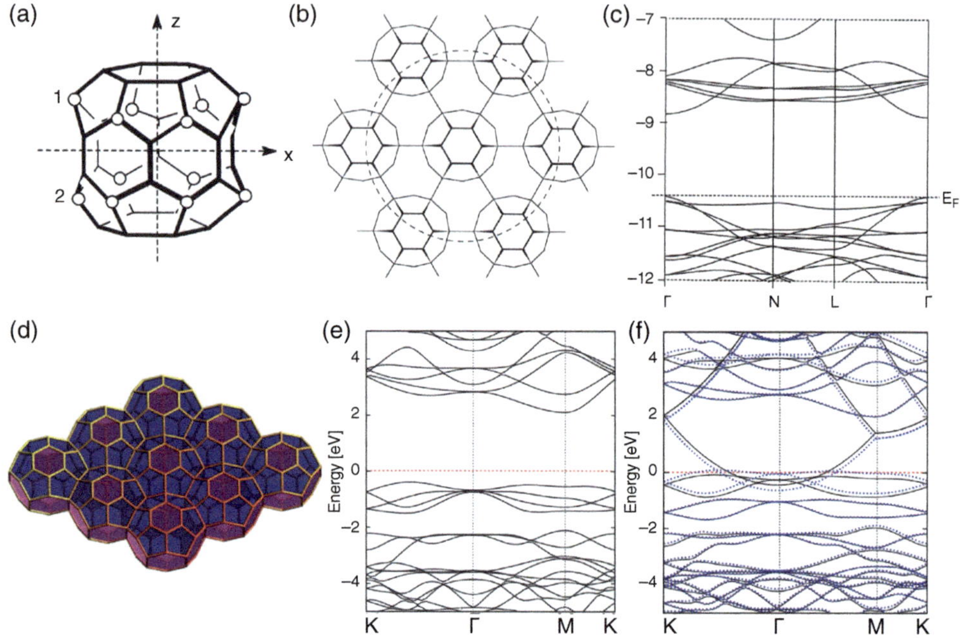

Figure 8.9 (a) Geometric structures of fullerene C_{36}, (b) top view of 2D C_{36} layer with trigonal lattice, and (c) the electronic band structure of trigonal C_{36} sheet. *Source:* pern et al. [130]. © Elsevier. (d) The 2D network of the fused C_{36} cluster, (e) electronic band structure of 2D C_{36}(I) phase, and (f) electronic band structure of 2D C_{36}(I) phase under parallel electric fields of 2.0 V/Å. *Source:* Maruyama et al. [121]. © The Physical Society of Japan.

blocks [130, 131]. Its electronic band structure obtained by the crystalline orbital method in the extended Hückel theory (EHT) approximation [132] shows that it is a semiconductor with a direct bandgap of ~1.5 eV [130] (Figure 8.9c). Maruyama et al. [121] also explored three possible 2D geometries of C_{36}-assembled phases by fusing the hexagonal rings. Among them, the most stable C_{36}(I) phase (Figure 8.9d) possesses total energy of 0.44 eV/atom with respect to graphene, which is comparable with the energy of C_{60} (0.4 eV/atom). Furthermore, such phase is confirmed to be both dynamically and thermally (up to 3000 K) stable by phonon dispersion calculation and MD simulation, respectively. Unlike the C_{36} assembly connected by covalent bonds, the fused C_{36}(I) phase is an indirect semiconductor with a larger bandgap of ~2.49 eV (Figure 8.9e). Interestingly, under parallel electronic field of 2.0 V/Å, the C_{36}(I) phase can be turned into a spin-polarized metal (Figure 8.9f) with a magnetic moment (MM) of 1.74 μ_B/nm^2. It was proved that such unusual spin states could be derived from the cooperation and competition between the triangularly arranged antiferromagnetic (AFM) polarized spins and the extended free electrons [121].

8.2.3 Fullerene Monolayers on Substrates

For realistic applications, a C_{60}-ML should be supported by a substrate. Therefore, the accurate characterization of the structural and electronic properties of 2D C_{60}-MLs on different substrates has become the topic of many studies [84, 89, 91, 133–149]. Compared with that of the above-mentioned freestanding C_{60}-MLs, the configurations of substrate-supported C_{60}-MLs are more complex because of (i) the elusive competition between C_{60}-substrate interactions and inter-C_{60} interactions and (ii) the possible C_{60}-induced surface reconstruction. However, using different techniques, including STM, STS, surface X-ray photoelectron diffraction (SXRD), low-energy electron diffraction (LEED) experiments [91, 139, 144, 145, 148–151], and first-principles calculations [89, 137, 138, 143, 152–156], many unique geometries of C_{60}-MLs on substrates have been determined, which reveal interesting electronic and transport properties for applications in fullerene-derived electronic devices.

In the case of C_{60} deposited on metal substrates, C_{60} cluster under thermal activation at high temperature can move to find the optimum adsorption sites. During this process, the matching of the high-symmetry sites between C_{60} and the substrate plays an important role in lowering the total energy of the C_{60}-metal composite system [89]. For the ideal clean close-packed metal surface fcc (111), the highest combined symmetry can be obtained when a threefold axis of C_{60} coincides with a threefold axis of the surface, namely, a hexagon of C_{60} being centered and aligned with on top or hollow site of the surface. This is the most commonly observed adsorption interface geometry in experiment, such as C_{60}-ML on Cu (111) [91, 157–159], Au(111) [160], Ag(111) [141], and Pt(111) [161–163], which have also been confirmed by DFT calculations. As for fourfold rotation symmetrical fcc (100) surfaces, the adsorbed C_{60} with a twofold symmetry is confirmed to the best match, such as that on Ag(100) [138, 164].

Except for the symmetry matching, size matching also plays an important role for determining the C_{60}-ML structure on metal substrate. The STM studies reveal that most of the ordered C_{60}-ML phases adopt a close-packed structure having an inter-cluster spacing very close to that of C_{60} bulk (~10 Å) due to the existence of in-plane C_{60}-C_{60} vdW interaction [89]. However, to reach size matching, there exist different C_{60}-ML unit cells and substrate supercell lattices. For instance, the unit cell of C_{60}-ML with one C_{60} cluster can closely match different supercells of various metal surface lattices, such as 4×4 Cu(111) [91], $\sqrt{13}\times\sqrt{13}$R13.9° Pt(111) [161], and $(2\sqrt{3}\times2\sqrt{3})$ R30° Ag(111) [141]. In some cases, C_{60}-MLs on the same kind of metal surface can also form different models of size matching. LEED, surface X-ray photoelectron diffraction, and STM measurements have uncovered that the close-packed C_{60}-MLs on Au(111) [145] have three prototypical superlattice phases, namely, $(2\sqrt{3}\times2\sqrt{3})$R30°, (7×7) R0°, and $(\sqrt{589}\times\sqrt{589})$R14.5°, which harbor 1, 4, and 49 C_{60} clusters per unit cell, respectively (see Figure 8.10).

Although C_{60}-ML on most metal substrates adopts similar close-packed structure, the orientation of C_{60} with respect to substrate, which is determined by a balance between cluster–substrate and inter-cluster vdW interactions, often affects the periodic pattern and properties of C_{60}-ML. Recently, Xu et al. [135] revealed that when C_{60} molecules are deposited on Cu(111) surface, there are two orientation-related configurations (see Figure 8.11a and b) that C_{60} molecules showed: one is a round shape (C_{60}-oriented pentagon-down), and the other is a three-lobe shape (C_{60}-oriented hexagon-down). If the C_{60}-covered Cu (111) surface is annealed at above 473 K, a well-ordered (4×4) close-packed C_{60}-ML structure is formed, in which the conformations of all C_{60} molecules are converted to the three-lobe structure. Additionally, there are two adsorption orientations where one is up-three-lobe and the other is down-three-lobe (Figure 8.11c). Using a controlled operation, the down-three-lobe conformation of C_{60} in a close-packed C_{60}-ML can be converted to the round

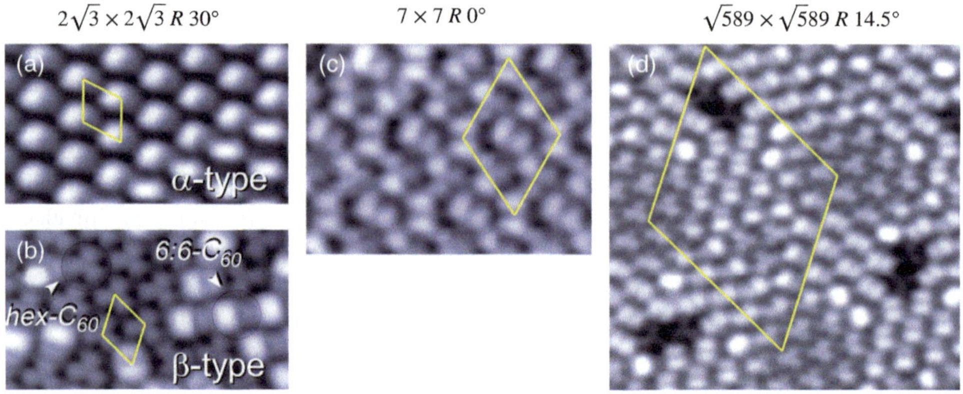

Figure 8.10 STM images recorded at sample voltages between 2 and 2.2 V of C_{60}-ML on Au(111). (a, b) $6.3 \times 3.6\,nm^2$ images of two different domains of a $(2\sqrt{3} \times 2\sqrt{3})$ R30° superlattice. (c) $6.3 \times 4.7\,nm^2$ image of the (7×7) R0° structure. (d) $11.2 \times 11.2\,nm^2$ image of a (589×589) R14.5° structure. Yellow rhombuses indicate the unit cells of the superstructures. *Source:* Adapted with permission from Shin et al. [145]. © American Physical Society.

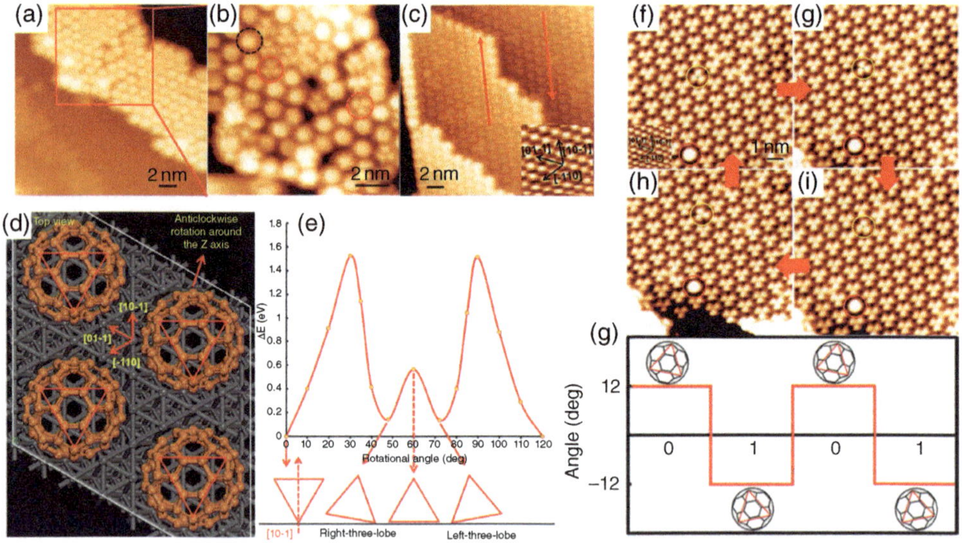

Figure 8.11 (a) STM images of the C_{60}-ML on the Cu(111) surface. (b) High-resolution STM image of the C_{60} domain marked by a red square in (a). There are two conformations of C_{60} in this domain, one has a round shape and the other has a three-lobe shape. (c) A very well-ordered (4×4) structure was formed when the annealing of the sample was over 473 K. The conformations of all C_{60} molecules were converted to the up-three-lobe or down-three-lobe conformation. The red arrow lines represent the direction of one lobe of three lobe. (d) Model of the unit cell for the (4×4) structure. There is only one C_{60} molecule with anticlockwise rotation around the z-axis among four C_{60} molecules. (e) Total energy of four C_{60} molecules as a function of one C_{60} molecular orientation. 48° and 72° are two angles of lower energy and represent the right-three-lobe and left-three-lobe conformation, respectively. Reversible and repeatable switching process between right-three-lobe and left-three-lobe conformation. (f) Initial status of C_{60} marked by a yellow circle within a molecular monolayer. (g–i) Conformation of C_{60} after the first, second, and third manipulation. (g) Simplified diagram for the working principle of a reversible and repeatable molecular switch. *Source:* Xu et al. [135]. © American Chemical Society.

conformation. Then, the down-three-lobe conformation is converted to the left-three-lobe conformation. On the basis of these, a reversible and repeatable switching between the left-three-lobe and right-three-lobe conformation can be realized (Figure 8.11d–g). Such a reversible C_{60} switch may potentially form building elements for molecular logic circuits.

8.3 Si-based Cluster Assembled 2D Materials

After the discovery of C_{60} [79], silicon (Si) clusters have also attracted a wide range of attention from both theorists and experimentalists due to the prominent role of Si in modern microelectronics. However, pure Si clusters are unsuitable as building blocks because they are chemically reactive due to the existence of dangling bonds. Although Si and C belong to the same group in the periodic table, their cluster structures are quite different. In Si clusters, it is difficult to obtain a stable hollow fullerene-like cage structure formed by large π bonds because the Si atom prefers to form σ bonds through sp^3 hybridization. By encapsulating small magic cluster [165], inner core [166], or transition metal (TM) atoms [167–170], closed-shell electronic structures can lead to species that are remarkably stable as a new class of endohedral silicon cage clusters. Using the external quadrupole static attraction ion trap, Hiura et al. [168] reported the synthesis of a series of metal-containing hydrogenated silicon clusters TM@Si_n (TM = Hf, Ta, W, Re, Ir for $n = 14, 13, 12, 11, 9$, respectively) and confirmed that the TM atom is endohedral and stabilizes the Si polyhedral cage. Inspired by this experiment, more and more TM-encapsulating Si clusters were obtained in following experiments [70], such as Mn@Si_n ($n = 12$~16) [170], V@Si_{12} [167], and TM@Si_{16} (TM = Ti, Zr, Hf, Ta) [171]. With these stable TM@Si_n cage clusters serving as the building blocks, one can construct some novel 1D, 2D, and 3D assembled materials with desirable properties [172–182]. In this section, we focus on the 2D materials assembled by TM@Si_{12} and TM@Si_{16} encapsulated clusters.

8.3.1 V@Si_{12} Assembled 2D Monolayer

For 2D spintronic materials, introducing TM atoms is a popular method to induce spin polarization and enhance magnetic interactions. However, the introduced TM atoms tend to cluster with each other because of the strong d–d or f–f interactions, resulting in inhomogeneous distribution of the TM atoms or chemical phase separation into alternating regions with higher and lower concentration of TM cations in the host matrix. In this context, Liu et al. [173] proposed an approach where endohedral Si clusters, used as building blocks, may bypass the long-standing problem because the outside Si cage can provide an appropriate potential well to trap foreign TM atoms and hinder the aggregation of TM atoms.

8.3.1.1 Structure and Stability

The magic cluster V@Si_{12} has a distorted hexagonal prism cage with a V atom located at the center, which has been characterized in both theory and experiment [167]. The stability of this structure can be understood by its compact geometrical shape and the saturated dangling bonds of Si_{12} with the endohedral V atom. Based on the C_6 symmetry, two sheets, i.e.

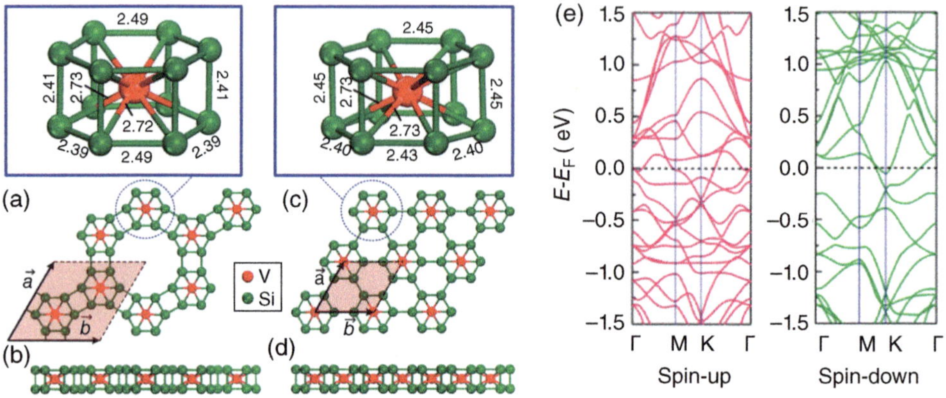

Figure 8.12 Atomic structures of the fully optimized $V@Si_{12}$-assembled sheets. (a) A top and (b) a side view of the hexagonal porous sheet (*P6/mmm*); (c) a top and (d) a side view of the honeycomb-like sheet (*P6/mmm*). (e) Spin-resolved band structures for the ferromagnetic state of the porous sheet. *Source:* Liu et al. [173]. © American Chemical Society.

hexagonal porous and honeycomb-like framework with regularly and separately distributed V atoms, can be built from $V@Si_{12}$ clusters (Figure 8.12a–d) [173]. The stability of these two suggested $V@Si_{12}$ sheets have been tested by analyzing their binding energies compared to n individual $V@Si_{12}$ clusters. From first-principles calculations, the binding energies are 2.04 and 3.09 eV per $V@Si_{12}$ for the porous and honeycomb-like sheet, respectively. This indicates that both of the assembled sheets are more stable than a collection of individual $V@Si_{12}$ clusters. Moreover, the two sheets are found to be dynamically stable as confirmed from phonon spectra. The analysis of electron density difference with respect to the sum of atomic densities further reveals that "V-modified sp^3 hybridization of the Si atoms in the inter-cluster linkage" is the assembled mechanism for these two stable 2D materials.

8.3.1.2 Electronic and Ferromagnetic Properties

Based on the calculations of energies in different magnetic states, the ferromagnetic (FM) state is found to be energetically the most stable state, with energy 0.207 and 0.026 eV lower than that of the AFM state for porous and honeycomb-like sheet, respectively. Compared with honeycomb-like sheet, the porous one has stronger FM coupling. The spin-resolved band structure (Figure 8.12e) reveals that the charge transport in porous sheet is mainly dominated by the spin-up electron. The measured spin polarization at the Fermi level is as large as 18.25%. It should be noted that the similar 2D structure of $Zr@S_{12}$ is a nonmagnetic indirect semiconductor [179] with a bandgap of ~0.3 eV (Figure 8.13), rather than an FM metal.

The MM of the V atom in isolated $V@Si_{12}$ cluster is almost quenched because the formation of V-Si bond enhances the electronic kinetic energy over the exchange energy favoring parallel spins. On the contrary, the MMs of the endohedral V atoms are recovered to a certain extent in both of the assembled sheets (3.5 and 3.8 μ_B per $V@Si_{12}$ for the porous and honeycomb-like sheet, respectively). This can be understood from the fact that the formation of Si-Si bonds between building blocks weakens the Si-V coupling. In particular, the first-principles MD simulations show that, at room temperature, the porous sheet with

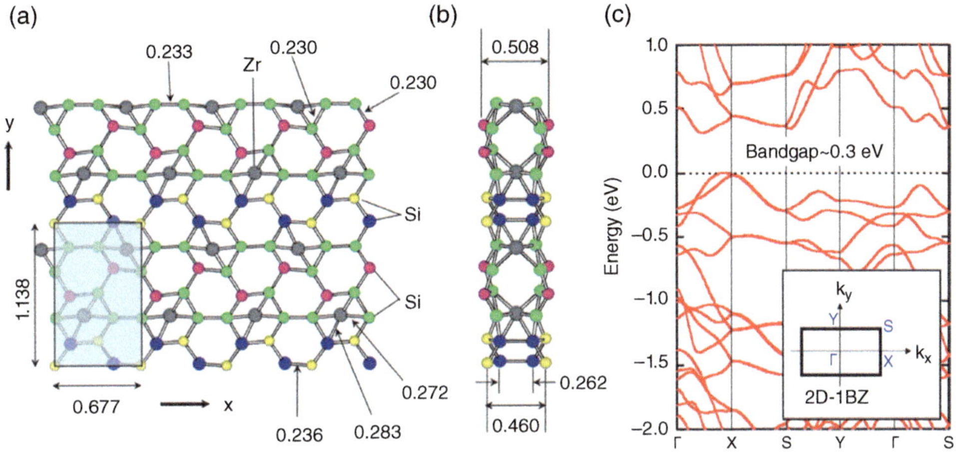

Figure 8.13 (a) Top and (b) side views of the optimized structure of Zr@Si_{12} monolayer. (c) Energy band structure of Zr@Si_{12} monolayer along some high-symmetry points. *Source:* Miyazaki et al. [179]. © The Japan Society of Applied Physics.

stronger FM coupling can remain magnetic with an average MM of around 21μ_B (2×2 supercell), indicating that the magnetism of the porous sheet is robust against thermal turbulence under ambient environment. Furthermore, to answer whether the magnetic states and MMs can be precisely modulated by external strain, the magnetic properties of the two proposed sheets under biaxial strain are also explored. Remarkably, the results show that both of the sheets have a robust FM state under compressive strain ($\varepsilon<0$). However, $\varepsilon>0$ strain leads to an AFM state. It is proved that the physical origin of this transition is from the competition between Si-V FM coupling through *p-d* hybridization and V-V AFM coupling through the Si-V super exchange interaction. For practical nano-spintronic applications such strain-induced magnetic transition between ferromagnetism and antiferromagnetism is very important. As for the MMs *m*, the increased strain reduces *m* from 7.0 to 5.5μ_B as a function of ε: $m = 6.96 + 0.24\varepsilon + 0.01\varepsilon^2$, because covalency is enhanced. For the honeycomb-like sheet, *m* is reduced from 15.3 to 12.9μ_B as a function of ε: $m = 15.29 - 0.13\varepsilon - 0.03\varepsilon^2$.

8.3.2 Other TM@Si_{12} Assembled 2D Monolayers

Inspired by the design of 2D V@Si_{12} sheet, Nie et al. [174] explored whether other 3*d* TM doped Si_{12} hexagonal prism clusters can also form a stable 2D material with magnetic order. For every TM@Si_{12} cluster, they constructed four 2D crystal structures: hexagonal honeycomb lattice (Figure 8.14a), hexagonal porous lattice (Figure 8.14b), square lattice (Figure 8.14c), and central rectangular lattice (Figure 8.14d). In these assembled 2D structures, the TM atoms are all separated and regularly distributed between the two layers of Si atoms, forming sandwich-like layered structures. Taking the Cr@Si_{12} cluster as an example, it is proved that the stability of these four structures is arranged as following: the honeycomb structure, the hexagonal porous structure, the central rectangular structure, and the square structure with binding energies of 3.73, 2.65, 2.06, and 1.22 eV, respectively. Owing to the D_{6h} symmetry of Cr@Si_{12}, the hexagonal lattices are relatively more stable against other

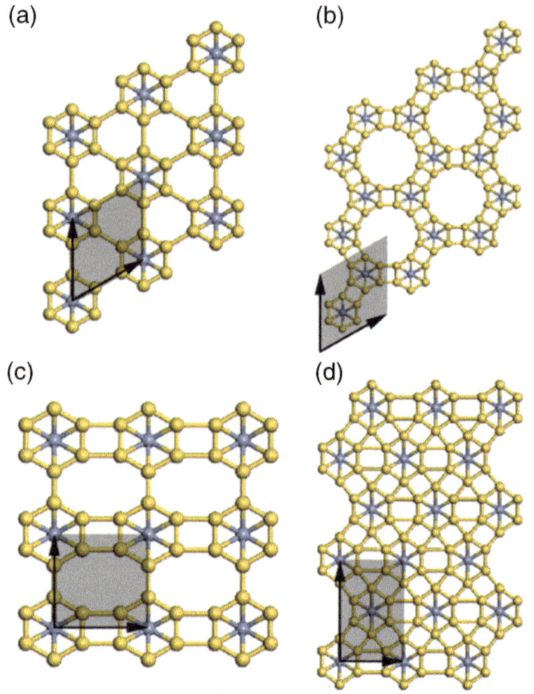

Figure 8.14 The atomic structures of four possible $TMSi_{12}$ assembled 2D crystals: (a) hexagonal honeycomb lattice, (b) hexagonal porous lattice, (c) square lattice, and (d) central rectangular lattice. *Source:* Nie et al. [174]. © Elsevier.

structures. Moreover, the first-principles MD simulations reveal that hexagonal honeycomb and porous structure can remain stable at room temperature, while the square and central rectangle lattices cannot. For all 3*d*-$TM@Si_{12}$ assembled hexagonal lattices, the optimized lattice constants, binding energies, geometric parameters, and relative energy of FM and AFM state with respect to the ground state are listed in Table 8.3. Except for the porous structure of $Ni@Si_{12}$ (with obvious structural deformation), the building blocks in all of other assembled 2D $TMSi_{12}$ sheets can be well preserved with small changes in bond lengths.

As in the case of $V@Si_{12}$, when other TM atoms are doped in Si_{12} hexagonal prism structure, most of their MMs are quenched (V, Cr, Mn, Fe, Co, see Figure 8.15). However, when $TM@Si_{12}$ clusters are assembled into a single cluster sheet, the MM is magically restored (except for Mn, see Figure 8.15). Further magnetic coupling studies uncover that $Ti@Si_{12}$ and $Fe@Si_{12}$ assembled structures exhibit intrinsic ferromagnetism, while $Mn@Si_{12}$ and $Co@Si_{12}$ sheets are AFM in both the honeycomb and the porous structure (MM = 0, see Figure 8.15). In particular, the honeycomb structure of $Cr@Si_{12}$ sheet is FM, while the porous structure is AFM. For $Ni@Si_{12}$ 2D structures, their honeycomb structure is FM and porous structure is nonmagnetic (NM). Based on these, one can find that the magnetic couplings depend on both the central atoms and cluster aggregation pattern.

8.3.3 $Ta@Si_{16}$ Assembled 2D Monolayer and That on Substrate

Aside from $TM@Si_{12}$ clusters, larger TM encapsulated Si_6 clusters were also found to form stable cage-like structures. Kumar, et al. [183] proposed two stable $TM@Si_{16}$ isomers [the fullerene-like (*f*) cage with D_{4d} symmetry and Frank–Kasper (FK) tetrahedral structure

Table 8.3 The optimized lattice constants (*a*); binding energies (E_b); geometric parameters (d0: Cr-Si bond, d1: Si-Si bonds of the hexagonal bottom edge parallel to the 2D plane, d2: Si-Si bonds of the hexagonal bottom edge perpendicular to the 2D plane, and d3: the Si-Si bond between adjacent clusters); energy difference between FM state, AFM state, and ground state; and MMs for the $TMSi_{12}$ assembled hexagonal honeycomb and porous structures.

$TMSi_{12}$	*a* (Å)	E_b (eV)	d0 (Å)	d1 (Å)	d2 (Å)	d3 (Å)	Δ (Å)	MM (μ_B)
Honeycomb structure								
Sc	7.62	5.096	2.85	2.58	2.43	2.45–2.46	0	0
Ti	7.53	3.432	2.80–2.81	2.52	2.46	2.49	413	10.1
V	7.48	3.046	2.78	2.5	2.43–2.44	2.48	56	16.4
Cr	7.47	3.733	2.78	2.5	2.43	2.48	12	20.9
Mn	7.49	3.104	2.78	2.51	2.43	2.48	123	20.0
Fe	7.51	2.664	2.76–2.77	2.48–2.49	2.42	2.54	104	12.9
Co	7.51	2.68	2.71–2.81	2.48	2.41	2.55	54	8.1
Ni	7.49	2.986	2.72–2.73	2.45	2.40–2.41	2.6	24	1.6
Porous structure								
Sc	11.84	4.499	2.82	2.51–2.59	2.42	2.44	0	0
Ti	11.7	2.756	2.76–2.77	2.44–2.53	2.41–2.42	2.47–2.48	471	4.2
V	11.64	2.193	2.73–2.75	2.42–2.50	2.41–2.42	2.47–2.48	244	7.4
Cr	11.63	2.652	2.74	2.44–2.48	2.42	2.46–2.47	100	10.0
Mn	11.62	2.279	2.74	2.44–2.48	2.41	2.46	168	10.0
Fe	11.57	1.878	2.66–2.76	2.41–2.46	2.41	2.47–2.50	73	7.1
Co	11.5	1.793	2.67–2.73	2.39–2.44	2.41–2.43	2.45–2.46	95	4.5
Ni	11.52	2.418	—	—	—	—	0	0

Source: Nie et al. [174]. © Elsevier.

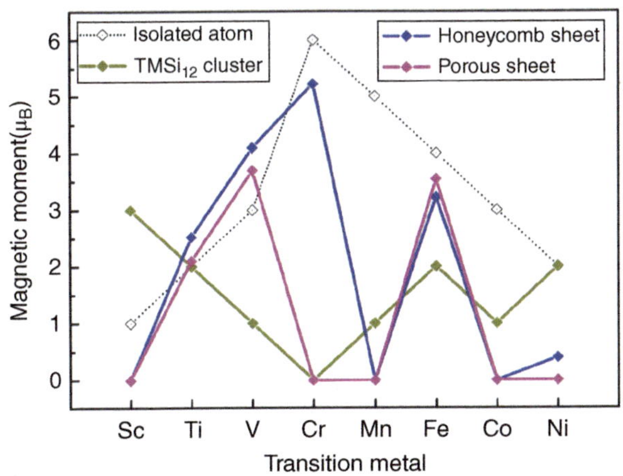

Figure 8.15 MM variation of the isolated TM atoms, $TM@Si_{12}$ clusters as well as their honeycomb and porous 2D sheets. *Source:* Nie et al. [174]. © Elsevier.

with C_{3v} symmetry] based on *ab initio* pseudopotential plane wave calculations, which received support from experiment [184]. As for the electronic properties, it has been found that there are three kinds of superatom-like TM@Si_{16} clusters: (i) halogen-like TM@Si_{16}^- (TM = Sc, Y, Lu), (ii) rare-gas-like TM@Si_{16} (TM = Ti, Zr, Hf), and (iii) alkali-like TM@Si_{16}^+ (TM = V, Nb, Ta), which had been selectively synthesized and analyzed using mass spectrometry and anion photoelectron spectroscopy (PES) [169, 185]. In particular, an integrated dry-wet synthesis method has recently been developed for large-scale synthesis of TM@Si_{16} (TM = Ti and Ta) cage clusters [186]. In this context, one is hopeful that mass selected clusters would indeed be synthesized on a large scale to form cluster assembled materials. With the neutral TM@Si_{16} (TM = Ti, Zr, Hf) FK structures as building blocks, Reis and Pacheco [187] confirmed that all of them can be aggregated to form a stable 3D bulk material with HCP structures. Since the cohesive energy per cluster is quite small (0.14–0.20 eV per cluster), the inter-cluster bonds should belong to weak vdW interaction. In contrast, cluster assembled multimers, nanowires, and bulk crystals can be obtained by the halogen-like TM@Si_{16}^- and alkali-like TM@Si_{16}^+ superatoms through strong binding between halogen-like and alkali-like clusters, alkali-like cluster and F, halogen-like cluster and K, or alkali-like cluster and C_{60} [182, 188–192].

All these results indicate the great potential of using TM@Si_{16} as a building block for cluster-assembled materials. Therefore, one natural question is whether TM@Si_{16} can self-assemble into a stable 2D crystal? To explore this, Liu et al. [175] designed possible self-assembled 2D sheets with the synthesized D_{4d} Ta@Si_{16} serving as the building blocks. Figure 8.16 shows six stable self-assembled 2D hexagonal structures (without structural collapse at 300 K proved by first-principles molecular dynamic [FPMD]), which share two types of lattices: hexagonal compact simple lattices (Figure 8.16a) and honeycomb complex

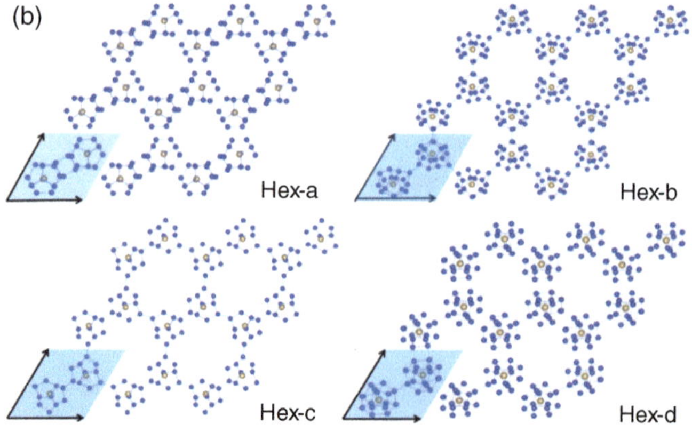

Figure 8.16 Optimized self-assembled 2D hexagonal structures of Ta@Si_{16}. (a) Compact simple lattice: P-0° and P-60° structures. (b) Honeycomb complex lattice: Hex-a, Hex-b, Hex-c, and Hex-d structures. *Source:* Liu et al. [175]. © American Chemical Society.

Table 8.4 Lattice constant (a), the range of Ta-Si distances ($d_{Ta\text{-}Si}$) and Si-Si distances ($d_{Si\text{-}Si}$) in a single Ta@Si_{16} cluster, the Si-Si distance [$d_{Si\text{-}Si(nearest)}$] between $TaSi_{16}$ clusters, the cohesive energy (E_c), and the energy difference [ΔE (meV)] between different magnetic states and ground state of the 2D crystal assembly.

2D	a (Å)	$d_{Ta\text{-}Si}$ (Å)	$d_{Si\text{-}Si}$ (Å)	$d_{Si\text{-}Si}$ (Å)	E_c (eV)	ΔE (meV)		
				(Nearest)		FM	AFM	NM
P-0°	8.127	2.69–3.71	2.30–2.59	—	2.92	705.5	0.04	0
P-60°	8.126	2.69–3.72	2.30–2.59	—	2.92	718.8	0.02	0
Hex-a	13.04	2.68–3.62	2.31–2.51	2.37	2.39	316.7	0	0
Hex-b	14.67	2.63–3.80	2.28–2.63	2.37	1.77	274.7	0.05	0
Hex-c	14.74	2.69–3.49	2.29–2.46	2.38	1.76	531.3	0.08	0
Hex-d	12.98	2.68–4.10	2.31–2.61	2.46	1.88	568.2	0.02	0
on C_{60}				$d_{\perp}$ (Å)				
Dot-C_{60}	10	2.80–3.36	—	2.01	1.24	24	0	167.2
Line-C_{60}	10.02	2.71–3.30	—	1.9	1.31	0	110.7	195.7
Face-C_{60}	9.98	2.90–3.00	—	3.52	1.3	0.3	0	197.4

Source: Liu et al. [175]. © American Chemical Society.

lattices (Figure 8.16b). The geometric optimization shows that the Ta@Si_{16} clusters in these assembled 2D crystals can maintain their fullerene-like structure very well, having only small bond length changes (see Table 8.4). Similar to the case of V@Si_{12} sheets [173], the hexagonal simple lattice generally has much larger cohesive energy E_c than that of the hexagonal honeycomb lattice (Table 8.4). Among the honeycomb structures, Hex-a has the largest E_c of 2.39 eV due to the strong cluster–cluster interaction with the edge-to-edge linkage.

Spin-polarized DFT calculations show that all these stable 2D sheets are nonmagnetic (see ΔE in Table 8.4), which is quite different from the situation of TM@Si_{12} monolayers [173, 174]. This is because the Ta–Si interaction (Ta-Si bond is 2.63 Å) in assembled sheets is stronger than that (Ta-Si bond is 2.87 Å) in the Ta@Si_{16} monomer. The electron density of state reveals that most of the assembled Ta@Si_{16} 2D structures are metallic, but the Hex-d phase exhibits direct semiconducting behavior with a narrow bandgap of 0.89 eV at the HSE06 level.

As mentioned above, the Ta@Si_{16} cluster is an alkali-like superatom, which provides a unique option for obtaining complex assembled materials. For instance, a metastable fcc bulk phase has been designed by DFT calculations, which takes the Ta@Si_{16}F supermolecule as the unit cell [188]. Similar to F, C_{60} has an electron affinity of 2.68 eV. In this regard, the order C_{60} films may provide an appropriate substrate that can immobilize the Ta@Si_{16} to form a 2D framework by superatom-molecule complex formation without surface diffusion or aggregation [190, 192]. Scanning tunneling microscopy and spectroscopy had demonstrated that Ta@Si_{16} cations could be densely immobilized onto ordered electron-acceptable C_{60} molecular film (Figure 8.17) [192]. Importantly, the Ta@Si_{16}

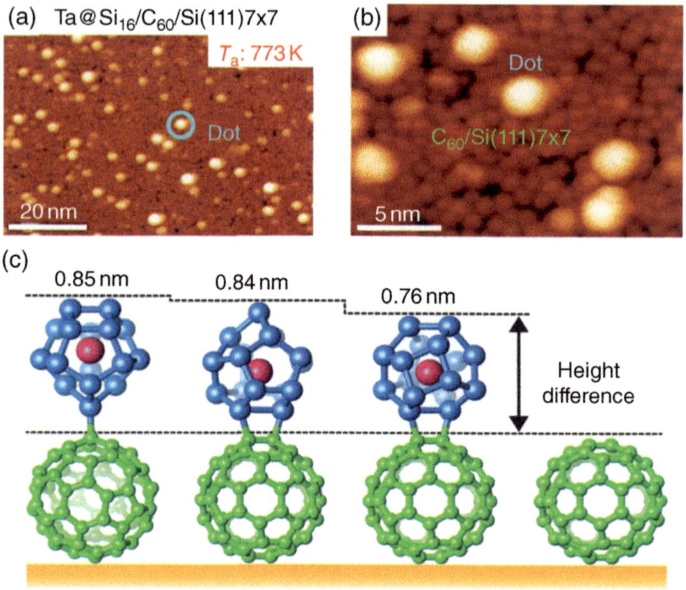

Figure 8.17 Stable immobilization of $Ta@Si_{16}$ clusters onto C_{60}-terminated surfaces. (a) Wide- and (b) molecular-scale STM images of the $Ta@Si_{16}^{+}$-deposited C_{60}/Si(111) 7 × 7 surface obtained after annealing at 773 K. (c) Examples of theoretical motifs of neutral $Ta@Si_{16}$-C_{60} complexes. *Source:* Nakaya et al. [192]. © Royal Society of Chemistry.

clusters in their monolayer exhibit excellent thermal stability with cage shape and positive charge, similar to that in the gas phase [192]. To provide a better understanding for this, the chemical states of $Ta@Si_{16}$ deposited on C_{60} fullerene film were further investigated by X-ray and ultraviolet photoelectron spectroscopies (XPS and UPS, respectively) [189]. From the XPS results of Si, Ta, and C, it is found that the superatomic charge transfer complex $(Ta@Si_{16})^{+}C_{60}^{-}$ can be formed between $Ta@Si_{16}$ and a single C_{60} molecule. Moreover, the XPS and UPS measurements reveal that such superatomic complex has high thermal and chemical stability: even when it is heated to 720 K or subjected to ambient oxygen, the cage-like $Ta@Si_{16}$ retains its original framework, forming oxides of $Ta@Si_{16}$ superatom.

The structural, electronic, and magnetic properties of $Ta@Si_{16}$ on C_{60} monolayer have been studied by first-principles calculations [175]. By placing a $Ta@Si_{16}$ on the pentagonal or hexagonal facet of C_{60} in different ways, three low-lying 2D interface structures were considered, including the dot-C_{60}-2D, line-C_{60}-2D, and face-C_{60}-2D (Figure 8.18). Owing to the interfacial interaction, the Ta-Si interactions are weaker than that in the freestanding $Ta@Si_{16}$ sheets. Consequently, the $Ta@Si_{16}/C_{60}$ interfaces present an MM of 1.0 μ_B per unit, and diverse magnetic couplings (FM or AFM) appear between $Ta@Si_{16}$ clusters. As can be seen from Table 8.4, the dot-C_{60}-2D and face-C_{60}-2D structures have the AFM ground state, while the line-C_{60}-2D structure is FM coupled, which indicates that the magnetic order of the $Ta@Si_{16}/C_{60}$ film depends on the contact configuration between $Ta@Si_{16}$ and C_{60}. Remarkably, the intrinsic FM state of the line-C_{60}-2D is robust with a Curie temperature of 294 K, estimated by the Monte Carlo simulation [175].

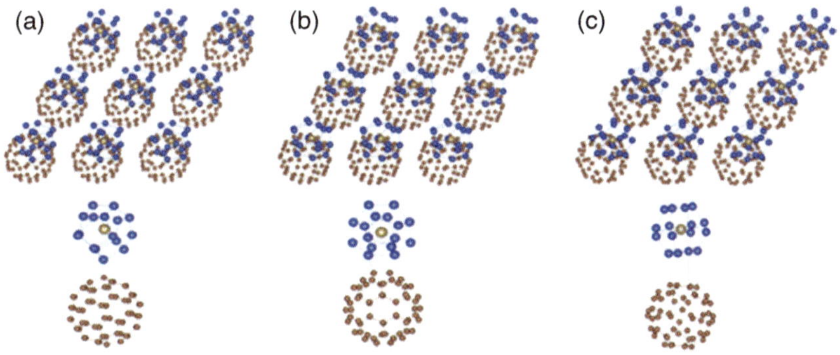

Figure 8.18 Ta@Si_{16}/C_{60} structures: (a) dot-contact structure [dot-C_{60}-2D (top) and dot-C_{60}-dim (bottom)]; (b) line-contact structure [line-C_{60}-2D (top) and line-C_{60}-dim (bottom)]; (c) face-contact structure [face-C_{60}-2D (top) and face-C_{60}-dim (bottom)]. *Source:* Liu et al. [175]. © American Chemical Society.

Beside the magnetic properties, the potential photocatalytic applications of Ta@Si_{16}/C_{60} sheets have also been explored [175]. Among the considered structures, face-C_{60}-2D is confirmed to be a photocatalytic semiconductor with an appropriate bandgap of 1.32 eV at the HSE06 level. As is known, a good photocatalyst should meet three basic requirements: (i) proper absorption spectrum for sunlight, (ii) easy to separate electrons and holes, and (iii) suitable band edges to promote a specific redox reaction [175]. Take water decomposition as an example: the calculated absorption spectra, work function, as well as the schematic diagram of photocatalytic hydrogen production of the Ta@Si_{16}/C_{60} interfaces are shown in Figure 8.19. The face-C_{60}-2D structure has a perfect absorption spectrum that is within the visible range of sunlight (Figure 8.19a). Furthermore, the electrons and holes in the Ta@Si_{16}/C_{60} interface are easily separated because the orbital composition of valence band maximum (VBM) and conduction band minimum (CBM) comes from nonoverlapping Ta atom *d* orbital and Si atom *p* orbital. Last, the band edges of the face-C_{60}-2D structure are also confirmed by the work function, which is an important physical parameter in studying the energy band alignment and charge transfer of the semiconductor heterostructure. As shown in Figure 8.19b, the calculated work function of the face-C_{60}-2D structure is 6.29 eV. Combined with the band structure, its VBM and CBM should be, respectively, located at 6.50 and 5.18 eV from the vacuum level (Figure 8.19c). The red line at 5.67 eV is the reduction potential position of H_2O, which is just below the CBM of the Ta@Si_{16}/C_{60} system, This makes the electron on CBM easily transferable to the reduction potential position of H_2O, promoting water reduction [175].

8.4 Binary Semiconductor Cluster Assembled 2D Materials

The discovery of fullerene C_{60} cluster [79] brought widespread attention to the valence isoelectronic III-V and II-VI binary semiconductor clusters X_nY_n because they can also form a stable fullerene-like cage in the medium size. For instance, the experimental and theoretical reports proved that the Cd_6Se_6 [193], $B_{12}N_{12}$ [194], $Zn_{12}S_{12}$ [195], $Zn_{12}O_{12}$ [196],

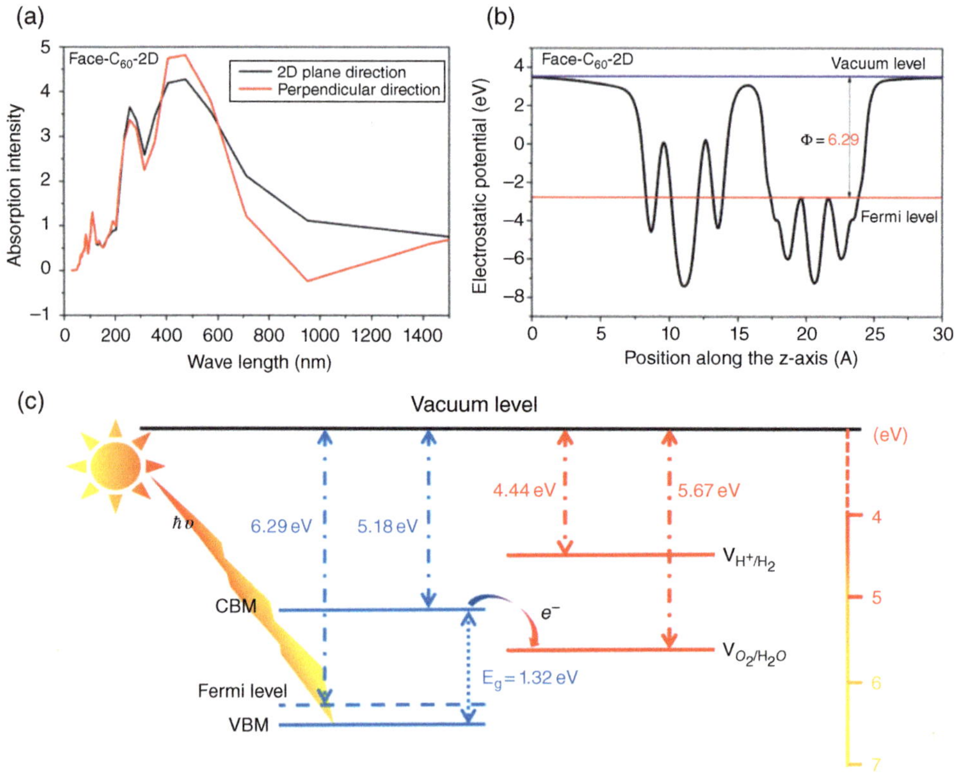

Figure 8.19 Schematic diagram of (a) optical absorption spectra and (b) work function as well as (c) photocatalytic water decomposition of the face-C_{60}-2D Ta@Si_{16}/C_{60} structure. *Source:* Liu et al. [175]. © American Chemical Society.

$B_{24}N_{24}$ [197], and $B_{36}N_{36}$ [198] have stable hollow cage ground state structures with an alternating X and Y arrangement. More interestingly, when these cage clusters serve as building blocks, they can lead to abundant new cluster assembled materials [199–205] with promising applications in microelectronics, spintronics, photovoltaic cell, molecular sieving, and gas storage. In this section, we focus on several 2D materials assembled from binary semiconductor clusters including Cd_6Se_6, $Zn_{12}S_{12}$, and $In_{12}As_{12}$.

8.4.1 Cd_6Se_6 Assembled Sheets

Since the CdSe semiconductor nanocrystals or clusters with narrow size distributions can be routinely synthesized, they have been considered as one of the most promising building blocks for new nanoscale materials [193, 207]. By combining experimental absorption, photoluminescence, and X-ray diffraction spectroscopies as well as *ab initio* DFT calculations, Jose et al. [193] observed the smallest wurtzite cage, Cd_6Se_6, which is obtained from an antiparallel stacking of the two Cd_3Se_3 layers. The diameter of Cd_6Se_6 determined from its molecular volume is 1.50 nm, which is consistent with the diameter determined from the experimental UV-vis absorption spectra (1.58 nm) at five minutes and at 120 °C in the absence of hexadecylamine. Motived by these, Saraf et al. [206] recently designed three configurations (see Figure 8.20a–c) by arranging the stable Cd_6Se_6 clusters as building blocks. Upon relaxation, the corresponding atomic structures are shown in Figure 8.20d–f. Among the three types, type-2 structure (Figure 8.20e) is the most stable one with binding energy of −2.79 eV/atom, which has the highest symmetry and shows geometry similar to that of A′A-type stacking in *h*-BN bilayers [208]. The calculated phonon dispersion (Figure 8.21a) shows that there are no unstable phonon modes with the exception of small imaginary frequencies (<10 THz) around the Γ point, which should be attributed to the limitations of the numerical procedure. This means that the type-2 structure of Cd_6Se_6 assembled sheets is dynamically stable. Moreover, by performing the *ab initio* MD simulations (Figure 8.21b), the structure was also confirmed to be thermally stable up to 300 K.

Figure 8.21c shows the electronic band structure of Cd_6Se_6 assembled type-2 phase, which has an indirect bandgap of 1.28 eV from M → Γ. The PDOS analysis reveals that the VBM of type-2 phase is mainly Se-*p* states with small proportion of Cd-*d* states, while the CBM is made up of Cd-*s* orbitals (Figure 8.21d). Using the Bardeen, Tersoff, and Hamann (BTH) formalism [209], the tunneling characteristics of type-2 Cd_6Se_6 bilayer were calculated in a low-bias range of −1.0 to 1.0 V. The I–V characteristics were computed at Cd and Se sites on the bilayer. By positioning the tip exactly above these sites, the spatially resolved tunneling spectra were obtained. As illustrated in Figure 8.22c, the I–V characteristics show rectification nature since no states are available in the gap region (1.28 eV) to

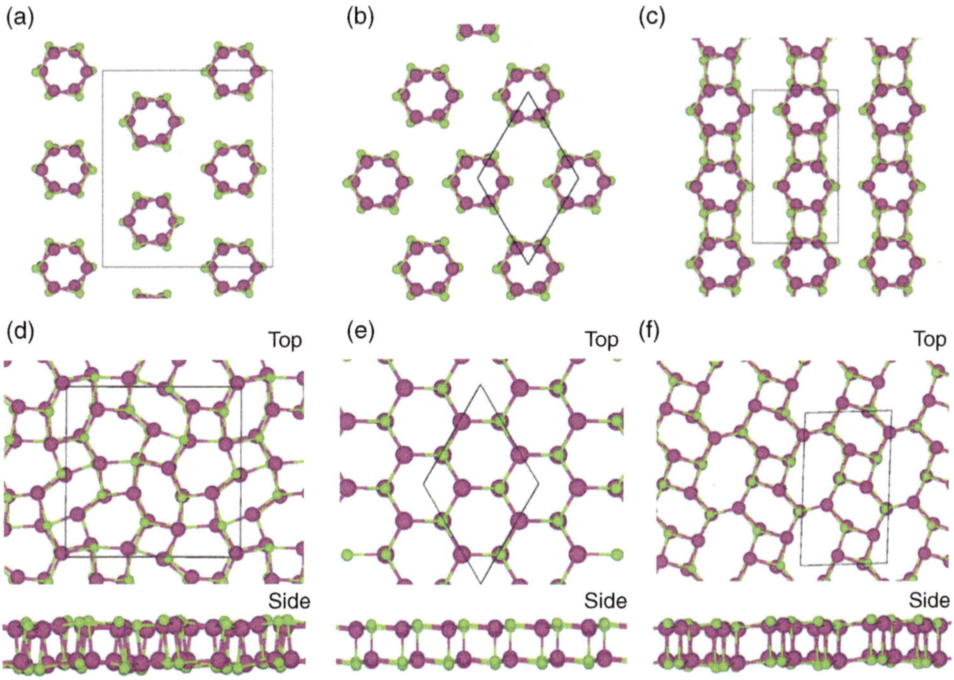

Figure 8.20 Initial and optimized geometries of Cd_6Se_6 cluster assembly, for type-1 (a) and (d), type-2 (b) and (e), and type-3 (c) and (f) structures, respectively. *Source:* Saraf and Kshirsagar [206]. © IOP Publishing.

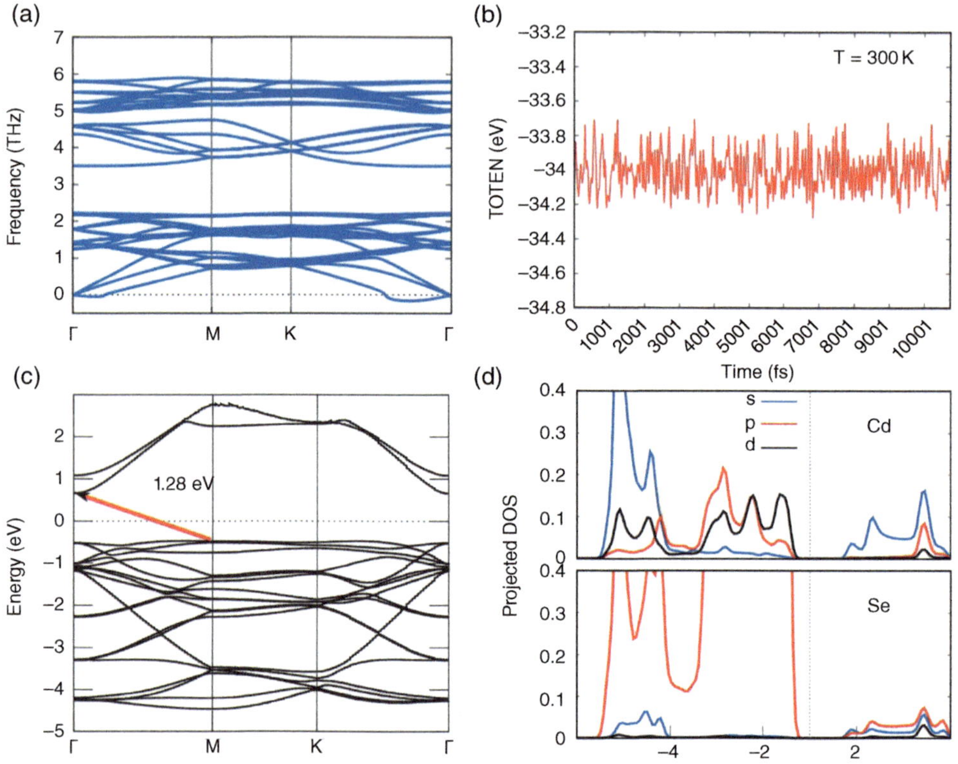

Figure 8.21 Dynamical and thermal stability of Cd_6Se_6 cluster assembly for type-2 structure shown using (a) phonon band structure and (b) total energy fluctuations obtained from *ab initio* MD simulations. (c) Electronic band structure of type-2 Cd_6Se_6 cluster assembly. (d) Total density of state (DOS) and projected density of state (PDOS) of the type-2 bilayers with Co adsorption at 50%. Se-*p* and Co-*d* states are indicated in blue and black, respectively. *Source:* Saraf and Kshirsagar [206]. © IOP Publishing.

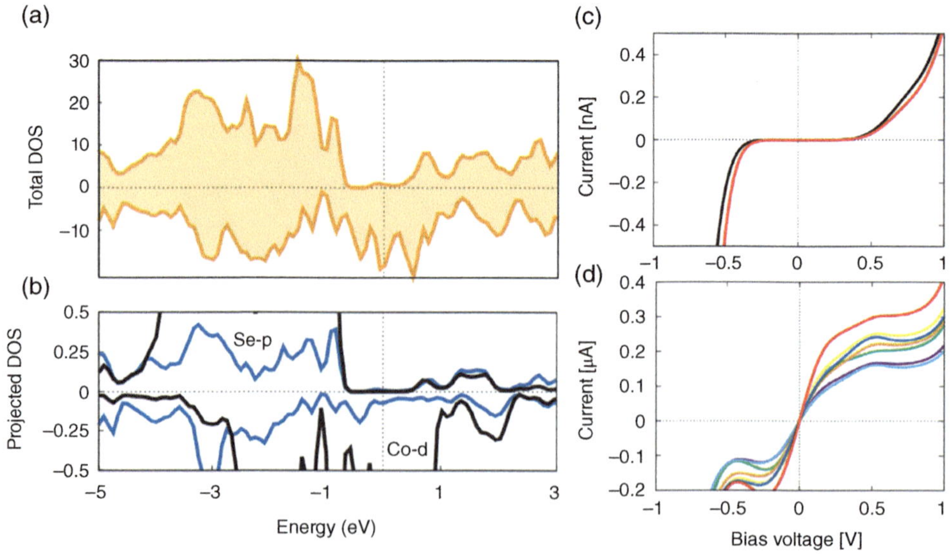

Figure 8.22 (a) Total DOS and (b) PDOS of type-2 Cd_6Se_6 cluster-assembled bilayers with Co adsorption at 50% per Cd-Se pair. Tunneling properties of I-V for (c) Cd_6Se_6 cluster-assembled bilayer and (d) that with Co adsorption at 50% per Cd-Se pair. *Source:* Saraf and Kshirsagar [206]. © IOP Publishing.

contribute to the tunneling current, which is similar to a classic Schottky diode [206]. Furthermore, using TM atoms to functionalize the pristine type-2 Cd_6Se_6 bilayer, some interesting magnetic and tunneling properties can be found: (i) the DFT calculations show that after the adsorption of Co atoms the bilayer is reorganized to form an FM ordered state. In particular, upon increasing the concentration (50% per Cd-Se pair) of Co adatoms, the structure turns into quasi-half-metal with near 100% spin polarizations (see Figure 8.22a and b). (ii) The exciting effect of doping is also seen in the I–V characteristics. Although the pristine Cd_6Se_6 bilayer shows classic Schottky diode-like features, it was found that the functionalized system shows the characteristics of a tunnel diode via negative differential conductance. Namely, the tunneling current decays up to a certain bias voltage range and begins to rise again as bias increases further (see Figure 8.22d).

8.4.2 $X_{12}Y_{12}$ Cage Cluster Assembled Monolayer

Among the III–V and II–VI binary semiconductor clusters X_nY_n with medium size, the most famous one should be $X_{12}Y_{12}$, which holds the truncated octahedral (sodalite) cage structure, sharing the same shape as the first Brillouin zone of fcc and the only solution to the Kelvin problem. Importantly, some of these have been detected in experiment, including $B_{12}N_{12}$ [194], $Zn_{12}S_{12}$ [195], and $Zn_{12}O_{12}$ [196]. As for $In_{12}As_{12}$, the similar cage structure was also confirmed to be the ground state in the DFT calculations at the hybrid Becke-3 Lee-Yang-Perdew (B3LYP) level of theory. On this basis, Liu et al. [205] explored the possible isomers of $In_{12}As_{12}$ dimer. From the stable inter-cluster link structure, they further designed different periodic crystals. Figure 8.23a displays the atomic structure of $In_{12}As_{12}$-assembled single sheet, which possesses graphene-like hexagonal symmetry. After optimizing, the sodalite cage can be well preserved. Moreover, two types of pores can be found on the assembled sheet. The diameter of the smaller one in the $In_{12}As_{12}$ cage is 0.508 nm and the larger one surrounded by six $In_{12}As_{12}$ cages is 1.091 nm. This special structure may make it a particularly promising molecular sieve membrane.

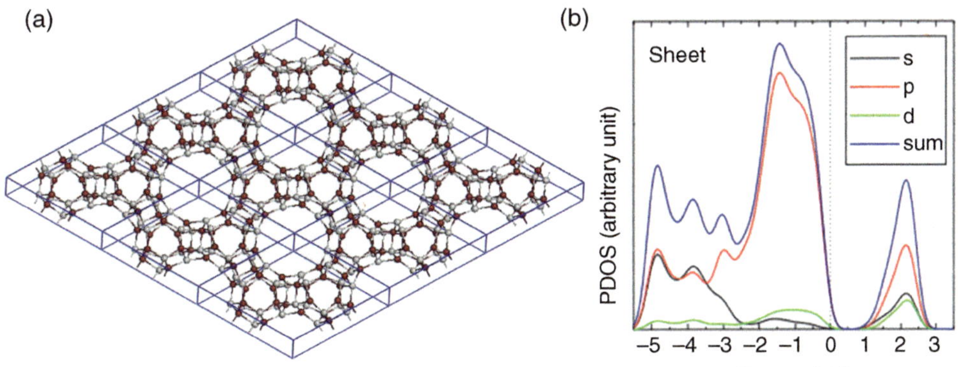

Figure 8.23 (a) The atomic structure of 2D graphene-like sheet assembled by $In_{12}As_{12}$ cage clusters. (b) The corresponding total and partial density of states. The vertical dashed line indicates the Fermi level. *Source:* Liu et al. [205]. © Royal Society of Chemistry [205].

To assess the viability of the proposed 2D $In_{12}As_{12}$-sheet, the binding energy was computed [205]. The results reveal that the binding energy (3.749 eV/InAs) of $In_{12}As_{12}$-sheet is larger than that of individual $In_{12}As_{12}$ cluster (3.256 eV/InAs). Although, the binding energy of $In_{12}As_{12}$-sheet lies 0.632 eV/InAs below that of the bulk zinc-blende InAs, it lies 0.207 eV/InAs above that of the synthesized rock-salt InAs phase. In this regard, the energy cost for forming $In_{12}As_{12}$-sheet does not seem prohibitive. Like the case of 2D Cd_6Se_6-sheet, the $In_{12}As_{12}$-sheet is also a semiconductor with a bandgap ratio of 3.22 with respect to that of zinc blende InAs bulk, as is seen in Figure 8.23b. The PDOS reveals that its electronic states of the valence bands near Fermi level are mainly dominated by As 4*p* and In 5*p* states, while the contribution from the As 4*s* and In 5*s* states is small. Note that the In 4*d* states only have a small contribution to the valence bands but a large contribution to the conduction bands. Therefore, p-p hybridization should be responsible for the interaction between the In and As atoms.

Following the design of 2D $In_{12}As_{12}$ monolayer, a similar 2D graphene-like cluster sheet, assembled by the experimentally synthesized $Zn_{12}S_{12}$ cage cluster [195], was proposed by Lu et al. [210] The DFT calculations show that in the assembled monolayer, all the $Zn_{12}S_{12}$ cage can be well preserved after full structural relaxation. The diameter of the existing pores is calculated to be 0.454 nm for the smaller one and 0.989 nm for the larger one, which is surrounded by six $Zn_{12}S_{12}$ cluster cages. From the density of state, such assembled sheet was confirmed to be a semiconductor with a bandgap ratio of 1 : 10 with respect to that of zinc blende ZnS bulk. As the binding energy of 6.4 eV/ZnS is larger than the isolated $Zn_{12}S_{12}$ cluster, the synthesis of the assembled $Zn_{12}S_{12}$ sheet should be feasible [210].

8.5 Simple and Noble Metal Cluster-assembled 2D Materials

Atomic clusters, composed of simple and noble metal atoms, are among the first clusters to be studied, and they exhibit unique geometry and electronic and magnetic properties [211–213]. Their stabilities are usually governed by two factors: (i) atomic packing and (ii) electron shell closure in the context of the jellium model [64]. Up to date, a lot of stable metal clusters have been identified as magic clusters or superatoms, which can serve as building blocks of new cluster-assembled materials [58, 59]. In this section, we focus on two types of 2D materials designed from metal clusters.

8.5.1 Mg_7 Assembled Monolayer

As a simple metal, Mg is especially appropriate for investigation of size-induced transition to metallicity. Like alkali metal cluster, Mg_n clusters have also attracted considerable attention from both experimentalists and theorists. Motivated by the experimental mass spectra [214] of Mg_n^+ ($n \leq 80$) and PES measurements [215] of Mg_n^- (n = 3–35), Jellinek et al. [216] reexamined the structural and electronic properties of neutral and anionic Mg_n (n = 2–22) clusters using gradient-corrected DFT calculation. Among Mg_n clusters, the Mg_7 cluster is found to form a pentagonal bipyramid configuration with D_{5h} symmetry [217], which has a relatively large highest occupied molecular orbital (HOMO)–LUMO gap. The binding energy of the Mg_7 cluster is calculated to be −0.367 eV [217] (−0.47 eV [216]) per Mg atom, indicating that it is stable and can be formed in an exothermic manner.

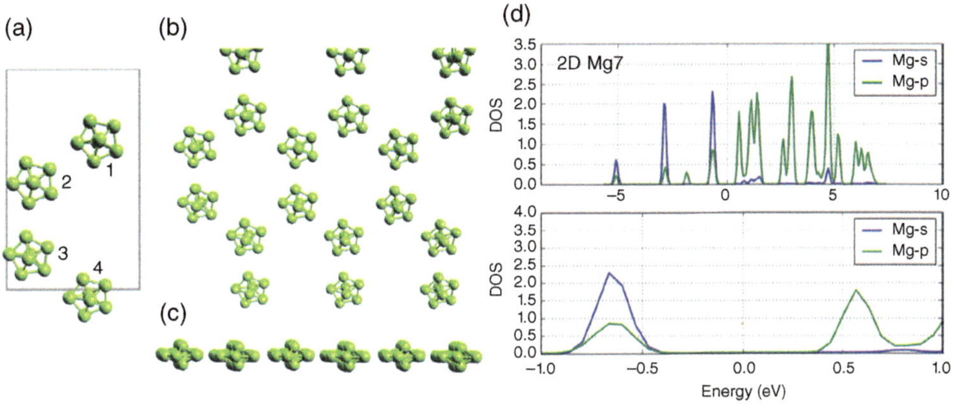

Figure 8.24 The atomic structure of 2D Mg_7 monolayer. (a) The corresponding unit cell having four Mg_7 clusters. Top (b) and side (c) view of the repeated lattice of 2D Mg_7 monolayer. (d) PDOS of 2D Mg_7 monolayer. *Source:* Takahashi and Takahashi [218]. © Elsevier.

Based on these facts, Takahashi et al. [218] used the Mg_7 cluster as the building block for a 2D material. By placing it into a hexagonal frame, a 2D periodic graphene-like monolayer was constructed. In every unit cell, there are four Mg_7 clusters (see Figure 8.24a). The ground state lattice constants of 2D Mg_7 sheet were calculated to be 18.929 and 34.79 Å by shrinking and expanding the lattice. By repeating the lattice, a flat hexagonal structure was found as shown in Figure 8.24b and c. The binding energy (E_b) in the assembled Mg_7 monolayer was calculated by $E_b = E[\text{2D-Mg}_7] - E[\text{Mg}_7\text{(cluster)}]$ to be −0.02 eV per Mg_7, much smaller than that of the Mg-Mg binding in Mg_7 cluster, suggesting that the bonding between Mg_7 clusters is characterized by vdW physisorption. However, the formation of Mg_7 monolayer is an exothermic process. Furthermore, when the effect of vdW forces is considered, the binding energy between Mg_7 clusters increases to −0.34 eV per Mg_7 cluster [218]. Therefore, one can conclude that the 2D Mg_7 assembled sheet is stabilized due to vdW forces.

Figure 8.24d presents the electronic structure of Mg_7 assembled monolayer. Interestingly, although the Mg_7 monolayer is composed of all metal atoms, it is a semiconductor with a bandgap of 0.8 eV at the Fermi level, similar to the case of Mg_7 isolated cluster. This is due to two factors: (i) the structure of Mg_7 in the assembled monolayer is almost the same as its cluster counterpart, and (ii) the interaction between Mg_7 clusters is weak vdW forces. It should be noted that the bandgap in 2D Mg_7 sheet is unusual because semiconductor characteristics are expected from elements within groups 13–15 in the periodic table, rather than a member of group 2. In this regard, the proposed all metal atom semiconductor should greatly expand the family of traditional semiconductors.

8.5.2 Au_9 and Pt_9 Assembled Square Monolayer

Among noble metal clusters, gold (Au) and platinum (Pt) clusters have attracted a lot of attention due to their high catalytic activities [220–222]. In particular, it was shown that small Pt and Au clusters grow into 2D planar structures [223, 224]. For instance, the ground state structure of Pt_9 is a perfect square [223]. To explore whether the planar Pt/Au clusters can expand toward a 2D periodic crystal, Takahashi et al. [219] designed three kinds of planar cluster-assembled monolayers, i.e. 2D-Au_9, -Pt_9, and -$Au_{18}Pt_{18}$ sheets. When the Au_9 and Pt_9 clusters are placed into a 2D unit cell with periodic boundary condition (see Figure 8.25), the corresponding infinite monolayer of square Au and Pt is formed, respectively. After optimizing by involving the stress tensor calculations, the ground state lattice constants of 2D-Au_9 and -Pt_9 are calculated to be 8.10 and 7.71 Å, respectively. Note that unlike the case of 2D-Pt_9, the geometry of 2D-Au_9 is 7° tilted against the unit cell. The average bond distance of Au-Au (Pt-Pt) in 2D-Au_9 (Pt_9) is calculated to be 2.77 Å (2.57 Å), which is 0.12 Å (0.11 Å) elongated with respect to the bond distances in the Au_9 (Pt_9) cluster, but is still shorter than the average bond length of Au(111) [Pt(111)] surface [219]. As for the binary 2D-$Au_{18}Pt_{18}$, the 2D framework is designed by combining and alternating two Au_9 and two Pt_9 clusters as shown in Figure 8.25c, where the ratio of Au_9 and Pt_9 is set to 1 : 1. The calculated ground state lattice constant is 15.82 Å. The average bond lengths of Au-Au and Pt-Pt are calculated to be 2.82 and 2.59 Å, respectively, slightly larger than that in the corresponding 2D-Au_9 and -Pt_9 due to the extra bonds with surrounding Au and Pt atoms satisfying the periodic boundary condition. However, the elongation of Au-Au and Pt-Pt bonds in $Au_{18}Pt_{18}$ is strong enough to keep the square atomic configuration.

Using the following definitions:

$$E_f\left[2\text{DAu}_9\left(\text{Pt}_9\right)\right]=\frac{NE\left[2\text{DAu}_9\left(\text{Pt}_9\right)\right]-NE\left[\text{Au}_9\left(\text{Pt}_9\right)\right]}{9N} \tag{8.1}$$

$$E_f\left[2\text{DAu}_{18}\text{Pt}_{18}\right]=\frac{NE\left[2\text{DAu}_{18}\text{Pt}_{18}\right]-2NE\left[\text{Au}_9\right]-2NE\left[\text{Pt}_9\right]}{36N} \tag{8.2}$$

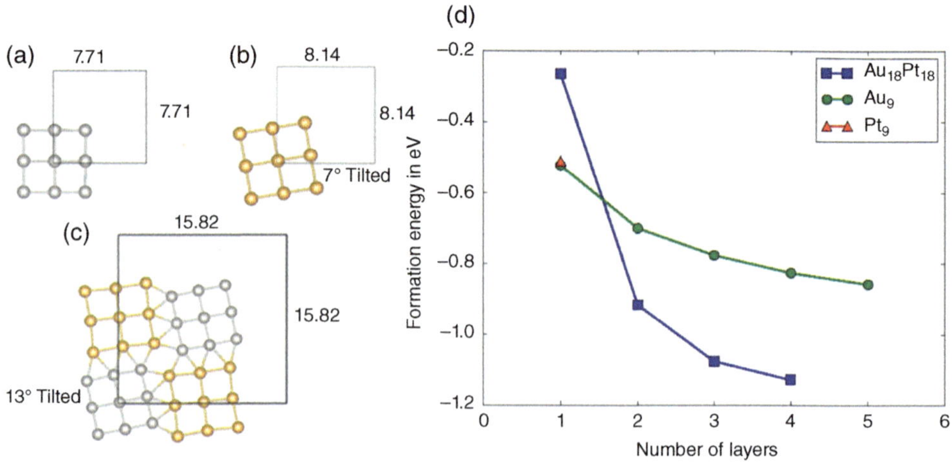

Figure 8.25 Structure of 2D (a) Pt_9, (b) Au_9, and (c) $Au_{18}Pt_{18}$. Black box indicates the unit cell in angstroms. Atomic color code: Au, gold; Pt, silver. (d) Formation energy per atom of multilayers of 2D Au_9, Pt_9, and $Au_{18}Pt_{18}$ from Au_9 and Pt_9 clusters. *Source:* Takahashi et al. [219]. © American Chemical Society.

2D-$Au_9(Pt_9)$ and 2D-$Au_{18}Pt_{18}$, with different layers were respectively calculated. The corresponding results are presented in Figure 8.25b. When the number of layer $N = 1$, all of 2D-Au_9, -Pt_9, and -$Au_{18}Pt_{18}$ monolayers are relatively stable because of the negative E_f, which indicates that the formation of these cluster-assembled single sheets is an exothermic process. Therefore, their synthesis from planar Au_9 and Pt_9 clusters is in principle achievable. Moreover, the multilayers of 2D-Au_9 (up to five layers) and 2D-$Au_{18}Pt_{18}$ (up to four layers) are also stable with increased negative formation energies. However, the multilayered 2D-Pt_9 is unstable because major structural relaxation occurs upon the adsorption of another 2D-Pt_9 layer, a result of the strong interaction between the Pt atoms [219].

To examine the electronic properties, the PDOSs of 2D-Au_9, 2D-Pt_9, and 2D-$Au_{18}Pt_{18}$ were calculated as shown in Figure 8.26a–d. All of the above 2D materials are metallic with bands crossing the Fermi level. The electronic states near the Fermi energy are mainly composed of d-orbitals of the Pt or Au atoms. In addition, spin up and down channels

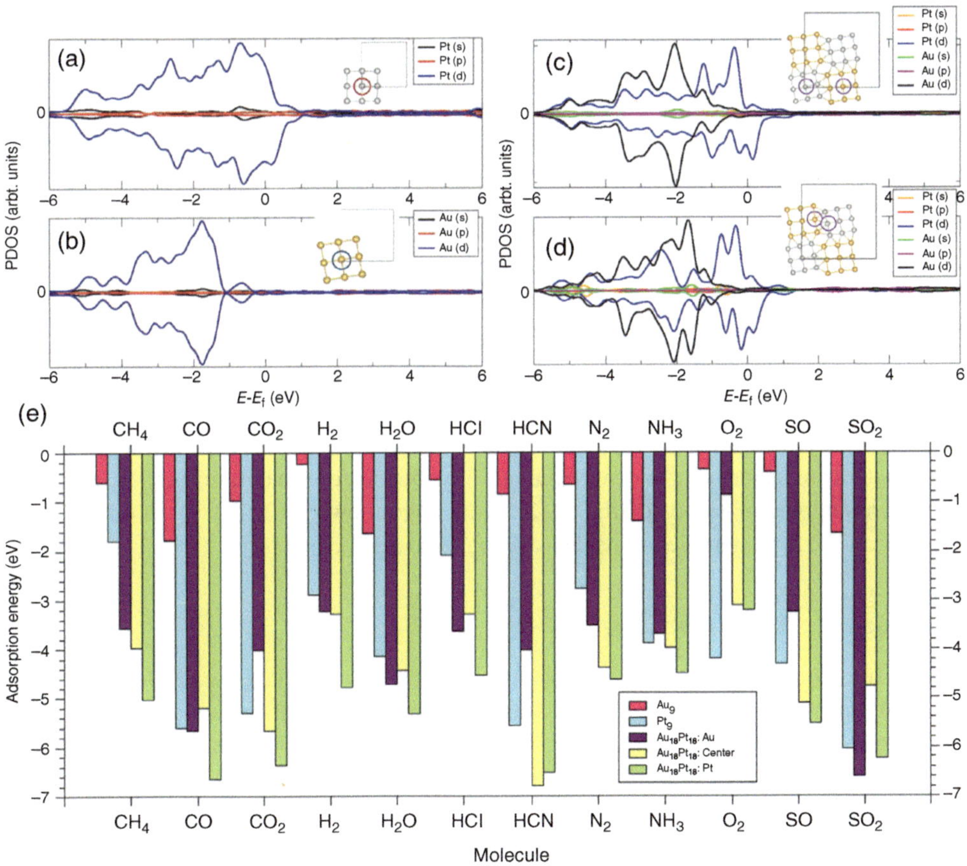

Figure 8.26 Spin-resolved PDOSs of (a) red-circled Pt atom in 2D-Pt_9 and (b) blue-circled Au atom in 2D-Au_9. (c) Nonbonded purple-circled Pt and Au; (d) bonded purple-circled Pt and Au in 2D-$Au_{18}Pt_{18}$. Atomic color code: Au, gold; Pt, silver. (e) Adsorption energy of various molecules when adsorbed onto different adsorption sites of 2D Au_9, Pt_9, and $Au_{18}Pt_{18}$ monolayers. *Sources:* (a–d) Takahashi et al. [219] and (e) Takahashi and Takahashi [225]. © American Chemical Society.

differ in the case of the 2D-Pt_9 (Figure 8.26b) and 2D-$Au_{18}Pt_{18}$ (Figure 8.26c and d), whereas they are the same in the case of the 2D-Au_9. Therefore, 2D-Au_9 is a nonmagnetic metal while 2D-Pt_9 and 2D-$Au_{18}Pt_{18}$ are magnetic metals.

Owing to the high surface-to-volume ratio, the cluster-assembled 2D-Au_9, 2D-Pt_9, and 2D-$Au_{18}Pt_{18}$ monolayers are expected to show high reactivities toward gas molecules. Using first-principles calculations, Takahashi et al. [225] investigated the commonly found molecules CH_4, CO, CO_2, H_2, H_2O, HCl, N_2, NH_3, O_2, SO, SO_2, and HCN to be adsorbed onto 2D-Au_9, 2D-Pt_9, and several locations on a 2D-$Au_{18}Pt_{18}$ surface. The calculated adsorption energies are presented in Figure 8.26e. The results show that 2D-Pt_9 and 2D-$Au_{18}Pt_{18}$ exhibit low adsorption energies where $Au_{18}Pt_{18}$ preserves the structural integrity of the molecule and surface. Particularly, adsorption of molecules onto 2D-$Au_{18}Pt_{18}$ frequently results in low adsorption energies and high reactivity with minor surface reconstruction of $Au_{18}Pt_{18}$ and small change of average bond lengths of the molecules. The decrease in adsorption energy can be attributed to the presence of Pt atoms, while Au atoms can be considered responsible for the structural stability. These results demonstrate that the high reactivities of small clusters can be achieved through the use of 2D assembled materials. Thus, $Au_{18}Pt_{18}$ can successfully utilize the low adsorption energies associated with Pt atoms while preserving the integrity of the surface structure using gold atoms, making it possible to adsorb or dissociate desired molecules using select areas of the 2D-$Au_{18}Pt_{18}$ [225].

8.6 Zintl-ion Cluster-assembled 2D Materials

Zintl clusters are the anionic units of Zintl phase structures [226, 227] in which the valence electrons are transferred from the more electropositive component to the more electronegative partner leading to the formation of salt-like compounds, such as the intermetallic compounds between alkaline or alkaline earth metals and *p*-block (semi)metals [228]. Zintl clusters are fascinating and versatile building blocks for the bottom-up fabrication of different assembled materials in 0D, 1D, 2D, and 3D framework [229, 230]. In this section, we focus on the 2D assembled materials arising from Zintl clusters.

8.6.1 Ge_9 Ion Cluster Monolayer

As the homoatomic Zintl cluster ions, group 14 Zintl clusters $[E_9]^{4-}$ (E = Si, Ge, Sn, Pb) are promising for preparation of new nanoscale architectures [230]. Particularly, it was confirmed that when the right experimental conditions are found, the $[Ge_9]^{4-}$ clusters can be oligomerized to form $[Ge_9\text{-}Ge_9]^{6-}$ dimer [231, 232], $[Ge_9=Ge_9=Ge_9]^{6-}$ trimer [233, 234], and $[Ge_9=Ge_9=Ge_9=Ge_9]^{6-}$ tetramer [235, 236] and 1D polymer [237] composed of interconnected Ge_9-clusters. After the preparation of the 1D polymeric Ge_9 chain [238], it was suggested that further oxidative coupling of the 1D chains could possibly result in a neutral 2D Ge_9 sheet [239, 240]. Based on this, Karttunen et al. [229] derived an initial structural model with four C_{4v}-symmetric Ge_9 clusters in the unit cell. The corresponding optimized structure (cell parameter a = 11.3 Å) forms a network of four- and eight-membered rings, which is related to a semi-regular tiling known as truncated square tiling.

The harmonic frequency calculation shows that the obtained Ge_9 sheet is a true local minimum. The theoretical IR spectrum of the sheet is dominated by two intense modes located at 119 and 239 cm^{-1}.

The calculated relative energies of various germanium modifications reveal that the predicted Ge_9 sheet is energetically quite unfavorable in comparison to diamond-like α-Ge, being 0.65 eV/atom higher in energy. However, for elemental nanostructures, energy differences of this magnitude are not unexpected. To compare with the experimentally known 1D Ge_9 chain, Karttunen et al. [229] further optimized the structure of a simple 1D Ge_9 chain, which can be derived from *closo*-type $[Ge_9]^{2-}$ clusters interlinked into a neutral 1D chain. The results show that Ge_9 sheet is 0.13 eV per atom less strained than 1D chain. Therefore, the proposed Ge_9 assembled sheet should be attainable in future experiment. Moreover, the analysis of electronic properties using B3LYP functional indicates that the predicted 2D Ge_9 sheet is a semiconductor with a direct bandgap of 1.3 eV at the Γ point [229].

Inspired by the high-yield synthesis of $[Ge_9\{Si\text{-}(SiMe_3)_3\}_3]^-$, Muñoz-Castro et al. [241] explored theoretically the formation of an extended 2D structure starting from a single monoanion $[Ge_9\{Si(SiMe_3)_3\}_3]^-$ (Figure 8.27a) acting as the source of Ge_9 building blocks. A periodic model (Figure 8.27b) for $[M_3\{Ge_9\}_3]_\infty$ (M = Li and Cs) sheet was constructed by involving three-connected Ge_9 cores resulting from the potential removal of the single ligand per unit in the hexameric species. In such structure, every Ge_9 cluster holds a charge of −1, and the counterions lithium or cesium cations have positive charge to ensure the charge neutrality in the periodic regime. After optimization, the ground state structure exhibits large pores with radius of 5.23 Å between the three-connected Ge_9 clusters for both of $[Cs_3\{Ge_9\}_3]_\infty$ and $[Li_3\{Ge_9\}_3]_\infty$. They also have the same lattice constants with 11.97 and 20.72 Å in the x and y directions, respectively [241]. The bonding between different Ge_9 cores is proved to be favorable and covalent in nature as a localized 2c-2e Ge-Ge exobond. Specifically, the binding energies of Ge-Ge bond in the extended sheets $[Cs_3\{Ge_9\}_3]_\infty$ and $[Li_3\{Ge_9\}_3]_\infty$ are calculated to be −1.32 and −1.27 eV, respectively. Although they are smaller than that in Ge bulk (−2.55 eV), they are similar to that obtained for 2D germanene (1.30 eV) [33]. Therefore, the proposed porous $[M_3\{Ge_9\}_3]_\infty$ sheet has similar stability as that of germanene.

8.6.2 Ti@Au_{12} Ion Cluster Monolayer

Generally, Zintl ion clusters are composed of group 14 and 15 post-transition metal elements [227], such as E_9^{4-} (E = Ge-Pb), E_9^{3-} (E = Si-Sn), Sn_8^{6-}, E_7^{3-} (E = P, As), E_{11}^{3-} (E = P-Sb), and As_{14}^{4-}. Based on this fact, an interesting question was proposed [242]: whether a Zintl ion cluster can be composed of only TM atoms? Using gradient-corrected DFT, the authors show that the 18-electron rule can be applied to design a new class of Zintl clusters, which are composed of all TM atoms. Noting that the existence of the FeH_6^{4-} anion in the experimentally synthesized Zintl phase material Mg_2FeH_6 [243] is consistent with the 18-electron rule, they replaced the H atoms with isovalent Au atoms. In the counterpart Mg_2FeAu_6, it was found that the large electron transfer from the Mg to the $FeAu_6$ unit still persists. This indicates that it is entirely possible to design Zintl ions composed of all TM atoms.

With the 18-electron rule in mind, Zhou et al. [242] proposed that the dianions $[Ti@Au_{12}]^{2-}$ is a stable all TM Zintl cluster because the total number of valence electrons of Ti@

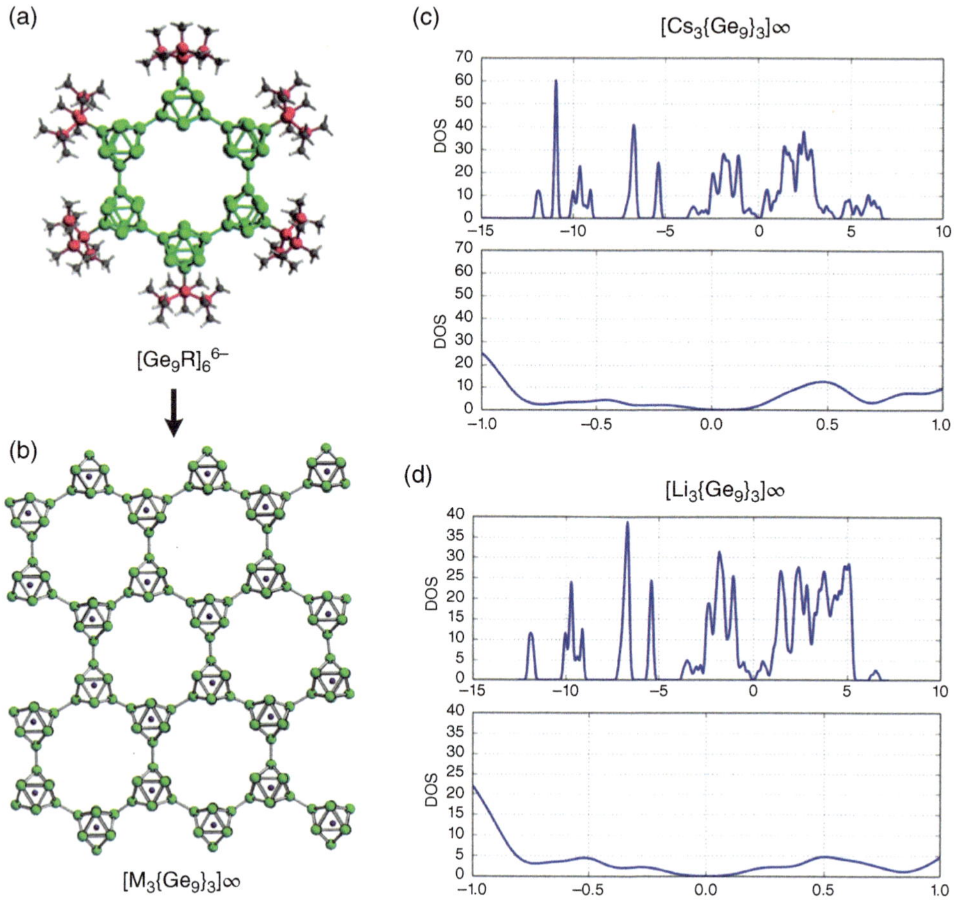

Figure 8.27 (a) Optimized structure of the hexameric species $[Ge_9R]_6^{6-}$ [R = $Si(SiMe_3)_3$]. (b) The extended 2D array involving three Ge-Ge exobonds per Ge_9 building block. (c) Density of state of 2D $[Cs\{Ge_9\}_3]_\infty$ and (d) $[Li\{Ge_9\}_3]_\infty$, between −15.0 and +10.0 eV, and between −1.0 and +1.0 eV in relation to the Fermi level. *Source:* Muñoz-Castro et al. [241]. © American Chemical Society.

Au_{12} cluster is 16 [Ti($3d^24s^2$) + 12 Au($6s^1$)]. Taking two Na atoms as countercations, the formed $Na_2[Ti@Au_{12}]$ complex displays high stability, which was confirmed by positive vibrational frequencies, as well as a large HOMO–LUMO gap (1.2 eV from the generalized gradient approximation (GGA)-Perdew–Burke–Ernzerhof (PBE) method [242]). To examine if this conclusion still holds in periodic systems, they further designed a 2D $Na_2[Ti@Au_{12}]$ structure, as is shown in Figure 8.28a. The relaxed geometry belongs to the *P4/mm* plane symmetry group. The bond length of Na-Au was calculated to be 2.92 Å, a little shorter than that in the cluster case. The cohesive energy, defined as $E_{coh} = 2E(Na^+) + E(Ti@Au_{12}^{2-}) - E(Na_2[Ti@Au_{12}])$, is 3.58 eV per unit cell. Bader's analysis shows that each Na transfers 0.77 electrons to the $Ti@Au_{12}$ moiety, confirming its Zintl-like character. Interestingly, electronic band structure calculations reveal that such an all-TM cluster assembled sheet is semiconducting rather than metallic. As shown in Figure 8.28b, the obtained bandgap is indirect with a value of 0.55 eV at the GGA-PBE level, in which the VBM and conduction band minimum are located at the M and Γ points, respectively.

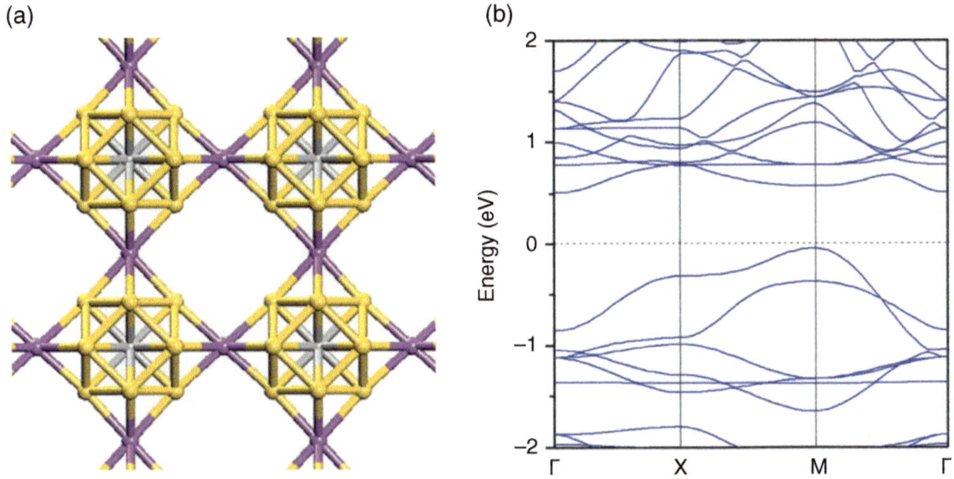

Figure 8.28 (a) Atomic structure and (b) electronic band structure of 2D $Na_2[Ti@Au_{12}]$ along high-symmetry points $\Gamma(0,0,0) \rightarrow X(0.5,0,0) \rightarrow M(0.5,0.5,0) \rightarrow \Gamma$. *Source:* Zhou et al. [242]. © Royal Society of Chemistry.

8.7 Chevrel Cluster-Assembled 2D Materials

Since the discovery of Chevrel phase in the ternary molybdenum chalcogenides [244], its material family has been getting bigger and bigger due to the versatile crystal structure [245]. In these Chevrel compounds, there exist a type of stable building block, octahedral chalcogenide M_6X_8, known as Chevrel cluster or superatom. Here, the X anion denotes chalcogen (S, Se, or Te) and can also be substituted by halogen elements, even by oxygen; the main cation, M, is Mo, which can be partially or totally replaced by another transition metal, like W, Re, Co, Nb, Ta, or others [245]. Essentially, Chevrel phase compounds belong to cluster-assembled materials. Recently, a new family of Chevrel materials with layered vdW hierarchical structure attracted research interest [55, 246–250] due to their novel multifunctionality and tunability, such as heavily doped superconductivity [250], strong polaronic effect [247], and strong in-plane anisotropy [249]. More importantly, they provide an ideal platform for producing 2D materials via mechanical exfoliation. In this section, we focus on two synthesized cluster-assembled 2D materials from layered Chevrel phases.

8.7.1 Re_6Se_8 Cluster-based Monolayer

Motivated by the lack of structural diversity and complexity in 2D atomic materials, Zhong et al. [55] reported a novel 2D semiconductor from $Re_6Se_8Cl_2$, a vdW layer material derived from the Chevrel phase, in which Re_6Se_8 clusters (isolated Re_6 octahedra enclosed in Se_8 cubes) are covalently linked into layers capped by terminal Cl atoms (see Figure 8.29a). Because of the strong in-plane inter-cluster bonding and weak interlayer interactions, the mechanical exfoliation of $Re_6Se_8Cl_2$ proved to be possible. Using the Scotch tape method, 2D rectangular flakes with thickness of ~15 nm have been produced [55], as determined by atomic force microscopy in Figure 8.29b. Based on experimental measurement and DFT calculations, the electronic bandgap, optical bandgap, and exciton binding energy of these

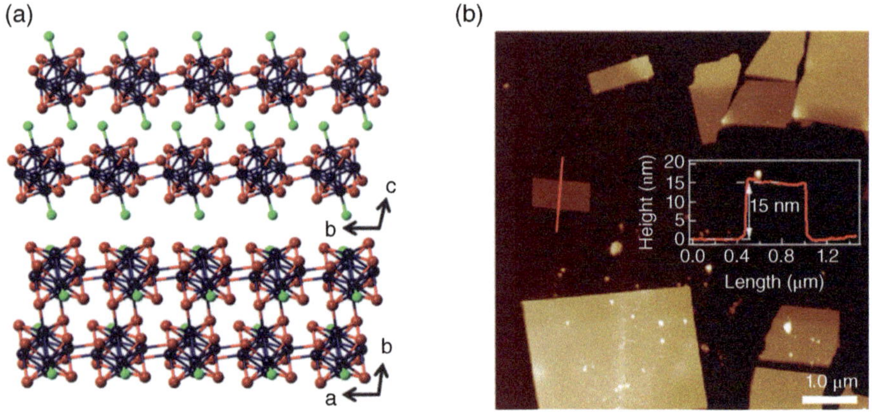

Figure 8.29 Crystal structure and exfoliation of the 2D vdW solid $Re_6Se_8Cl_2$. (a) Side view (top) of the ab plane and top view (bottom) of a single layer. Color code: Re, blue; Se, red; Cl, green. (b) Exfoliated $Re_6Se_8Cl_2$ flakes on SiO_2/Si. The height profile in red (insert) shows the thickness of a ~15 nm thin flake. *Source:* Zhong et al. [55]. © American Chemical Society.

exfoliated multilayer $Re_6Se_8Cl_2$ flakes were confirmed to be 1.58 eV, 1.48 eV(indirect), and 100 meV, respectively, which closely resemble those of the bulk solid.

To reach the monolayer limit, a new technique (intercalation of Li^+ and liquid exfoliation in *N*-methylformamide) was further used for the exfoliation of layered $Re_6Se_8Cl_2$ [248]. When solvated $Re_6Se_8Cl_2$ monolayers are drop-cast onto a sapphire substrate and the solvent is evaporated, the structure of the $Re_6Se_8Cl_2$ monolayers is verified by the AFM image. In Figure 8.30a, one can find that on the substrate a high coverage density of plate-like micrometer-sized $Re_6Se_8Cl_2$ nanosheets are formed. A closer examination of these nanosheets (Figure 8.30b) indicates that the thickness of the first layer is ~1.7 nm. Furthermore, the thickness of the subsequent layers is identified to be ~1.5 nm in the step height histograms, which corresponds to the height of one monolayer. The ordered oblique array of individual Re_6Se_8 cluster can be seen in the high-resolution transmission electron microscopy, which appears as white dots in the high-resolution high-angle annular dark field micrograph, as is shown in Figure 8.30c. The calculations of the electronic band structure (Figure 8.30f) at the HSE06-SOC level reveal that the $Re_6Se_8Cl_2$ monolayer is a semiconductor with an indirect bandgap of 1.72 eV, which is larger than that (1.49 eV) of the corresponding solid bulk.

As the Chevrel Re_6Se_8 clusters can be decorated with a wide range of ligands using ligand substitution chemistry [251], the prospect of functionalizing has also been investigated for the surface of $Re_6Se_8Cl_2$ monolayers. As the first demonstration, the $Re_6Se_8Cl_2$ monolayers on a sapphire substrate are reacted with a solution of trimethylsilyl cyanide (TMSCN). Then, the labile Cl atoms are substituted by the cyanide groups, which have a distinctive and easily identifiable vibrational signature (Figure 8.30d). Remarkably, the monolayer Raman modes in the 100–400 cm^{-1} spectral region reveal that the characteristics of the Re_6Se_8 bonding are essentially unchanged before and after functionalization (Figure 8.30e). That is to say, the ligand substitution strategy can be used to modify the surface of the 2D $Re_6Se_8Cl_2$ monolayers while preserving its internal structure, which establishes a promising route toward multifunctional 2D materials with tunable physical and chemical properties.

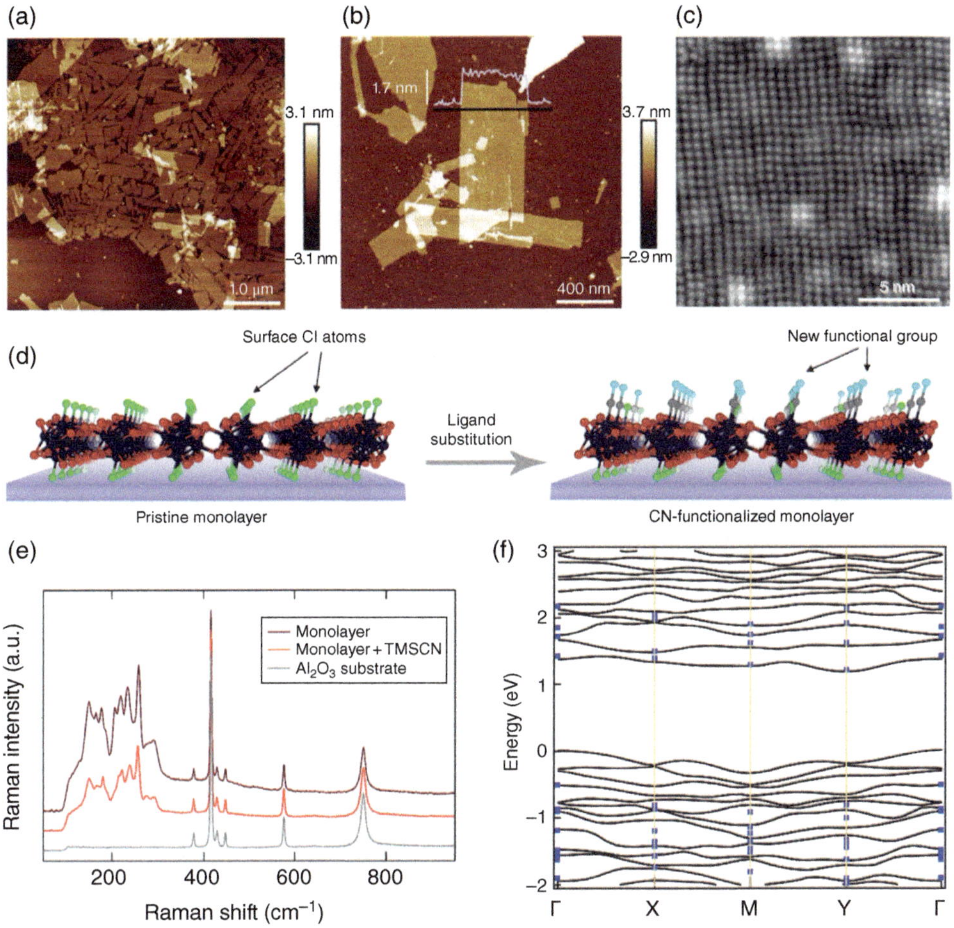

Figure 8.30 (a) AFM image of monolayers drop-cast on a sapphire substrate. (b) AFM image of a few monolayers and height profile along the black line. (c) High-resolution transmission electron microscopy (TEM) micrograph of a monolayer. (d) Schematic of the surface functionalization reaction of monolayers supported on a substrate. (e) Raman spectrum in the 100–400 cm^{-1} region of the monolayers before and after treatment with TMSCN. (f) Calculated DFT band structure with PBE-SOC (solid black line) and HSE06-SOC (blue dots) for the 2D $Re_6Se_8Cl_2$ monolayer. *Source:* Choi et al. [248]. © American Chemical Society.

8.7.2 Co_6Se_8 Cluster-based Monolayer

Like the case of Re_6Se_8, the Co_6Se_8 cluster linked with different ligands also behave as superatoms [252, 253] and can serve the building blocks role to form diverse solids [254, 255]. Recently, Champsaur et al. [256] synthesized the $Co_6Se_8[PEt_2(4\text{-}C_6H_4COOH)]_6$ superatom and proved that it can aggregate into a 3D solid, in which the carboxylic acids form inter-cluster hydrogen bonds. By replacing the reversible hydrogen bonds that hold the superatoms together with zinc carboxylate bonds via the solvothermal reaction between $Co_6Se_8[PEt_2(4\text{-}C_6H_4COOH)]_6$ and $Zn(NO_3)_2$, they further obtained two other types of crystalline solids, a trigonal 3D solid (**Trig**$_{3D}$) and a tetragonal 2D solid (**Tet**$_{2D}$). From single-crystal X-ray diffraction (SCXRD) measurements, the **Trig$_{3D}$** is confirmed to be a 3D network of superatoms held together with zinc carboxylate bonds. In contrast, the **Tet$_{2D}$** form 2D sheets that then stack through noncovalent forces into a 3D solid. Specifically, each 2D layer of **Tet$_{2D}$** is a square arrangement of Co_6Se_8 superatoms with four phosphine ligands residing in the 2D plane and bonding to four-bladed Zn-carboxylate paddlewheels (see Figure 8.31a). In the perpendicular direction of the 2D sheet, the axial carboxylate ligands coordinate an additional Zn^{2+} ion that is positioned above or below the square sheet (Figure 8.31b).

Owing to the layered structure, the **Tet$_{2D}$** crystals are naturally expected to be exfoliated so as to obtain the corresponding 2D material. When the **Tet$_{2D}$** is immersed in a solution of a weak acid (e.g. benzoic acid), the layers of **Tet$_{2D}$** would be chemically dissociated. Drop-casting this solution on a silicon substrate (SiO_2 on Si), thin sheets with a thickness of 7.5 nm can be detected from optical microscopy and AMF (see Figure 8.31c). Since thickness of a single 2D sheet in **Tet$_{2D}$** is 1.5 nm (i.e. the Zn-Zn distance between stacked mononuclear Zn atoms in adjacent layers), the 7.5 nm sheet should correspond to five distinct Co_6Se_8 superatom layers. Except for 7.5 nm sheet, other thinner sheets with step sizes of 3.8 and 5.3 nm (corresponding to three and four layers, respectively) have also been observed. To probe the redox activity of these exfoliated Co_6Se_8 sheets, they were drop-casted onto a

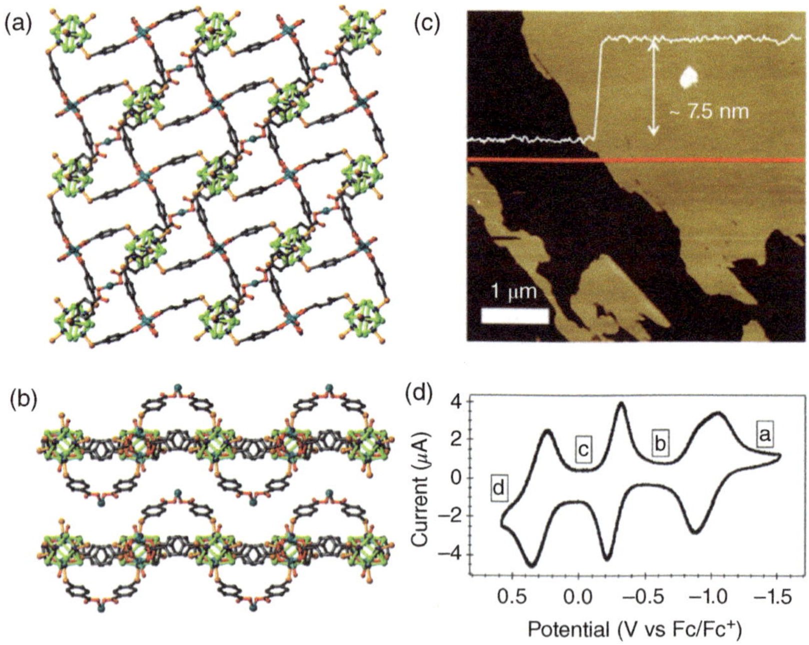

Figure 8.31 Structure of **Tet$_{2D}$** from SCXRD: square sheets in the crystalline state. (a) Top-down view of a single layer within **Tet$_{2D}$** along the b-axis. (b) Side-on view of **Tet$_{2D}$** layers along the c-axis. Noncovalent forces hold the layers together along the third dimension. (c) AFM topographic image of exfoliated sheets. Sheets that are about 7.5 nm in thickness remain, with distinct step sizes being apparent. (d) Solid-state cyclic voltammogram of exfoliated **Tet$_{2D}$** sheets in 0.1 M TBAPF6 in tetrahydrofuran with a 50 mV/s scan rate. The solution of exfoliated sheets was drop-cast onto a glassy carbon electrode. *Source:* Copyright 2017 American Chemical Society.

glassy carbon electrode. Then, the cyclic voltammogram was measured, as shown in Figure 8.31d. Similar to the case of superatom $Co_6Se_8[PEt_2(4\text{-}C_6H_4COOH)]_6$, one can find three reversible oxidations relative to Fc/Fc^+. Therefore, the redox properties of the component superatoms have well persisted in the assembled 2D sheets. For future applications, the preserved redox activity and porous and ultrathin features in structure may make the sheets find utility in a number of fields, such as modified electrodes for catalysis, batteries, and nanoscale electronic sieves.

8.8 Summary and Future Perspectives

In addition to the 2D atomic crystals with relatively simple lattice structures, there is a growing interest in 2D superatomic materials with complex hierarchical structures that could result in unique physical properties and functions. Nowadays, the theoretical design and experimental synthesis of new 2D superatomic materials, as well as the studies of their structure and properties, have become an important branch of both cluster science and 2D materials science. In this chapter, we have made an effort to review the current advances in this branch by focusing on six types of clusters (i.e. fullerenes, silicon-based TM@Si*n* clusters, binary X_nY_n semiconductor clusters, metal clusters, Zintl clusters, and Chevrel clusters) as superatomic building blocks of 2D materials. For most of the predicted or synthesized 2D superatomic materials, we discussed their geometries, relative stability, and electronic properties. In particular, several novel collective behaviors of these reported 2D assembled sheets were discussed, which include the ultrahigh thermal conductivity, linear band dispersion, room-temperature ferromagnetism and half-metallicity, photocatalytic properties, and all-metal-atom-based semiconductor. Furthermore, we also highlighted the experimental syntheses of some 2D superatomic materials on substrate.

Compared with the rapidly developing studies of 2D atomic materials, research on 2D superatomic materials is at the embryonic stage. Although there have been some theoretical designs and a few experimental reports for cluster-assembled sheets, there are at least three major challenges that should be overcome to make continued progress in this field:

First, "good" superatoms are required, which should meet the following important criteria: they (i) can be easily obtained and separated in experiment with low cost and (ii) are stable enough to preserve their identity in different environments. Guided by theory, some stable clusters in the gas phase have been identified to be superatoms, mimicking the properties of atoms. However, the experimental conditions to obtain them are often harsh, including ultrahigh vacuum and ultralow temperature. Therefore, it is highly desirable to find simpler ways to synthesize superatoms in sufficient quantities so that they can be used as building blocks of new 2D materials.

Second, new experimental approaches need to be developed for assembling superatoms into extended structures. In our opinion, starting with superatoms in the gas phase is not an ideal route for obtaining assembled 2D materials. The combination between bottom-up and top-down methods may be a promising route. The synthesis of Co_6Se_8-based sheet we discussed is a typical example. Using ligand-induced self-assembly, one can first obtain the vdW layered superatomic solid from the bottom-up and then 2D sheets can be exfoliated from the layered bulk in the top-down way. However, the synthesized layered vdW superatomic solids are scarce, and the controllability for the liquid or mechanical exfoliation of 2D sheets with desirable thickness needs to be improved. In this context, new approaches for the bottom-up and top-down route are highly required.

Third, as new 2D superatomic sheets are developed, more efforts are needed to explore their unique structure and collective properties, which cannot be found in traditional 2D atomic materials, or at least have superior properties. These efforts will be the key in realizing the promise of superatoms as building blocks for 2D materials. Also, the discovery of new collective properties can keep the vitality of the research in 2D superatomic sheets going.

Facing these challenges, we think that the following topics would be interesting to investigate in future studies:

a) With the help of high-speed computers, along with efficient DFT codes, high-throughput search for "good" superatoms with larger size and more complex geometry is possible. Superatoms can establish a fascinating 3D periodic table in which the number of elements is infinite. However, to date the superatom concept has not been generalized. And we also lack a specialized superatoms database. We believe that theoretical studies can not only guide experimental synthesis and characterization of new superatoms but also create more possibilities for the design of new 2D superatomic materials.
b) Within the past few years, some effective strategies have been proposed to assemble superatoms into periodic 2D crystals, which make use of inter-cluster and cluster-substrate charge transfer, bridging ligands, and exfoliating from vdW layered superatomic solid. For all of them, understanding of ligand–cluster, cluster–cluster, and cluster–substrate interactions is the key. In previous theoretical studies, much of the focus is on freestanding 2D assembled structures and their electronic properties. Although some of these exhibit attractive properties, there are many challenges for their realization in experiment and practical application. In this context, more efforts should be devoted to explore the ligand–cluster, inter-cluster, and cluster–substrate interactions. The corresponding studies can help us to understand if superatoms can strongly couple with each other to form stable 2D materials (survive above room temperature and withstand their own weight to keep the freestanding structure) and if the coupling (also including the cluster–substrate interaction) is weak enough to preserve the integrity of superatoms.
c) As the superatomic 2D materials are usually complex, hierarchical, and tunable structures, they are expected to exhibit special properties superior to atom-based materials. As is known, graphene is essentially a quantum spin Hall insulator. However, the bulk bandgap is too small. Can we obtain the desired large-gap quantum spin Hall insulator in superatomic 2D materials? On the other hand, the linear Dirac band has been observed in C_{26}-assembled sheet. Can other ideal topological semimetals (e.g. Weyl semimetals, nodal line semimetals, and multiple degeneracy point semimetals) also be found in 2D superatomic materials? Furthermore, can the 2D superatomic materials exhibit high-temperature FM half-metallicity, topological superconductivity, highly efficient catalytic and energy storage performance, and so on? We have a long way to go before we realize the potential applications of 2D superatomic materials. It is, however, clear that further research into superatom-based materials will lead to discovery of new phenomena.

References

1 Novoselov, K.S., Geim, A.K., Morozov, S.V. et al. (2004). *Science* 306: 666–669.
2 Castro Neto, A.H., Guinea, F., Peres, N.M.R. et al. (2009). *Rev. Mod. Phys.* 81: 109–162.
3 Zhu, Y., Murali, S., Cai, W. et al. (2010). *Adv. Mater.* 22: 3906–3924.
4 Geim, A.K. (2009). *Science* 324: 1530–1534.
5 Allen, M.J., Tung, V.C., and Kaner, R.B. (2010). *Chem. Rev.* 110: 132–145.
6 Abergel, D.S.L., Apalkov, V., Berashevich, J. et al. (2010). *Adv. Phys.* 59: 261–482.
7 Berman, D., Erdemir, A., and Sumant, A.V. (2014). *Mater. Today* 17: 31–42.
8 Wu, J., Pisula, W., and Müellen, K. (2007). *Chem. Rev.* 107: 718–747.
9 Choi, W., Lahiri, I., Seelaboyina, R., and Kang, Y.S. (2010). *Crit. Rev. Solid State Mater. Sci.* 35: 52–71.
10 Schwierz, F. (2010). *Nat. Nanotechnol.* 5: 487–496.
11 Pumera, M. (2011). *Energy Environ. Sci.* 4: 668–674.
12 Xiang, Q., Yu, J., and Jaroniec, M. (2012). *Chem. Soc. Rev.* 41: 782–796.
13 Shao, Y., Wang, J., Wu, H. et al. (2010). *Electroanalysis* 22: 1027–1036.
14 Kuila, T., Bose, S., Khanra, P. et al. (2011). *Biosens. Bioelectron.* 26: 4637–4648.
15 Liu, Y., Dong, X., and Chen, P. (2012). *Chem. Soc. Rev.* 41: 2283–2307.
16 Bhimanapati, G.R., Lin, Z., Meunier, V. et al. (2015). *ACS Nano* 9: 11509–11539.
17 Butler, S.Z., Hollen, S.M., Cao, L. et al. (2013). *ACS Nano* 7: 2898–2926.
18 Das, S., Robinson, J.A., Dubey, M. et al. (2015). *Annu. Rev. Mater. Res.* 45: 1–27.
19 Molle, A., Goldberger, J., Houssa, M. et al. (2017). *Nat. Mater.* 16: 163–169.
20 Kong, X., Liu, Q., Zhang, C. et al. (2017). *Chem. Soc. Rev.* 46: 2127–2157.
21 Zhang, S., Guo, S., Chen, Z. et al. (2018). *Chem. Soc. Rev.* 47: 982–1021.
22 Xu, M., Liang, T., Shi, M., and Chen, H. (2013). *Chem. Rev.* 113: 3766–3798.
23 Miro, P., Audiffred, M., and Heine, T. (2014). *Chem. Soc. Rev.* 43: 6537–6554.
24 Tao, W., Kong, N., Ji, X. et al. (2019). *Chem. Soc. Rev.* 48: 2891–2912.
25 Chaudhari, N.K., Jin, H., Kim, B. et al. (2017). *J. Mater. Chem. A* 5: 24564–24579.
26 Gao, G., Ding, G., Li, J. et al. (2016). *Nanoscale* 8: 8986–8994.
27 Mounet, N., Gibertini, M., Schwaller, P. et al. (2018). *Nat. Nanotechnol.* 13: 246–252.
28 Mannix, A.J., Kiraly, B., Hersam, M.C., and Guisinger, N.P. (2017). *Nat. Rev. Chem.* 1: 1–14.
29 Zhao, J., Liu, H., Yu, Z. et al. (2016). *Prog. Mater. Sci.* 83: 24–151.
30 Li, Y., Xu, L., Liu, H., and Li, Y. (2014). *Chem. Soc. Rev.* 43: 2572–2586.
31 Zhang, S., Zhou, J., Wang, Q. et al. (2015). *Proc. Natl. Acad. Sci. U. S. A.* 112: 2372–2377.
32 Guzmán-Verri, G.G. and Lew Yan Voon, L.C. (2007). *Phys. Rev. B* 76: 075131.
33 Acun, A., Zhang, L., Bampoulis, P. et al. (2015). *J. Phys. Condens. Matter.* 27: 443002.
34 Cahangirov, S., Topsakal, M., Aktürk, E. et al. (2009). *Phys. Rev. Lett.* 102: 236804.
35 Xu, Y., Yan, B., Zhang, H.J. et al. (2013). *Phys. Rev. Lett.* 111: 136804.
36 Liu, H., Neal, A.T., Zhu, Z. et al. (2014). *ACS Nano* 8: 4033–4041.
37 Zhang, S., Yan, Z., Li, Y. et al. (2015). *Angew. Chem. Int. Ed.* 127: 3155–3158.
38 Zhang, S., Xie, M., Li, F. et al. (2016). *Angew. Chem.* 128: 1698–1701.
39 Feng, B., Zhang, J., Zhong, Q. et al. (2016). *Nat. Chem.* 8: 563–568.
40 Mannix, A.J., Zhou, X.F., Kiraly, B. et al. (2015). *Science* 350: 1513–1516.
41 Coleman, J.N., Lotya, M., O'Neill, A. et al. (2011). *Science* 331: 568–571.
42 Zeng, Z., Yin, Z., Huang, X. et al. (2011). *Angew. Chem. Int. Ed.* 50: 11093–11097.

43 Gogotsi, Y. and Anasori, B. (2019). *ACS Nano* 13: 8491–8494.
44 Huang, B., Clark, G., Navarro-Moratalla, E. et al. (2017). *Nature* 546: 270–273.
45 Seyler, K.L., Zhong, D., Klein, D.R. et al. (2018). *Nat. Phys.* 14: 277–281.
46 Sun, Z., Liao, T., Dou, Y. et al. (2014). *Nat. Commun.* 5: 3813.
47 Ida, S., Ogata, C., Eguchi, M. et al. (2008). *J. Am. Chem. Soc.* 130: 7052–7059.
48 Topsakal, M., Aktürk, E., and Ciraci, S. (2009). *Phys. Rev. B* 79: 115442.
49 Giovannetti, G., Khomyakov, P.A., Brocks, G. et al. (2007). *Phys. Rev. B* 76: 073103.
50 Kim, K.K., Hsu, A., Jia, X. et al. (2012). *Nano Lett.* 12: 161–166.
51 Jin, C., Lin, F., Suenaga, K., and Iijima, S. (2009). *Phys. Rev. Lett.* 102: 195505.
52 Dean, C.R., Young, A.F., Meric, I. et al. (2010). *Nat. Nanotechnol.* 5: 722–726.
53 Liu, C.-C., Feng, W., and Yao, Y. (2011). *Phys. Rev. Lett.* 107: 076802.
54 Liu, C.-C., Jiang, H., and Yao, Y. (2011). *Phys. Rev. B* 84: 195430.
55 Zhong, X., Lee, K., Choi, B. et al. (2018). *Nano Lett.* 18: 1483–1488.
56 Dou, L., Wong, A.B., Yu, Y. et al. (2015). *Science* 349: 1518–1521.
57 Jariwala, D., Marks, T.J., and Hersam, M.C. (2017). *Nat. Mater.* 16: 170–181.
58 Khanna, S.N. and Jena, P. (1992). *Phys. Rev. Lett.* 69: 1664–1667.
59 Khanna, S.N. and Jena, P. (1995). *Phys. Rev. B* 51: 13705–13716.
60 Jena, P., Khanna, S.N., and Rao, B.K. (1996). *Mater. Sci. Forum* 232: 1–26.
61 Castleman, A.W. (2011). *J. Phys. Chem. Lett.* 2: 1062–1069.
62 Jena, P. (2013). *J. Phys. Chem. Lett.* 4: 1432–1442.
63 Castleman, A.W. Jr. and Khanna, S.N. (2009). *J. Phys. Chem. C* 113: 2664–2675.
64 Jena, P. and Sun, Q. (2018). *Chem. Rev.* 118: 5755–5870.
65 Doud, E.A., Voevodin, A., Hochuli, T.J. et al. (2020). *Nat. Rev. Mater.* 5: 371–387.
66 Krätschmer, W., Lamb, L.D., Fostiropoulos, K., and Huffman, D.R. (1990). *Nature* 347: 354–358.
67 Fischer, J.E., Heiney, P.A., McGhie, A.R. et al. (1991). *Science* 252: 1288–1290.
68 Claridge, S.A., Castleman, A.W., Khanna, S.N. et al. (2009). *ACS Nano* 3: 244–255.
69 Qian, M., Reber, A.C., Ugrinov, A. et al. (2010). *ACS Nano* 4: 235–240.
70 Zhao, J., Du, Q., Zhou, S., and Kumar, V. (2020). *Chem. Rev.* 120: 9021–9163.
71 Mandal, S., Reber, A.C., Qian, M. et al. (2013). *Acc. Chem. Res.* 46: 2385–2395.
72 Gadjieva, N.A., Champsaur, A.M., Steigerwald, M.L. et al. (2020). *Eur. J. Inorg. Chem.* 2020: 1245–1254.
73 Konarev, D.V., Khasanov, S.S., and Lyubovskaya, R.N. (2014). *Coord. Chem. Rev.* 262: 16–36.
74 Huang, Y.-G., Jiang, F.-L., and Hong, M.-C. (2009). *Coord. Chem. Rev.* 253: 2814–2834.
75 Pinkard, A., Champsaur, A.M., and Roy, X. (2018). *Acc. Chem. Res.* 51: 919–929.
76 Long, D.-L., Burkholder, E., and Cronin, L. (2007). *Chem. Soc. Rev.* 36: 105–121.
77 Mandal, S., Reber, A.C., Qian, M. et al. (2012). *Dalton. Trans.* 41: 12365–12377.
78 Tomalia, D.A. and Khanna, S.N. (2016). *Chem. Rev.* 116: 2705–2774.
79 Kroto, H.W., Heath, J.R., O'Brien, S.C. et al. (1985). *Nature* 318: 162–163.
80 Feng, M., Zhao, J., and Petek, H. (2008). *Science* 320: 359–362.
81 Zhao, J., Feng, M., Yang, J., and Petek, H. (2009). *ACS Nano* 3: 853–864.
82 Hebard, A.F., Rosseinsky, M.J., Haddon, R.C. et al. (1991). *Nature* 350: 600–601.
83 Schön, J.H., Kloc, C., Haddon, R.C., and Batlogg, B. (2000). *Science* 288: 656–658.
84 Modesti, S., Cerasari, S., and Rudolf, P. (1993). *Phys. Rev. Lett.* 71: 2469–2472.
85 Nakayama, T., Onoe, J., Takeuchi, K., and Aono, M. (1999). *Phys. Rev. B* 59: 12627–12631.

86 Cepek, C., Vobornik, I., Goldoni, A. et al. (2001). *Phys. Rev. Lett.* 86: 3100–3103.
87 Grobis, M., Lu, X., and Crommie, M.F. (2002). *Phys. Rev. B* 66: 161408.
88 Yuan, L.-F., Yang, J., Wang, H. et al. (2003). *J. Am. Chem. Soc.* 125: 169–172.
89 Shi, X.-Q., Van Hove, M.A., and Zhang, R.-Q. (2012). *J. Mater. Sci.* 47: 7341–7355.
90 Chen, W., Zhang, H., Huang, H. et al. (2008). *ACS Nano* 2: 693–698.
91 Pai, W.W., Jeng, H.T., Cheng, C.M. et al. (2010). *Phys. Rev. Lett.* 104: 036103.
92 Cho, J., Smerdon, J., Gao, L. et al. (2012). *Nano Lett.* 12: 3018–3024.
93 Nakamura, J., Nakayama, T., Watanabe, S., and Aono, M. (2001). *Phys. Rev. Lett.* 87: 048301.
94 Saito, S. and Oshiyama, A. (1991). *Phys. Rev. Lett.* 66: 2637–2640.
95 Reddy, C.D., Yu, Z.G., and Zhang, Y.-W. (2015). *Sci. Rep.* 5: 12221.
96 Stetzer, M.R., Heiney, P.A., Fischer, J.E., and McGhie, A.R. (1997). *Phys. Rev. B* 55: 127–131.
97 Leifer, S.D., Goodwin, D.G., Anderson, M.S., and Anderson, J.R. (1995). *Phys. Rev. B* 51: 9973–9978.
98 Shibuta, M., Yamamoto, K., Guo, H. et al. (2020). *J. Phys. Chem. C* 124: 734–741.
99 Rao, A.M., Zhou, P., Wang, K.-A. et al. (1993). *Science* 259: 955–957.
100 Nunez-Regueiro, M., Marques, L., Hodeau, J.L. et al. (1995). *Phys. Rev. Lett.* 74: 278–281.
101 Iwasa, Y., Arima, T., Fleming, R.M. et al. (1994). *Science* 264: 1570–1572.
102 Okada, S. and Saito, S. (1997). *Phys. Rev. B* 55: 4039–4041.
103 Okada, S. and Saito, S. (1999). *Phys. Rev. B* 59: 1930–1936.
104 Okada, S. and Oshiyama, A. (2003). *Phys. Rev. B* 68: 235402.
105 Berber, S., Osawa, E., and Tománek, D. (2004). *Phys. Rev. B* 70: 085417.
106 Pérez-Terrazas, J.E., Romero, A.H., and Terrones, M. (2008). *Chem. Phys. Lett.* 458: 128–133.
107 Goze, C., Rachdi, F., Hajji, L. et al. (1996). *Phys. Rev. B* 54: R3676–R3678.
108 Xu, C.H. and Scuseria, G.E. (1995). *Phys. Rev. Lett.* 74: 274–277.
109 Okotrub, A.V., Belavin, V.V., Bulusheva, L.G. et al. (2001). *J. Chem. Phys.* 115: 5637–5641.
110 Fowler, P.W. and Manolopoulos, D. (2007). *An Atlas of Fullerenes*. Courier Corporation.
111 Prinzbach, H., Weiler, A., Landenberger, P. et al. (2000). *Nature* 407: 60–63.
112 Dunk, P.W., Kaiser, N.K., Mulet-Gas, M. et al. (2012). *J. Am. Chem. Soc.* 134: 9380–9389.
113 Kietzmann, H., Rochow, R., Ganteför, G. et al. (1998). *Phys. Rev. Lett.* 81: 5378–5381.
114 Handschuh, H., Ganteför, G., Kessler, B. et al. (1995). *Phys. Rev. Lett.* 74: 1095–1098.
115 Wang, Z., Ke, X., Zhu, Z. et al. (2001). *Phys. Lett. A* 280: 351–356.
116 Piskoti, C., Yarger, J., and Zettl, A.J.N. (1998). *Nature* 393: 771–774.
117 Shen, Y., Wang, F.Q., Liu, J. et al. (2018). *Nanoscale* 10: 6099–6104.
118 Maruyama, M. and Okada, S. (2013). *J. Phys. Soc. Jpn.* 82: 043708.
119 Maruyama, M. and Okada, S. (2013). *Phys. Stat. Solid C* 10: 1620–1623.
120 Maruyama, M. and Okada, S. (2015). *Jpn. J. Appl. Phys.* 54: 06FF02.
121 Maruyama, M., Cuong, N.T., and Okada, S. (2015). *J. Phys. Soc. Jpn.* 84: 084706.
122 Okada, S., Miyamoto, Y., and Saito, M. (2001). *Phys. Rev. B* 64: 245405.
123 Chen, Z., Heine, T., Jiao, H. et al. (2004). *Chemistry* 10: 963–970.
124 Shen, Y., Wang, F.Q., Liu, J., and Wang, Q. (2019). *Phys. Chem. Chem. Phys.* 22: 306–312.
125 Spagnolatti, I., Bernasconi, M., and Benedek, G. (2002). *Europhys. Lett.* 59: 572–578.
126 Liu, H., Qin, G., Lin, Y., and Hu, M. (2016). *Nano Lett.* 16: 3831–3842.
127 Wang, F.Q., Yu, J., Wang, Q. et al. (2016). *Carbon* 105: 424–429.

128 Hu, M., Jing, Y., and Zhang, X. (2015). *Phys. Rev. B* 91: 155408.
129 Chang, Y., Jalbout, A.F., Zhang, J. et al. (2006). *Chem. Phys. Lett.* 428: 148–151.
130 Gal'pern, E.G., Stankevich, I.V., Chistyakov, A.L., and Chernozatonskii, L.A. (2001). *J. Mol. Graphics Modell.* 19: 189–193.
131 Gal'pern, E.G., Sabirov, A.R., Stankevich, I.V. et al. (2005). *Fullerenes, Nanotubes, Carbon Nanostruct.* 12: 455–461.
132 Hoffmann, R. (1963). *J. Chem. Phys.* 39: 1397–1412.
133 Wilson, R., Meijer, G., Bethune, D.S. et al. (1990). *Nature* 348: 621–622.
134 Kuk, Y., Kim, D.K., Suh, Y.D. et al. (1993). *Phys. Rev. Lett.* 70: 1948–1951.
135 Xu, H., Wang, Z., Lu, Y. et al. (2020). *J. Phys. Chem. C* 124: 11158–11164.
136 Zhang, X., Fan, X., Zhu, G. et al. (2020). *J. Phys. Chem. C* 124: 12589–12595.
137 Stengel, M., Vita, A.D., and Baldereschi, A. (2003). *Phys. Rev. Lett.* 91: 166101.
138 Brouet, V., Yang, W.L., Zhou, X.J. et al. (2004). *Phys. Rev. Lett.* 93: 197601.
139 Schull, G. and Berndt, R. (2007). *Phys. Rev. Lett.* 99: 226105.
140 Wang, Y.Y., Yamachika, R., Wachowiak, A. et al. (2007). *Phys. Rev. Lett.* 99: 086402.
141 Li, H.I., Pussi, K., Hanna, K.J. et al. (2009). *Phys. Rev. Lett.* 103: 056101.
142 Tang, L. and Guo, Q. (2012). *Phys. Chem. Chem. Phys.* 14: 3323–3328.
143 Torrelles, X., Pedio, M., Cepek, C., and Felici, R. (2012). *Phys. Rev. B* 86: 075461.
144 Jung, M., Shin, D., Sohn, S.D. et al. (2014). *Nanoscale* 6: 11835–11840.
145 Shin, H., Schwarze, A., Diehl, R.D. et al. (2014). *Phys. Rev. B* 89: 245428.
146 Ledieu, J., Gaudry, E., de Weerd, M.C. et al. (2015). *Phys. Rev. B* 91: 155418.
147 Smerdon, J.A., Darancet, P., and Guest, J.R. (2017). *J. Chem. Phys.* 146: 092328.
148 Zotov, A.V., Olyanich, D.A., Mararov, V.V. et al. (2018). *J. Chem. Phys.* 149: 034702.
149 Shang, Y., Wang, Z., Yang, D. et al. (2020). *Nanomaterials* 10: 1305.
150 Hou, J.G., Yang, J.L., Wang, H.Q. et al. (1999). *Phys. Rev. Lett.* 83: 3001–3004.
151 Kim, K., Lee, T.H., Santos, E.J.G. et al. (2015). *ACS Nano* 9: 5922–5928.
152 Wang, L.L. and Cheng, H.P. (2004). *Phys. Rev. B* 69: 165417.
153 Che, J.G. and Cheng, H.P. (2005). *Phys. Rev. B* 72: 115436.
154 Li, L., Wu, Y.N., and Cheng, H.P. (2010). *J. Chem. Phys.* 132: 074702.
155 Torrelles, X., Langlais, V., De Santis, M. et al. (2010). *Phys. Rev. B* 81: 041404.
156 Hamada, I. and Tsukada, M. (2011). *Phys. Rev. B* 83: 245437.
157 Pai, W.W., Hsu, C.L., Lin, M.C. et al. (2004). *Phys. Rev. B* 69: 125405.
158 Tamai, A., Seitsonen, A.P., Baumberger, F. et al. (2008). *Phys. Rev. B* 77: 075134.
159 Wang, L.L. and Cheng, H.P. (2004). *Phys. Rev. B* 69: 045404.
160 Tang, L., Zhang, X., Guo, Q.M. et al. (2010). *Phys. Rev. B* 82: 125414.
161 Felici, R., Pedio, M., Borgatti, F. et al. (2005). *Nat. Mater.* 4: 688–692.
162 Shi, X.Q., Pang, A.B., Man, K.L. et al. (2011). *Phys. Rev. B* 84: 235406.
163 Liu, C.D., Qin, Z.H., Chen, J.A. et al. (2011). *J. Chem. Phys.* 134: 044707.
164 Zhang, X., He, W., Zhao, A. et al. (2007). *Phys. Rev. B* 75: 235444.
165 Sun, Q., Wang, Q., Jena, P. et al. (2003). *Phys. Rev. Lett.* 90: 135503.
166 Röthlisberger, U., Andreoni, W., and Parrinello, M. (1994). *Phys. Rev. Lett.* 72: 665–668.
167 Claes, P., Janssens, E., Ngan, V. et al. (2011). *Phys. Rev. Lett.* 107: 173401.
168 Hiura, H., Miyazaki, T., and Kanayama, T. (2001). *Phys. Rev. Lett.* 86: 1733.
169 Koyasu, K., Akutsu, M., Mitsui, M., and Nakajima, A. (2005). *J. Am. Chem. Soc.* 127: 4998–4999.

170 Ngan, V.T., Janssens, E., Claes, P. et al. (2012). *Chem. - Eur. J.* 18: 15788–15793.
171 Furuse, S., Koyasu, K., Atobe, J., and Nakajima, A. (2008). *J. Chem. Phys.* 129: 064311.
172 Menon, M., Andriotis, A.N., and Froudakis, G. (2002). *Nano Lett.* 2: 301–304.
173 Liu, Z.F., Wang, X.Q., Cai, J.T., and Zhu, H.J. (2015). *J. Phys. Chem. C* 119: 1517–1523.
174 Nie, Z., Guo, P., Zheng, J. et al. (2018). *Comput. Mater. Sci.* 146: 134–142.
175 Liu, J., Guo, P., Zheng, J. et al. (2020). *J. Phys. Chem. C* 124: 6861–6870.
176 Dhaka, K. and Bandyopadhyay, D. (2016). *Dalton. Trans.* 45: 12432–12443.
177 Reis, C.L. and Pacheco, J.M. (2009). *J. Phys. Condens. Matter.* 22: 035501.
178 Gueorguiev, G.K., Stafström, S., and Hultman, L. (2008). *Chem. Phys. Lett.* 458: 170–174.
179 Miyazaki, T. and Kanayama, T. (2007). *Jpn. J. Appl. Phys.* 46: L28–L30.
180 Singh, A.K., Kumar, V., Briere, T.M., and Kawazoe, Y. (2002). *Nano Lett.* 2: 1243–1248.
181 Saranin, A.A., Zotov, A.V., Kotlyar, V.G. et al. (2004). *Nano Lett.* 4: 1469–1473.
182 Torres, M.B., Fernández, E.M., and Balbás, L.C. (2011). *Int. J. Quantum Chem.* 111: 444–462.
183 Kumar, V. and Kawazoe, Y. (2001). *Phys. Rev. Lett.* 87: 045503.
184 Ohara, M., Koyasu, K., Nakajima, A., and Kaya, K. (2003). *Chem. Phys. Lett.* 371: 490–497.
185 Nakajima, A. (2013). *Bull. Chem. Soc. Jpn.* 86: 414–437.
186 Tsunoyama, H., Akatsuka, H., Shibuta, M. et al. (2017). *J. Phys. Chem. C* 121: 20507–20516.
187 Reis, C.L. and Pacheco, J.M. (2010). *J. Phys. Condens. Matter.* 22: 035501.
188 Cantera-López, H., Balbás, L.C., and Borstel, G. (2011). *Phys. Rev. B* 83: 075434.
189 Ohta, T., Shibuta, M., Tsunoyama, H. et al. (2016). *J. Phys. Chem. C* 120: 15265–15271.
190 Nakaya, M., Iwasa, T., Tsunoyama, H. et al. (2015). *J. Phys. Chem. C* 119: 10962–10968.
191 Cantera-López, H., Fernández, E.M., Balbás, L.C., and Borstel, G. (2013). *Mater. Chem. Phys.* 139: 247–255.
192 Nakaya, M., Iwasa, T., Tsunoyama, H. et al. (2014). *Nanoscale* 6: 14702–14707.
193 Jose, R., Zhanpeisov, N.U., Fukumura, H. et al. (2006). *J. Am. Chem. Soc.* 128: 629–636.
194 Oku, T., Nishiwaki, A., and Narita, I. (2004). *Sci. Technol. Adv. Mater.* 5: 635–638.
195 Burnin, A., Sanville, E., and BelBruno, J.J. (2005). *J. Phys. Chem. A* 109: 5026–5034.
196 Łazarski, R., Sierka, M., Heinzelmann, J. et al. (2015). *J. Phys. Chem. C* 119: 6886–6895.
197 Oku, T., Nishiwaki, A., Narita, I., and Gonda, M. (2003). *Chem. Phys. Lett.* 380: 620–623.
198 Oku, T., Nishiwaki, A., and Narita, I. (2004). *Phys. B (Amsterdam, Neth.)* 351: 184–190.
199 Liu, Z.F., Wang, X.G., Cai, J.T. et al. (2013). *J. Phys. Chem. C* 117: 17633–17643.
200 Liu, Z.F., Wang, X.Q., Liu, G.B. et al. (2013). *Phys. Chem. Chem. Phys.* 15: 8186–8198.
201 Yong, Y., Song, B., and He, P. (2011). *Phys. Chem. Chem. Phys.* 13: 16182–16189.
202 Woodley, S.M., Watkins, M.B., Sokol, A.A. et al. (2009). *Phys. Chem. Chem. Phys.* 11: 3176–3185.
203 Carrasco, J., Illas, F., and Bromley, S.T. (2007). *Phys. Rev. Lett.* 99: 235502.
204 Liu, Z.F., Wang, X.Q., and Zhu, H.J. (2015). *Phys. Chem. Chem. Phys.* 17: 13117–13122.
205 Liu, Z.F., Wang, X.Q., and Zhu, H.J. (2013). *RSC Adv.* 3: 1450–1459.
206 Saraf, D. and Kshirsagar, A. (2020). *J. Phys. Condens. Matter.* 32: 015501.
207 Puzder, A., Williamson, A.J., Gygi, F., and Galli, G. (2004). *Phys. Rev. Lett.* 92: 217401.
208 Constantinescu, G., Kuc, A., and Heine, T. (2013). *Phys. Rev. Lett.* 111: 036104.
209 Tersoff, J. and Hamann, D.R. (1983). *Phys. Rev. Lett.* 50: 1998–2001.
210 Lu, J.-Z., Liu, Z.-F., Zhu, H.-J., and Xiang, M. (2017). *Comput. Theoret. Chem.* 1101: 74–82.

211 Alonso, J.A. (2000). *Chem. Rev.* 100: 637–677.

212 Lu, Y. and Chen, W. (2012). *Chem. Soc. Rev.* 41: 3594–3623.

213 Chakraborty, I. and Pradeep, T. (2017). *Chem. Rev.* 117: 8208–8271.

214 Diederich, T., Döppner, T., Braune, J. et al. (2001). *Phys. Rev. Lett.* 86: 4807–4810.

215 Thomas, O.C., Zheng, W., Xu, S., and Bowen, K.H. (2002). *Phys. Rev. Lett.* 89: 213403.

216 Jellinek, J. and Acioli, P.H. (2002). *J. Phys. Chem. A* 106: 10919–10925.

217 Acioli, P.H. and Jellinek, J. (2002). *Phys. Rev. Lett.* 89: 213402.

218 Takahashi, L. and Takahashi, K. (2017). *Flatchem* 1: 57–59.

219 Takahashi, K., Hussain, T., Takahashi, L., and Baran, J.D. (2016). *Cryst. Growth Des.* 16: 1746–1750.

220 Boronat, M., Leyva-Pérez, A., and Corma, A. (2014). *Acc. Chem. Res.* 47: 834–844.

221 Harding, D.J. and Fielicke, A. (2014). *Chem. - Eur. J.* 20: 3258–3267.

222 Liu, L., Lopez-Haro, M., Lopes, C.W. et al. (2019). *Nat. Mater.* 18: 866–873.

223 Xiao, L. and Wang, L. (2004). *J. Phys. Chem. A* 108: 8605–8614.

224 Häkkinen, H. (2008). *Chem. Soc. Rev.* 37: 1847–1859.

225 Takahashi, L. and Takahashi, K. (2016). *Inorg. Chem.* 55: 9410–9416.

226 Fässler, T.F. (2011). *Zintl Phases: Principles and Recent Developments*. Berlin: Springer.

227 Gärtner, S. and Korber, N. (2013). 1.09 - Zintl anions. In: *Comprehensive Inorganic Chemistry II*, 2e (eds. J. Reedijk and K. Poeppelmeier). Amsterdam: Elsevier.

228 Bentlohner, M.M., Waibel, M., Zeller, P. et al. (2016). *Angew. Chem. Int. Ed.* 55: 2441–2445.

229 Karttunen, A.J., Fassler, T.F., Linnolahti, M., and Pakkanen, T.A. (2010). *ChemPhysChem* 11: 1944–1950.

230 Liu, C. and Sun, Z.-M. (2019). *Coord. Chem. Rev.* 382: 32–56.

231 Hauptmann, R. and Fässler, T.F. (2003). *Z. Anorg. Allg. Chem.* 629: 2266–2273.

232 Xu, L. and Sevov, S.C. (1999). *J. Am. Chem. Soc.* 121: 9245–9246.

233 Yong, L., Hoffmann, S.D., and Fässler, T.F. (2005). *Z. Anorg. Allg. Chem.* 631: 1149–1153.

234 Ugrinov, A. and Sevov, S.C. (2002). *J. Am. Chem. Soc.* 124: 10990–10991.

235 Yong, L., Hoffmann, S.D., and Fässler, T.F. (2004). *Z. Anorg. Allg. Chem.* 630: 1977–1981.

236 Ugrinov, A. and Sevov, S.C. (2003). *Inorg. Chem.* 42: 5789–5791.

237 Downie, C., Mao, J.-G., Parmar, H., and Guloy, A.M. (2004). *Inorg. Chem.* 43: 1992–1997.

238 Ugrinov, A. and Sevov, S.C. (2005). *C. R. Chim.* 8: 1878–1882.

239 Fässler, T.F. (2001). *Angew. Chem. Int. Ed.* 40: 4161–4165.

240 Fässler, T.F. (2007). *Angew. Chem. Int. Ed.* 46: 2572–2575.

241 Muñoz-Castro, A. and Takahashi, K. (2017). *J. Phys. Chem. C* 121: 1934–1940.

242 Zhou, J., Giri, S., and Jena, P. (2014). *Phys. Chem. Chem. Phys.* 16: 20241–20247.

243 Miwa, K., Takagi, S., Matsuo, M., and Orimo, S.-i. (2013). *J. Phys. Chem. C* 117: 8014–8019.

244 Chevrel, R., Sergent, M., and Prigent, J. (1971). *J. Solid State Chem.* 3: 515–519.

245 Peña, O. (2015). *Phys. C (Amsterdam, Neth.)* 514: 95–112.

246 Lee, K., Maehrlein, S.F., Zhong, X. et al. (2019). *Adv. Mater.* 31: 1903209.

247 Li, Q., Liu, F., Russell, J.C. et al. (2020). *J. Chem. Phys.* 152: 171101.

248 Choi, B., Lee, K., Voevodin, A. et al. (2018). *J. Am. Chem. Soc.* 140: 9369–9373.

249 Zhong, X., Lee, K., Meggiolaro, D. et al. (2019). *Adv. Funct. Mater.* 29: 1902951.

250 Telford, E.J., Russell, J.C., Swann, J.R. et al. (2020). *Nano Lett.* 20: 1718–1724.

251 Kim, Y., Fedorov, V.E., and Kim, S.-J. (2009). *J. Mater. Chem.* 19: 7178–7190.
252 Champsaur, A.M., Velian, A., Paley, D.W. et al. (2016). *Nano Lett.* 16: 5273–5277.
253 Chauhan, V., Reber, A.C., and Khanna, S.N. (2017). *J. Am. Chem. Soc.* 139: 1871–1877.
254 Choi, B., Yu, J., Paley, D.W. et al. (2016). *Nano Lett.* 16: 1445–1449.
255 Turkiewicz, A., Paley, D.W., Besara, T. et al. (2014). *J. Am. Chem. Soc.* 136: 15873–15876.
256 Champsaur, A.M., Yu, J., Roy, X. et al. (2017). *ACS Cent. Sci.* 3: 1050–1055.

9

Superatom-Based Ferroelectrics

Menghao Wu[1] and Puru Jena[2]

[1] *School of Physics, Huazhong University of Science and Technology, Wuhan, Hubei, China*
[2] *Department of Physics, Virginia Commonwealth University, Richmond, VA, USA*

9.1 Introduction

Ferroelectric (FE) materials are polar substances with spontaneous electric polarization that can be switched by an external electric field. They are indispensable for electronics, micromechatronics, and electro-optics. Prevalent pervoskite ferroelectrics such $PbTiO_3$ [1] are among FE materials that have found a variety of commercial uses, such as nonvolatile memories [2], piezoelectric actuators, thermal sensors, light modulators, etc. However, these materials often are composed of toxic lead or rare metals such as bismuth, niobium, or tantalum. Moreover, in their ultrathin films, FE perpendicular to the film surface is often suppressed by the depolarizing field and disappears when the film thicknesses are below 24 Å in $BaTiO_3$ and 12 Å in $PbTiO_3$ [3, 4]. In addition, the displacements of metal ions during FE switching are within 0.5 Å and can be diminished by the depolarizing field. To resolve these issues, especially for the high-density integration of nonvolatile memories, a variety of new FE systems have been recently explored, including a class of ferroelectrics composed of superatoms.

Superatoms are clusters with specific size and composition that mimic the properties of atoms [5]. Among these, superhalogens [6–9] and superalkalis [10–13] are two of the most studied superatoms. The former mimics halogens but has electron affinities (EAs) that are larger than halogen atoms (~3.6 eV for Cl), while the latter mimics alkalis but has ionization potentials (IPs) lower than that of alkali atoms (~3.9 eV for Cs). In addition, superatoms are anisotropic and, unlike atoms, lack spherical symmetry. These unique features enable superatoms to form ideal building blocks of a new generation of FE materials with improved properties. The higher EAs of superhalogens and lower IPs of superalkalis may enhance the amount of charge transfer, while the larger size of superatoms compared with single atoms may also increase the charge transfer distance, giving rise to higher polarization. In addition, the anisotropic feature of superatoms may also lead to the breaking of centrosymmetry or even ferroelectricity as long as a switchable polarization can be formed.

Superatoms: Principles, Synthesis and Applications, First Edition. Edited by Purusottam (Puru) Jena and Qiang Sun.
© 2022 John Wiley & Sons Ltd. Published 2022 by John Wiley & Sons Ltd.

In this chapter, we review some recent works on the design of superatom-based ferroelectrics, including organic ferroelectrics, hybrid organic-inorganic perovskites (HOIPs), and supersalt-based ferroelectrics.

9.2 Organic Ferroelectrics

Nonpolar low-dimensional materials may become polar via surface functionalization that breaks the centrosymmetry, e.g. transition-metal molecular sandwich nanowires (SNWs). Ferrocene, $Fe(C_5H_5)_2$, is a molecule composed of an Fe atom sandwiched between two parallel cyclopentadienyl (Cp) rings. Since its discovery in 1950 [14], zero-dimensional (0D) ferrocene-like molecular complexes [15–18] and one-dimensional (1D) transition-metal molecular SNWs have motivated a vast body of research due to their potential applications in spintronics [19–21]. Various ferrocene-like complexes such as VCp_2, $CrCp_2$, $MnCp_2$, $CoCp_2$, $NiCp_2$, and $ZnCp_2$ where the central Fe atom is replaced by other metal atoms and 1D SNWs with Cp or benzene (Bz) rings, e.g. seven-decker $V_7(Bz)_8$ sandwich clusters, have been successfully synthesized in the laboratory [15–17].

These pristine SNWs are all nonpolar. However, it is known that –OH is a superhalogen, isoelectronic with F, and when the hydrogen atoms on the SNWs are substituted by –OH, those hydroxyl groups can form into hydrogen-bonded chains, giving rise to polarizations along the SNW direction that are switchable under an applied electric field, as shown in Figure 9.1a [22]. If every Bz or Cp in SNWs is decorated by hydroxyl groups, they will possess dipole moments up to 7.3D. Meanwhile, they are ferroelectrics only when the SNWs are insulating along the SNW axis, and CrBz, MnCp, and TiBz SNWs are insulating candidates. In fully hydroxylized MnCp SNWs three-dimensional (3D) crystal where SNWs are connected by hydrogen bonds, a high polarization of $23.0\,\mu C/cm^2$ is obtained, which is even larger compared with croconic acid ($21\,\mu C/cm^2$) at room temperature.

The first design of two-dimensional (2D) van der Waals ferroelectrics was obtained by a similar functionalization method, which endows nonpolar graphene with a switchable polarization [23]. Although pristine graphene is a semimetal and chemically inert, it has been demonstrated experimentally that graphene can be chemically converted into insulating graphene/fluorographene by saturating all carbon atoms with atomic hydrogen/fluorine, which is still nonpolar. If the hydrogen/fluorine atoms on one side are substituted by hydroxyl groups, as shown in Figure 9.1b, which is denoted as semi-hydroxylized-graphane (SHLGA), a 2D hydrogen-bonded network with an in-plane polarization of $41.1\,\mu C/cm^2$ can be formed due to symmetry breaking by the dipoles of –OH. With the help of an external electric field, the polarization can be switched through the rotation of –OH as the energy barrier is only 0.042 eV per proton. If we substitute the hydrogen atoms in a semi-hydrogenated graphone by –OH groups, which is denoted as SHLGO, the system will occupy less space. So, if SHLGOs are stacked layer by layer, the resulting spontaneous polarization $43.7\,\mu C/cm^2$ is even larger compared with bulk SHLGAs. Moreover, the system will become magnetic and semiconducting like graphone, with a magnetic moment of $2.0\,\mu_B$ per unit cell (every unit cell contains two layers), resulting in a 2D multiferroics. For a graphene layer with all the hydrogen atoms substituted by hydroxyl groups, denoted as a fully hydroxylized graphene sheet (HLG), with the formation of interlayer hydrogen bonds,

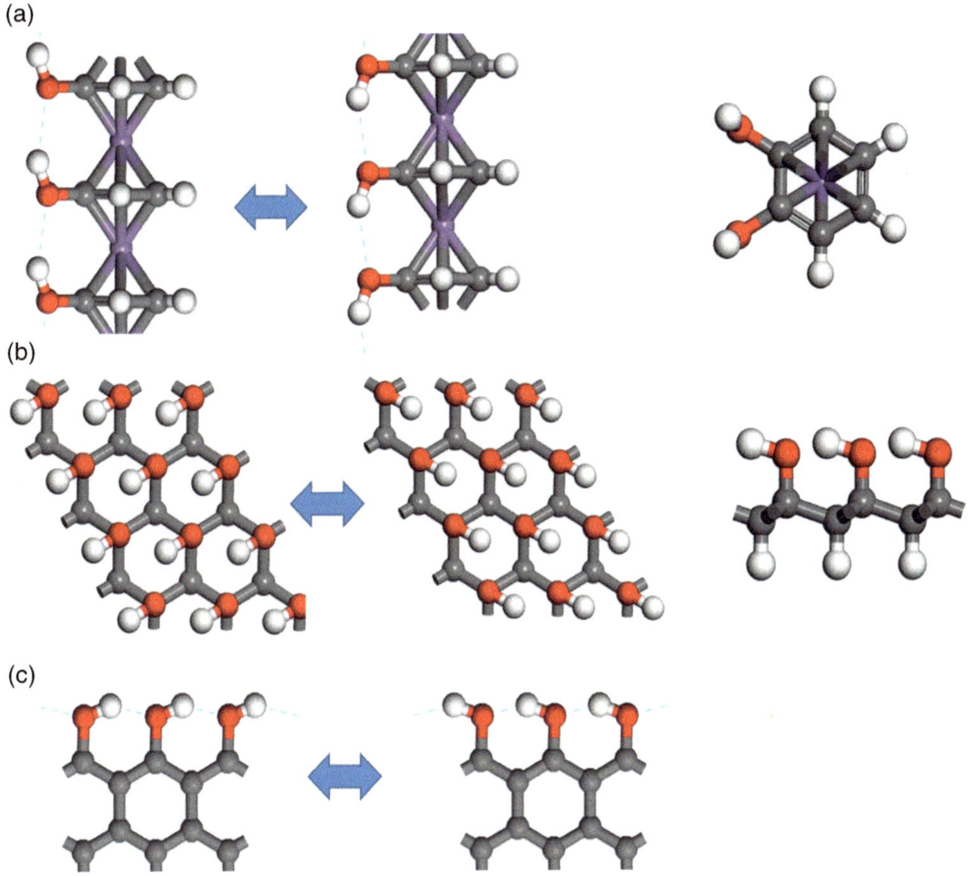

Figure 9.1 FE switching of hydroxylized (a) transition-metal molecular SNWs, (b) graphene, and (c) graphene nanoribbons (GNR). Gray, red, white, and purple spheres denote C, O, H, and transition metal atoms, respectively.

a spontaneous polarization of 67.7 $\mu C/cm^2$ that is switchable under an applied electric field is obtained. This is almost twice that of the value for bulk SHLGA.

1D ferroelectricity can also be formed in hydroxylized zigzag-edge GNR (ZGNR), where –OH groups decorated on zigzag edges will form hydrogen bonds with adjacent –OH groups, giving rise to a spontaneous polarization where the displacements of proton are switchable through the rotation of –OH bond along the direction of GNR, as shown in Figure 9.1c. To achieve the maximum magnitude of ferroelectricity, both edges of the narrowest ZGNRs are hydroxylized and a high polarization of 27.0 $\mu C/cm^2$ is obtained. Suppose that there are two proton vacancies, i.e. two =O bonds at the end of a hydrogen-bonded chain. Under an external electric field along the ZGNR, the protons can simultaneously hop and bind to adjacent O atoms, rotate to the other side, and hop to the next O atoms, crossing an energy barrier of 0.34 eV per proton. Here, the displacements of every proton during hopping and rotating process are, respectively, 0.58 and 1.88 Å, and the changes of dipole moment are both 1.23 eÅ. Therefore, a polarization of $(1.23+1.23+1.23)/2 = 1.85$ eÅ is switched during such transfer. If there are N proton vacancies at the end of every zigzag edge, the 1D polarization will be $0.62+(N-1)1.23\,\mathrm{e\AA}$ per proton.

If the switching barrier is large enough against thermal fluctuation, the data storage density may be further enhanced, even in 0D. For an isolated 0D single-molecule magnet, its magnetism with a spin switching barrier (~meV) cannot survive even at ultralow temperature of a few Kelvin, whereas for a single-molecule FE, the switching barrier can still be more than hundreds of meV [24]. Even though the adjacent dipole–dipole interaction is negligible, the sizable energy barrier would allow FE to survive at ambient conditions. Actually, negligible adjacent dipole–dipole interaction is indispensable for independent bit storage of each dipole, where ultrahigh data storage density can be obtained as each molecule can store 1 bit.

The well-known superatom C_{60} can be transformed into a 0D ferroelectrics [25]. The nonpolar fullerene can become polar by various chemical modifications that have been experimentally realized, as the symmetry can be easily broken by functionalization of dopants, covalent ligands, or endohedral complex. A series of endohedral metallofullerenes $M@C_{60}$ have been isolated experimentally, such as $La@C_{60}$ [26], $Li@C_{60}$ [27, 28], $Er@C_{60}$ [29], $Dy@C_{60}$ [30], $Gd@C_{60}$, $Ca@C_{60}$, $Nd@C_{60}$, and $Y@C_{60}$, etc. [31, 32]. Here M is usually located away from the center of the fullerene cage, giving rise to a dipole moment, while the screening of cages may also greatly reduce the adjacent dipole interactions. In previous experimental reports, fullerene may also become polar via radical functionalization [33, 34], boron/nitrogen doping [35–38], or intercalation of polar diatomic molecules like MX (M = metal, X = nonmetal) inside [39, 40]. However, these single polar molecules, without fixed polarization axis, are still not ferroelectrics that would possess several equivalent polar ground state polarization directions. Their polarizations with zero switching barrier would not be stable against thermal fluctuation, even at low temperatures. The barrier of FE switching should be within the desirable range for single-molecule ferroelectrics; high enough to be robust against thermal fluctuations at ambient conditions while low enough to be tuned by applicable voltage. The desirable switching barrier may be achieved by applying proper substrates interacting with those functionalized fullerenes.

If a fullerene is fixed on the substrate with strong bonding, the dipole moment of the fullerene, functionalized by dopants or ligands on the surface, will be difficult to switch. If the ionic radius of the metal atom, M in $M@C_{60}$ is smaller than that of the fullerene radius, M will be located away from the cage center. The direction of the dipole moment will then be altered upon the movement of the metal ion. The intercalated M ion usually prefers to reside over the center of a hexagon, while its position over a pentagon is metastable. C_{60} is composed of 20 hexagons and 12 pentagons, where the pentagons are all isolated from each other while hexagons are interconnected. The polarization can be switched as the M ion continuously hops to adjacent hexagonal or pentagonal rings along the inner wall of the C_{60} cage from the top toward the bottom. Since the barrier from hexagon-to-hexagon hopping is lower in energy compared with from hexagon-to-pentagon hopping, the most favorable pathway from the top to the bottom should be composed of five equivalent hopping pathways, from a hexagon to another adjacent one. According to previous estimation, the ideal range of switching barrier should approximately range from 0.1 to 1 eV, which renders negligible probability of quantum tunneling and moderate required switching voltage [24]. Here, the switching pathway of M from one hexagon to an adjacent one using the

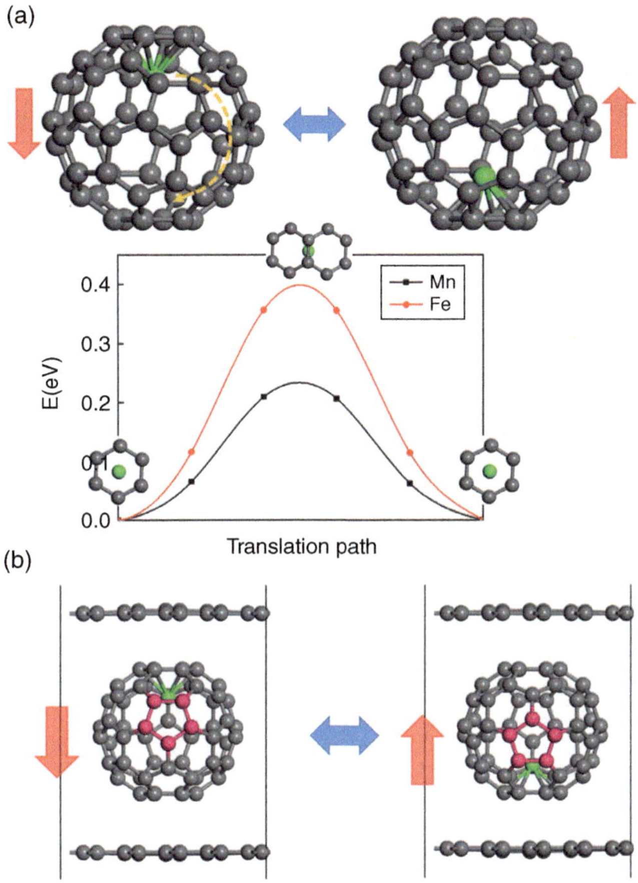

Figure 9.2 FE switching pathway of (a) isolated $M@C_{60}$ fixed on the substrate, and (b) $M@C_{60}$ intercalated between two graphene layers, with a C5 ring in pink to mark the rotation. The polarization directions are denoted by red arrows. Gray and green spheres denote carbon and transition metal atoms, respectively.

nudge elastic band (NEB) method is shown in Figure 9.2a. For M = Li where $Li@C_{60}$ can be deemed as a superalkali atom, the barrier of 0.02 eV can be easily overcome by thermal fluctuation at ambient conditions. If we choose 3*d* magnetic ions M = Fe or Mn as paradigmatic cases, with relatively smaller ion radius compared with alkali ions, their switching barriers are, respectively, 0.40 and 0.23 eV, which are within the desirable range. Thus, each metallofullerene can be used for data storage of 1 bit. Moreover, each Mn and Fe ion carries a magnetic moment of 3 and 2 μ_B, respectively, which may render multiferroicity, although the magnetism may not be as robust as the ferroelectricity.

If the rotation barrier for $M@C_{60}$ is lower than the migration barrier of M ion inside due to the weak interaction with the substrate, the rotation of overall endohedral metallofullerenes will be a favorable path for FE switching with the aid of an electric field. For example, if $M@C_{60}$ molecules are intercalated between two graphene layers with van der Waals interactions, the degeneracy for M residing over different hexagons will be slightly lifted.

From our calculation, the adsorption energy of each Fe@C_{60} on one side of graphene is ~0.03 eV. In the ground state, the upper and lower layers, respectively, form AB stacking with a hexagon of fullerene. The M ion can be located over either of the top or bottom hexagons, giving rise to a vertical polarization. The polarization along the vertical direction in M@C_{60} (M = Fe, Mn) is approximately 30 meV lower in energy than that of the in-plane polarization. In addition, the polarization in Fe@C_{60} is slightly enhanced over that in Mn@C_{60} while each Fe and Mn ion maintain a magnetic moment of 3 and 2 μ_B, respectively. The barrier for ion migration from one hexagon to an adjacent one is slightly influenced by the attached graphene layers. In comparison, due to weak van der Waals interaction, the rotation of fullerene with much lower barrier is more favorable. The pathway for rotation involves a series of metastable intermediate states where the top and bottom hexagons of fullerene form AB stacking with graphene layers. This is revealed by changes in the position of the pink pentagonal ring on the cage shown in Figure 9.2b, where the overall barriers for M = Fe and Mn are, respectively, 0.12 and 0.11 eV.

For the design of FE tunneling junction with different top and bottom electrodes, two states can be differentiated by the variation of conductance upon FE switching. If such a structure can be utilized as FE hard disk data storage system for perpendicular recording, it is expected to increase the areal storage density of hard disk from the superparamagnetic limit of around 40 Gbit/in^2 to ~10^5 Gbit/in^2, which is also around two orders of magnitude higher compared to the current record of FE data storage (~Tbit/in^2) [41]. In addition, compared to bulk ferroelectrics where polarization switching involves domain nucleation and growth, the polarization of M@C_{60} is switched via simple flipping of dipoles. As a result, the switching speed could be substantially enhanced. Such speculation is further confirmed by Born–Oppenheimer molecular dynamics (BOMD) simulation of Fe@C_{60} intercalated between two fixed graphene layers [25], where a canonical ensemble, the algorithm of Nosè, and a timestep of 1 fs are adopted. Snapshot of the equilibrium structure of Fe@C_{60} at the end of 5 ps of BOMD simulation at 300 K in Ref. 25 confirms the thermal stability of its vertical polarization. Upon the application of a vertical electric field of 0.15 V/Å (corresponding to a voltage of ~1.8 V between graphene layers), the polarization in Fe@C_{60} is reversed in about 5 ps due to its rotation. This corresponds to a switching frequency of ~10^2 GHz. Compared to prevalent ferroelectrics like $PbTiO_3$ and $BiFeO_3$ thin films with switching time around 100–400 ns [42, 43], the writing speed of Fe@C_{60} is enhanced by about three to four orders of magnitude.

9.3 Hybrid Organic-Inorganic Perovskites

Hybrid organic-inorganic perovskites (HOIPs), with a record of power conversion efficiency higher than 22% within several years [44, 45], have attracted intensive attention for their high performance in photovoltaics [46–49]. ABX_3 is the general formula of most studied HOIPs, where A is an organic cation like methylammonium ($CH_3NH_3^+$, MA^+) or formamidinium ($[NH_2]_2CH^+$, FA^+), B is a metal cation like Pb^{2+} or Sn^{2+}, and X is a halide anion [46, 50–53]. HOIP was first synthesized by substituting Cs cations in 3D $CsPbI_3$ with nitrogen-containing organic groups [54] like MA or FA. These organic groups possess low ionization energy (IE), mimicking superalkalis [5]. They are usually polar and can give rise to ferroelectricity, thus playing a critical role in the efficient separation of photoinduced electron-hole pairs. Consequently, power conversion efficiency of HOIPs can be high. Previous first-principles computational study [55] has showed that MA cations in tetragonal $MAPbI_3$ have a preferential alignment along the c-axis, and the polarization intensity is estimated to be 4.42 $\mu C/cm^2$, which is mainly contributed by the MA dipoles.

$MAPbI_3$ can be synthesized by the reaction $PbI_2 + MAI \rightarrow MAPbI_3$ with the formation of new Pb-I bonds and 3D perovskite structure. However, due to the long diffusion pathways and considerable structural reconstruction, such insertion of MAI species into 2D layered PbI_2 can be sluggish. And, this can give rise to nonuniformity and incomplete film coverage. This can be improved by using so-called $HPbI_3$ as a new precursor to replace lead halide [56, 57]. However, the claimed existence of $HPbI_3$ has not yet been clarified, with no structural or spectroscopic validation to date. It is not chemically stable and, according to theory, it readily dissociates to PbI_2 and HI [58, 59]. The IE of H atom is 13.6 eV, so its behavior is distinct from alkali metal or MA whose IEs are less than 6 eV. According to our calculations, the cubic perovskite structure given in the Materials Project database [60] turns out to be highly unstable, where a high energy of 3.31 eV/f.u. can be released if it decomposes into PbI_2 and HI. This instability is expected due to the small radius and high EA of hydrogen. This problem might be solved when protons attach to water molecules in the solution and generate superalkali cations with lower EA.

For example, two superalkali cations can be formed when H^+ binds with one and two water molecules: H_3O^+ and $H_5O_2^+$, where their EAs are reduced to 6.4 and 3.1 eV, respectively. The ground state geometry of $H_5O_2^+$ cation is shown in Figure 9.3a, where a proton is located at the center of two water molecules [61]. The distances between the central proton and two O atoms are both 1.21 Å, right between the covalent O-H bond length (usually 1.0 ~ 1.1 Å) and O..H hydrogen bond length (usually 1.4~1.6 Å). The electron of central H in $H_5O_2^+$ is found to be highly delocalized, judging from the electron deformation density. In addition, the non-centrosymmetric structure of $H_5O_2^+$ should give rise to a dipole moment. Aside from the reduced EA, the radius of cation is also greatly enlarged, which may stabilize the perovskite structure with PbI_2. For two structures formed by replacing H in orthorhombic and cubic $HPbI_3$ by H_5O_2, the energy costs ΔE of their formation $PbI_2 + HI + 2H_2O \rightarrow (H_5O_2)PbI_3$ are, respectively, reduced to −0.043 and 0.167 eV/f.u. for orthorhombic and cubic $(H_5O_2)PbI_3$. Here, the lattice parameters ($a = b = 8.64$ Å, $c = 8.13$ Å) are consistent with the reported values fitting the X-ray diffraction pattern [57]. The formation of $(H_5O_2)PbI_3$ is exothermic, which may explain the previously claimed existence of $HPbI_3$. We also note previous report of phase transitions of $MAPbI_3$ from tetragonal to cubic structure at 327.4 K and from orthorhombic to tetragonal structure at 162.2 K [62]. Previous density-functional-theory (DFT) calculations [63–65] show that the energy costs for the reaction $PbI_2 + MAI \rightarrow MAPbI_3$ are, respectively, 100 and −3 meV/f.u. for the formation of cubic phase and orthorhombic phase, respectively. If the high-temperature cubic structure of $(H_5O_2)PbI_3$ is replaced by the low-temperature orthorhombic structure that is 0.170 eV/f.u. lower in energy, the reaction can be deemed as exothermal with

$$\Delta E = -3\,\mathrm{meV\,/\,f.u.}$$

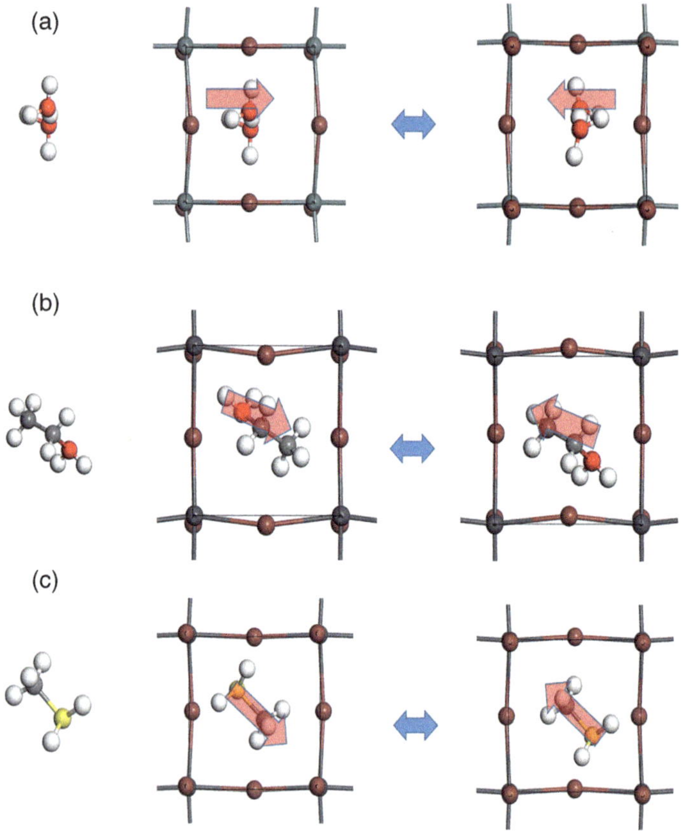

Figure 9.3 Geometric structures of super-cations and FE switching of (a) $(H_5O_2)PbI_3$, (b) $(C_2H_5OH_2)PbI_3$, and (c) $(CH_3SH_2)PbI_3$ [61]. The polarization directions are denoted by red arrows. Gray, white, red, brown, yellow, and dark-gray spheres denote C, H, O, I, S, and Pb atoms, respectively.

It also turns out that the formation of the same group of compounds with lighter elements, such as $(H_5O_2)SnI_3$ and $(H_5O_2)PbBr_3$, can be energetically more favorable, even for the high-temperature cubic phase:

$$SnI_2 + HI + 2H_2O \rightarrow (H_5O_2)SnI_3, \Delta E = -0.03\,eV / f.u.$$

$$SnBr_2 + HBr + 2H_2O \rightarrow (H_5O_2)SnBr_3, \Delta E = -0.12\,eV / f.u.$$

$$PbBr_2 + HBr + 2H_2O \rightarrow (H_5O_2)PbBr_3, \Delta E = -0.09\,eV / f.u.$$

It is likely that their facile solution-processed fabrication can be feasible via codeposition of metal dihalides MX_2 and halide acids HX mixture. The band structures of these perovskites in Ref. 61 calculated by using HSE + SOC method possess direct bandgaps of 0.7, 2.0, and 1.4 eV in $(H_5O_2)SnI_3$, $(H_5O_2)SnBr_3$, and $(H_5O_2)PbBr_3$, respectively. To our knowledge, $(H_5O_2)SnI_3$ possesses one of the lowest bandgaps among current HIOPs. To achieve high photovoltaic performance, direct bandgaps within the optimal range of 0.9–1.6 eV are desirable to achieve a Shockley–Queisser efficiency of ~25%. Herein, $(H_5O_2)PbBr_3$ seems to be the ideal candidate with modest bandgap and low ΔE. However, the photovoltaic performance of the most current HOIPs is still poor in the infrared region [26]. This is where the low bandgap HOIPs like $(H_5O_2)SnI_3$ can play a role. The computed optical absorption spectrum reveals even stronger absorption compared with $MAPbI_3$, making it a promising candidate for being a solar-absorber material. If the alloy of a low-bandgap ABX_3 and a high-bandgap $A'BX_3$ can be fabricated (i.e. $AxA'_{1-x}BX_3$), a wide range of bandgaps can be obtained by tailoring x, which may overwhelm the Shockley–Queisser detailed-balance limit of photovoltaic conversion efficiency for mono-bandgap semiconductors. Here, the low bandgap of $(H_5O_2)SnI_3$ and wide bandgap of $(H_5O_2)SnBr_3$ may achieve such a wide range. It is worth noting that the dipole moment of $H_5O_2^+$ gives rise to a switchable FE polarization of 22.7 $\mu C/cm^2$ for $(H_5O_2)PbI_3$, as shown in Figure 9.3a, which is much higher than previously predicted polarization of $MAPbI_3$ [55]. Theoretically, ferroelectricity offers additional advantage for photovoltaics as high photovoltage can be induced by polarization, while photogenerated electrons and holes can be separated by the built-in electric field [66, 67]. Moreover, it also renders a desirable combination of high-mobility low-bandgap semiconductor and nonvolatile memory [68, 69].

During the formation of these perovskites, halide anions are dissociated from halide acids and join the skeleton of the framework, while the protons are attached to the oxygen atoms of water molecules, giving rise to a low-EA cation mimicking MA^+. Aside from water molecules, proton may also bind with oxygen-containing organic molecules, resulting in cations that may give rise to new stable HOIPs. Ethanol, for example, can become a cation with EA of 5.27 eV when the proton is bound near the site of the oxygen atom. A stable perovskite structure can then be constructed:

$$PbI_2 + HI + C_2H_5OH \rightarrow (C_2H_5OH_2)PbI_3, \Delta E = -0.09\,eV / f.u.$$

$$SnI_2 + HI + C_2H_5OH \rightarrow (C_2H_5OH_2)SnI_3, \Delta E = -0.28\,eV / f.u.$$

$$PbBr_2 + HBr + C_2H_5OH \rightarrow (C_2H_5OH_2)PbBr_3, \Delta E = -0.24\,eV / f.u.$$

It is possible that these can be formed spontaneously in a solution of metal dihalides, halide acids, and ethanol. The thermal stability of $(C_2H_5OH_2)PbI_3$ was investigated via molecular dynamics (MD) simulations in Ref. 61, which shows that the cubic structure is stable at 300 K. Aside from oxygen, sulfur atom in the molecule can also be the binding site for proton. For example, methanethiol can be transformed to a cation with EA of 5.33 eV when bound to a proton. The formation reaction of perovskite containing such cations is also exothermal: $PbI_2 + HI + CH_3SH \rightarrow (CH_3SH_2)PbI_3$, $\Delta E = -0.34\,eV/f.u.$ Such reaction can be processed by solid–gas interaction between PbI_2 and methanethiol. Note that methanethiol should be in the gas phase at ambient conditions. These structures may be much more stable than $MAPbI_3$, considering their much larger energy release during formation. The band structures of $(C_2H_5OH_2)PbI_3$, $(C_2H_5OH_2)SnI_3$, $(C_2H_5OH_2)PbBr_3$, and $(CH_3SH_2)PbI_3$ possess direct bandgaps of 1.48, 1.47, 2.31, and 1.03 eV, respectively. Except for $(C_2H_5OH_2)PbBr_3$, the bandgaps of all these materials lie within the optimal range of 0.9–1.6 eV. According to Mulliken charge analysis, the average charges on organic cation, Pb, and I ions are, respectively, 0.65 e, 0.49 e, and −0.38 e for $(C_2H_5OH_2)PbI_3$ and 0.62 e, 0.40 e, and −0.34 e for $(CH_3SH_2)PbI_3$. $(C_2H_5OH_2)PbI_3$ is found to be more ionic, which may explain why it possesses a wider bandgap compared with $(CH_3SH_2)PbI_3$. The case is similar for $(C_2H_5OH_2)PbBr_3$ with an even wider bandgap as Br is more electronegative than I. The computed optical absorption spectra exhibit even stronger absorption compared with that of $MAPbI_3$. The major factor that causes the difference in the optical absorption may be attributed to different bandgaps, as shown by the redshift of the spectrum of $(CH_3SH_2)PbI_3$ due to its smaller bandgap compared with other HOIPs. Similarly, the absence of centrosymmetry of the cations (as shown in Figure 9.3b and c) would also manifest in a polar material. With the reorientations of these polar cations, where the red arrows denote the polarization direction of the cations and the whole crystal, considerable polarization would emerge both for $(C_2H_5OH_2)PbI_3$ and $(CH_4SH)PbI_3$. Distinct from $(H_5O_2)SnI_3$, here, their polarizations possess projection along both $-x$-axis and $-y$-axis. The computed Px and Py are, respectively, 11.08 and 10.98 $\mu C/cm^2$ for $(C_2H_5OH_2)PbI_3$ and 7.65 and 1.68 $\mu C/cm^2$ for $(CH_4SH)PbI_3$.

9.4 Supersalts

Supersalts are charge transfer complexes composed of superalkali cations and superhalogen anions, akin to traditional ionic salts containing alkali cations and halogen anions [70, 71]. Superalkali and superhalogen species have been previously used to design perovskites for photovoltaics [72, 73], lithium superionic conductors [74], etc. Recently, a number of superatomic solids assembled from molecules like nickel telluride clusters and fullerenes [75] have been successfully synthesized in experiments [76] and interesting properties like controllable phase transitions have been observed [77–80]. The bonding between adjacent superalkali cations and superhalogen anions in supersalts should be highly ionic. Ionic solids usually adopt structures with higher coordination numbers so as to maximize ionic bonding, enhance the Madelung constant, and lower the lattice energy. As a result, typical structures for ionic binary systems such as alkali halides are nonpolar rock salt (RS) or cesium chloride (CsCl) types where the coordination numbers are, respectively, six and eight and are higher than the coordination number of four for covalent sp^3 hybridization. Due to directionless ionic bonding interactions, the structures such as RS or CsCl types are all nonpolar centrosymmetrical structures. Hence, the absence of current reports on highly ionic ferroelectricity should be attributed to the conflict between polarity and such features. In contrast, the directionality and saturation of covalent bonding may facilitate the formation of crystalline polarization. According to the Philips scale, when the ionicity declines below the critical value 0.785, marking the idealized boundary between predominantly "covalent" and "ionic" systems [81], zinc-blende (ZB) structure with a lower coordination number will become energetically more favorable. For example, BN, AlP, and ZnS, with iconicity of 0.256, 0.307, and 0.623, respectively, prefer to form ZB structures. Those structures are polar, although their polarizations are not switchable due to the brittle nature associated with covalent bonding. Mixed ionic-covalent bonding may facilitate the formation of switchable polarizations with moderate barriers (i.e. ferroelectricity), which has been revealed in well-known perovskite ferroelectrics [82] ($BaTiO_3$, $PbTiO_3$, etc.). However, supersalt may not necessarily adopt the nonpolar structures for most ionic systems. Compared with isotropic single atoms, the anisotropy of superatoms may give rise to breaking of centrosymmetry. If such breaking can induce switchable polarizations, highly ionic ferroelectrics may be formed via facile and solution-processed [66] fabrication. This offers a route for large-scale production that cannot be applied to current prevailing ferroelectrics and can find promising applications in a broad range of fields including nonvolatile memories, sensor and actuators, nonlinear optical materials, etc.

For example, superalkali PH_4 and superhalogen BCl_4 can be used to construct a supersalt PH_4BCl_4, where both superatoms are non-centrosymmetric compared with isotropic single atoms, as shown in Figure 9.4a [83]. The ground state structure of supersalt PH_4BCl_4, obtained after an unbiased swarm-intelligence structural search using the Crystal structure AnaLYsis by Particle Swarm Optimization(CALYPSO) code [84, 85], is displayed in Figure 9.4b. The structure is a distorted zinc-blende (dZB) structure with lattice constants ($|a| : |b| : |c| = \sqrt{2}:\sqrt{2}:1$) relatively doubled in $|a|$ and $|b|$ (or contracted by half in $|c|$) compared with a four-atom unit cell of standard ZB ($|a| : |b| : |c| = 1:1:\sqrt{2}$). It is non-centrosymmetric with superalkali PH_4 and superhalogen BCl_4 ions remaining in their pristine forms. The total energy of the derived structure is lower than that of the corresponding RS and CsCl structures. For the dZB supersalt structure, similar to sp^3 hybridization of covalent bonding in prevalent ZB structures, each superalkali cation is tetrahedrally bonded with four adjacent superhalogen anions. Meanwhile, each superhalogen anion is

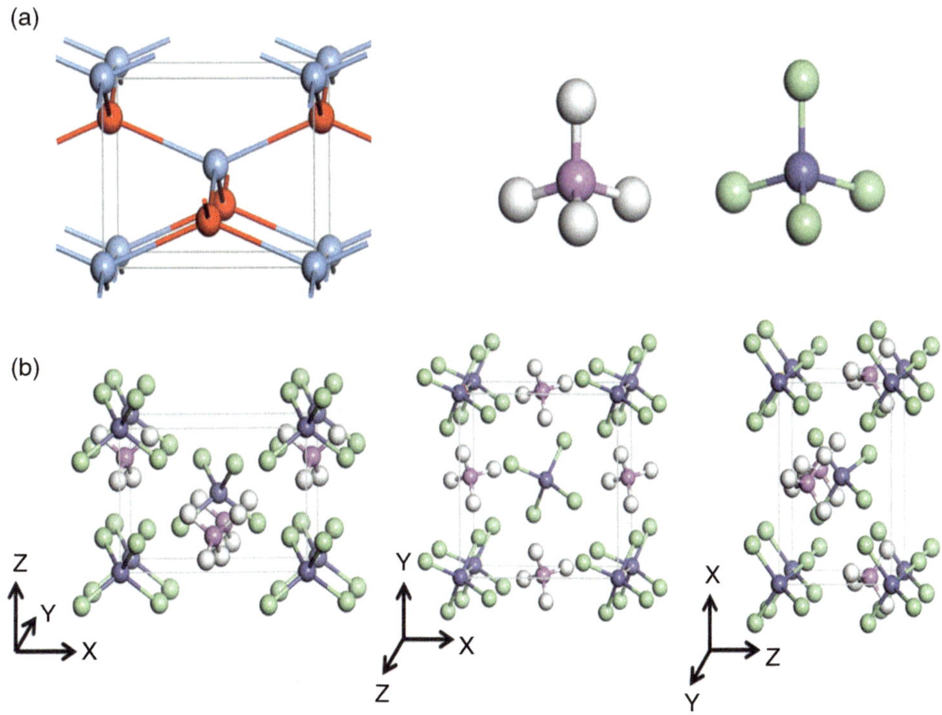

Figure 9.4 (a) ZB structure, superalkali cations, and superhalogen anions; (b) geometric structures of supersalts NH_4MX_4/PH_4MX_4. *Source:* Gao et al. [83]. © John Wiley & Sons.

also tetrahedrally bonded with four adjacent superalkali cations, which are also connected by weak H. . .Cl hydrogen bonds. According to our calculations, the energy release in the reaction $BP + 4HCl \rightarrow PH_4BCl_4$ is 0.57 eV/f.u. (i.e. $\Delta E = -0.57$ eV/f.u.), which may enable its solution-processed synthesis via etching of boron pnictides by acid.

The energy, ΔE associated with decomposition, $PH_4BCl_4 \rightarrow PH_4Cl + BCl_3$, is 0.23 eV/f.u. Thus, the reaction is also endothermal and confirms the chemical stability of PH_4BCl_4, hindering the formation of other products like PH_4Cl during the synthesis. However, the strong covalent B-P bonding may give rise to high energy cost to break those bonds. Consequently, the formation $BP + 4HCl \rightarrow PH_4BCl_4$ may not spontaneously take place at ambient conditions. In comparison, the reaction $PH_4Cl + BCl_3 \rightarrow PH_4BCl_4$ might be more feasible for its fabrication. For the same group pnictides where the bonding is weaker and the hollow spaces are larger, protons or halogen anions may be able to permeate from the surface to the interior. This would facilitate similar formation of supersalts. To see if there are any same-group low-energy structures, we carried out an extensive search for NH_4MX_4/PH_4MX_4 supersalts, where M = Al and B and X = Cl, Br. The ground states of PH_4AlCl_4, PH_4BBr_4, and PH_4BBr_4 share the same dZB-type structure with PH_4BCl_4, which is similar to NH_4BBr_4 and NH_4BCl_4, with a slightly different relative angle between the cation tetrahedra and anion tetrahedra. A 3*d* metal like Fe may also be used to construct superhalogens, such as $FeCl_4$. Their corresponding supersalts like PH_4FeCl_4 and PH_4FeBr_4 turn out to possess similar dZB ground state structures, and their formations are all energetically favorable with positive energy release. There are a few exceptions like NH_4BF_4 and NH_4AlCl_4, which prefer to form structures where each superalkali cation is hydrogen bonded to eight adjacent superhalogen anions and vice versa, while NH_4AlF_4 prefers to form a layered perovskite structure.

The polar dZB structures in Figure 9.4b may be utilized as FE materials as long as their polarizations are switchable. In previous theoretical reports [86, 87], the calculated switching barriers for ZB structures like CuCl and ZnO are over 0.30 eV/f.u., which make their polarizations difficult to reverse at ambient conditions. Compared with the breaking of covalent bonds in these ZB structures during polarization switching, the energy barrier for ions to migrate in highly ionic supersalt systems should be much lower. Taking NH_4BCl_4 as an example, if all the NH_4 cations at the initial ground state are simultaneously displaced by 2.63 Å along −z direction, as shown in Figure 9.5a, with the rotation of all superatoms by a certain angle along −z-axis, we can get an equivalent structure that is energetically degenerate with a reversed polarization. The reversed polarization turns out to be of considerable value, around 14 $\mu C/cm^2$. As revealed by its band structure, the system is insulating with a large bandgap of over 5 eV, which should indicate a high dielectric breakdown voltage. The FE switching pathway calculated by using the solid-state NEB method indicates a switchable barrier of only 0.13 eV/f.u. (0.013 eV/atom), which is impressive for such a high ion displacement. For traditional ferroelectrics such as $PbTiO_3$, the ion displacement during FE switching is much smaller than 1 Å while the switching barrier is even larger (~0.15 eV/f.u., i.e. 0.03 eV/atom) compared with NH_4BCl_4 [88]. By performing BOMD simulations for NH_4BCl_4 at 300 K, the robustness of its ferroelectricity at ambient conditions can be verified. Such mechanism of FE switching can be applied to other supersalts like PH_4BBr_4 via the displacement of superalkali cations and superhalogen anions (2.86 Å along −y direction), where the switching barrier is also around 0.13 eV/f.u. (i.e. 0.013 eV/atom).

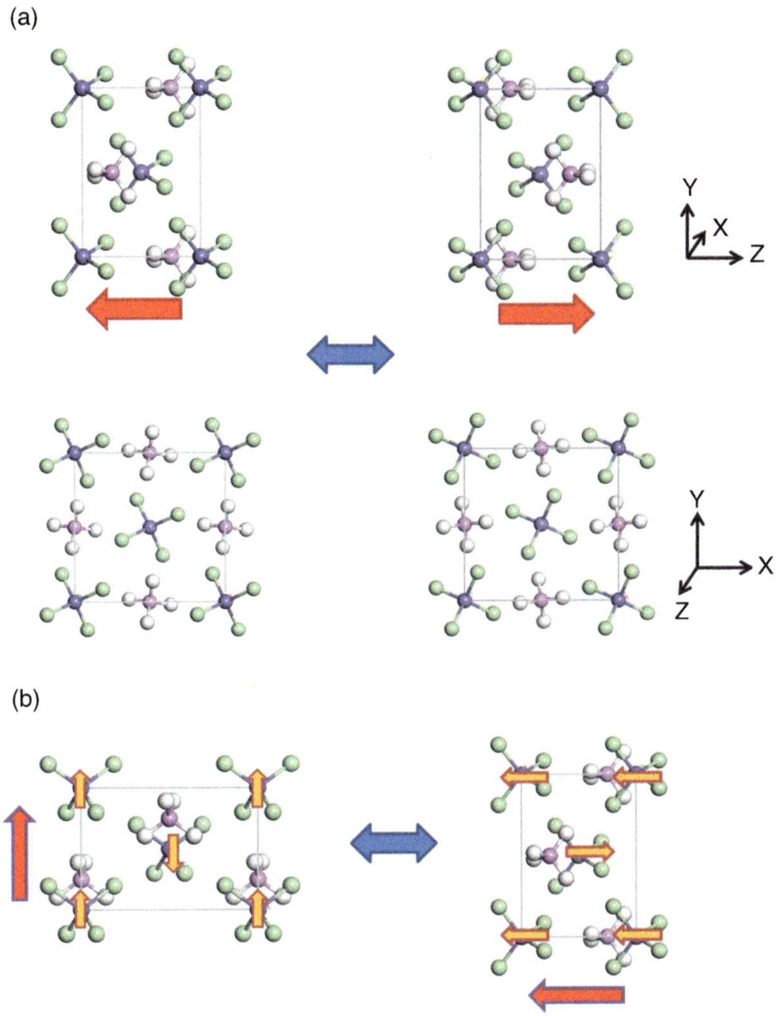

Figure 9.5 (a) FE switching of NH_4MX_4/PH_4MX_4. (b) Triferroic switching of PH_4FeX_4 [83], where the directions of both spin and electrical polarization are altered upon ferroelastic switching. Yellow and red arrows denote spin and electrical polarization directions, respectively. *Source:* (b) Gao et al. [83]. © John Wiley & Sons.

It is noteworthy that compared with standard ZB phase, the dZB structure in Figure 9.4 is relatively prolonged in two directions and contracted in the third direction. Such distortion may give rise to ferroelasticity, with a spontaneous strain that can be reversed via stress. During such ferroelastic pathway switching for NH_4BCl_4, a symmetrical nonpolar state appears as an intermediate state as its polarization is switched from −z direction at the initial state to −x/−y direction at the final state. The switching barrier of such ionic ferroelasticity with long-range Coulomb interaction between ions is unimaginable, which is as low as 0.065 eV/f.u. (i.e. 0.0065 eV/atom) for reversing such a high ferroelastic strain (defined as $(|b|/|c| - 1) \times 100\%$) more than 40%. In previous reports [89–91], the calculated switching barriers (up to 0.5 eV/atom) for high-strain (>20%) ferroelasticity were almost orders of magnitude higher, which can be attributed to the distinct features of covalent bonds in those systems. Meanwhile, the electric polarization is also reversed by 90° upon ferroelastic switching. Such 90° FE switching can also be achieved by applying an external electric field along −x or −y direction. As a result, the ferroelectricity and ferroelasticity are strongly coupled, making it possible to use strain to control the polarization direction as well as electric field to control strain. Due to the low barrier, a much smaller electric field or stress is required for reversing such a large strain.

Similar coupling between ferroelectricity and ferroelasticity can be applied to supersalts containing 3*d* magnetic ions like PH_4FeBr_4. It would be interesting to investigate whether their magnetism can also be coupled or not. After checking different possible spin configurations, the obtained ground state for PH_4FeBr_4 is antiferromagnetic as displayed in Figure 9.5b [83], where the spins of two Fe ions in a unit cell are along opposite directions, while the spins on the same lines along −x, −y, or −z direction are along the same direction. Calculations involving noncollinear spins show that the magnetic anisotropy energy is around 0.4 meV for the spin of each Fe atom, where the spins are aligned along −y direction (the "contracted" direction). However, upon ferroelastic switching, i.e. 90° FE switching, as the "contracted" direction is switched to −x direction, the magnetization easy axis is also switched from −y to −x direction. Equivalently, it might also be switched to −z direction via a strain or electric field along −z direction. The switching barrier of PH_4FeBr_4 is larger (~0.13 eV/f.u.) compared with NH_4BCl_4, with its bandgap reduced to only ~1 eV. This is within the desirable range for nanoelectronics. As a result, the magnetism, ferroelectricity, and ferroelasticity in PH_4FeBr_4 are all coupled, so the system is a multiferroic material with "triferroic" coupling. Electrical control of giant anisotropic magnetoresistance for nonvolatile memories may be realized by utilizing such triferroic coupling, which is long sought but still challenging in the field of antiferromagnetic spintronics [92].

9.5 Conclusion

In summary, we reviewed some of the recent theoretical designs of superatom-based ferroelectrics, including organic ferroelectrics, HOIPs, and supersalt ferroelectrics. In these reports, the anisotropic features of superatoms give rise to ferroelectricity, where the higher EAs of superhalogens and lower IP of superalkalis as well as their larger sizes compared with single atoms can greatly enhance the polarization. The theory-driven design of a new class of FE materials based on superatoms may stimulate experimental efforts in this field.

References

1 Dawber, M., Rabe, K.M., and Scott, J.F. (2005). Physics of thin-film ferroelectric oxides. *Reviews of Modern Physics* 77 (4): 1083–1130.
2 Hong, S., Auciello, O., and Wouters, D. (2014). *Emerging Non-Volatile Memories*, 103–166. Springer US.

3 Junquera, J. and Ghosez, P. (2003). Critical thickness for ferroelectricity in perovskite ultrathin films. *Nature* 422 (6931): 506–509.
4 Fong, D.D., Stephenson, G.B., Streiffer, S.K. et al. (2004). Ferroelectricity in ultrathin perovskite films. *Science* 304 (5677): 1650–1653.
5 Jena, P. (2013). Beyond the periodic table of elements: the role of superatoms. *Journal of Physical Chemistry Letters* 4 (9): 1432–1442.
6 Gutsev, G. and Boldyrev, A. (1981). DVM-Xα calculations on the ionization potentials of $MX_{k+1}-$ complex anions and the electron affinities of MX_{k+1} "superhalogens". *Chemical Physics* 56 (3): 277–283.
7 Gutsev, G. and Boldyrev, A. (1985). The theoretical investigation of the electron affinity of chemical compounds. *Advances in Chemical Physics* 61: 169–221.
8 Wang, X.-B., Ding, C.-F., Wang, L.-S. et al. (1999). First experimental photoelectron spectra of superhalogens and their theoretical interpretations. *The Journal of Chemical Physics* 110 (10): 4763–4771.
9 Elliott, B.M., Koyle, E., Boldyrev, A.I. et al. (2005). MX_3-superhalogens (M= Be, Mg, Ca; X= Cl, Br): a photoelectron spectroscopic and ab initio theoretical study. *The Journal of Physical Chemistry A* 109 (50): 11560–11567.
10 Rehm, E., Boldyrev, A.I., and Schleyer, P.v.R. (1992). Ab initio study of superalkalis. First ionization potentials and thermodynamic stability. *Inorganic Chemistry* 31 (23): 4834–4842.
11 Zakrzewski, V.G., von Niessen, W., Boldyrev, A.I., and von Ragué Schleyer, P. (1993). Green function calculation of ionization energies of hypermetallic molecules. *Chemical Physics* 174 (2): 167–176.
12 Gutsev, G.L. and Boldyrev, A. (1982). DVM Xα calculations on the electronic structure of "superalkali" cations. *Chemical Physics Letters* 92 (3): 262–266.
13 Alexandrova, A.N. and Boldyrev, A.I. (2003). σ-Aromaticity and σ-antiaromaticity in alkali metal and alkaline earth metal small clusters. *The Journal of Physical Chemistry A* 107 (4): 554–560.
14 Kealy, T. and Pauson, P. (1951). A new type of organo-iron compound. *Nature* 168 (4285): 1039–1040.
15 Hedberg, L. and Hedberg, K. (1970). Molecular structure of dicyclopentadienylnickel (C5H5) 2Ni. *The Journal of Chemical Physics* 53 (3): 1228–1234.
16 Gard, E., Haaland, A., Novak, D.P., and Seip, R. (1975). The molecular structures of dicyclo pentadienylvanadium,(C5H5) 2V, and dicyclopentadienylchromium,(C5H5) 2Cr, determined by gas phase electron diffraction. *Journal of Organometallic Chemistry* 88 (2): 181–189.
17 Almenningen, A., Gard, E., Haaland, A., and Brunvoll, J. (1976). Dynamic Jahn–Teller effect and average structure of dicyclopentadienylcobalt,(C5H5) 2Co, studied by gas phase electron diffraction. *Journal of Organometallic Chemistry* 107 (2): 273–279.
18 Kandalam, A.K., Rao, B., Jena, P., and Pandey, R. (2004). Geometry and electronic structure of V n (Bz) m complexes. *The Journal of Chemical Physics* 120 (22): 10414–10422.
19 Koleini, M., Paulsson, M., and Brandbyge, M. (2007). Efficient organometallic spin filter between single-wall carbon nanotube or graphene electrodes. *Physical Review Letters* 98 (19): 197202.
20 Wu, M. and Zeng, X.C. (2011). Transition-metal-molecular sandwich nanowires as magnetic on/off switch. *Applied Physics Letters* 99 (5): 053121.

21 Xiang, H., Yang, J., Hou, J.G., and Zhu, Q. (2006). One-dimensional transition metal–benzene sandwich polymers: possible ideal conductors for spin transport. *Journal of the American Chemical Society* 128 (7): 2310–2314.

22 Wu, M., Burton, J.D., Tsymbal, E.Y. et al. (2012). Multiferroic materials based on organic transition-metal molecular nanowires. *Journal of the American Chemical Society* 134 (35): 14423–14429.

23 Wu, M., Burton, J.D., Tsymbal, E.Y. et al. (2013). Hydroxyl-decorated graphene systems as candidates for organic metal-free ferroelectrics, multiferroics, and high-performance proton battery cathode materials. *Physical Review B* 87 (8): 081406.

24 Yang, Q., Zhong, T., Tu, Z. et al. (2019). Design of single-molecule multiferroics for efficient ultrahigh-density nonvolatile memories. *Advanced Science (Weinheim, Baden-Wurttemberg, Germany)* 6 (1): 1801572.

25 Zhong, T. and Wu, M. (2020). Fullerene-based 0D ferroelectrics/multiferroics for ultrahigh-density and ultrafast nonvolatile memories. *Physical Chemistry Chemical Physics* 22 (21): 12039–12043.

26 Heath, J.R., O'Brien, S.C., Zhang, Q. et al. (1985). Lanthanum complexes of spheroidal carbon shells. *Journal of the American Chemical Society* 107 (25): 7779–7780.

27 Aoyagi, S., Nishibori, E., Sawa, H. et al. (2010). A layered ionic crystal of polar Li@ C 60 superatoms. *Nature Chemistry* 2 (8): 678.

28 Chandler, H.J., Stefanou, M., Campbell, E.E.B., and Schaub, R. (2019). Li@C60 as a multi-state molecular switch. *Nature Communications* 10 (1): 2283.

29 Ogawa, T., Sugai, T., and Shinohara, H. (2000). Isolation and characterization of Er@ C60. *Journal of the American Chemical Society* 122 (14): 3538–3539.

30 Kanbara, T., Kubozono, Y., Takabayashi, Y. et al. (2001). Dy@C60: evidence for endohedral structure and electron transfer. *Physical Review B* 64 (11): 113403.

31 Kubozono, Y., Ohta, T., Hayashibara, T. et al. (1995). Preparation and extraction of Ca@ C60. *Chemistry Letters* 24 (6): 457–458.

32 Kubozono, Y., Maeda, H., Takabayashi, Y. et al. (1996). Extractions of Y@ C60, Ba@ C60, La@ C60, Ce@ C60, Pr@ C60, Nd@ C60, and Gd@ C60 with Aniline. *Journal of the American Chemical Society* 118 (29): 6998–6999.

33 Ding, Y., Chen, C.-C., Zeng, Q. et al. (2014). Novel high-pressure monoclinic metallic phase of V2O3. *Physical Review Letters* 112 (5): 056401.

34 Guan, J. and Tomanek, D. (2017). Can CF3-functionalized La@C60 be isolated experimentally and become superconducting? *Nano Letters* 17 (6): 3402–3408.

35 Cao, B., Zhou, X., Shi, Z. et al. (1998). Synthesis of C60-NBN and C70-NBN(N=L, 2) by DC arc burning method. *Fullerene Science and Technology* 6 (4): 639–648.

36 Yu, R., Zhan, M., Cheng, D. et al. (1995). Simultaneous synthesis of carbon nanotubes and nitrogen-doped fullerenes in nitrogen atmosphere. *The Journal of Physical Chemistry* 99 (7): 1818–1819.

37 Zou, Y., Chen, G., and Yan, H. (2000). Formation and infrared studies of boron-doped C60. *Journal of Physics and Chemistry of Solids* 61 (7): 1051–1054.

38 Muhr, H.-J., Nesper, R., Schnyder, B., and Kötz, R. (1996). The boron heterofullerenes C59B and C69B: generation, extraction, mass spectrometric and XPS characterization. *Chemical Physics Letters* 249 (5–6): 399–405.

39 Wang, Y., Shi, Y., Fan, X. et al. (2019). Encapsulation of an ionic bond in fullerenes: what is the difference? *Inorganic Chemistry* 58 (6): 3601–3605.

40 Jaros, A., Bonab, E.F., Straka, M., and Foroutan-Nejad, C. (2019). Fullerene-based switching molecular diodes controlled by oriented external electric fields. *Journal of the American Chemical Society* 141 (50): 19644–19654.

41 Aoki, T., Hiranaga, Y., and Cho, Y. (2016). High-density ferroelectric recording using a hard disk drive-type data storage system. *Journal of Applied Physics* 119 (18): 184101.

42 Kim, W.-H., Yoon, S.M., and Son, J.Y. (2014). Ferroelectric domain wall motion in epitaxial PbTiO3 and BiFeO3 thin films. *Materials Letters* 124: 47–49.

43 Chen, X.B., Yan, F., Li, C.H. et al. (2000). Switching properties of SrBi2Ta2O9 thin films produced by metalorganic decomposition. *Applied Physics Letters* 76 (3): 369–371.

44 Burschka, J., Pellet, N., Moon, S.-J. et al. (2013). Sequential deposition as a route to high-performance perovskite-sensitized solar cells. *Nature* 499 (7458): 316.

45 Yang, W.S., Park, B.-W., Jung, E.H. et al. (2017). Iodide management in formamidinium-lead-halide–based perovskite layers for efficient solar cells. *Science* 356 (6345): 1376–1379.

46 Kojima, A., Teshima, K., Shirai, Y., and Miyasaka, T. (2009). Organometal halide perovskites as visible-light sensitizers for photovoltaic cells. *Journal of the American Chemical Society* 131 (17): 6050–6051.

47 Etgar, L., Gao, P., Xue, Z. et al. (2012). Mesoscopic CH3NH3PbI3/TiO2 heterojunction solar cells. *Journal of the American Chemical Society* 134 (42): 17396–17399.

48 Green, M.A., Ho-Baillie, A., and Snaith, H.J. (2014). The emergence of perovskite solar cells. *Nature Photonics* 8 (7): 506.

49 Yablonovitch, E. (2016). Lead halides join the top optoelectronic league. *Science* 351 (6280): 1401–1401.

50 Eperon, G.E., Stranks, S.D., Menelaou, C. et al. (2014). Formamidinium lead trihalide: a broadly tunable perovskite for efficient planar heterojunction solar cells. *Energy & Environmental Science* 7 (3): 982–988.

51 Noel, N.K., Stranks, S.D., Abate, A. et al. (2014). Lead-free organic–inorganic tin halide perovskites for photovoltaic applications. *Energy & Environmental Science* 7 (9): 3061–3068.

52 Kulkarni, S.A., Baikie, T., Boix, P.P. et al. (2014). Band-gap tuning of lead halide perovskites using a sequential deposition process. *Journal of Materials Chemistry A* 2 (24): 9221–9225.

53 Lee, M.M., Teuscher, J., Miyasaka, T. et al. (2012). Efficient hybrid solar cells based on meso-superstructured organometal halide perovskites. *Science* 338 (6107): 643–647.

54 Weber, D. (1978). CH3NH3PbX3, ein Pb (II)-system mit kubischer perowskitstruktur/ CH3NH3PbX3, a Pb (II)-system with cubic perovskite structure. *Zeitschrift für Naturforschung B* 33 (12): 1443–1445.

55 Stroppa, A., Quarti, C., De Angelis, F., and Picozzi, S. (2015). Ferroelectric polarization of CH3NH3PbI3: a detailed study based on density functional theory and symmetry mode analysis. *Journal of Physical Chemistry Letters* 6 (12): 2223–2231.

56 Wang, F., Yu, H., Xu, H., and Zhao, N. (2015). HPbI3: a new precursor compound for highly efficient solution-processed perovskite solar cells. *Advanced Functional Materials* 25 (7): 1120–1126.

57 Pang, S., Zhou, Y., Wang, Z. et al. (2016). Transformative evolution of organolead triiodide perovskite thin films from strong room-temperature solid–gas interaction between HPbI3-CH3NH2 precursor pair. *Journal of the American Chemical Society* 138 (3): 750–753.

58 Ke, W., Spanopoulos, I., Stoumpos, C.C., and Kanatzidis, M.G. (2018). Myths and reality of HPbI3 in halide perovskite solar cells. *Nature Communications* 9 (1): 4785.

59 Daub, M. and Hillebrecht, H. (2018). On the demystification of "HPbI3" and the peculiarities of the non-innocent solvents H2O and DMF. *Zeitschrift für Anorganische und Allgemeine Chemie* 644 (22): 1393–1400.

60 Jain, A., Ong, S.P., Hautier, G. et al. (2013). Commentary: the materials project: a materials genome approach to accelerating materials innovation. *APL Materials* 1 (1): 011002.

61 Chang, K., Kuster, F., Miller, B.J. et al. (2020). Microscopic manipulation of ferroelectric domains in SnSe monolayers at room temperature. *Nano Letters* 20 (9): 6590–6597.

62 Poglitsch, A. and Weber, D. (1987). Dynamic disorder in methylammoniumtrihalogenoplumbates (II) observed by millimeter-wave spectroscopy. *The Journal of Chemical Physics* 87 (11): 6373–6378.

63 Ju, M.G., Dai, J., Ma, L., and Zeng, X.C. (2017). Lead-free mixed tin and germanium perovskites for photovoltaic application. *Journal of the American Chemical Society* 139 (23): 8038–8043.

64 Becker, M. and Wark, M. (2018). Organic cation substitution in hybrid perovskite CH3NH3PbI3 with hydroxylammonium (NH3OH+): a first-principles study. *The Journal of Physical Chemistry C* 122 (6): 3548–3557.

65 Giorgi, G., Fujisawa, J.-I., Segawa, H., and Yamashita, K. (2015). Organic–inorganic hybrid lead iodide perovskite featuring zero dipole moment guanidinium cations: a theoretical analysis. *The Journal of Physical Chemistry C* 119 (9): 4694–4701.

66 Li, L., Wu, M.H., and Zeng, X.C. (2019). Facile and versatile functionalization of two-dimensional carbon nitrides by design: magnetism/multiferroicity, valleytronics, and photovoltaics. *Advanced Functional Materials* 29 (43): 1905752.

67 Tu, Z. and Wu, M. (2019). 2D diluted multiferroic semiconductors upon intercalation. *Advanced Electronic Materials* 5 (6): 1800960.

68 Wu, M. and Zeng, X.C. (2017). Bismuth oxychalcogenides: a new class of ferroelectric/ferroelastic materials with ultra high mobility. *Nano Letters* 17 (10): 6309–6314.

69 Wu, M., Dong, S., Yao, K. et al. (2016). Ferroelectricity in covalently functionalized two-dimensional materials: integration of high-mobility semiconductors and nonvolatile memory. *Nano Letters* 16 (11): 7309–7315.

70 Giri, S., Behera, S., and Jena, P. (2014). Superalkalis and superhalogens as building blocks of supersalts. *The Journal of Physical Chemistry A* 118 (3): 638–645.

71 Srivastava, A.K. and Misra, N. (2014). Novel (Li2X)+(LiX2)− supersalts (X= F, Cl) with aromaticity: a journey towards the design of a new class of salts. *Molecular Physics* 112 (19): 2621–2626.

72 Fang, H. and Jena, P. (2016). Super-ion inspired colorful hybrid perovskite solar cells. *Journal of Materials Chemistry A* 4 (13): 4728–4737.

73 Zhou, T., Zhang, Y., Wang, M. et al. (2019). Tunable electronic structures and high efficiency obtained by introducing superalkali and superhalogen into AMX3-type perovskites. *Journal of Power Sources* 429: 120–126.

74 Fang, H. and Jena, P. (2017). Li-rich antiperovskite superionic conductors based on cluster ions. *Proceedings of the National Academy of Sciences* 114 (42): 11046–11051.

75 Lee, C.-H., Liu, L., Bejger, C. et al. (2014). Ferromagnetic ordering in superatomic solids. *Journal of the American Chemical Society* 136 (48): 16926–16931.

76 Ong, W.-L., O'Brien, E.S., Dougherty, P.S. et al. (2017). Orientational order controls crystalline and amorphous thermal transport in superatomic crystals. *Nature Materials* 16 (1): 83.

77 Pinkard, A., Champsaur, A.M., and Roy, X. (2018). Molecular clusters: nanoscale building blocks for solid-state materials. *Accounts of Chemical Research* 51 (4): 919–929.

78 Turkiewicz, A., Paley, D.W., Besara, T. et al. (2014). Assembling hierarchical cluster solids with atomic precision. *Journal of the American Chemical Society* 136 (45): 15873–15876.

79 Zhong, X., Lee, K., Choi, B. et al. (2018). Superatomic two-dimensional semiconductor. *Nano Letters* 18 (2): 1483–1488.

80 O'Brien, E.S., Trinh, M.T., Kann, R.L. et al. (2017). Single-crystal-to-single-crystal intercalation of a low-bandgap superatomic crystal. *Nature Chemistry* 9 (12): 1170.

81 Phillips, J.C. (1970). Ionicity of the chemical bond in crystals. *Reviews of Modern Physics* 42 (3): 317–356.

82 Cohen, R.E. (1992). Origin of ferroelectricity in perovskite oxides. *Nature* 358 (6382): 136–138.

83 Gao, Y.X., Wu, M., and Jena, P. (2021). A family of ionic supersalts with covalent-like directionality and unconventional multiferroicity. *Nature Communications* 12: 1331.

84 Wang, Y., Lv, J., Zhu, L., and Ma, Y. (2012). CALYPSO: a method for crystal structure prediction. *Computer Physics Communications* 183 (10): 2063–2070.

85 Wang, Y., Lv, J., Zhu, L., and Ma, Y. (2010). Crystal structure prediction via particle-swarm optimization. *Physical Review B* 82 (9): 094116.

86 Gao, Y., Wu, M., and Zeng, X.C. (2019). Phase transitions and ferroelasticity–multiferroicity in bulk and two-dimensional silver and copper monohalides. *Nanoscale Horizons* 4 (5): 1106–1112.

87 Li, L. and Wu, M. (2017). Binary compound bilayer and multilayer with vertical polarizations: two-dimensional ferroelectrics, multiferroics, and nanogenerators. *ACS Nano* 11 (6): 6382–6388.

88 Tadmor, E.B., Waghmare, U.V., Smith, G.S., and Kaxiras, E. (2002). Polarization switching in PbTiO3: an ab initio finite element simulation. *Acta Materialia* 50 (11): 2989–3002.

89 Wu, M., Fu, H., Zhou, L. et al. (2015). Nine new phosphorene polymorphs with non-honeycomb structures: a much extended family. *Nano Letters* 15 (5): 3557–3562.

90 Li, W. and Li, J. (2016). Ferroelasticity and domain physics in two-dimensional transition metal dichalcogenide monolayers. *Nature Communications* 7: 10843.

91 Zhang, C., Nie, Y., Sanvito, S., and Du, A. (2019). First-principles prediction of a room-temperature ferromagnetic janus VSSe monolayer with piezoelectricity, ferroelasticity, and large valley polarization. *Nano Letters* 19 (2): 1366–1370.

92 Wadley, P., Howells, B., Železný, J. et al. (2016). Electrical switching of an antiferromagnet. *Science* 351 (6273): 587–590.

10

Cluster-based Materials for Energy Harvesting and Storage

Puru Jena[1], Hong Fang[1], and Qiang Sun[2,3]

[1]*Department of Physics, Virginia Commonwealth University, Richmond, VA, USA*
[2]*School of Materials Science and Engineering, Peking University, Beijing, China*
[3]*Center for Applied Physics and Technology, Peking University, Beijing, China*

10.1 Introduction

Imagine that a superatomic cluster is so stable that it can retain its structure and properties when assembled. Such a cluster-assembled material (CAM) is expected to exhibit properties very different from those where individual atoms form the building blocks. This is because: (i) unlike conventional crystals, CAMs have two characteristic length scales, an intracluster distance between atoms that make up the superatomic cluster and intercluster distance corresponding to the large lattice constant. (ii) The band structure of CAMs results from the overlap of molecular orbitals formed by the superatoms, while the band structure of conventional crystals is determined by overlap of atomic orbitals. (iii) The phonon spectrum of CAMs is governed by both intracluster and intercluster vibrations. Consequently, the electron–phonon interaction in CAMs can exhibit features commonly not seen in crystals composed of individual atoms. If the superatom is chemically inert, the cohesive energy of the corresponding CAM is expected to be low. On the other hand, if the designed superatoms are chemically active, they are likely to form ionically or covalently bound CAMs. Since this idea was first introduced by Khanna and Jena in 1992 [1, 2], many studies have been carried out to explore these possibilities [3].

A classic example of a cluster-assembled crystal is fullerite [4], which is an face centered cubic (fcc) crystal composed of C_{60} carbon fullerenes [5]. In Figure 10.1a and b, we give the geometry of the C_{60} fullerene and the corresponding fullerite crystal. Note that the geometry of C_{60} remains intact in the fullerite. The lattice constant of fullerite under ambient conditions is 14.17 Å. The interaction between C_{60} fullerenes, which are 10.02 Å apart, is governed by weak van der Waals force.

The highest occupied molecular orbital (HOMO) and the lowest unoccupied molecular orbital (LUMO) of C_{60} are shown in Figure 10.2. C_{60} is insulating due to its filled fivefold degenerate HOMO and the empty threefold degenerate LUMO. The richness of its property can be demonstrated by the electron-doped variants, e.g. $C_{60}{}^{3-}$ and $C_{60}{}^{4-}$. For instance, in

Superatoms: Principles, Synthesis and Applications, First Edition. Edited by Purusottam (Puru) Jena and Qiang Sun.
© 2022 John Wiley & Sons Ltd. Published 2022 by John Wiley & Sons Ltd.

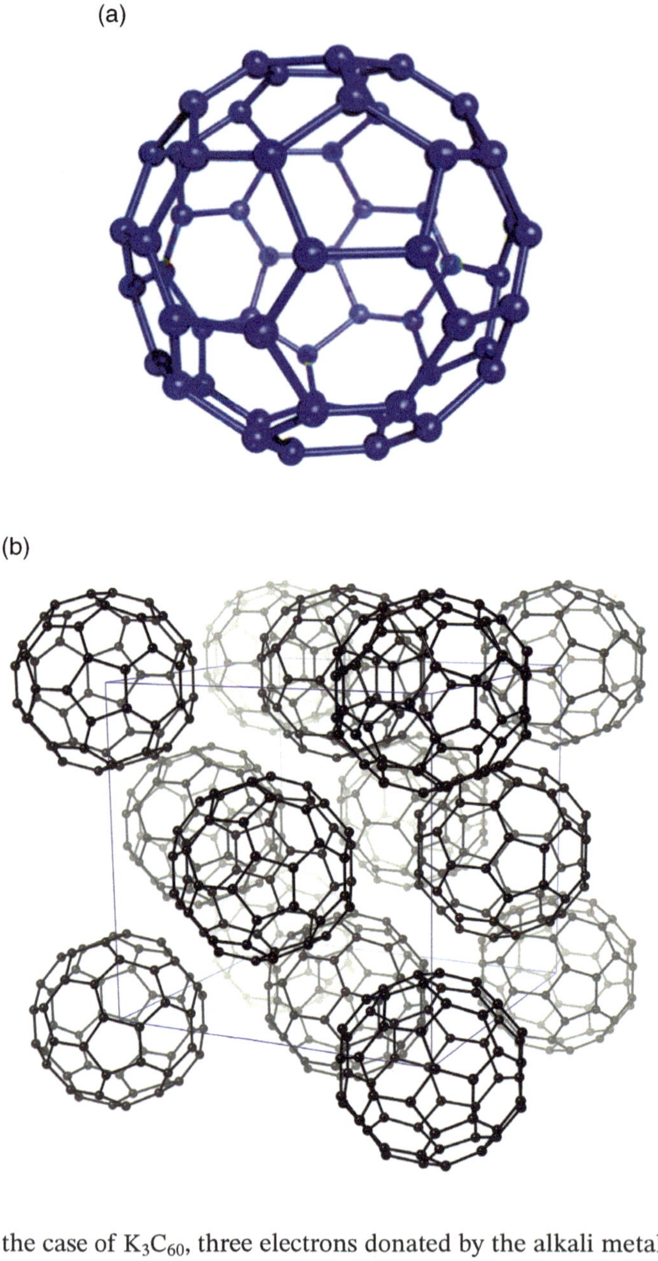

Figure 10.1 (a) Geometry of C_{60} fullerene. (b) Geometry of fcc-C_{60} solid. *Source:* Compounds formed from C_{60} and potassium: K_6C_{60} and K_3C_{60}. ChemTube3D. © The University of Liverpool.

the case of K_3C_{60}, three electrons donated by the alkali metal atoms transfer to the empty LUMO, making it half-filled (Figure 10.2). Therefore, K_3C_{60} is metallic. On the other hand, for the case of K_4C_{60}, where four electrons are transferred to the fullerene cage, there will be a spontaneous molecular distortion (Jahn–Teller effect), which lifts the degeneracy of LUMO. This leads to a set of lower levels that are fully occupied and a completely empty upper level. Therefore, K_4C_{60} becomes nonmetallic.

The band structure of the fullerite, which evolves as the molecular orbitals of C_{60} overlap, is given in Figure 10.3 using two different methods. The electronic structures of carbon

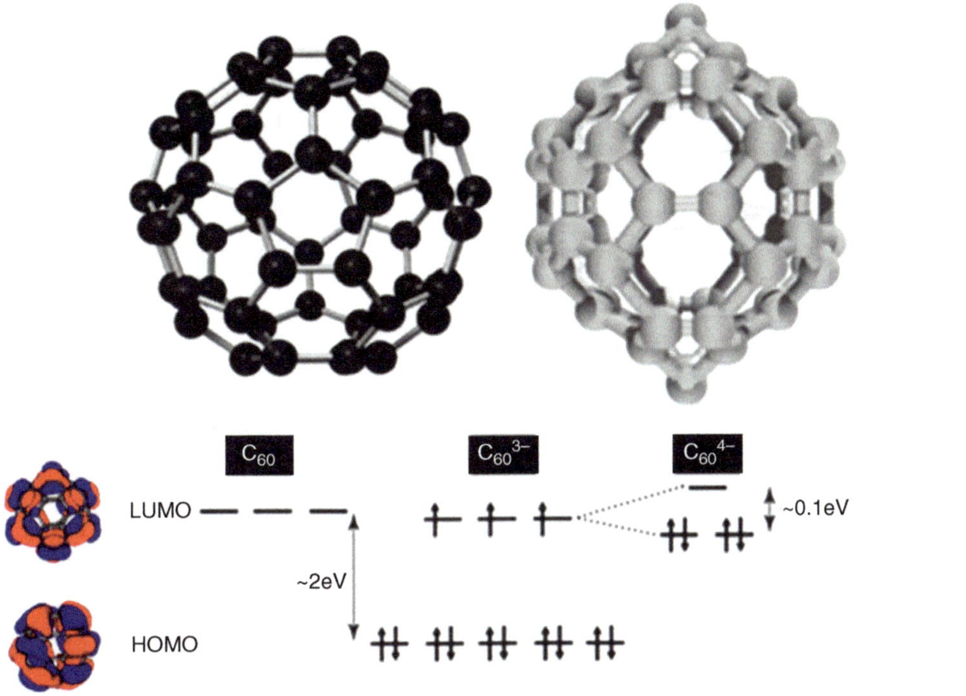

Figure 10.2 The HOMO and LUMO (c.a. 2 eV above) of neutral C_{60} (*Left*) and two of its electron-doped variants $C_{60}{}^{3-}$ and $C_{60}{}^{4-}$. The HOMO is fivefold degenerate filled with 10 electrons. The LUMO becomes half-filled for $C_{60}{}^{3-}$. $C_{60}{}^{4-}$ has a distorted structure (*Right*) due to Jahn–Teller effect, leading to the split of the LUMO into a twofold degenerate level that is fully occupied and an empty level about 0.1 eV above. *Source:* Adapted from O'Shea [6]. © John Wiley & Sons.

materials are often determined by the property of their cluster/molecular/atomic building blocks. Assembled by C_{60} molecules via weak dispersive interactions, crystalline C_{60} is insulating (Figure 10.3) as anticipated from the above-mentioned nature of the molecular orbitals of the fullerene cage. By comparison, zero energy gaps appear at certain special wave vectors in the band structure of graphite, since its building block is the conducting graphene layer. On the other hand, diamond is an insulator due to its atomic building block of sp^3 carbon with localized bonding.

Materials based on C_{60} can also be synthesized by noting that the electron affinity (EA) of C_{60} is 2.68 eV. Thus, C_{60} can act as an electron acceptor and form a cluster solid by interacting with electron-donating species. A_3C_{60} (A = K, Rb) is such a class of crystals, which form fcc structure while Cs_3C_{60} exhibits both fcc and body centered cubic (bcc) structure. Unlike intercalated graphite, when fullerites are doped with alkali atoms (K, Rb, Cs), the resulting fullerides display rich properties [8]. As can be seen from the phase diagram in Figure 10.4, fullerides with doped alkali atoms can be tuned to be metals, superconductors (SC), antiferromagnetic insulators (AFI), or Mott insulators [9]. A high transition temperature (T_c) s-wave superconducting state emerges next to a Mott insulating phase as a function of lattice spacing. The unconventional s-wave superconductivity results from the cooperation between the

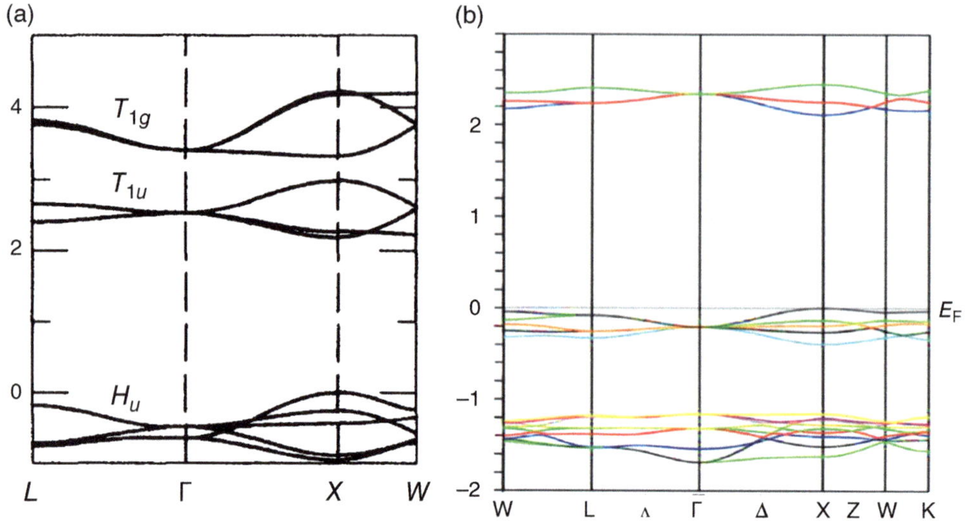

Figure 10.3 (a) Calculated band structures of fcc-C_{60} using (a) GW and (b) nonregular Tran-Blaha-modified Becke–Johnson potentials. *Source:* Jalali-Asadabadi et al. [4]. © Springer. Shirley and Louie [7]. © American Physical Society.

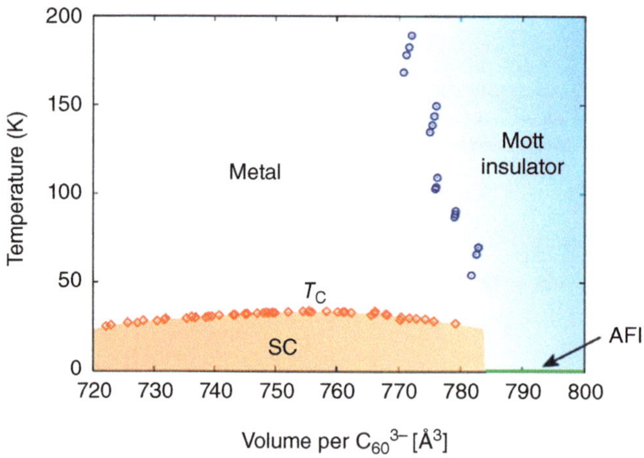

Figure 10.4 Experimental phase diagram for fcc A_3C_{60} (A = K, Rb, Cs). Copyright 2016 IOP Publishing. *Source:* Yusuke et al. [9]. Licenced under CC BY-3.0. © IOP Publishing.

electron–phonon coupling and the electron–electron interactions, which arise due to a competition between the Jahn–Teller coupling and the Hund's rule coupling within the C_{60} cluster. The above results illustrate that crystals composed of C_{60} have properties very different from those of graphite and diamond where individual carbon atoms form the building block. Other methods for synthesizing C_{60}-based materials include fullerene-rich dendrimers, fullerene hexa-adduct as building block, and protein-directed C_{60} assembly. Examples of synthesized fullerene-based materials are given in Table 10.1 [10].

Table 10.1 Unit cell parameters (a, c, Å), densities (ρ, g/cm^3), bulk moduli of elasticity (B, GPa), bandgap widths (E_g, eV), and bond energies for fullerene C_n molecules (E_{mol}, eV/atom) and their crystalline forms (E_{cr}, eV/atom) as compared to diamond.

Phase	a	c	ρ	B	E_g	E_{mol}	E_{cr}
bcc-fullerite C_{20}	6.897		2.43	201.87	3.51	8.01	8.07
hyperdiamond C_{28}	15.953		1.10	39.26	2.04	8.29	8.50
hyperlonsdaleite C_{28}	11.277	18.415	1.10	41.20	1.99	8.29	8.50
graphite-like C_{36}	11.725	8.462	1.43	64.03	1.85	8.23	8.61
fcc-fullerite C_{60}	~14.1		~1.7	<0.01	1.66	8.85	8.85
Diamond	0.3576		3.49	478.20	7.48		9.22

Source: Ivanovskii [10]. © Springer Nature.

Silicon, although isoelectronic with carbon, does not exhibit a fullerene-like structure. However, metal-encapsulated stable silicon clusters such as $M@Si_{12}$ (M = W, V), alkali-like $M@Si_{16}$ (M = V, Nb, Ta), halogen-like $M@Si_{16}$ (M = Sc, Y, Lu), and neutral magic $M@Si_{16}$ (M = Ti, Zr, Hf) clusters have been synthesized. The neutral magic clusters $M@Si_{16}$ (M = Ti, Zr, Hf) have Frank–Kasper cage structures and form weakly bound CAMs. Using a dry–wet combination synthesis technique Nakajima and coworkers [11] have recently synthesized $TiSi_{16}$ clusters on a large scale and confirmed its structure to be that of metal-encapsulated tetrahedral Si cage with Frank–Kasper structure – in agreement with earlier theoretical prediction (see Chapter 4). Strongly bound MSi_{16}-based CAMs can be formed by combing alkali-like $M@Si_{16}$ (M = V, Nb, Ta) clusters with F and C_{60}. Recently, experiments have shown that $TaSi_{16}$ combines with a single C_{60} molecule to form the superatomic charge transfer (CT) complex, $(TaSi_{16})^+C_{60}^-$, which remains stable up to 720 K. Scanning tunneling microscopy and spectroscopy experiments have demonstrated that $TaSi_{16}$ cations could be densely immobilized onto C_{60}-terminated surfaces, while retaining their cage shape and positive charge [12] (see Figure 10.5).

In addition to creating CAMs by depositing clusters on substrates, considerable amount of work is being done on ligated clusters and assembling these to form into bulk materials. These include $Co_6Se_8(PEt_3)_6$, $Cr_6Te_8(PEt_3)_6$, and $Ni_9Te_6(PEt_3)_8$ cluster-assembled ionic solids; ZnO-based clusters; ligated Al-, Cu-, Ag-, and Au-based CAMs; Zintl cluster-based CAMs; Chevrel-like cluster-based assemblies; as well as clusters confined inside nanotubes and porous materials. The reader is directed to Chapters 6 and 7, which are devoted to ligated clusters and assemblies. Note that CAMs synthesized by first isolating a cluster from the gas phase and then depositing these on surfaces or aggregating these after coating with appropriate ligands is not an easy process, and it is difficult to synthesize large quantities of these materials.

In this chapter we discuss a different class of materials we term cluster-based materials where lessons learnt from cluster science help to design materials where selected atoms/ions are replaced by superatoms/superions. Note that there are many compounds and complex crystals that can be viewed as cluster-based materials. Consider for example a well-known disinfectant, potassium permanganate $KMnO_4$. It is composed of K^+

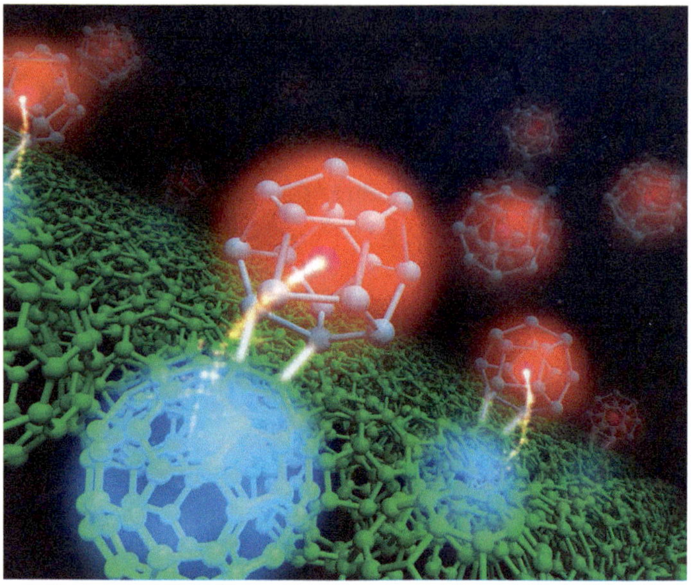

Figure 10.5 $TaSi_{16}$ superatom monolayer on C_{60}-terminated surface. *Source:* Adapted with permission from Nakaya et al. [12]. Licenced under CC BY-3.0. © Royal Society of Chemistry.

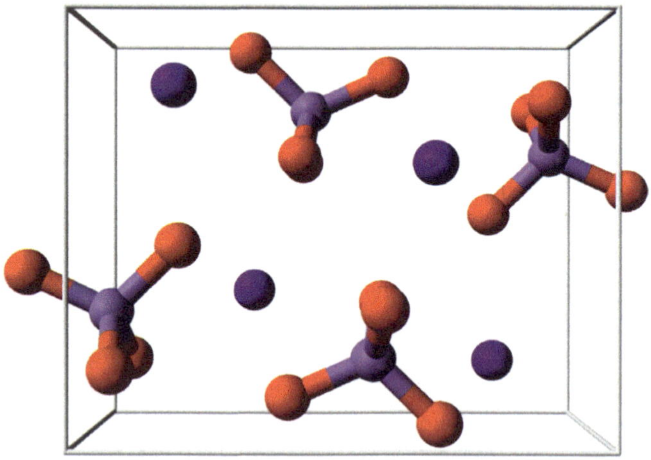

Figure 10.6 Ball and stick model of $K^+MnO_4^-$ crystal.

and MnO_4^- (see Figure 10.6) ions. MnO_4 can be regarded as a superhalogen cluster. With a valence electron configuration of $3d^54s^2$, Mn can donate at most seven electrons, while four oxygen atoms, with valence electron configuration of $2s^22p^4$, require eight electrons for octet shell closure. However, the extra electron is distributed over four oxygen atoms. As discussed in Chapter 2, the EA of MnO_4 is expected to be larger than that of a halogen atom, making it a superhalogen. Indeed, Gutsev et al. [13] were the first ones to show that a transition metal atom can be used to design and synthesize superhalogens. Its predicted EA of 4.96 eV agrees well with the experimental value of

4.80 ± 0.10 eV. There are numerous ionically bound compounds that can be viewed as cluster-based materials. Here, the cations can be replaced by superalkali clusters and/or the anions can be replaced by superhalogen clusters, thus forming cluster-based supersalts [14].

In the following, we show how knowledge gained from cluster science is used to rationally design materials for energy harvesting and storage as well as those capable of promoting unusual reactions. Note that solar cell materials and electrolytes in batteries are essentially salts composed of cations and anions. By replacing these ions with cluster ions of suitable size and composition, one can design new materials with improved properties [15]. We focus on the design of moisture-resistant hybrid perovskite solar cells for energy harvesting, halogen-free electrolytes in Li-ion batteries (LIBs) for energy storage, and metal-coated C_{60} fullerenes for hydrogen storage based on lessons learnt from cluster science. We also discuss how clusters can be used to promote unusual reactions such as enabling noble gas atoms like argon to form covalent bonds at room temperature (RT) and Zn to exist in +III oxidation state.

10.2 Cluster-based Materials for Moisture-resistant Hybrid Perovskite Solar Cells

This section deals with the role clusters can play in harvesting solar energy. While fossil fuels have been instrumental in vast economic and social development of the world, it has come at a cost. In addition to its limited supply, fossil fuels contribute to global warming and environmental pollution. Thus, it is a matter of time before we will face an energy and environmental crisis. To mitigate this problem, one must rely on clean and renewable energy sources such as the sun, wind, and hydro. Note that the amount of the solar energy that falls on Earth in a day can meet world's need for a year. However, this energy needs to be harvested and stored. Although technologies have been developed for solar cells for energy harvesting and batteries for energy storage, they have limitations in meeting the growing world demand.

We first discuss the challenges in solar cells. Although more than 50 years of research has led to the development of many solar cell materials, for commercial application, Si has emerged as the material of choice. However, as shown in Figure 10.7, the efficiency of Si solar cells has not improved much over the past 30 years [16]. On the other hand, because of their easy solution-based synthesis, excellent light emission properties, and tunable bandgaps from visible to infrared regions, organic–inorganic hybrid perovskites $A^+(MX_3)^-$, [$A = CH_3NH_3$, $HC(NH_2)_2$; M = Pb, Sn, Ge; X = Cl, Br, I], and in particular $A^+(PbI_3)^-$ have emerged as promising candidates for a new generation of solar cells. In the past 10 years, their efficiency has increased from about 3% to over 20%. However, there are practical problems associated with the stability of these materials when exposed to intense light and moisture. They degrade fast and Pb is toxic.

To understand the underlying reason for its affinity toward water, Fang and Jena [17] noted that the geometries of the organic cations, A^+ in the perovskite crystal (Figure 10.8) [18], are same as their structure in the gas phase. Indeed, CH_3NH_3, with an ionization potential of 4.13 eV, can be thought of as a superalkali cluster. Similarly, PbI_3

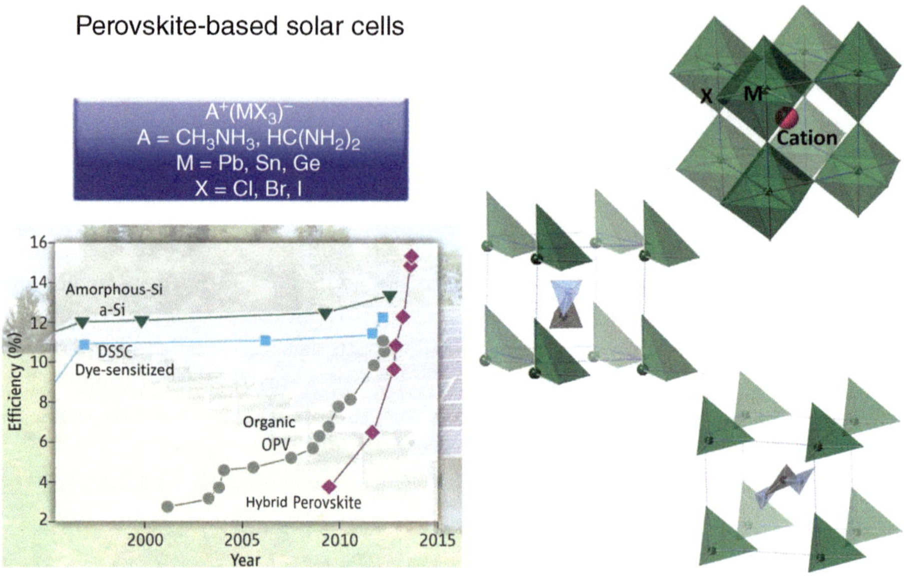

Figure 10.7 Efficiency of different solar cells. *Source:* Modified from Hodes [16]. © John Wiley & Sons.

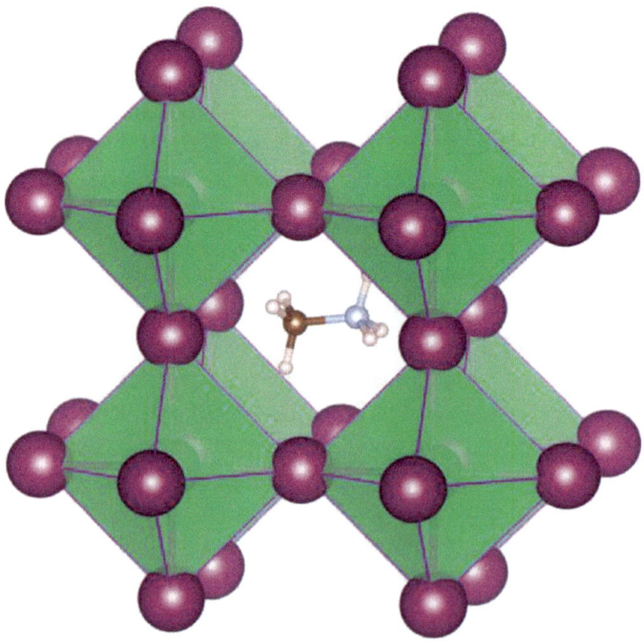

Figure 10.8 Crystal structure of $CH_3NH_3PbX_3$ perovskites (X = I, Br, and/or Cl). The methylammonium cation ($CH_3NH_3^+$) is surrounded by PbX_6 octahedra. *Source:* Eames et al. [18]. © Springer Nature.

can be thought of a superhalogen cluster, as its EA in the gas phase is 4.42 eV. Thus, the hybrid perovskites can be visualized as supersalts [14], where the alkali atom is replaced by organic cations ($CH_3NH_3^+$, $HC(NH_2)_2^+$) and the halogen anion is replaced by an inorganic anion PbI_3^-.

Fang and Jena [17] carried out a molecular dynamics study of the interaction between CH_3NH_3 molecule and PbI_3 cluster in the presence of a water molecule. The results are shown in Figure 10.9. Note that the water molecule enters into the bond between $CH_3NH_3^+$ and PbI_3^- with the H^+ pointing toward I^-. This results in the formation of HI. Similarly, O^- attracts H^+ from $CH_3NH_3^+$, forming H_2O and CH_3NH_2. Both HI and CH_3NH_2 are volatile, leaving the system with PbI_2 and one less hybrid perovskite moiety. The process continues until all that is left is H_2O and PbI_2. Note that here H_2O acts like an undesirable catalyst. To overcome this problem, it is necessary to prevent the formation of HI. Realizing that some clusters have the same chemistry as the halogens, Fang and Jena replaced the iodine in PbI_3 with a cluster BH_4, which is a superhalogen with an EA of 4.50 eV. In addition, its size is similar to that of I. Thus, $Pb(BH_4)_3$, with an EA of 4.98 eV, can be regarded as a hyperhalogen [19]. In PbI_3, each I atom carries one extra electron, while in $Pb(BH_4)_3$, the extra electron is distributed over four H atoms, thus reducing electrostatic interaction between $H\delta^+$ and $H\delta^-$. Indeed, the hybrid

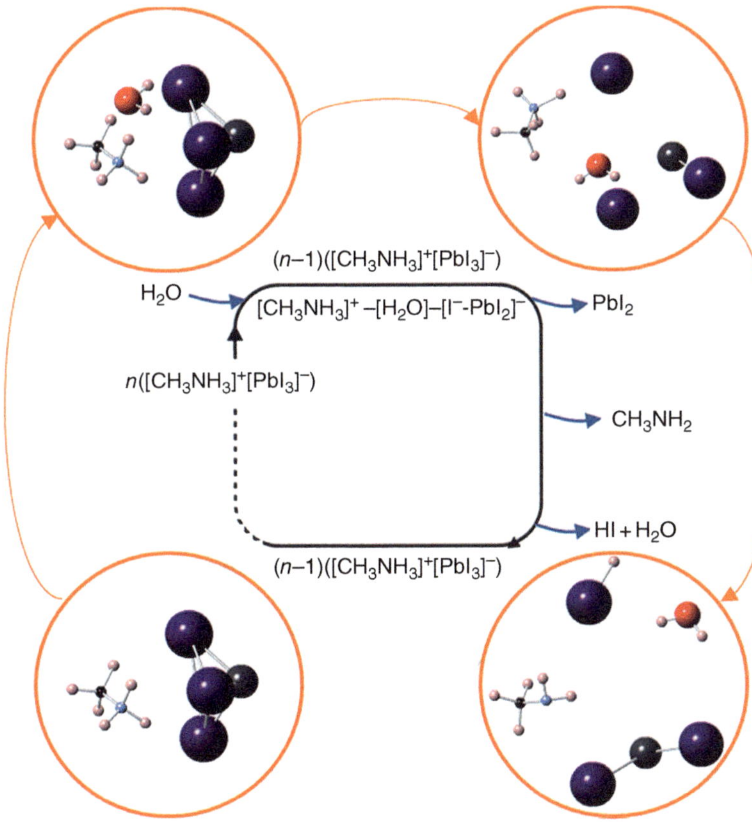

Figure 10.9 Degradation mechanism of $CH_3NH_3PbI_3$ exposed to moisture. The ball-and-stick plots (Pb, lead black; I, purple; N, cyan; C, black; O, red; H, pink) in the circles demonstrate the process in sequence as indicated by the arrows. It starts with the reactant as one water molecule is trapped between the superalkali $[CH_3NH_3]^+$ and superhalogen $[PbI_3]^-$ and ends up with products of PbI_2, CH_3NH_2, HI, and H_2O. *Source:* Fang and Jena [17]. © Royal Society of Chemistry.

perovskites consisting of hyperhalogens [Ge/Sn(BH_4)$_3$] show significantly reduced (about 20%) binding energy with the water molecule, as compared to the materials with superhalogens [Ge/SnX_3] (X being a halogen). It should be noted that Schouwink et al. [20] have successfully replaced halogens in perovskites with BH_4 and synthesized more than 30 new perovskite crystals.

It is important to realize that any replacement of I by other anion moieties should not adversely affect the efficiency of the solar cell. In this case, the bandgap of the solar cell material is an important parameter. Fang and Jena noted that the ionic radii and the bonding ionicity of the cations and anions control the bandgap. The ionic radii of the cations and the anions in hybrid perovskites are larger than the radii of the corresponding atomic ions. They showed that with the superions one can achieve lower ionicity and lower radius ratio than with the ordinary ions in the periodic table, and, hence, much smaller bandgaps for the superalkali halide crystals as compared to the normal alkali halides. In Figure 10.10, we plot the bandgaps as a function of the ionicity of hydride-based hybrid perovskites. The results show that not only the bandgaps can be further reduced with superion-based hybrid perovskites but also they can be tailored by choosing the size of the hyperhalogens. In a later paper [21], these authors showed that water-resistant hybrid perovskites can also be designed by replacing halogens by

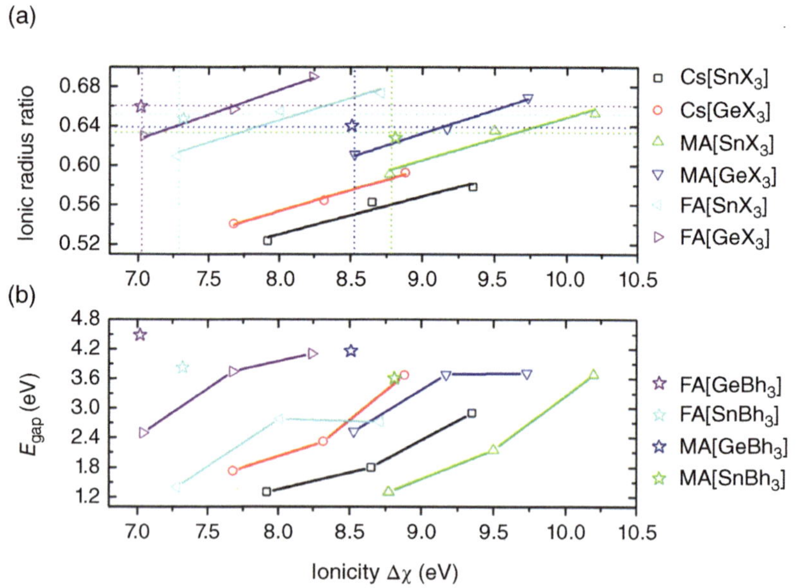

Figure 10.10 Ionic radius ratio and the bandgaps of hybrid perovskites against the bond ionicity. (a) Cation versus anion radii ratio, R^+/R^-, for the hybrid perovskites composed of super- and hyperhalogens. MA and FA stand for CH_3NH_3 and $HC(NH_2)_2$, respectively. (b) The bandgaps of hybrid perovskites obtained either from experiments or calculations using functional against ionicity. In each case with superhalogens, the data points correspond to X = Cl, Br, and I from right to left. "Bh" stands for (BH_4). The dotted lines show that the hypersalts show almost identical ionicity and radius ratio with their supersalt counterparts with X = I and Br, respectively. *Source:* Fang and Jena [17]. © Royal Society of Chemistry.

pseudohalogens such as SCN, SeCN, and PCO and that their electronic properties including the bandgap, the gap deformation potential, the exciton binding energy as well as the chemical stability of organic–inorganic perovskites may be traced back to their corresponding molecular motifs. This understanding helps in estimating the properties of bulk semiconductors from their building blocks.

10.3 Cluster-based Materials for Optoelectronic Devices

Superatomic clusters can also play a role in energy efficiency. Considerable amount of energy can be saved if incandescent light bulbs can be replaced by energy-efficient light-emitting diodes (LED). In the United States alone, use of LEDs can save about 75% of this energy. Recently, Dou et al. [22] have studied two-dimensional (2D) hybrid perovskite $(C_4H_9NH_3)_2PbBr_4$ for LEDs. In Figure 10.11 we show the structure of this material where the corner-sharing $[PbBr_6]^{4-}$ octahedral is capped by intervening layers of organic cations. The insulating long organic chains and semiconducting inorganic layer give rise to a quantum well (QW) superlattice. Due to the reduced dimensionality, the bandgap and the exciton binding energy are enhanced over their corresponding values in the three-dimensional (3D) bulk phase. In addition, the color of the emitted light can be readily tuned by changing the thickness and/or halogen composition of the 2D materials. For this reason, the 2D hybrid perovskites are preferred for LED and photoluminescent applications.

Yao et al. [23] showed that the bandgap and exciton binding energies of 2D hybrid perovskites can be tailored by sequentially replacing the halogens (I) with BH_4 superhalogens in $(C_4H_9NH_3)_2PbI_4$, namely, $(C_4H_9NH_3)_2MI_{4-x}(BH_4)_x$ [M = Pb, Sn, Ge]. In Figure 10.12 we show their results of relative bandgaps measured with respect to $x = 0$. Note that the bandgaps and exciton binding energies for Pb-, Sn-, and Ge-based hybrid 2D perovskites increase with the replacement of halogens with superhalogens, achieving the same purpose as lowering the dimensionality. Combining quantum confinement with a compositional change of hybrid perovskites can, therefore, produce materials with improved optoelectronic properties.

10.4 Cluster-based Materials for Solid-state Electrolytes in Li- and Na-ion Batteries

Because of the intermittent nature of solar and wind energy, development of safe, efficient, and cost-effective energy storage devices for large-scale applications is essential for addressing the impending energy crisis. LIBs, due to their light weight and high energy densities, have been the storage medium of choice for decades. However, there are major challenges with LIBs that need to be addressed to meet the growing demand for energy storage. First, rarity of Li combined with its uneven distribution in Earth's crust and high costs implies that it cannot single-handedly meet the demand of the future widespread commercialization of electric vehicles. Second, the lithium extraction process is very harmful to the

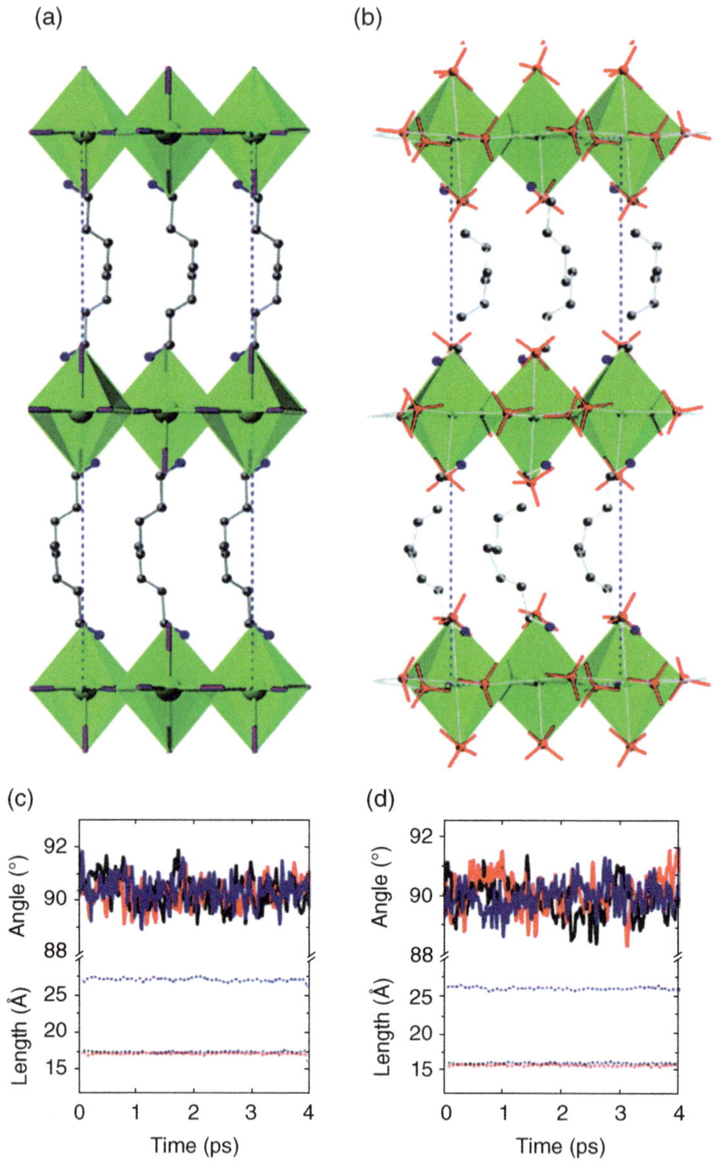

Figure 10.11 Crystal structures of (a) BA_2MI_4 and (b) $BA_2M(BH_4)_4$ (M = Ge, Sn, Pb). The gray and blue balls denote C and N atoms. $[MI_6]^{4-}$ are in green octahedra. BH_4^- tetrahedra are in red. I and H atoms are removed for clarity. Ab initio molecular dynamics simulations (AIMD)-simulated structure parameters of (c) $BA_2Sn(BH_4)_4$ and (d) $BA_2Ge(BH_4)_4$. Lattice angles α, β, and γ are in black, red, and blue solid lines, respectively. Lattice lengths *a*, *b*, and *c* of the 2 × 2 × 1 supercell are in black, red, and blue dotted lines, respectively. *Source:* Yao et al. [23]. © Royal Society of Chemistry.

environment and consumes an enormous amount of water per ton of lithium. There are also challenges with the current LIBs. For example, the electrolytes in LIBs contain halogens that are toxic and corrosive. The graphite-based anode has low specific capacity (372 mAh/g) and poor cycle life. In the following we discuss how use of cluster ions can mitigate some of these problems.

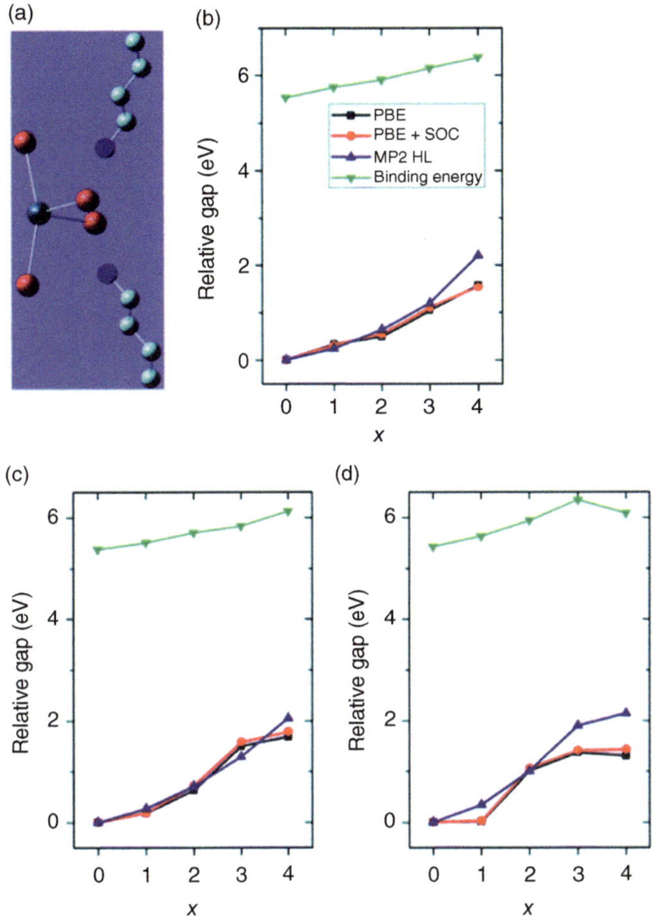

Figure 10.12 (a) Cluster model of 2D hybrid perovskite. The organic chains of $C_4H_9NH_3$ are in blue and gray stick-and-ball representation. The red balls denote the halogen atoms or BH_4 unit, surrounding the black metal atom (Pb, Sn, or Ge). Calculated relative gaps of $(C_4H_9NH_3)_2MI_{4-x}(BH_4)_x$ when M is (b) Pb, (c) Sn, and (d) Ge for both bulk materials and cluster models. The exciton binding energies of these admixtures are for the cluster model. *Source:* Yao et al. [23]. © Royal Society of Chemistry.

10.4.1 Halogen-free Electrolytes

Some of the electrolytes currently used in LIBs are $LiAsF_6$, $LiBF_4$, $LiPF_6$, $LiFePO_4$, $LiClO_4$, $LiN(SO_2F)_2$, and $LiN(SO_2CF_3)_2$. Most of these contain halogens and can be viewed as complex salts where Li is the cation and AsF_6, BF_4, PF_6, $FePO_4$, ClO_4, $N(SO_2F)_2$, and $N(SO_2CF_3)_2$ are the negative ions. A simple counting of valence electrons shows that all these anions lack one electron to form closed shells. Because these anions are larger in size than the halogens, as discussed in Chapter 2, the EAs of these clusters are expected to be larger than those of halogens, i.e. these are superhalogens. Indeed, the calculations by Giri et al. [24] revealed this to be true. In Table 10.2 we present the calculated vertical detachment energies (VDEs), i.e. the energies needed to remove an electron from these anions without changing their geometries. The VDEs range between 4.32 and 8.91 eV.

Table 10.2 Vertical detachment energies (VDEs), energies to dissociate the Li salt into individual ions (ΔE), and binding affinity (E_b) to H_2O of currently used and predicted electrolytes[a].

Negative ions	VDE in eV	ΔE in eV	E_b in eV
$[AsF_6]^-$	8.91	5.65	1.09
$[BF_4]^-$	7.66	6.08	1.41
$[PF_6]^-$	8.55	5.73	1.07
$[FePO_4]^-$	4.32	7.38	1.04
$[ClO_4]^-$	5.83	5.96	1.02
$[N(SO_2F)_2]^-$	6.89	5.82	1.02
$[N(SO_2CF_3)_2]^-$	7.01	6.01	0.99
NO_3^-	4.22	6.53	0.96
BH_4^-	4.50	6.62	0.92
$B_3^-H_8^-$	4.72	6.25	0.93
$CB_{11}^-H_{12}^-$	5.99	5.08	1.08

Note that BH_4, which does not contain a single halogen atom, is a superhalogen with EA of 4.50 eV. Thus, $LiBH_4$ can be regarded as a halogen-free electrolyte. Orimo and coworkers found that $LiBH_4$, having a hexagonal phase, is a superionic conductor at high temperature (~390 K) with ionic conductivity on par with or even in some instances better than that reported for other solid-state electrolytes in LIBs [25]. To lower the temperature to a more practical level, the authors partially replaced BH_4^- anion (0.205 nm) with an anion of larger size, such as iodine (0.211 nm), as transition temperature in alkaline borohydride tracked with the distances between the BH_4^- anions [26, 27]. Because the ionic conduction was found to be related to the anion configuration, mixed phases such as $Li_2BH_4NH_2$ and $Li_4(BH_4)(NH_2)_3$ were reported [28]. These exhibited orders of magnitude higher conductivities than that in the low temperature $LiBH_4$ phase. Note that the ionic size of currently used electrolytes containing halogens is larger than that of BH_4^-.

To design a halogen-free electrolyte with superionic conductivity, Giri et al. [24] used a rational approach to find a halogen-free anion with size larger than that of BH_4^-. They also suggested that, in addition to a large anion, the energy to dissociate the electrolyte into Li cation and superhalogen anion [$\Delta E = -\{E(LiX) - E(Li^+) - E(X^-)\}$] as well as the binding energy [$E_b = -\{E(LiXH_2O) - E(LiX) - E(H_2O)\}$] of the salt with H_2O are important factors and should be on par with that for the currently used electrolytes. They calculated these quantities by approximating the electrolytes by their respective molecular units. The results are given in Table 10.2. Note that the dissociation energies, ΔE range between 5.65 and 7.38 eV, while the binding energies, E_b with H_2O range between 0.92 and 1.41 eV. Giri et al. [24] considered halogen-free anions NO_3, B_3H_8, and $CB_{11}H_{12}$ and compared their properties with BH_4 as well as with halogen-containing anions in Table 10.2. The geometries of these clusters are given in Figure 10.13.

The VDE of $CB_{11}H_{12}$ is larger than those of NO_3, BH_4, and B_3H_8, which is consistent with their ionic size. The choice of $CB_{11}H_{12}$ followed from the earlier discussion in Chapter 2

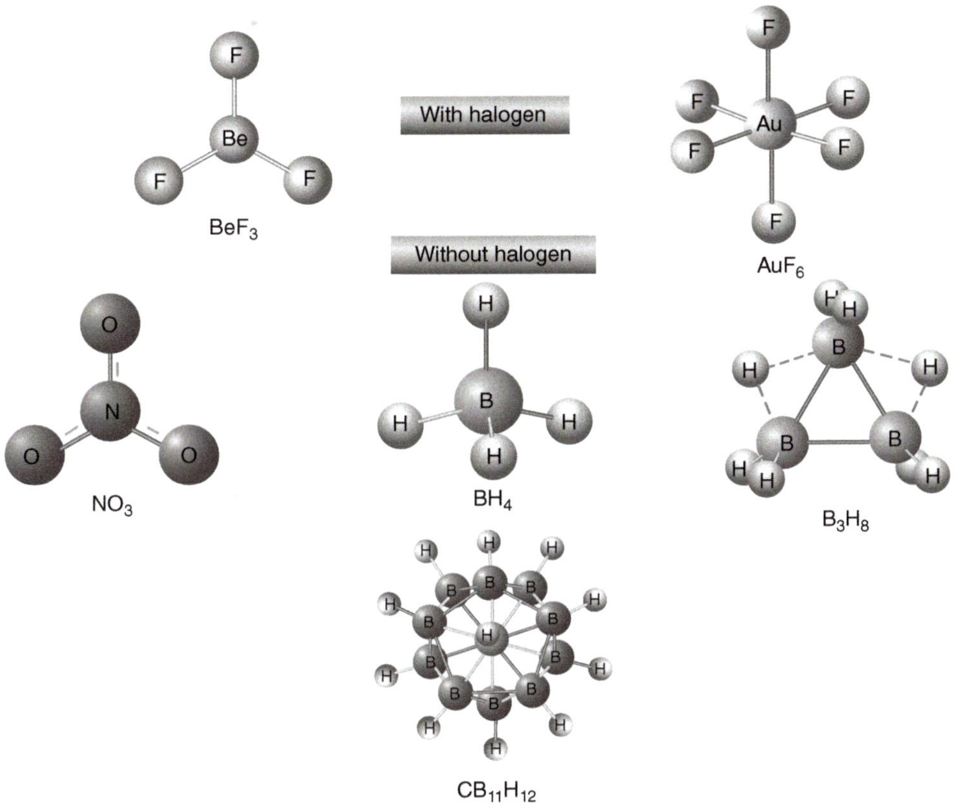

Figure 10.13 Optimized geometries of anions of four potential candidates for halogen-free electrolytes. Also given for comparison are two other halogen-containing anions that are not currently used in commercial applications. *Source:* Giri et al. [24]. © John Wiley & Sons, Inc.

where the stability of $B_{12}H_{12}^{2-}$ as a dianion was explained to be due to Wade–Mingos rule. By replacing one of the B atoms by C, one then needs only one extra electron to satisfy the Wade–Mingos rule. Hence, $CB_{11}H_{12}$ should behave like a superhalogen. This is indeed what was established by Pathak et al. [29] before its potential as an electrolyte component was illustrated by Giri et al. [24] While the binding energies of all the molecular salts, Li^+X^- in Table 10.2 with H_2O are about 1 eV, the energy to dissociate $LiCB_{11}H_{12}$ to ionic components is the smallest of all the electrolytes. Thus, Giri et al. suggested that $LiCB_{11}H_{12}$ should be an ideal halogen-free electrolyte for LIBs. This prediction has been verified by two subsequent experiments. Tutusaus et al. [30] found that $CB_{11}H_{12}$ serves as an effective anion for the electrolyte in Mg-ion batteries, which has been a challenge. To quote the authors "This achievement is a turning point in the research and development of Mg electrolytes that has deep implications on realizing practical rechargeable Mg batteries." In a later experiment, Orimo and coworkers [31] used $CB_{11}H_{12}$ anion for electrolyte components in Li- and Na-ion batteries and observed "unparalleled lithium and sodium superionic conduction in solid electrolytes" with "...truly stellar ionic conductivities (>0.1 S/cm) unmatched by any other known polycrystalline materials..."

Table 10.3 Dissociation energies in lithium salt (ΔE_{Li1}, ΔE_{Li2}) and in magnesium salt (ΔE_{Mg}).

	ΔE_{Li1} (eV)	ΔE_{Li2} (eV)		ΔE_{Mg} (eV)
$Li_2B_{12}H_{12}$	5.91	9.03	$MgB_{12}H_{12}$	21.31
$Li_2B_{12}(CN)_{12}$	4.66	6.83	$MgB_{12}(CN)_{12}$	17.46
$LiCB_{11}(CN)_{12}$	4.09		$MgCB_{11}(CN)_{12}$	18.30
$Li_2CB_{11}(CN)_{12}$	4.67	6.89		

Realizing that the size of the anion as well as its EA are important factors in the performance of an electrolyte, Zhao et al. [32] studied the potential of $CB_{11}(CN)_{12}$ and compared it to that of $CB_{11}H_{12}$. Recall that CN with an EA of 3.86 eV is a superhalogen. This needs to be compared with the EA of H, which is only 0.75 eV. In addition, the size of $CB_{11}(CN)_{12}$ is significantly larger than that of $CB_{11}H_{12}$. These factors are reflected in the EAs of $CB_{11}H_{12}$ and $CB_{11}(CN)_{12}$, namely, 5.99 eV and 8.72 eV, respectively. The larger size as well as the larger EA of $CB_{11}(CN)_{12}$ compared to that of $CB_{11}H_{12}$ make it a better candidate for halogen-free electrolyte; the dissociation energy of $LiCB_{11}(CN)_{12}$ is 4.09 eV compared to 5.99 eV for $LiCB_{11}H_{12}$ (see Table 10.3). Another interesting property of $CB_{11}(CN)_{12}$ is that it can bind a second electron with an energy of 1.07 eV, even if $CB_{11}(CN)_{12}{}^{2-}$ has one electron more than it needs to satisfy the Wade–Mingos rule. Consequently, there are two Li atoms in $Li_2B_{12}H_{12}$, $Li_2B_{12}(CN)_{12}$, and $Li_2CB_{11}(CN)_{12}$ electrolytes. However, the energy cost to remove the second Li is significantly higher than that to remove the first Li.

Despite the improved properties of $Li_2B_{12}(CN)_{12}$, and $LiCB_{11}(CN)_{12}$ as halogen-free electrolytes, their practical application may be hindered due to the toxicity of CN. Fang and Jena [33] studied the potential of $Li_2B_{12}(SCN)_{12}$ and $LiCB_{11}(SCN)_{12}$ as alternative candidates for halogen-free electrolytes as S is used in the detoxification of CN. They calculated the geometry and the energy needed to dissociate $Li_2B_{12}(SCN)_{12}$ and $LiCB_{11}(SCN)_{12}$ into individual ions. The energies to remove the first and second Li^+ ion from $Li_2B_{12}(SCN)_{12}$ are 5.07 and 6.89 eV, respectively. Similarly, the energy to dissociate $LiCB_{11}(SCN)_{12}$ into Li^+ and $CB_{11}(SCN)_{12}{}^-$ is 4.40 eV. These energies are significantly lower than their borane counterparts and close to those when CN is used as a ligand. Thus, $Li_2B_{12}(SCN)_{12}$ and $LiCB_{11}(SCN)_{12}$ are expected to outperform $LiCB_{11}H_{12}$ as an electrolyte.

10.4.2 Cluster-based Antiperovskites for Electrolytes in Li-ion Batteries

Recently, a family of low cost, Li-rich antiperovskite (LiRAP) conductors Li_3OA (A = halogen) have shown great promise for solid-state electrolytes. Li_3OA is an ionic solid that can be viewed as $[Li_3O]^+A^-$ where $[Li_3O]^+$ is a superalkali as its ionization potential of 3.88 eV is smaller than that of a Li atom, namely, 5.39 eV. The high RT ionic conductivity, low activation energy, large electrochemical stability window (ESW), low electronic conductivity, and excellent electrochemical stability against lithium metal anode make LiRAPs good candidates for solid-state electrolytes in LIBs [34–36]. The ionic conductivity can be enhanced by mixing the halogens and/or by doping divalent cations such as Ba at the Li site [37]. For example, RT ionic conductivity of Li_3OCl is 0.85×10^{-3} S/cm, which can be increased to 1.94×10^{-3} S/cm by replacing half of the Cl with Br, namely, $Li_3OCl_{0.5}Br_{0.5}$. Similarly, the ionic conductivities of $Li_{3-2x}Ba_xOCl$ ($x = 0.005$) and $Li_{3-2x}Ba_xOCl_{0.5}I_{0.5}$ ($x = 0.005$) are 25×10^{-3} S/cm at 25 °C and 121×10^{-3} S/cm at 50 °C, respectively. These results show that the composition of LiRAPs play an important role in their ionic conductivity. Fang and Jena [38] have shown that replacing halogens by superhalogens such as BH_4 and BF_4 can further improve the ionic conductivity of cluster-based antiperovskites. The underlying reason is that clusters offer additional degrees of freedom that atoms do not. First is the size. By replacing halogens by larger size superhalogens, one can create larger open channels for the Li ions to migrate. In addition, a cluster, due to its rotational degree of freedom, can alter the potential energy landscape and reduce the activation barrier. The atoms being spherical in shape do not offer this freedom. Using density functional theory, the authors calculated the lattice structure and ionic conductivity of $Li_3O(BH_4)$ and $(Li_3O)(BH_4)_{0.5}Cl_{0.5}$. In Figure 10.14 we show the optimized lattice structure of $Li_3O(BH_4)$ and its phonon spectrum, the radial distribution function, and the electronic density of states (DoS). The geometry of BH_4^- in Figure 10.14 is same as that in isolated gas phase, but Li_3O^+ in $Li_3O(BH_4)$ has a pyramidal structure while in the gas phase it is planar.

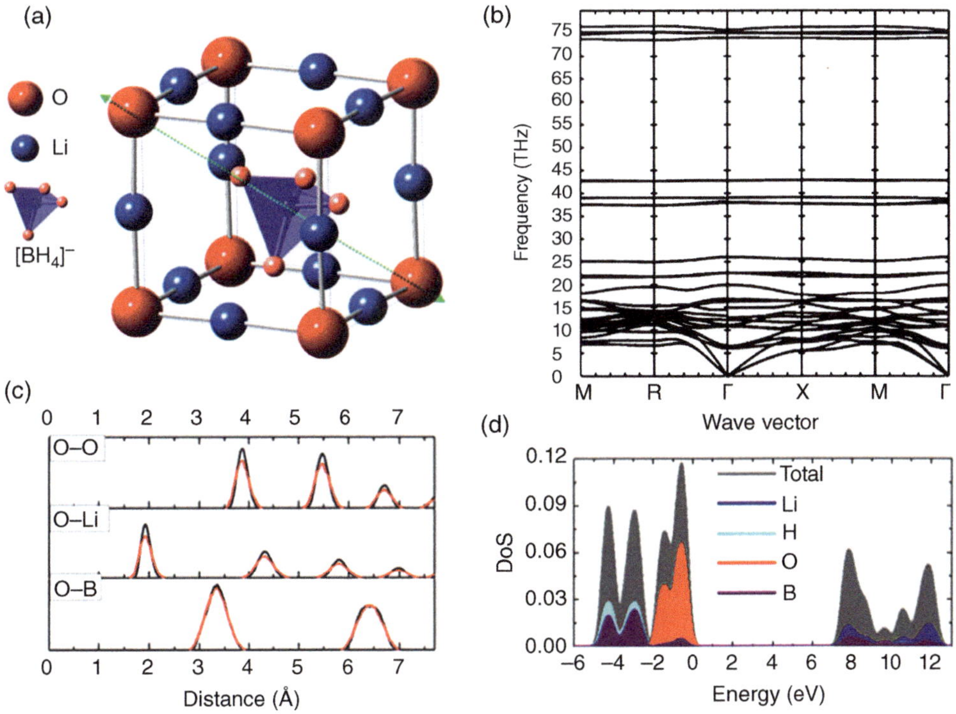

Figure 10.14 (a) Unit cell of the optimized ground state of $Li_3O(BH_4)$. The green arrow indicates the C_3 axis of BH_4^- tetrahedron inside the cubic cell. (b) Calculated phonon spectrum of $Li_3O(BH_4)$ along high-symmetry directions. (c) Radial distribution functions between atom pairs computed from the trajectory data of molecular dynamics simulations at 400 K (in black) and 600 K (in red). (d) Calculated electronic DoS and partial DoS using Heyd–Scuseria–Ernzerhof functional (HSE06) method. *Source:* Fang et al. [38]. © Royal Society of Chemistry.

The phonon spectrum does not show any imaginary modes, confirming the dynamical stability of $Li_3O(BH_4)$. Molecular dynamics simulation at 600 K and the radial distribution function between atom pairs further show that the lattice is thermally stable. The calculated electronic DoS using HSE06 functional yields a large bandgap of 7 eV, which is much larger than the 5-eV bandgap of Li_3OA. These results suggest that $Li_3O(BH_4)$ would have a larger ESW than that of Li_3OA. The calculated RT ionic conductivity of 0.10×10^{-3} S/cm is mediated by the Li-ion vacancy as is the case with Li_3OA. The ionic conductivity can be increased by introducing disorder either chemically or structurally. For example, the ionic conductivity of $Li_3O(BH_4)_{0.5}Cl_{0.5}$ of 0.21×10^{-3} S/cm is twice that of $Li_3O(BH_4)$. The theory used by Fang and Jena underestimates the ionic conductivity as the calculated value for Li_3OCl is 0.12×10^{-3} S/cm compared to the experimental value of 0.85×10^{-3} S/cm [34, 35]. Being light weight and with good mechanical properties, $Li_3O(BH_4)$ has the added advantage that it can be used in wearable electronics and implantable medical devices.

Realizing that the ion size is an important factor in determining the ionic conductivity, Fang and Jena [39] considered AlH_4^- and BF_4^- anions, which are isoelectronic with BH_4^-, as components of antipervoskite family of salts. The ionic sizes of AlH_4^- and BF_4^- are 2.66 and 2.43 Å, respectively, compared to 2.03 Å of BH_4^-. As anion size is an important factor in determining the ionic conductivity, it is expected that BF_4^- may have advantages over AlH_4^-. Note that for the stability of antiperovskites, the ionic radii have to satisfy the Goldschmidt tolerance factor t, which for the antiperovskite Li_3OX, is given as

$$t = \sqrt{2}\left(r_A + r_C\right) / \left(r_B + r_C\right)$$

Here, r_A, r_B, and r_C are the ionic radii of oxygen, X, and the lithium ion, respectively. To stabilize the antiperovskite structure with large superhalogen ions, such as X = AlH_4^- and BF_4^-, the oxygen atom should be replaced by a larger group element, such as sulfur. The resulting materials Fang and Jena studied are Li_3SAlH_4 and Li_3SBF_4. In Table 10.4, we list the corresponding Goldschmidt tolerance factors, which suggest that Li_3SAlH_4 and Li_3SBF_4 should be stable.

Fang and Jena optimized the antiperovskite lattice structure of Li_3SAlH_4 and Li_3SBF_4 and confirmed that it is lattice-dynamically stable by calculating their phonon spectra. The results are shown in Figure 10.15b and compared with that of Li_3OBH_4. The calculated formation energy of Li_3SBF_4 antiperovskite for the reaction $LiBF_4 + Li_2S \rightarrow Li_3SBF_4$ is 39.4 meV per atom, which is significantly lower than 58.8 meV per atom of the antiperovskite Li_3OBH_4 ($LiBH_4 + Li_2O \rightarrow Li_3OBH_4$) and is close to those of Li_3OA ($LiA + Li_2O \rightarrow Li_3OA$) of 13.9 and 25.8 meV per atom for A = Cl and Br, respectively. Note that the geometry of Li_3O^+/Li_3S^+ in the crystal is a tetrahedron indicated by red lines in Figure 10.15a, while it is planar in the gas phase. This results from the interaction of these clusters with their surroundings.

Li_3SX (X = AlH_4, BH_4, BF_4) may be viewed as super-LiRAP as both Li_3S cation and X anions are clusters with superalkali and superhalogen properties, respectively. From earlier studies of alkali halide crystals, it is known that the bandgap increases with the bonding ionicity, which is defined as the VDEs of the anions minus the ionization potential of the cation. Realizing that the antiperovskites are ionic crystals like alkali halides, we expect that the bandgaps of Li_3SX (X = AlH_4, BH_4, BF_4) would be larger than that when X is a

Table 10.4 Key information about the cluster ions in the super-LiRAP family. In each case, the channel size (CS) is calculated using the volume of the unit cell minus the volumes of all the ions inside. The used ionic radii (*R*) of Li, O, S, B, Al, H, and F are 0.90, 1.26, 1.70, 0.41, 0.68, 0.81, and 1.01 Å, respectively. The volume of a superhalogen ion is defined as the volume of the cube enclosing the superhalogen tetrahedron. The value in the parenthesis is the calculated bonding ionicity (BI), measured by the value of the VDE of the superhalogen ions $AlH_4^-/BH_4^-/BF_4^-$ minus the ionization potential (IP) of the superalkali ions Li_3O^+/Li_3S^+ (3.88/3.72 eV) in $(Li_3S)(AlH_4)$, $(Li_3O)(BH_4)$, $(Li_3S)(BF_4)$, $(Li_3O)Cl$, and $(Li_3O)Br$. $(Li_3S)(BF_4)$ has the largest bonding ionicity of all. The calculated Goldschmidt's tolerance factor is denoted by "*t*." BL is the bond length inside the superhalogens. LP is the calculated lattice parameters of the corresponding crystal.

VDE(BI) (eV)		BL (cluster) (Å)	BL (crystal) (Å)	R (Å)	CS (Å³/unitcell)	LP (Å)	*t*
AlH_4^-	4.46 (0.73)	1.64	1.67	2.66	56.27	4.73	1.03
BH_4^-	4.50 (0.56)	1.24	1.22	2.03	42.74	4.00	1.04
BF_4^-	7.42 (3.69)	1.44	1.42	2.43	62.48	4.72	1.10
Cl^-	3.71 (−0.17)	—	—	1.67	31.89	3.91	1.19
Br^-	3.58 (−0.30)	—	—	1.82	31.33	4.02	1.12

halogen. Indeed, the electronic DoS of Li_3SBF_4 in Figure 10.15c, calculated using the HSE06 functional, shows the largest bandgap of 8.5 eV compared with the 5.0 eV of Li_3SAlH_4, 7.0 eV of Li_3OBH_4, and those of Li_3OA (A = halogen). This is consistent with the anomalously large bonding ionicity of Li_3SBF_4 given in Table 10.4. No known crystalline electrolyte with high Li^+ conductivity exhibits such a large bandgap. As the bandgap provides a measure of the ESW, Li_3SBF_4 has the potential to be an ideal electrolyte.

Thermal stability of the super-LiRAP, including Li_3SAlH_4, Li_3OBH_4, and Li_3SBF_4, is tested by MD simulations at constant temperature and pressure. The radial distribution functions are calculated using the MD trajectory data collected over 100 ps. As shown in Figure 10.15d for Li_3SBF_4, no melting is observed up to 600 K. The calculated linear thermal expansion coefficient of Li_3SBF_4 from the MD data is 1.6×10^{-5}/K, which is smaller than 3.0×10^{-5}/K of Li_3OBH_4. The value is also smaller than 2.1×10^{-5}/K and 1.8×10^{-5}/K of Li_3OA for A = Cl and Br, respectively. This should be considered as another advantage of Li_3SBF_4 – smaller thermal expansion means higher compatibility with other parts inside a composite device, subject to large temperature fluctuations.

The conductivity of the Li^+ ion depends upon the channel size as well as the energy barrier for Li^+ migration. From Table 10.4, we note that Li_3SBF_4 provides a much larger channel size compared with the rest. This is caused by the larger charge on the F atoms (−1.9e) compared to that on the H atoms (−0.8e), thus making more room between sulfur and BF_4^- in the crystal and keeping the sulfur atoms further apart. In addition, the weaker interaction between sulfur and Li^+ compared to that between oxygen and Li^+ and the tetrahedral geometry of the BF_4^- anion plays an important role. Unlike a spherical ion, at elevated temperatures the translational and rotational modes of the BF_4^- superhalogen ions can constantly change the orientations of the superhalogen tetrahedra from their C_{3v} ground-state symmetry. This, in turn, generates a shifting and varying potential surface throughout the crystal, which can then facilitate fast ion migration of Li^+ between different sites. The RT conductivity of Li_3SBF_4 is found to be 1.0×10^{-2} S/cm, which is 12 times

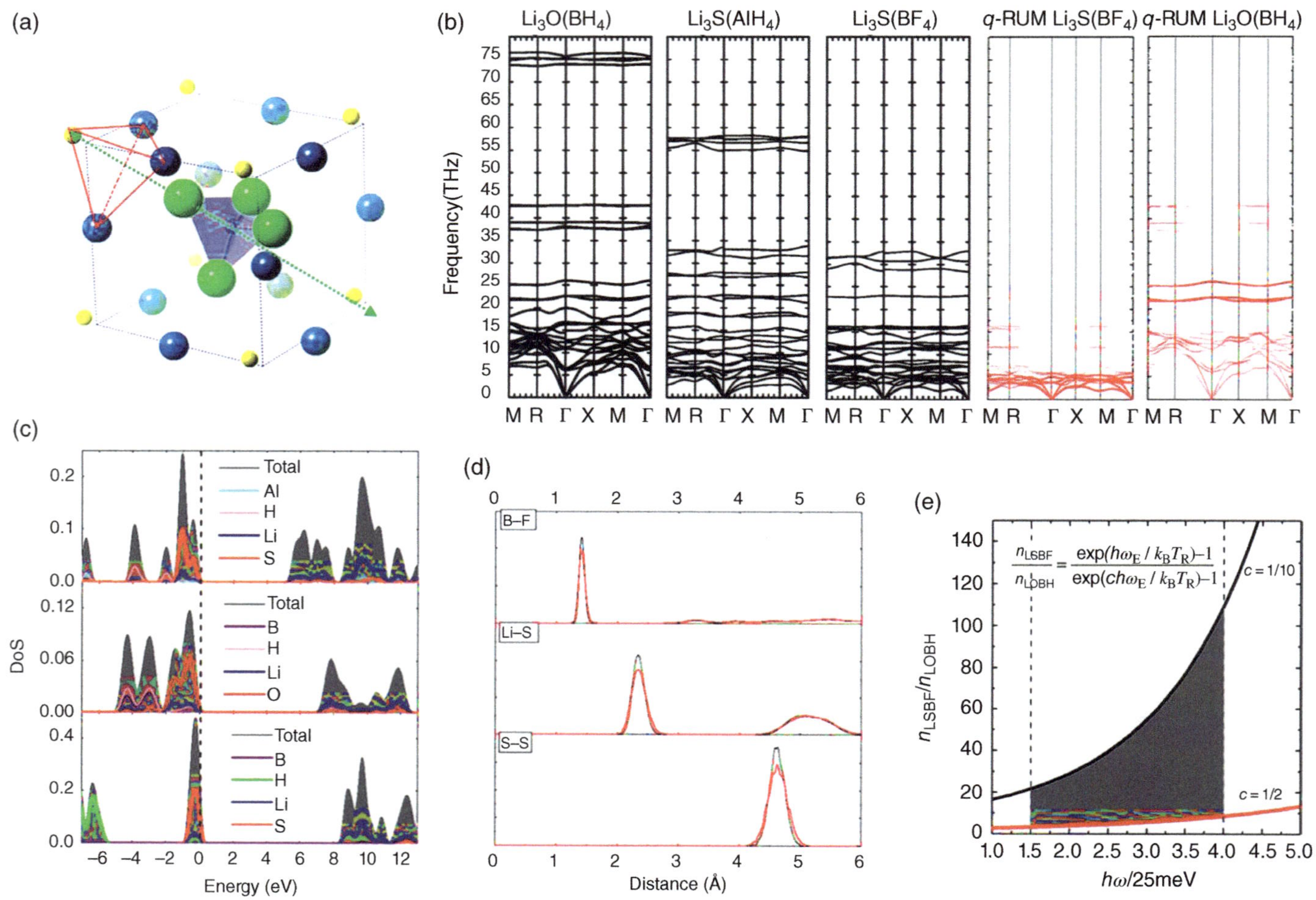

Figure 10.15 (a) Optimized unit cell of Li_3SBF_4, with Li in blue, S in yellow, B in gray, and F in green. The green arrow indicates the C_3v orientational symmetry adopted by the BF_4^- tetrahedral unit in the cubic cell. The red outline highlights the pyramidal configuration of Li_3S^+. (b) Calculated phonon spectra of the studied super-LiRAP. The red-colored spectra of Li_3SBF_4 and Li_3OBH_4 show the positions of the q-RUMs (rigid unit modes) in these materials. The modes with large distortion of superhalogen ions are colored in white. (c) HSE06 calculated (partial) DoS of the studied super-LiRAP. The Fermi level is set to zero in each case. (d) Calculated radial distribution functions of Li_3SBF_4 using Molecular dynamics (MD) trajectory data over 100 ps at 400 and 600 K under ambient pressure. (e) Estimated range of ratio between the ionic conductivity of Li_3SBF_4 and that of Li_3OBH_4 at RT by using the energy of the q-RUMs of these materials in a simple Einstein model. The mean phonon

higher than the experimentally observed RT conductivity of Li_3OCl, namely, 0.85×10^{-3} S/cm. This value is same as that of the organic liquid electrolyte used in practical batteries. The superionic conductivity can be further increased by partially replacing BF_4^- with Cl^-. The calculated RT conductivity of $Li_3S(BF_4)_{0.5}Cl_{0.5}$ is 1.9×10^{-2} S/cm, which is about 14 times higher than the calculated value of Li_3SBF_4 and is already higher than the conductivity of the organic liquid electrolytes. The high melting point over 600 K and the low activation energy allow these materials to operate over a wide range of temperatures from below −30 °C with conductivity above 10^{-3} S/cm to over 300 °C with conductivity well above 10^{-1} S/cm.

10.4.3 Cluster-based Antiperovskites for Electrolytes in Na-ion Batteries

LIBs have undoubtedly brought one of the greatest benefits to humankind by enabling the development of miniaturized and portable electronics and opening the gateway toward a fossil fuel free economy. However, its rarity, combined with its uneven distribution in Earth's crust and high costs, implies that it cannot single-handedly meet the demand of the future widespread commercialization of electric vehicles [40]. In addition, the lithium extraction process is very harmful to the environment and consumes an enormous amount of water per ton of lithium [41]. Sodium, on the other hand, is abundant and Na-ion batteries can indeed meet the world's demand for energy storage. However, here too there are many challenges. In addition to finding the ideal anodes, problems that plague all solid-state LIBs also exist for Na-ion batteries, i.e. the absence of solid electrolytes.

Recently, there has been considerable interest [42–46] in developing RT ionic conductivity in the Na-rich antiperovskites. However, it is rather low, only of the order of 10^{-8} S/cm in Na_3OBr. The experimental activation energies of Na_3OX (X = Cl or Br) are around 0.7 eV, which is much higher than the regular value below 0.3 eV. The calculated Young's modulus of the sodium antiperovskites is above 50 GPa with Poisson's ratio around 0.22 [47]. These values correspond to ceramics-like materials, which make them hard to densify as electrolytes having close contact with electrodes. It seems natural to apply the same strategy as has been done for LiRAPs, to boost the ionic conductivity of the material by using cluster ions. Fang and Jena [48] started by replacing Br^- in Na_3OBr with superhalogens AlH_4^-, BF_4^-, and BCl_4^-. Note that for the stability of antiperovskites, the Goldschmidt's tolerance factor, t, should be around 1.0. In Table 10.5 we list the tolerance factor, t, for a range of cluster-based antiperovskites. For superionic conductivity, the channel size for Na^+ ion migration should also be high. While all cluster-based antiperovskites satisfy the Goldschmidt's tolerance criteria, we see from Table 10.5 that $Na_3S(BCl_4)$ offers the largest channel size.

The optimized structures of $Na_3O(AlH_4)$, $Na_3O(BF_4)$, and $Na_3S(BCl_4)$ are given in Figure 10.16. These are pseudocubic structures with LPs of 4.74, 4.74, and 5.88 Å, respectively, and angles are close to 90°. All the three structures have no imaginary frequencies and, thus, are dynamically stable. Molecular dynamics study of $Na_3S(BCl_4)$, which has the largest channel size, showed that the crystal is stable up to 600 K. The calculated formation energy of $Na_3S(BCl_4)$ from the reaction $Na + NaSH + Na(BCl_4) \rightarrow Na_3S(BCl_4) + (1/2)H_2$ is negative (−82.4 meV/atom), suggesting that this may be a preferred route for its synthesis. The bandgaps of $Na_3O(AlH_4)$, $Na_3O(BF_4)$, and $Na_3S(BCl_4)$ are all about 4.0 eV, which are typical of a solid electrolyte in sodium-ion battery and similar to those of Na_3OX (X = halogen).

Table 10.5 The calculated Goldschmidt's tolerance factor, $t = (r_B + r_{Na}) / \sqrt{2}(r_A + r_{Na})$, of the studied Na-rich antiperovskite structures, where r_A, r_B, and r_{Na} are the ionic radii of the divalent anion, monovalent anion, and Na, respectively.

Crystal	BL (Å)	r (Å)	t	LP(Å)	CS (Å^3/unitcell)	V_{Na}/CS^a
Na_3OBr	—	1.82	0.87	4.57^b	68.92	0.28
$Na_3O(BF_4)$	1.41	2.43	1.05	4.74	76.03	0.26
$Na_3O(AlH_4)$	1.62	2.57	1.09	4.74	71.98	0.27
$Na_3S(BCl_4)$	1.85	3.29	1.10	5.88	127.89	0.15
$Na_{2.875}S(BCl_4)^c$	1.85	3.29	1.10	5.96	136.30	0.14 (4)
$Na_{2.875}S(BCl_4)_{0.5}I_{0.5}{}^c$	1.85	2.67^d	0.95	5.72	137.26	0.14 (3)

The ionic radii of B, Al, Li, Na, O, S, Br, and I are 0.41, 0.67, 0.90, 1.16, 1.26, 1.70, 1.82, and 2.06 Å, respectively. BL is the bond length inside the cluster ion. In each case, the channel size (CS) is calculated by using the volume of the unit cell minus the volumes of all the ions (apart from Na ions) inside. The volume of a cluster ion is defined as the volume of the cube enclosing the cluster. The volume of the unit cell is computed from the optimized lattice parameter (LP) of the material with pseudocubic symmetry. V_{Na}/CS is the ratio between the total volume of the three Na atoms inside a unit cell versus the channel size that is available for Na to migrate in each material.

a These values are compared to that of $V_{Li}/CS = 0.15$ for $Li_3S(BF_4)$.

b Experimental value.

c Nonstoichiometric system with one Na^+ vacancy in a 2×2×2 supercell used in the simulations.

d Calculated as the average of ionic radii of I^- and BCl_4^-. *Source:* Fang and Jena [48]. © American Chemical Society.

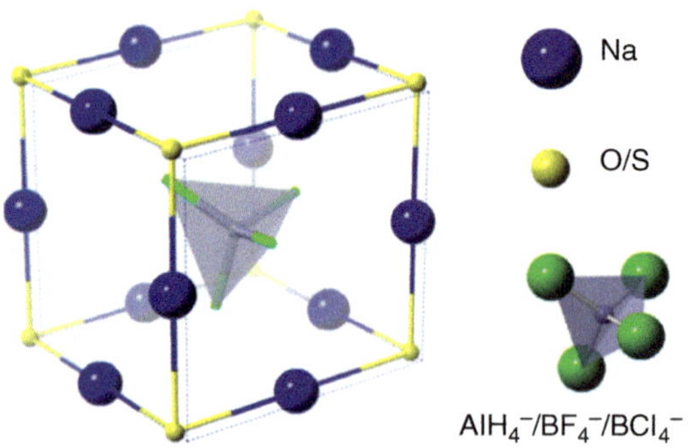

Figure 10.16 Optimized pesudocubic structures of $Na_3O(AlH_4)$, $Na_3O(BF_4)$, and $Na_3S(BCl_4)$ with cluster ions AlH_4^-, BF_4^-, and BCl_4^-, respectively. The tetrahedral cluster with the cube adopts a C_3v symmetry in the ground state, where the threefold axis of the tetrahedral unit is along the diagonal of the pseudocubic cell. *Source:* Fang and Jena [48]. © American Chemical Society.

The Na^+ ion conductivity, studied using *ab initio* molecular dynamics, shows that Na^+ vacancies are necessary for its diffusion. Results are shown in Figure 10.17. The supercell where a Na^+ vacancy is introduced to trigger fast-ion hopping is shown in Figure 10.17a. The average mean-square displacement of Na^+ is extracted in Figure 10.17c from the molecular dynamics data. The Na^+ conductivity is then computed using the Nernst–Einstein relation. Among the materials studied, $Na_3S(BCl_4)$ shows the highest 3D Na^+ conductivity of

2.8×10^{-3} S/cm at RT, which is significantly higher than that of Na_3OBr (1.2×10^{-5} S/cm). Similarly, the activation energy (Ea) of $Na_3S(BCl_4)$ is 0.17 eV, which is much lower than that of Na_3OBr (0.35 eV). The Na^+ conductivity can be further increased by comixing the cluster ion with elementary halogen inside the antiperovskite lattice, making $Na_3S(BCl_4)_{1-x}A_x$ (A = Cl, Br, or I; $0<x<1$). This was demonstrated by using $Na_3S(BCl_4)_{0.5}I_{0.5}$ in Figure 10.17b. The calculated channel size and the ratio of V_{Na}/CS for the material are more favorable than those of $Na_3S(BCl_4)$, as shown in Table 10.5. Indeed, the RT conductivity, 3.8×10^{-3} S/cm, of $Na_3S(BCl_4)_{0.5}I_{0.5}$ is higher than (2.8×10^{-3} S/cm) that of $Na_3S(BCl_4)$. These results confirm that replacing elemental ions by superions can lead to better electrolytes.

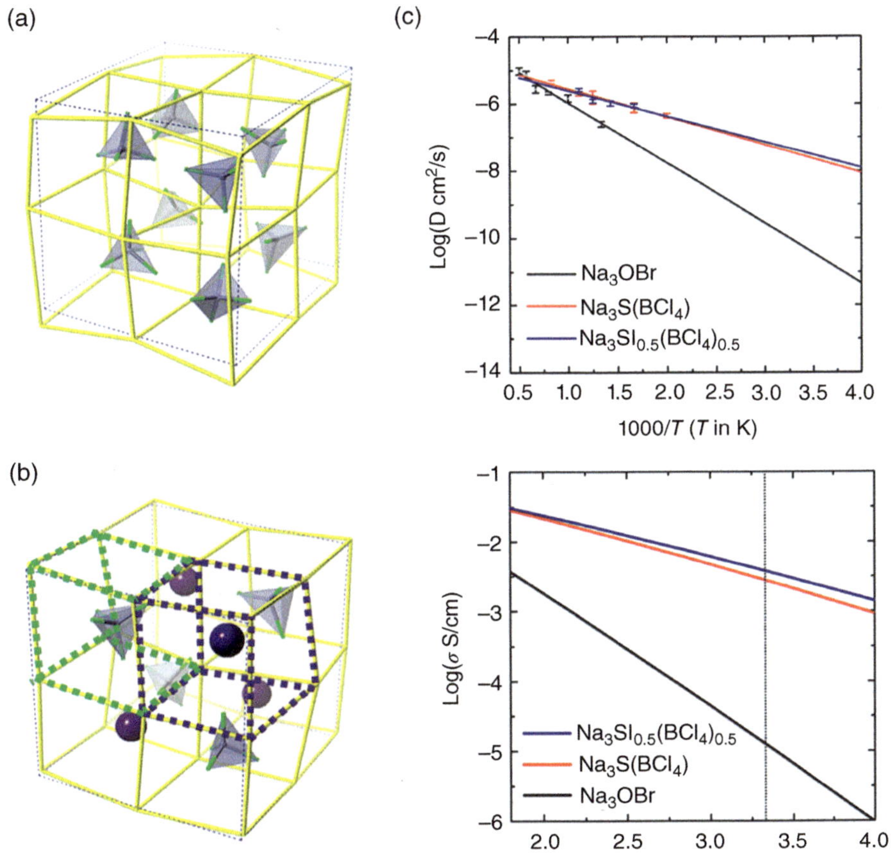

Figure 10.17 Snapshots of (a) $Na_3S(BCl_4)$ and (b) $Na_3S(BCl_4)0.5I_{0.5}$ containing one Na^+ vacancy after 250 ps of AIMD simulations. BCl_4 clusters are gray tetrahedra with Cl in green. Iodine atoms are in purple. Sulfur atoms are shown in yellow bonds. Sodium atoms are neglected for clarity. The supercell can be equipartitioned into unit cells containing either BCl_4^- or I^-, as shown by the green outline and the purple outline, respectively. (c) The calculated Na diffusion coefficients and conductivities of Na_3OBr, $Na_3S(BCl_4)$, and $Na_3S(BCl_4)_{0.5}I_{0.5}$. The error of the diffusion coefficient (D) is transferred from the error of the slope of the linear fitting to the mean square displacement (MSD) data. The error of the logarithm of D is then estimated by $\log(D+\Delta D) = \log[D(1+\Delta D/D)] \approx \log \Delta D + \Delta D/D$. $Na_3S(BCl_4)$ and $Na_3S(BCl_4)_{0.5}I_{0.5}$ exhibit much higher conductivities and lower activation energies than that of Na_3OBr. The dotted line shows the ionic conductivity at RT. *Source:* Fang and Jena [48]. © American Chemical Society.

10.5 Cluster-based Materials for Hydrogen Storage

Note that about 75% of the oil used in the United States is spent on transportation. Another solution to the energy problem is, therefore, to find alternate sources to power the mobile industry that does not depend on fossil fuels. If hydrogen can be used in fuel cells, not only will it reduce our dependence on fossil fuels, but also it can benefit the environment as water is the only byproduct. However, hydrogen is not an energy source as it has to be produced and stored. The biggest hurdle to a successful hydrogen economy is the safe and economical storage of hydrogen with high gravimetric and volumetric density under ambient conditions. While liquid hydrogen or hydrogen stored under high pressure can meet some of these requirements, they are not practical for widespread commercial use. This is because hydrogen liquifies at cryogenic temperatures and pressure of 10 000 psi is needed to store enough hydrogen in a tank to drive automobiles over a 500 km range. Even at these extreme conditions, the energy density of hydrogen in liquid or under high pressure is only 4.4 and 8.4 MJ/l, respectively, compared to the energy density of gasoline, namely, 31.6 MJ/l. The alternate method is to store hydrogen in solid materials. Here too, there are challenges. The solid-state materials should be capable of storing hydrogen with a gravimetric and volumetric density of 7.5 wt% and 70 g/l, respectively. While some metals like Pd meet the volumetric density requirement, the gravimetric density is very low due to the heavy mass of Pd. For larger gravimetric density, the materials ideally should be lighter than Al. Unfortunately, many light elements do not store hydrogen reversibly. Thus, to find the ideal hydrogen storage materials, it is first necessary to understand how hydrogen interacts with atoms in a crystal lattice.

10.5.1 Hydrogen Interaction Mechanism

Typically, hydrogen as it approaches a material is either physiosorbed or chemisorbed. In the former, hydrogen remains in molecular form and its binding energy to the surface is rather low, of the order of meV. In such materials, hydrogen cannot be retained at RT and thus, they are not effective as hydrogen storage materials for vehicles. In the latter, hydrogen is chemisorbed where the hydrogen molecules dissociate into atomic hydrogen, diffuse into the materials, and bind strongly. The typical binding energy in this case is about 2–4 eV. This implies that hydrogen can only be released at high temperatures, again making it impractical as a hydrogen storage material. Thus, the ideal binding energy of hydrogen should be intermediate between physisorption and chemisorption, namely, between 0.2 and 0.8 eV. These three mechanisms are depicted in Figure 10.18 [49]. Indeed, such a process was demonstrated earlier by Kubas [50, 51] and Niu et al. [52, 53] where hydrogen is bonded in quasi-molecular form, with H–H bond stretched but not broken.

In the Kubas mechanism, hydrogen is bound to a transition metal atom through a charge donation and back donation process. Hydrogen donates electron to the unfilled *d*-orbitals of the transition metal atom, while the transition metal atom back donates charge to the antibonding orbital of the H_2 molecule, weakening the H–H bond. In the case studied by Niu et al., a positively charged atom polarizes the H_2 molecule, which then binds in quasi-molecular form. In either case, the binding energy of hydrogen lies in the ideal range, i.e. 0.2–0.8 eV. In Figure 10.19, we show schematically a Li^+ ion binding to as many as six H_2 molecules through the polarization mechanism [53]. Zhou et al. [54] later showed that an

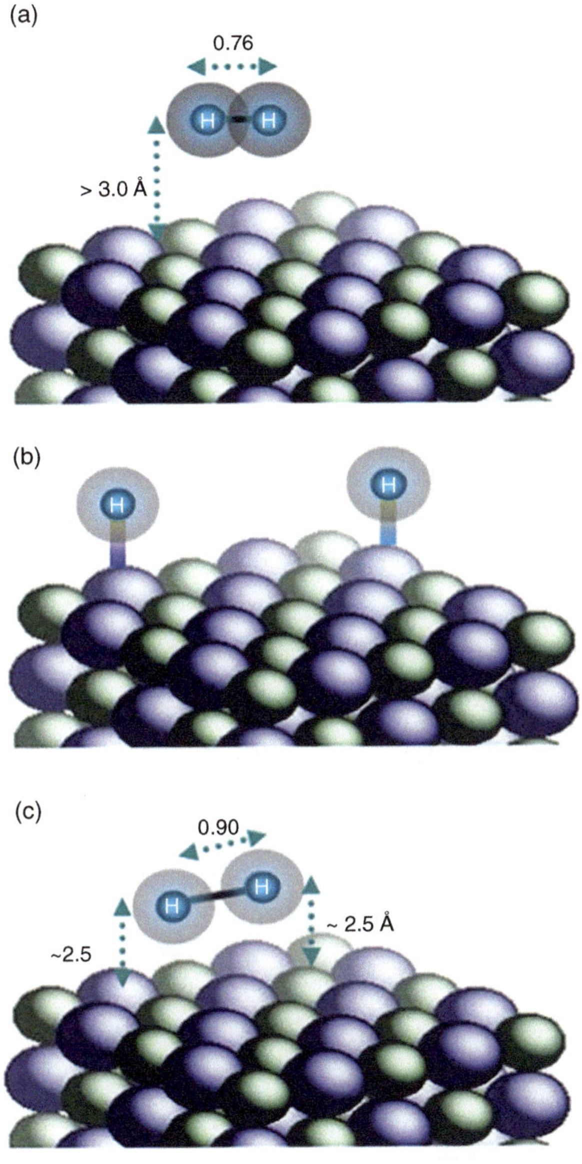

Figure 10.18 Adsorption of hydrogen on substrates, (a) physisorption, (b) chemisorption, and (c) quasi-molecular bonding. *Source:* Jena [49]. © American Chemical Society.

external electric field can also polarize the hydrogen molecules and store them in quasi-molecular form.

Zhang and coworkers [55] were the first to show that the Kubas mechanism can be used to design an ideal hydrogen storage material by focusing on Sc-decorated $C_{48}B_{12}$ fullerene. They showed that up to 11 hydrogen atoms can be trapped by each Sc atom, with 10 of these hydrogen atoms bound in quasi-molecular form (see Figure 10.20). Later, Yildirim

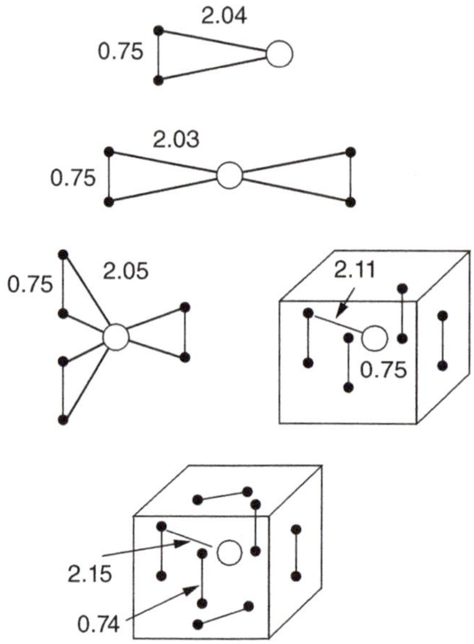

Figure 10.19 Geometries of $Li^+(H_2)n$ clusters with $n \leqslant 6$. The bond lengths are given in Å. *Source:* Niu et al. [53]. © American Chemical Society.

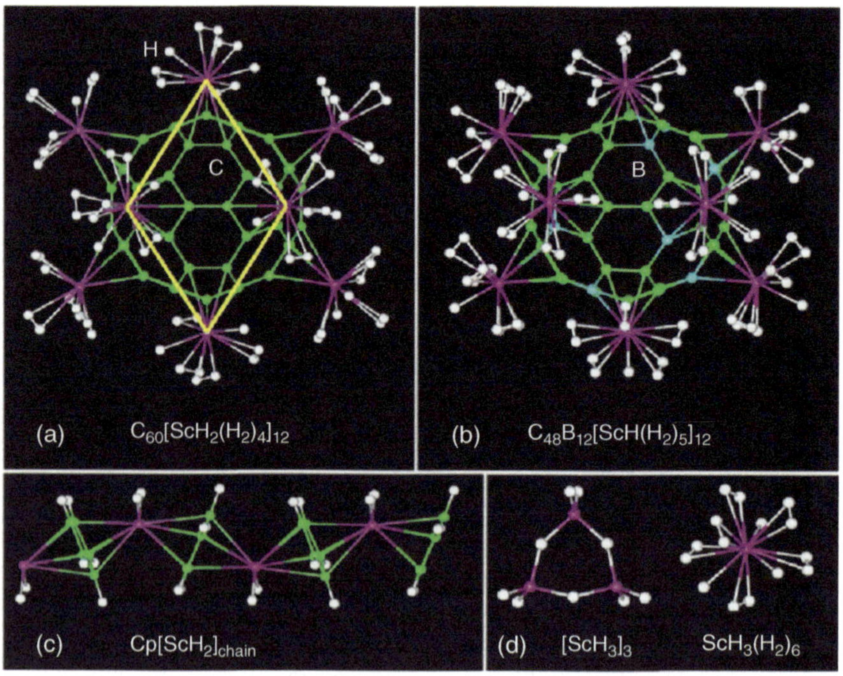

Figure 10.20 (a) Optimized atomic structures of $C_{60}[ScH_2(H_2)_4]_{12}$, (b) $C_{48}B_{12}[ScH(H_2)_5]_{12}$, (c) $Cp[ScH_2]$chain, and (d) $[ScH_3]_3$ (left) and $ScH_3(H_2)_6$ (right). For clarity, only part of the buckyball is shown. A close-packed van der Waals interaction (vdW) solid formed from $C_{48}B_{12}[ScH(H_2)_5]_{12}$ would have a volumetric storage density of 43 kg H_2/m^3 even without consideration of the possibility of interlocking using large open spaces indicated by the rhombus in (a). *Source:* Zhao et al. [55]. © American Physical Society.

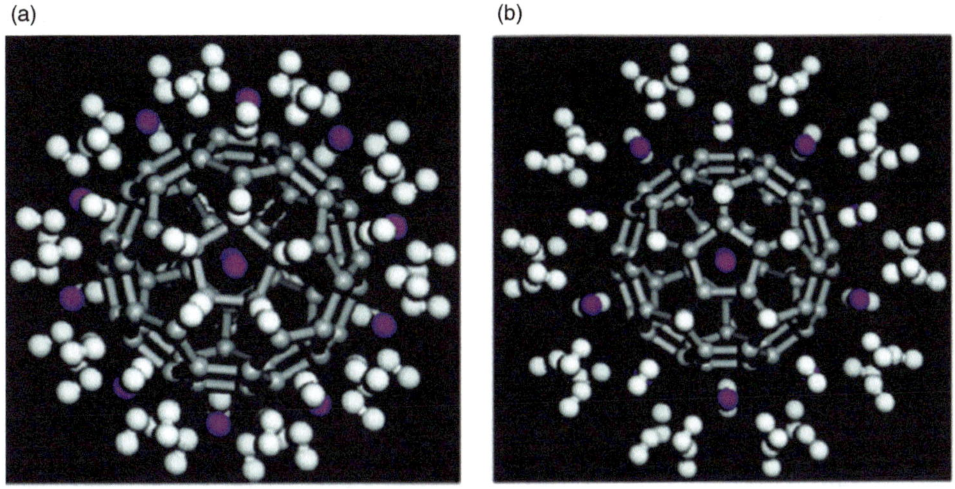

Figure 10.21 (a) Initial and (b) optimized geometries of $Li_{12}C_{60}(H_2)_{60}$. *Source:* Sun et al. [58]. © American Chemical Society.

and Ciraci [56] arrived at a similar conclusion by decorating carbon nanotube with Ti. However, Sun et al. [57] showed that these atoms would cluster when deposited on a C_{60} surface, thus undermining the amount of hydrogen that can be stored. Instead, Li atoms were found not to cluster as the Li–C bonding is stronger than the Li–Li bonding [58]. These authors found that up to 13 wt% hydrogen can be stored reversibly on $Li_{12}C_{60}$ cluster (Figure 10.21).

Teprovich et al. [59] studied the hydrogenation of Li_6C_{60} by using solvent-assisted mixing and found that up to 40 hydrogen atoms can be stored reversibly. The fullerene cage remained almost intact during the hydrogenation/dehydrogenation process. Using density functional theory, Wang and Jena [60] showed that Li atoms preferably occupy the adjacent fivefold sites on C_{60}. This leaves 30 C atoms that are not forming a bond with Li and 30 H atoms that are radially bonded to these C atoms. When additional hydrogen atoms are introduced, one of the Li atoms assumes an endohedral position. The remaining 5 Li atoms bind 10 hydrogen atoms in quasi-molecular form. Later experiment by Zidan's group confirmed the presence of the endohedral Li (R. Zidan, private communication). Several studies involving metal atoms embedded in clusters, organometallic complexes, metal-organic frameworks, and porous carbon have since been carried out.

10.5.2 Intermediate States

Another important property of hydrogen storage material is that they must be reversible. Unfortunately, some of the light metal hydrides lack this property. For example, light metal borohydrides $M(BH_4)_n$ where M is a light metal (Li, Na, Mg, Al) with valence n are an important class of materials that store hydrogen with high gravimetric density. Among these, $LiBH_4$ has a gravimetric and volumetric density of 18.4% and 102 kg/m^3, respectively, and exceeds the storage requirement set forth by the Department of Energy [61]. Unfortunately, the hydrogen desorption temperature is high and hydrogen cannot be stored reversibly. This is because during dehydrogenation, $LiBH_4$ forms a very stable intermediate phase, $Li_2B_{12}H_{12}$ [62, 63]. Similarly, $Mg(BH_4)_2$ and $Y(BH_4)_3$ yield $Mg(B_3H_8)_2$ and $Y(B_3H_8)_3$ during decomposition [64, 65]. Understanding the reason behind the formation of these intermediate states is important to best utilize these materials for hydrogen storage.

Studies using clusters as models have helped to understand this reason [66–68]. Note that $M(BH_4)_n$, $Li_2B_{12}H_{12}$, $Mg(B_3H_8)_2$, and $Y(B_3H_8)_3$ are ionically bonded solids where BH_4^-, $B_{12}H_{12}^{2-}$, and $B_3H_8^-$ are very stable negative ions. Their stability arises due to electronic shell closure. To illustrate this, we show in Figure 10.22 the equilibrium geometry of BH_4^-, $B_{12}H_{12}^{2-}$, and $B_3H_8^-$ [68]. BH_4^- is a tetrahedral cluster where an extra electron is needed for octet shell closure. $B_{12}H_{12}$ is an icosahedral cluster whose stability is governed by the Wade–Mingos rule. According to this rule, $(n+1)$ pairs of electrons are needed to form cage bonding where n is the number of vertices of the polyhedron. $B_{12}H_{12}$ being an icosahedron, $n = 12$. Therefore, to stabilize $B_{12}H_{12}$, 26 electrons are needed. Since each BH pair can contribute two electrons to the cage bonding, two extra electrons are needed to satisfy the Wade–Mingos rule. Consequently, $B_{12}H_{12}^{2-}$ is a very stable dianion. Similarly, B_3H_8 has $3\times3+8\times1 = 17$ electrons. Of these, 12 electrons (3×4) are occupied in covalent bonds between 2 H atoms and 1 B atom. This leaves five electrons, but six are needed to satisfy the aromatic $(4n+2)$ rule, n being an integer (here $n = 1$). Thus, BH_4, $B_{12}H_{12}$, and B_3H_8 are all superhalogens. Indeed, the energy to remove one electron from BH_4 and B_3H_8 and two electrons from $B_{12}H_{12}$ are 4.50, 4.72, and 13.84 eV, respectively. In Table 10.6 we compare the binding energies of the intermediate phases with those of the original compounds. Note that the intermediate phase $Na_2(B_{12}H_{12})$ is more stable than the parent phase $NaBH_4$, clearly showing why the material is irreversible. Similarly, the binding energies of intermediate phases $Mg(B_3H_8)_2$ and $Y(B_3H_8)_3$ are comparable to those of their parent borohydrides.

Recently, Yan et al. [69] proposed a novel strategy for reversible hydrogen storage in $Ca(BH_4)_2$ where the decomposition pathway could be effectively controlled by the reaction temperature. Knowing that CaB_2H_6 is the crucial intermediate step on the way to CaB_6, they efficiently avoided the formation of boron sinks by controlling the decomposition temperature, leading to CaB_6 and CaH_2 as the final products. This facilitated the reversibility of $Ca(BH_4)_2$ under much milder conditions.

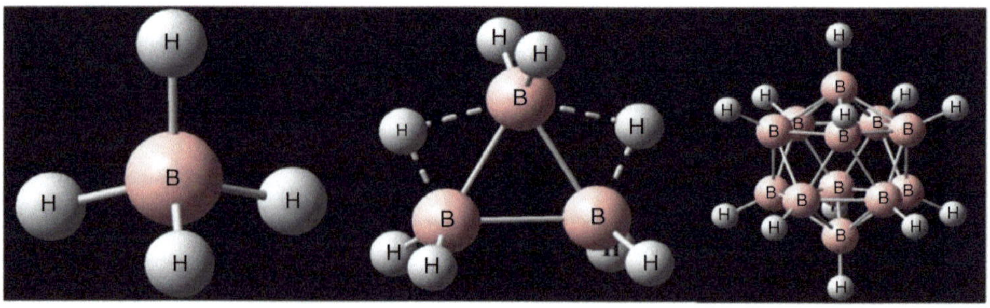

Figure 10.22 Geometries of negative ion components of intermediate phases involving BH_4^-, $B_3H_8^-$, and $B_{12}H_{12}^{2-}$. *Source:* Mohtadi et al. [68]. © IOP Publishing.

Table 10.6 Binding energies (BE, in eV) between the metal cation and the borohydride anion of the $M[BH_4]_n$, $M[B_3H_8]_n$, and $M_2[B_{12}H_{12}]_n$ (M = Na, Mg, Y) clusters.

Na compound	BE	Mg compound	BE	Y compound	BE
$NaBH_4$	5.72	$Mg(BH_4)_2$	23.46	$Y(BH_4)_3$	44.25
NaB_3H_8	5.28	$Mg(B_3H_8)_2$	21.95	$Y(B_3H_8)_3$	43.87
$Na_2B_{12}H_{12}$	6.58	$MgB_{12}H_{12}$	20.97	$Y_2(B_{12}H_{12})_3$	43.79

Source: Liu et al. [67]. © American Chemical Society.

10.5.3 Catalysts for Lowering the Dehydrogenation Temperature

One of the other reasons why light metal hydrides have not been suitable for hydrogen storage is that the hydrogen desorption temperature is high. This is because the bonding of hydrogen to the metal atom is covalent. Bogdanovic and Schwickardi [70] showed that the hydrogen desorption temperature of $NaAlH_4$ could be lowered from 265 to 150 °C by using a Ti-based catalyst. However, the role of the catalyst was difficult to understand as the catalyst is ball milled with the sample, which inadvertently introduces defects. Study based on clusters have shed light on this issue. For example, it was shown earlier from cluster calculations that neutral AlH_4 cluster is unstable and fragments to $AlH_2 + H_2$ [71]. However, once an electron is added, AlH_4^- is very stable. This is consistent with the fact that in $NaAlH_4$ the stability comes from CT from Na to AlH_4, and $Na^+AlH_4^-$ becomes a salt with ionic bonding. Thus, dehydrogenation can be made possible if the electron that Na donates to AlH_4 can be less available. This is possible if the catalyst is more electronegative than AlH_4. This indeed was found to be the case by Berseth et al. [72] when the authors supported $NaAlH_4$ on graphene, carbon nanotube with varying radii, and C_{60} fullerene and compared those results with $TiCl_3$ catalyst. The results are shown in Figure 10.23. Note that C_{60} acts as good a catalyst as $TiCl_3$.

In Figure 10.24 we plot the energy to remove a hydrogen atom from a $NaAlH_4$ cluster supported on graphene, nanotubes of varying radii, and C_{60} versus their EA. Note that, as expected, higher the EA, lower is the hydrogen removal energy.

10.6 Clusters Promoting Unusual Reactions

Chemical reactions involve CT or charge sharing. The former leads to an ionic bond while the latter leads to a covalent bond. Consider, for example, formation of NaCl. Here, an electron is transferred from Na to Cl, resulting in an ionic bond. For H_2, on the other hand, the hydrogen atoms share their electrons, forming a covalent bond. In the former case, the quantities that are important are the ionization potential of the atom donating the electron and the EA of the atom receiving it. CT is facilitated if the ionization potential is small and the EA is large. As we have discussed before, clusters that mimic alkalis (halogens) can be designed to have ionization potential (EA) smaller (larger) than that of alkali (halogen) atoms. In this section, we present two examples where such clusters can promote reactions, commonly thought not possible.

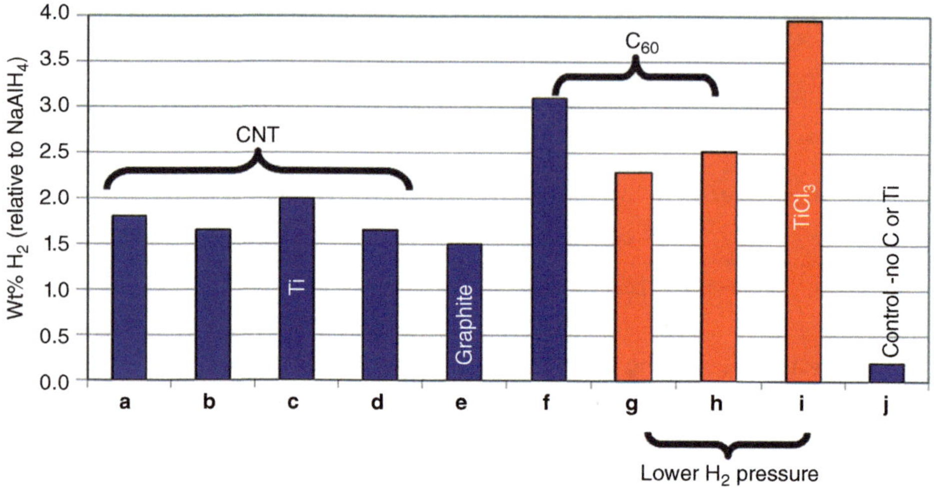

Figure 10.23 Screening study results of $NaAlH_4$/carbon mixtures. These values are for hydrogen desorbed on the second desorption cycle. Samples a–f and j were heated to 280 °C for three hours, and the previous absorption step was performed at ~2.1 × 107 Pa H_2 and 150 °C. Samples g–i were heated to 350 °C for three hours, and the previous absorption step was performed at ~1.2 × 107 Pa H_2 and 150 °C. Sample key: (a) 8 nm Carbon nanotubes (CNT), (b) 10–20 nm CNT, (c) 10–20 nm CNT with 4 mol % Ti, (d) 50 nm CNT, (e) graphite, (f) C_{60}[1], (g) C_{60}[2], (h) C_{60}[3], (i) control no carbon, ball mill 4 mol % $TiCl_3$, and (j) control no carbon or Ti. The two pressures used for the rehydriding step (which affects the amount of hydrogen desorbed in the second cycle) are highlighted by color: high pressure experiments are blue, while lower pressure experiments are red. *Source:* Berseth et al. [72]. © American Chemical Society.

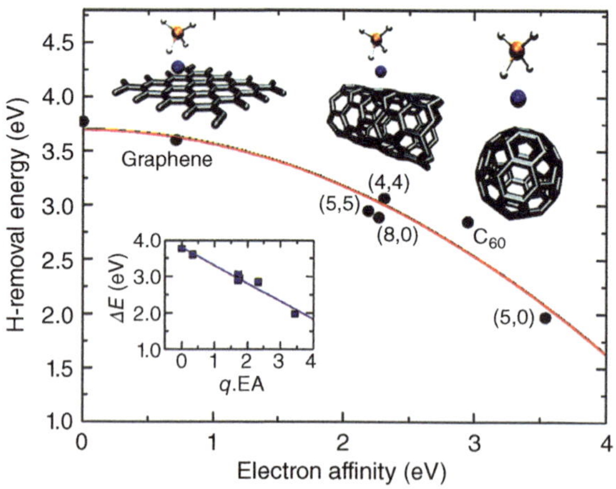

Figure 10.24 Correlation of the carbon substrate EA and the hydrogen removal energy. Black circles are ab initio results for H-removal energy as a function of EA of the carbon substrate. The red curve is a fit of the data with a quadratic function. The inset below the red curve shows the linear relationship between the H-removal energy (ΔE) to the product of transferred charge (q) and EA. The inset above the red curve displays the equilibrium configurations for $NaAlH_4$ interacting with a graphene sheet, (5,0) zigzag carbon nanotubes, and fullerene. Na atoms are shown as blue spheres, Al in gold, C in dark gray, and H in white. *Source:* Berseth et al. [72]. © American Chemical Society.

10.6.1 Zn in +III Oxidation State

The oxidation state of an atom is the charge it would have in a compound if the bonding were completely ionic. This is a fundamental quantity for understanding redox reactions, reaction mechanisms, catalysis, etc. Group 12 elements Zn, Cd, and Hg have an oxidation state of +II. While, due to relativistic effects, Hg has been known to exist in +IV oxidation state, finding Zn in +III state has been challenging. This is because Zn has an outer electron configuration of $3d^{10}4s^2$. For Zn to have an oxidation state of +III, the filled $3d$-core orbitals have to be involved. This is difficult as the third ionization potential of Zn is 39.7 eV. It is needless to mention that the discovery of Zn in +III oxidation state will open a new branch in Zn chemistry as zinc has many applications in the pure metallic state in forming alloys, as salts used in white pigments, as biocomplexes in metalloenzymes, and as organometallic reagents in organic synthesis. Realizing that superhalogens can have large EAs, Samanta and Jena [73] studied the interaction of a Zn atom with F, BO_2, and AuF_6 whose EAs are progressively large, i.e. 3.4, 4.5, and 8.4 eV, respectively. For Zn to be in +III oxidation state, it would need to bind to three atoms and there must be involvement of $3d$ electrons in the bonding. The authors calculated the equilibrium geometries, electronic structure, and nature of bonding of neutral and anionic form of ZnX_3 (X = F, BO_2, AuF_6). The results are shown in Figure 10.25. Note that in ZnF_2 and $Zn(BO_2)_2$, Zn has an oxidation state of +II. In neutral ZnF_3, the Zn-F bond lengths are different, one F atom is at a distance of 1.76 Å while the other two are at 1.93 Å. This implies that two of the F atoms are bound in a quasi-molecular form while the third one is bound in atomic form. In the case of $Zn(BO_2)_3$, there are also two bonds with Zn and two of the BO_2 moieties dimerize and bind as one unit to Zn. In addition, Zn clearly makes three bonds in ZnF_3^- and $Zn(BO_2)_3^-$ and, as expected, both are superhalogens with EAs of 6.2 and 5.63 eV, respectively. Thus, it is clear that neither of these ligands can make Zn to achieve a +III oxidation state.

$Zn(AuF_6)_3$ on the other hand has C_2 symmetry where Zn is surrounded by three AuF_6 units. However, the three Au atoms are not equivalent, with one Au atom being at a distance of 3.15 Å from the Zn atom while the other two Au atoms are at a distance of 3.26 Å. Zn is hexa-coordinated with six F atoms. We note that as in the case of ZnF_3, here also two AuF_6 moieties are close to each other, but they do not dimerize. In addition, in contrast to ZnF_3, $Zn(AuF_6)_3$ is stable with respect to all of the fragmentation channels studied. An analysis of the natural bond orbitals (NBOs) shows that the $3d$ orbital of Zn in completely full. While, ideally, one would have liked to see $3d$ orbital participation in bonding, the fact that three AuF_6 units maintain their structural identity and the optimized $Zn(AuF_6)_3$ is a stable molecule, one could argue that Zn in this case is in a +III oxidation state. For a more conniving evidence, it is worthwhile to search for different ligands that may be even more electronegative or a ligand that is stable as a trianion [74, 75].

10.6.2 Covalent Binding of Noble Gas Atoms

Noble gas (NG) atoms seldom form chemical bonds because of their very large ionization potentials and zero to negative EAs. Nearly a century ago, Pauling [76] speculated that heavier NG atoms such as Xe and Kr may form stable compounds. In 1962 Bartlett [77] confirmed this speculation by synthesizing the first Xe-containing compound, $Xe^+PtF_6^-$. Note that PtF_6 is a superhalogen. With an EA of 7 eV, it is capable of oxidizing oxygen

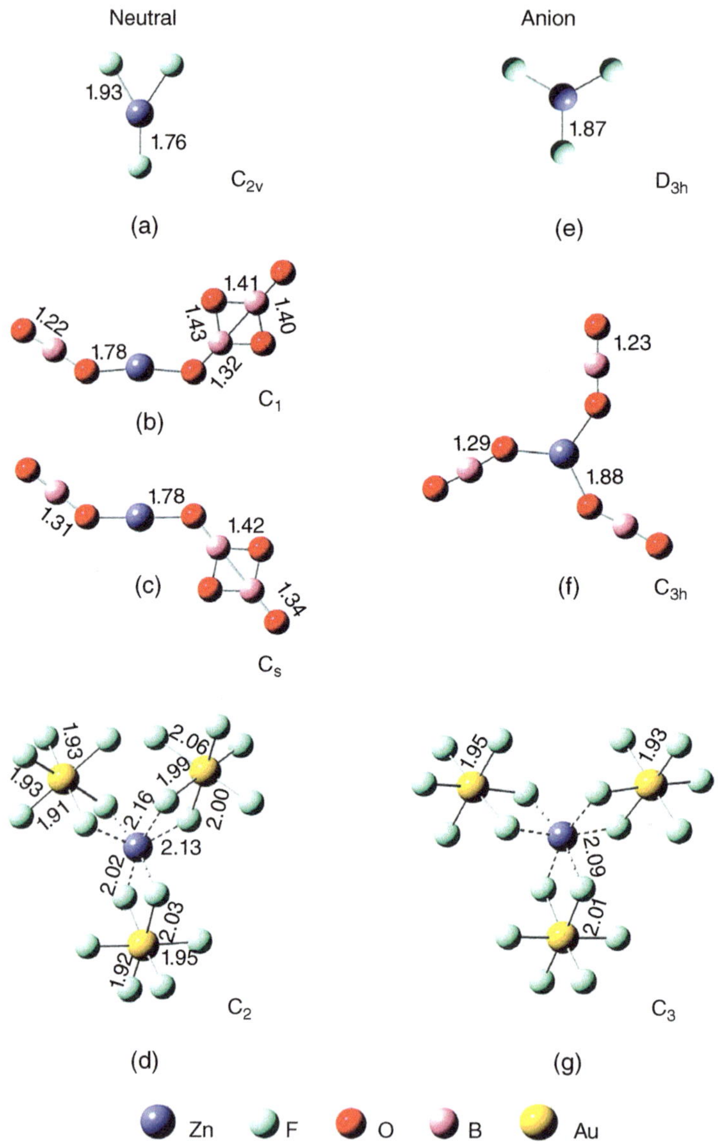

Figure 10.25 Optimized structures of (a–d) neutral and (e–g) anionic ZnX_3 (X = F, BO_2, AuF_6) clusters. Bond lengths are in Å. *Source:* Samanta and Jena [73]. © American Chemical Society.

molecule as well as a xenon atom whose ionization potentials are 12.07 and 12.13 eV, respectively. Since then, several compounds of Xe and Kr have been found [78–86]. However, compounds containing lighter NG atoms are rare.

In particular, argon, the most abundant NG on Earth, binds to molecules at near absolute zero. Discovery of a single neutral Ar-containing compound HArF [87], isolated in a low-temperature matrix, was a seminal achievement and opened a new chapter in NG chemistry. These kinds of compounds are known as "insertion compounds" where HAr moiety carries a partial positive charge and F carries an equal but opposite charge. Since then, some other NG-containing compounds HKrF, HKrCl, and HXeNC have been synthesized [81–86]. Realizing that the EAs of Cl and CN are higher than that of F, Samanta [88] systematically studied the role of superhalogens, Y, in forming more strongly bound H(NG)Y (NG = Ar, Kr; Y = F, BO_2, BF_4) molecules. Note that the EAs of F, BO_2, and BF_4 are 3.4, 4.46, and 6.86 eV, respectively. All these molecules were found to be metastable. However, the ionic characters as well as the dipole moments of HArY and HKrY are found to increase with the increasing EA of the superhalogens, Y.

Strongest interaction of Ar have been reported with transient cationic species [89–94]. RT Ar chemistry is currently limited to a small number of isolated cations, whose observation and characterization are rather challenging. Because of the negative EA of Ar, its binding to negative ions is seldom considered. Recently, Mayer et al. [95] reported a counterintuitive method where Ar can be bound covalently at RT to a negative ion. This involves a strong electrophilic center firmly embedded in a negatively charged framework of exceptional stability. The starting point is the extremely sable percyano-dodecaborate $B_{12}(CN)_{12}^{2-}$ whose second electron is bound by 5.5 eV, making it the most stable dianion in nature. This molecule, predicted earlier by Jena and coworkers [96], is discussed in more detail in Chapter 2. It is stabilized by satisfying two electron-counting rules simultaneously, the Wade–Mingos rule for the boron cage and the octet rule for the CN^- moiety. Removal of CN^- from $B_{12}(CN)_{12}^{2-}$ leaves behind a mono-anion $B_{12}(CN)_{11}^-$ with a strong electrophilic B site (Figure 10.26). This design was based on an earlier work by Rohdenburg et al. [97] who

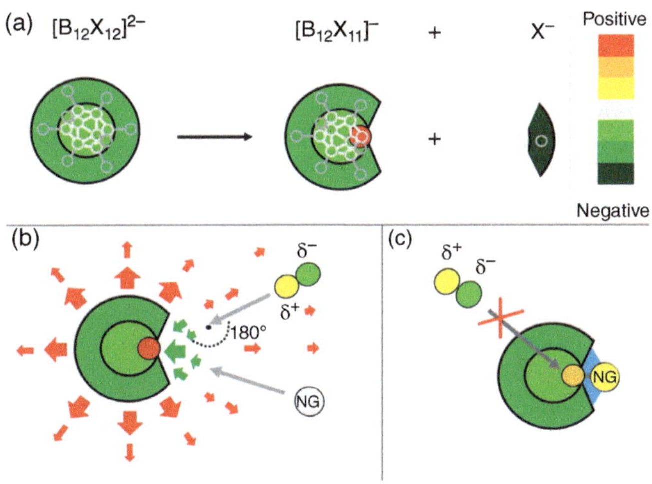

Figure 10.26 (a) Fragmentation reaction that yields the electrophilic anion $[B_{12}X_{11}]^-$ depicting the differently charged areas within the reactant and product. Note that the color coding is just for illustration purposes and is neither quantitative nor linear. (b) Visualization of the electric field resulting from the molecular charge. Colored arrows depict the force on a negative point charge, which is large and attractive near the positively charged binding site (green arrows) and becomes repulsive (red arrows) at larger distances due to the overall negative charge. The change in direction of the electric field close to the binding site leads to reversal of the preferred orientation of a dipole. (c) The NG atom binds to the positive binding site, provides electron density (dative bond, dark gray), and is charged slightly positive, allowing for an attractive electrostatic interaction with the surrounding negatively charged substituent shell. Nucleophilic substitution (SN)2-type substitution of the bound NG atom is prevented by the dodecaborate scaffold (reaction path crossed out in red). Additionally, a large contact surface is available for dispersion interactions. *Source:* Mayer et al. [95]. © National Academy of Sciences.

abstracted Cl^- from $B_{12}Cl_{12}^{2-}$ to form a super electrophilic $B_{12}Cl_{11}^-$, which binds to Xe and Kr at RT. Here, the free B coordination site carries a positive charge and is electrophilic, even though the entire cluster carries a net negative charge. The strongly electrophilic (positive) binding site within the overall anionic framework generates an unusual electric field in the vicinity of the reactive boron atom. In addition to the dative NG-B bond, dispersion interactions and electrostatics contribute to the stabilization of NG complexes. The charge on the free B site in $B_{12}(CN)_{11}^-$ was found to be +0.82*e*. The Ar attachment energy to $B_{12}(CN)_{11}^-$ was calculated to be 61 kJ/mol with a B-Ar distance of 1.98 Å. These values show that Ar is covalently bonded to B and open the door to new NG chemistry.

10.7 Conclusions

In this chapter, we discussed the merits of materials where clusters, instead of individual atoms, form the building blocks. The concept of CAMs was inspired by the fact that fullerites and fullerides, which are crystals composed of C_{60} clusters, have very different properties than diamond and graphite where individual carbon atoms form the building blocks. However, no other cluster created in the gas phase from a single element has had the same success as C_{60} in producing bulk materials. Early promise that Ti_8C_{12} and their analogues, commonly known as metallo-carbohedrenes or "met-cars" [98–100], could be used to synthesize cluster-assembled crystals did not materialize as these clusters were found to be reactive and coalesced when brought together. Significant progress has been made by either protecting the bare clusters with ligands or depositing clusters on substrates. In either case, the properties of bare clusters are modified by their interaction with the ligands or substrate atoms. In addition, large quantities of these materials have not been synthesized to be useful for practical applications.

In this chapter we took a different view. Instead of first producing stable clusters in the gas phase and then putting them together to form a CAM, we propose to synthesize a new class of what we term as cluster-based materials. Here, we replace an atom in a conventional crystal by a cluster that has the same chemistry as the atom being replaced. Instead of creating this cluster in the gas phase, we look for a conventional route for its synthesis. The advantage is that the size and the symmetry of the cluster are different from the atom and these flexibilities could lead to different properties. To illustrate this procedure, consider two crystals that already exist – KCl and $KMnO_4$. Here, the MnO_4 is a superhalogen that has the same chemistry as Cl, but due to the size and composition, properties of $KMnO_4$ are different from that of KCl.

We discussed two different classes of cluster-based materials; one for energy applications and the other for promoting unusual reactions. The former includes materials for harvesting solar energy, electrochemical storage in LIBs and Na-ion batteries, and hydrogen storage. The latter class corresponds to highly electronegative clusters that can enable Zn to achieve +III oxidation state and allow NG atoms to form chemical bonds.

For solar cells, we considered the hybrid-perovskite solar cells where the cations are organic and the anions are inorganic. We demonstrated that replacing iodine with cluster ions, such as BH_4, one can reduce the sensitivity of these materials to moisture. In the case of LIBs, we showed that the conventional halogen-based electrolytes can be replaced by halogen-free electrolytes such as $LiCB_{11}H_{12}$ with remarkable enhancement in safety and ionic conductivity. Similarly, all solid-state LIBs and Na-ion batteries with superionic conductivity can be made possible by replacing halogens in antiperovskites Li_3OCl with cluster ions such as Li_3OBH_4, Li_3SBF_4, and Na_3SBF_4 whose ionic conductivities are orders of magnitude higher than that of Li_3OCl. Lessons learnt from cluster science has also helped to develop efficient hydrogen storage materials. Based on mechanisms introduced by Kubas and by Jena and coworkers, it has been shown that hydrogen molecules can bind to clusters in quasi-molecular form and large quantities of hydrogen with high gravimetric density can be stored under ambient conditions.

Knowledge gained from cluster science has helped to design highly stable multiply charged cluster ions. In addition, size and composition of clusters can be tailored to produce superhalogens and hyperhalogens with large EAs. The species can help Zn to achieve an oxidation state of +III and NG atoms like argon to form chemical bonds at RT. In Chapter 12 of this book, the use of clusters in the sequestration and conversion of CO_2 is discussed.

While identifying very stable clusters in the gas phase and putting them together to synthesize bulk materials like fullerites and fullerides still remain a dream of cluster science, much can be achieved in developing novel materials by replacing elemental ions with cluster ions using a top-down approach. A paradigm shift in materials synthesis with clusters as building blocks can usher a new era in materials synthesis.

References

1. Jena, P. and Sun, Q. (2018). Super atomic clusters: design rules and potential for building blocks of materials. *Chem. Rev.* 118: 5755–5870.
2. Jena, P. (2013). Beyond the periodic table of elements: the role of superatoms. *J. Phys. Chem. Lett.* 4: 1432.
3. Khanna, S.N. and Jena, P. (1992). Assembling crystals from clusters. *Phys. Rev. Lett.* 69: 1664–1667.
4. Jalali-Asadabadi, S., Ghasemikhah, E., Ouahrani, T. et al. (2016). Electronic structure of crystalline buckyballs: fcc-C60. *J. Electron. Mater.* 45: 339–348.
5. Kroto, H.W., Heath, J.R., O'Brien, S.C. et al. (1985). C60: buckminsterfullerene. *Nature* 318: 162–163.
6. O'Shea, J.N. (2005). Molecular orbitals tell the story. *Science* 310: 453–454.
7. Shirley, E.L. and Louie, S.G. (1993). Electron excitations in solid C_{60}: energy gap, band dispersions, and effects of orientational disorder. *Phys. Rev. Lett.* 71: 133–136.
8. Hebard, A.F., Rosseinsky, M.J., Haddon, R.C. et al. (1991). Superconductivity at 18 K in potassium-doped C_{60}. *Nature (London)* 350: 600–601.
9. Yusuke, N., Shiro, S., Massimo, C., and Ryotaro, A. (2016). Exotic s-wave superconductivity in alkali-doped fullerides. *J. Phys.: Condens. Matter* 28: 153001.
10. Ivanovskii, A.L. (2008). Assemblies of carbon and boron–nitrogen nanotubes and fullerenes: structure and properties. *Russ. J. Inorg. Chem.* 53: 2083–2102.
11. Tsunoyama, H., Akatsuka, H., Shibuta, M. et al. (2017). Development of integrated dry–wet synthesis method for metal encapsulating silicon cage superatoms of $M@Si_{16}$ ($M = Ti$ and Ta). *J. Phys. Chem. C* 121: 20507–20516.

12 Nakaya, M., Iwasa, T., Tsunoyama, H. et al. (2014). Formation of a superatom monolayer using gas-phase-synthesized Ta@Si_{16} nanocluster ions. *Nanoscale* 6: 14702–14707.

13 Gutsev, G.L., Rao, B.K., Jena, P. et al. (1999). Origin of the unusual stability of MnO_4−. *Chem. Phys. Lett.* 312 (5–6): 598–605.

14 Giri, S., Behera, S., and Jena, P. (2014). Superalkalis and sueprhalogens as building blocks of supersalts. *J. Phys. Chem. A* **118**: 638.

15 Jena, P. (2015). Superhalogens: a bridge between complex metal hydrides and Li-ion batteries. *J. Phys. Chem. Lett.* **6**: 1119.

16 Hodes, G. (2013). Perovskite-based solar cells. *Science* 342: 317–318.

17 Fang, H. and Jena, P. (2016). Super-ion inspired colorful hybrid Perovskite solar cells. *J. Mat. Chem. A* 4: 4728.

18 Eames, C., Frost, J.M., Barnes, P.R. et al. (2015). Ionic transport in hybrid lead iodide perovskite solar cells. *Nat. Commun.* 6: 7497.

19 Willis, M., Gotz, M., Kandalam, A.K. et al. (2010). Hyperhalogens: discovery of a new class of electro-negative species. *Angew. Chem. Int. Ed.* 49: 8966.

20 Schouwink, P., Ley, M.B., Tissot, A. et al. (2014). Structure and properties of complex hydride perovskite materials. *Nat. Commun.* 5: 5705.

21 Fang, H. and Jena, P. (2016). Molecular origin of the properties of organic–inorganic hybrid perovskites: the big picture from small clusters. *J. Phys. Chem. Lett.* 7: 1596–1603.

22 Dou, L., Wong, A.B., Yu, Y. et al. (2015). Atomically thin two dimensional organic-inorganic hybrid perovskites. *Science* 349: 1518–1521.

23 Yao, Q., Fang, H., Deng, K. et al. (2016). Superhalogens as building blocks of two-dimensional organic–inorganic hybrid perovskites for optoelectronics applications. *Nanoscale* 8: 17836–17842.

24 Giri, S., Behera, S., and Jena, P. (2014). Superhalogens as building blocks of halogen-free electrolytes in Li-ion batteries. *Angew. Chem. Int. Ed.* **53**: 13916.

25 Matsuo, M., Nakamori, Y., Orimo, S.-i. et al. (2007). Lithium superionic conduction in lithium borohydride accompanied by structural transition. *Appl. Phys. Lett.* 91: 224103.

26 Stephenson, C.C., Rice, D.W., and Stockmayer, W.H. (1955). Orderdisorder transitions in the alkali borohydrides. *J. Chem. Phys.* 23: 1960–1960.

27 Oguchi, H., Matsuo, M., Hummelshøj, J.S. et al. (2009). Experimental and computational studies on structural transitions in the $LiBH_4$—LiI pseudobinary system. *Appl. Phys. Lett.* 94: 141912.

28 Matsuo, M., Remhof, A., Martelli, P. et al. (2009). Complex hydrides with $(BH_4)^-$ and $(NH_2)^-$ anions as new lithium fast-ion conductors. *J. Am. Chem. Soc.* 131: 16389–16391.

29 Pathak, B., Samanta, D., Ahuja, R., and Jena, P. (2011). Borane derivatives: a new class of super and hyper halogens. *Chem. Phys. Chem.* 12: 2422.

30 Tutusaus, O., Mohtadi, R., Arthur, T.S. et al. (2015). An efficient halogen-free electrolyte for use in rechargeable magnesium batteries. *Angew. Chem., Int. Ed.* 54: 7900–7904.

31 Tang, W.S., Unemoto, A., Zhou, W. et al. (2015). Unparalleled lithium and sodium superionic conduction in solid electrolytes with large monovalent cage like anions. *Energy Environ. Sci.* 8: 3637–3645.

32 Zhao, H., Zhou, J., and Jena, P. (2016). Stability of $B_{12}(CN)_{12}{}^{2-}$: implications for lithium and magnesium ion batteries. *Angew. Chem. Int. Ed.* 55: 3704–3708.

33 Hong, F. and Jena, P. (2017). $B_{12}(SCN)_{12}^{2-}$: an ultra-stable weakly-coordinating dianion for electrolyte salts in metal-ion batteries. *J. Phys. Chem. C.* 121: 7697.

34 Zhang, Y., Zhao, Y., and Chen, C. (2013). Ab initio study of the stabilities of and mechanism of superionic transport in lithium-rich antiperovskites. *Phys. Rev. B: Condens. Matter Mater. Phys.* 87: 134303.

35 Zhao, Y. and Daemen, L.L. (2012). Superionic conductivity in lithium-rich anti-perovskites. *J. Am. Chem. Soc.* 134: 15042–15047.

36 Emly, A., Kioupakis, E., and Van der Ven, A. (2013). Phase stability and transport mechanisms in antiperovskite Li_3OCl and Li_3OBr superionic conductors. *Chem. Mater.* 25: 4663–4670.

37 Braga, M.H., Ferreira, J.A., Stockhausen, V. et al. (2014). Novel Li_3ClO based glasses with superionic properties for lithium batteries. *J. Mater. Chem. A* 2: 5470–5480.

38 Fang, H., Wang, S., Liu, J. et al. (2017). Superhalogenbased lithium superionic conductors. *J. Mater. Chem. A* 5: 13373–13381.

39 Fang, H. and Jena, P. (2017). Li-rich antiperovskite superionic conductors based on cluster ions. *Proc. Natl. Acad. Sci.* 14: 1047.

40 Chen, L., Fiore, M., Wang, J.E. et al. (2018). Readiness level of sodium-ion battery technology: a materials review. *Adv. Sustain. Syst.* 2 (3): 1700153. https://doi.org/10.1002/adsu.201700153.

41 Wanger, T.C. (2011). The lithium future-resources, recycling, and the environment. *Conserv. Lett.* 4 (3): 202–206. https://doi.org/10.1111/j.1755-263X.2011.00166.x.

42 Zhu, Z., Chu, I.-H., Deng, Z., and Ong, S.P. (2015). Role of Na^+ interstitials and dopants in enhancing the Na+ conductivity of the cubic Na_3PS_4 superionic conductor. *Chem. Mater.* 27: 8318–8325.

43 Wang, Y., Wang, Q., Liu, Z. et al. (2015). Structural manipulation approaches towards enhanced sodium ionic conductivity in Na-rich antiperovskites. *J. Power Sour.* 293: 735–740.

44 Zhao, Y., Wang, Y., Zou, R. (2016). Sodium anti-perovskite solid electrolyte compositions. WO Application WO2016026130A1.

45 Zhu, J., Wang, Y., Li, S. et al. (2016). Sodium ion transport mechanisms in antiperovskite electrolytes Na_3OBr and Na_4OI_2: an in-situ neutron diffraction study. *Inorg. Chem.* 55: 5993–5998.

46 Dawson, J.A., Chen, H., and Islam, M.S. (2018). Composition screening of lithium- and sodium-rich anti-perovskites for fast-conducting solid electrolytes. *J. Phys. Chem. C* 122: 23978–23984.

47 Ramanna, J., Yedukondalu, N., Babu, K.R., and Vaitheeswaran, G. (2013). Ab initio study of electronic structure, elastic and optical properties of anti-perovskite type alkali metal oxyhalides. *Solid State Sci.* 20: 120–126.

48 Fang, H. and Jena, P. (2019). Sodium superionic conductors based on clusters. *ACS Appl. Mat. Interf.* 11: 963.

49 Jena, P. (2011). Materials for hydrogen storage: Past, present, and future. *J. Phys. Chem. Lett.* 2: 206–211.

50 Kubas Gregory, J. (1988). Molecular hydrogen complexes: coordination of a sigma bond to transition metals. *Acc. Chem. Res.* 21: 120–128.

51 Kubas, G.J. (2001). Metal–dihydrogen and σ-bond coordination: the consummate extension of the Dewar–Chatt–Duncanson model for metal–olefin π bonding. *J. Organomet. Chem.* 635: 37–68.

52 Niu, J., Rao, B.K., and Jena, P. (1992). Binding of hydrogen molecules by a transition-metal ion. *Phys. Rev. Lett.* 68: 2277–2280.

53 Niu, J., Rao, B.K., Jena, P., and Manninen, M. (1995). Interaction of H_2 and He with metal atoms, clusters, and ions. *Phys. Rev. B: Condens. Matter Mater. Phys.* 51: 4475––4484.

54 Zhou, J., Wang, Q., Sun, Q. et al. (2010). Electric field enhanced hydrogen storage on polarizable materials substrates. *Proc. Natl. Acad. Sci. USA* 107: 2801–2806.

55 Zhao, Y., Kim, Y.-H., Dillon, A.C. et al. (2005). Hydrogen storage in novel organometallic buckyballs. *Phys. Rev. Lett.* 94: 155504.

56 Yildirim, T. and Ciraci, S. (2005). Titanium-decorated carbon nanotubes as a potential high-capacity hydrogen storage medium. *Phys. Rev. Lett.* 94: 175501.

57 Sun, Q., Wang, Q., Jena, P., and Kawazoe, Y. (2005). Clustering of Ti on a C60 surface and its effect on hydrogen storage. *J. Am. Chem. Soc.* 127: 14582–14583.

58 Sun, Q., Jena, P., Wang, Q., and Marquez, M. (2006). First-principles study of hydrogen storage on $Li_{12}C_{60}$. *J. Am. Chem. Soc.* 128: 9741–9745.

59 Teprovich, J.A., Wellons, M.S., Lascola, R. et al. (2012). Synthesis and characterizationof a lithium-doped fullerane (Lix-C60-Hy) for reversible hydrogen storage. *Nano Lett.* 12: 582–589.

60 Wang, Q. and Jena, P. (2012). Density functional theory study of the interaction of hydrogen with Li_6C_{60}. *J. Phys. Chem. Lett.* 3: 1084–1088.

61 Satyapal, S., Petrovic, J., Read, C. et al. (2007). Department of Energy's National Hydrogen Storage Project: progress towards meeting hydrogen-powered vehicle requirements. *Catal. Today* 120: 246–256.

62 Orimo, S.-I., Nakamori, Y., Ohba, N. et al. (2006). Experimental studies on intermediate compound of $LiBH_4$. *Appl. Phys. Lett.* 89: 021920.

63 Ohba, N., Miwa, K., Aoki, M. et al. (2006). First-principles study on the stability of intermediate compounds of $LiBH_4$. *Phys. Rev. B: Condens. Matter Mater. Phys.* 74: 075110.

64 Chong, M., Karkamkar, A., Autrey, T. et al. (2011). Reversible dehydrogenation of magnesium borohydride to magnesium triborane in the solid state under moderate conditions. *Chem. Commun.* 47: 1330–1332.

65 Yan, Y., Remhof, A., Rentsch, D. et al. (2013). Is Y2(B12H12)3 the main intermediate in the decomposition process of $Y(BH_4)_3$? *Chem. Commun.* 49: 5234–5236.

66 Li, S., Willis, M., and Jena, P. (2010). Reaction intermediates during the dehydrogenation of metal borohydrides: a cluster perspective. *J. Phys. Chem. C* 114: 16849–16854.

67 Liu, Y., Giri, S., Zhou, J., and Jena, P. (2014). Intermediate phases during decomposition of metal borohydrides, $M(BH_4)_n$ (M = Na, Mg, Y). *J. Phys. Chem. C* 118: 28456–28461.

68 Mohtadi, R., Remhof, A., and Jena, P. (2016). Complex metal borohydrides: multifunctional materials for energy storage and conversion. *J. Phys.: Condens. Matter* 28: 353001.

69 Yan, Y., Remhof, A., Rentsch, D. et al. (2015). A novel strategy for reversible hydrogen storage in $Ca(BH_4)_2$. *Chem. Commun.* 51: 11008–11011.

70 Bogdanović, B. and Schwickardi, M. (1997). Ti-doped alkali metal aluminium hydrides as potential novel reversible hydrogen storage materials. *J. Alloys Compd.* 253–254: 1–9.

71 Rao, B.K., Jena, P., Burkart, S. et al. (2001). AlH_3 and Al_2H_6: magic clusters with unmagical properties. *Phys. Rev. Lett.* 86: 692–695.

72 Berseth, P.A., Harter, A.G., Zidan, R. et al. (2009). Carbon nanomaterials as catalysts for hydrogen uptake and release in $NaAlH_4$. *Nano Lett.* 9: 1501–1505.

73 Samanta, D. and Jena, P. (2012). Zinc in +III Oxidation State. *J. Am. Chem. Soc. (Commun.)* 134: 8400.

74 Zhao, T., Zhu, J., Wang, Q., and Jena, P. (2017). Colossal stability of gas-phase tri-anions: super-pnictogens. *Angew. Chem. Int. Ed.* 56: 13421.

75 Fang, H. and Jena, P. (2019). Stable tetra- and penta-anions in the gas phase. *Angew. Chem. Int. Ed.* 58: 1.

76 Pauling, L. (1933). The formulas of antimonic acid and the antimonates. *J. Am. Chem. Soc.* 55: 1895–1900.

77 Bartlett, N. (1962). Xenon hexafluoroplatinate (V) $Xe^+[PtF_6]^-$. *Proc. Chem. Soc.* 218: 197–236.

78 Grochala, W. (2007). Atypical compounds of gases, which have been called 'Noble'. *Chem. Soc. Rev.* 36: 1632–1655.

79 Christe, K.O. (2001). A renaissance in noble gas chemistry. *Angew. Chem., Int. Ed.* 40: 1419–1421.

80 Khriachtchev, L., Räsänen, M., and Gerber, R.B. (2009). Noble-gas hydrides: new chemistry at low temperatures. *Acc. Chem. Res.* 42: 183–191.

81 Patterson, M., Lundell, J., and Räsänen, M. (1995). Neutral rare-gas containing charge-transfer molecules in solid matrices. I. HXeCl, HXeBr, HXeI, and HKrCl in Kr and Xe. *J. Chem. Phys.* 102: 6423–6431.

82 Lundell, J., Khriachtchev, L., Pettersson, M., and Räsänen, M. (2000). Formation and characterization of neutral krypton and xenon hydrides in low-temperature matrices. *Low Temp. Phys.* 26: 680–690.

83 Khriachtchev, L., Isokoski, K., Cohen, A. et al. (2008). A small neutral molecule with two noble-gas atoms: HXeOXeH. *J. Am. Chem. Soc.* 130: 6114–6118.

84 Lignell, A., Khriachtchev, L., Lundell, J. et al. (2006). On theoretical predictions of noble-gas hydrides. *J. Chem. Phys.* 125: 184514.

85 Pettersson, M., Khriachtchev, L., Lignell, A. et al. (2002). HKrF in solid krypton. *J. Chem. Phys.* 116: 2508–2515.

86 Pettersson, M., Lundell, J., Khriachtchev, L., and Räsänen, M. (1998). Neutral rare-gas containing charge-transfer molecules in solid matrices. III. HXeCN, HXeNC, and HKrCN in Kr and Xe. *J. Chem. Phys.* 109: 618–625.

87 Khriachtchev, L., Pettersson, M., Runeberg, N. et al. (2000). A stable argon compound. *Nature* 406: 874––876.

88 Samanta, D. (2014). Prediction of superhalogen-stabilized noble gas compounds. *J. Phys. Chem. Lett.* 5: 3151–3156.

89 Koskinen, J.T. and Cooks, R.G. (1999). Novel rare gas ions BXe^+, BKr^+, and BAr^+ formed in a halogen/rare gas exchange reaction. *J. Phys. Chem. A* 103: 9565–9568.

90 Ascenzi, D., Tosi, P., Roithová, J., and Schröder, D. (2008). Gas-phase synthesis of the rare-gas carbene cation $ArCH_2^+$ using doubly ionized bromomethane as a superelectrophilic reagent. *Chem. Commun. (Camb.)*: 4055–4057.

91 Ascenzi, D., Tosi, P., Roithová, J. et al. (2008). Generation of the organo-rare gas dications $HCCRg^{2+}$ (Rg = Ar and Kr) in the reaction of acetylene dications with rare gases. *Phys. Chem. Chem. Phys.* 10: 7121–7128.

92 Roithová, J. and Schröder, D. (2009). Silicon compounds of neon and argon. *Angew. Chem. Int. Ed. Engl.* 48: 8788–8790.

93 Lockyear, J.F., Douglas, K., Price, S.D. et al. (2010). Generation of the $ArCF_2^{2+}$ dication. *J. Phys. Chem. Lett.* 1: 358–362.

94 Wagner, J.P., McDonald, D.C. 2nd, and Duncan, M.A. (2018). An argon-oxygen covalent bond in the $ArOH^+$ molecular ion. *Angew. Chem. Int. Ed. Engl.* 57: 5081–5085.

95 Mayer, M., Van Lessen, V., Rohdenburg, M. et al. (2019). Rational design of an argon-binding superelectrophilic anion. *Proc. Natl. Acad. Sci. USA* 116: 8167–8172. https://doi.org/10.1073/pnas.1820812116.

96 Zhao, H., Zhou, J., and Jena, P. (2016). Stability of $B_{12}(CN)_{12}^{2-}$: implications for lithium and magnesium ion batteries. *Angew. Chem. Int. Ed. (VIP)* 55: 3704.

97 Rohdenburg, M., Mayer, M., Grellmann, M. et al. (2017). Superelectrophilic behavior of an anion demonstrated by the spontaneous binding of noble gases to $[B_{12}Cl_{11}]^-$. *Angew. Chem. Int. Ed. Engl.* 56: 7980–7985.

98 Guo, B.C., Kerns, K.P., and Castleman, A.W. Jr. (1992). Ti_8C^{12+}-metallo-carbohedrenes: a new class of molecular clusters? *Science* 255: 1411.

99 Guo, B.C., Wei, S., Purnell, J. et al. (1992). Metallo-carbohedrenes $[M_8C_{12}{}^+$ (M = V, Zr, Hf, and Ti)]: a class of stable molecular cluster ions. *Science* 256: 515.

100 Wei, S., Guo, B.C., Purnell, J. et al. (1992). Metallo-carbohedrenes: formation of multi-cage structures. *Science* 256: 818.

11

Thermal and Thermoelectric Properties of Cluster-based Materials

Tingwei Li[1,2], *Qiang Sun*[1,2], *and Puru Jena*[3]

[1] *School of Materials Science and Engineering, Peking University, Beijing, China*
[2] *Center for Applied Physics and Technology, Peking University, Beijing, China*
[3] *Physics Department, Virginia Commonwealth University, Richmond, VA, USA*

11.1 Introduction

Heat is one of the most common forms of energy in our daily life. However, about two-thirds of the heat is released to the atmosphere, causing severe energy waste and environment problems. Conversion of this waste heat into usable energy is important not only to address the impending energy crisis but also to mitigate its adverse environmental impact. To this end, thermoelectric (TE) materials are of current interest where a temperature difference generates an electric field and vice versa, enabling direct conversion of heat energy to electrical energy [1] as evidenced from the Seebeck effect, Peltier effect, and Thomson effect since the nineteenth century [2]. A solid-state TE device is free of moving parts, noise, and greenhouse emission. In addition, it can work in tandem with other energy-conversion technologies [3]. However, due to the low performance of TE materials [4], the efficiency of current TE devices is lower than that of other energy-conversion technologies. Thus, development of new materials with high TE performance is a hotly pursued field of research [5].

A good strategy for developing new materials with high TE performance is to go beyond the conventional synthesis route by using clusters, instead of atoms, as the building blocks. In this chapter, we explore such an option. Note that the size, shape, and composition of clusters can be varied one atom at a time, lending diversity and flexibility in the design and synthesis of cluster-based materials. In recent years, it has been demonstrated that phonon scattering can be enhanced and lattice thermal conductivities can be reduced in materials composed of Mo_6 and Mo_9 cluster selenides [6], tetrahedrite Ag-based clusters [7], B-based clusters [8], Zintl-phase clusters [9], argyrodite compounds [10], and superatomic clusters [11]. In this chapter, we briefly review the progress made in using cluster-assembled materials for TE applications, aiming to promote and stimulate more effort in this direction.

Superatoms: Principles, Synthesis and Applications, First Edition. Edited by Purusottam (Puru) Jena and Qiang Sun.
© 2022 John Wiley & Sons Ltd. Published 2022 by John Wiley & Sons Ltd.

In the following, we first review the basic theory behind the TE effect and the material characteristics needed for increasing the efficiency of TE devices. In Section 11.3, we discuss the advantages inherent in cluster-based materials such as large unit cell of hetero-atomic composition, structural disorder, and orientational disorder. Examples of cluster-based materials and their TE properties are given in Section 11.4, and a summary of our conclusions is provided in Section 11.5.

11.2 Basic Theory

11.2.1 Thermoelectric Effect

The first TE effect discovered by Seebeck [12] and later explained by Oersted relates to the electric potential, ΔV, that appears at the junction of two materials a and b when exposed to a temperature gradient (Figure 11.1a). The well-known application of the Seebeck effect is the measurement of temperature by a thermocouple. The Seebeck coefficient is given as [14] $\alpha_{ab} = \Delta V/\Delta T$, where ΔV is the voltage difference and ΔT is the temperature difference. The subscript ab signifies two conductors, A and B, of the thermocouple. All materials can exhibit the TE effect to some extent. The known materials typically have a Seebeck coefficient between 100 and 1000 μV/K.

Later on, Peltier [15] discovered the second effect, which is opposite to the Seebeck effect, i.e. a temperature difference can be induced by a current I passing through the junction of two different materials (Fig. 11.1b). The magnitude of Peltier heat is directly proportional to the current, with the sign changing for cooling or heating effect, depending on the direction of the electric current. The rate of evolved or absorbed heat, q, is directly proportional to the electric current in the circuit and is given as [13] $q = (\pi a - \pi b)I$, where I denotes the electric current from conductor A–B and πa and πb are the proportionality constants known as Peltier coefficient for conductors A and B.

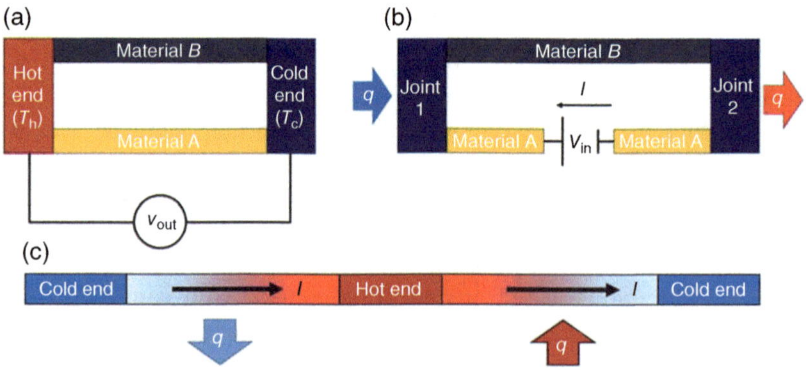

Figure 11.1 (a) Seebeck effect: Two TE materials joined at hot and cold junctions, which are maintained at temperatures T_h and T_c. A potential appears across the junctions. (b) Peltier effect: Two TE materials joined at the two ends. An electric current "I" is passed through the circuit, and heat q is evolved at one end and absorbed at the other. (c) Thomson effect induced by spatial gradient in temperature and current. *Source:* Kishore and Priya [13]. Licenced under CC BY-4.0. © MDPI.

The third effect [16] occurs when both a gradient of temperature and an electric current exist simultaneously. When current passes through a material under a temperature gradient, the material exchanges heat with the outside medium. Conversely, a current is produced when a heat flux passes through this material (Figure 11.3c). The thermal flux, $\mathrm{d}Q/\mathrm{d}x$, is given by $\tau I \mathrm{d}T/\mathrm{d}x$, with τ being the Thomson coefficient [17].

It is important to note that the Seebeck and Peltier effects occur in a thermocouple, whereas Thomson effect can exist in a single conductor. In addition, Seebeck coefficient (α), Peltier coefficient (π), and Thomson coefficient (τ) are related by the Kelvin relations [18]:

$$\pi_{ab} = \alpha_{ab} T \tag{11.1}$$

$$\tau_a - \tau_b = T \frac{\mathrm{d}\alpha_{ab}}{\mathrm{d}T} \tag{11.2}$$

where

$$\alpha_{ab} = \alpha_a - \alpha_b \text{ and } \pi_{ab} = \pi_a - \pi_b \tag{11.3}$$

11.2.2 Material Performance

The TE performance of a material is measured by three parameters. The primary parameter is the dimensionless figure of merit, ZT, which determines the "efficiency." The maximum efficiency of a power generation device (η) is given by [19]

$$\eta = \frac{T_{\text{hot}} - T_{\text{cold}}}{T_{\text{hot}}} \left[\frac{\sqrt{1 + \text{ZT}_{\text{m}}} - 1}{\sqrt{\left(1 + \text{ZT}_{\text{m}}\right)} + \left(\frac{T_{\text{cold}}}{T_{\text{hot}}} \right)} \right] \tag{11.4}$$

where the Carnot efficiency is the ratio of the temperature difference between the hot end and the cold end ($T_{\text{hot}} - T_{\text{cold}}$) and T_{hot}. To increase the conversion efficiency, it is essential to increase the dimensionless figure of merit (ZT) defined as

$$\text{ZT} = \frac{\alpha^2 T}{\rho\kappa} = \frac{\alpha^2 \sigma T}{\left(\kappa_{\text{L}} + \kappa_{\text{e}}\right)} = \frac{\text{PF}}{\kappa\sigma} \tag{11.5}$$

where α is the Seebeck coefficient (i.e. the thermopower), ρ is the electrical resistivity, σ is the electrical conductivity, PF is the power factor, T is the temperature in Kelvin, κ_{L} is the lattice thermal conductivity, and κ_{e} is the charge-carrier thermal conductivity. The value of α is positive for p-type (hole) conduction and negative for n-type (electron) conduction. A good TE material should have a large Seebeck coefficient (α), high electrical conductivity (σ), and poor thermal conductivity (κ). The interdependence of these three transport properties via the carrier concentration makes the optimization of TE response a challenging task. Among the various strategies used to identify promising systems, the search for materials that inherently behave as good thermal insulators can be driven by simple guiding rules based on their crystallographic characteristics. A complex unit cell containing a large number of heavy atoms is one of the guiding principles for discovering compounds that can exhibit low lattice thermal conductivity values.

The second combinatorial material parameter is the *B* factor [20] defined by microscopic band structure and transport parameters [21]. For a nondegenerate semiconductor

$$B \propto \frac{\mu_c m_*^{1.5} T^{2.5}}{\kappa_L} \tag{11.6}$$

where m^* is the carrier effective mass and μ_c is the carrier mobility at the nondegenerate limit. For a given scattering mechanism and at any Fermi energy E_F, the higher the *B* value, the higher the ZT value. The *B* factor reflects the importance of decoupling μ_c and m^* from κ_L. In practice, Eq. (11.6) remains a useful indicator for prescreening and optimizing more-complex materials [21].

The third combinatorial material parameter is the TE compatibility factor *S* [22] defined as

$$S = \frac{\sqrt{1 + ZT} - 1}{\alpha T} \tag{11.7}$$

If the compatibility factors differ by a factor of 2 or more, the maximum efficiency can decrease by segmentation [23]. It rationalizes how to extract the maximum efficiency out of a material by optimizing its geometrical shape, compositional gradient, or segmentation of a TE-device leg with different materials.

11.2.3 Tuning ZT by Carrier Concentration

Among all the parameters that affect the TE performance of materials, α, σ, and κ_e are directly related to carrier concentration *n* through the following relationships [24]

$$\alpha = \frac{8\pi^2 \kappa_B^2}{3eh^2} m^* T \left(\frac{\pi}{3n} \right)^{2/3} \tag{11.8}$$

$$\sigma = 1/\rho = ne\mu \tag{11.9}$$

$$\kappa_e = L\sigma T = ne\mu LT. \tag{11.10}$$

where *e* is the electrical charge of an electron, κ_B is the Boltzmann constant, μ is the carrier mobility, and *L* is the Lorenz number [24]. Figure 11.2 illustrates the variation of these parameters with carrier concentration for bismuth telluride, one of the most common TE materials. It can be noted that both electrical and thermal conductivities, σ and κ, increase with increasing carrier concentration, while Seebeck coefficient (α) has an opposite trend. This indicates that figure of merit, ZT, cannot be improved by simply increasing the carrier concentration. Maximizing ZT requires optimizing σ, α, and κ, simultaneously. Such interrelationships suggest that *n* needs to be optimized for high σ and α but low κ_e. Taking PbSe as an example, through controlled doping with K, *n* in $K_xPb_{1-x}Se$ can be increased as high as $1.5 \times 10^{20}\,cm^{-3}$, leading to the highest ZT value of 1.2 when $x = 0.0125$ [25].

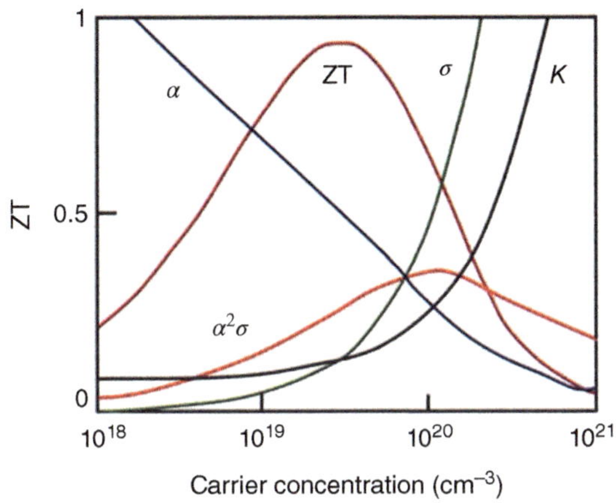

Figure 11.2 Schematic diagram showing how ZT and its related parameters (electrical conductivity σ, Seebeck coefficient α, power factor, and total thermal conductivity κ) change as a function of carrier concentration *n*. *Source:* Kishore and Priya [13]. Licenced under CC BY-4.0. © MDPI.

In general, the value of n depends strongly upon T and the nature of the materials themselves [26] and usually increases rapidly with rising temperature, roughly obeying the power law of $(T)^{3/2}$ [27]. Alternatively, based on the temperature-dependent solubility [28] of some specific dopants, one can create a gradient of n within a single material, controlled only by temperature and its gradient. A well-known example is that of Cu, Ag, and excess Pb, which have negligible solubility within PbTe at room temperature, but are highly soluble at elevated temperatures [28]. This temperature-dependent doping is reversible in the heating-cooling process.

11.2.4 Tuning ZT by Electronic Structure

The maximum ZT of a given TE material can generally be achieved by optimizing carrier concentration. However, to increase the maximum ZT to much higher levels, one needs to tailor the electronic structure as well as microstructure carefully. According to the B factor (Eq. (11.6)), the maximum ZT value is related to the carrier effective mass m^*, the carrier mobility μ_c, and the lattice thermal conductivity κ_L. The carrier effective mass m^* is related to the band effective mass m_b^* through the number of equivalent valleys of the band structure (N_V) [29]

$$m^* = \left(N_V\right)^{2/3} m_b^* \tag{11.11}$$

Thus, to increase the maximum ZT, one needs to maximize N_V, m_b^*, and μ_c and minimize κ_L concurrently to achieve higher α and σ but lower κ at a constant n according to Eqs. (11.8) and (11.10) as discussed below.

11.2.4.1 Carrier Effective Mass, m^*

Increasing the number of band extrema N_V and the band effective mass m_b^* are two effective ways for increasing carrier effective mass. N_V is closely related to the crystal structure symmetry – a crystal structure with high symmetry has a high value of N_V. Taking tetragonal chalcopyrite compounds [30] for instance, as shown in Figure 11.3, going from a non-cubic

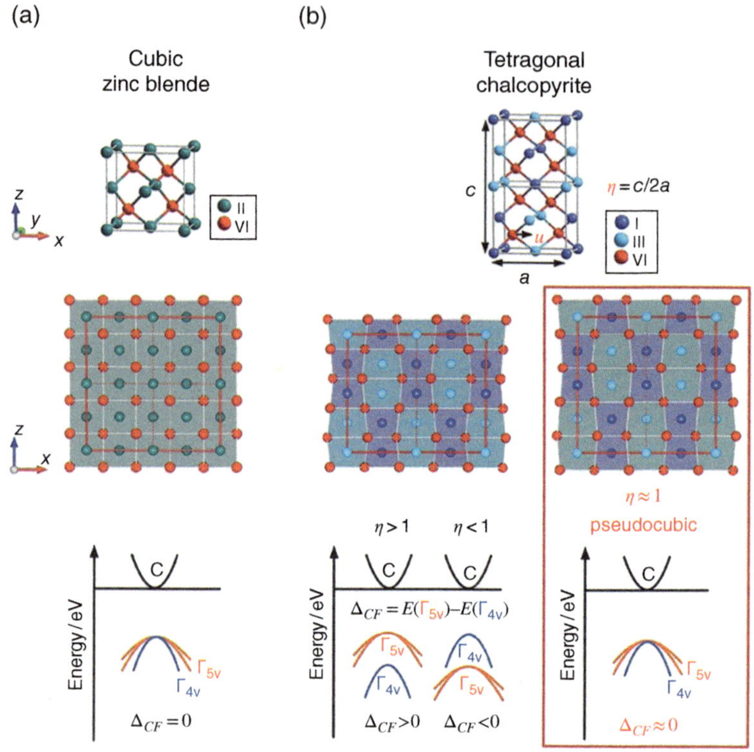

Figure 11.3 The pseudocubic band-convergence scheme. (a) Crystal structure and schematic electron-band structure of cubic zinc blende structure. (b) Crystal structure and schematic electronic bands of ternary chalcopyrites. The *c* and *a* are the lattice constants along the *c*- and *a*-axis, respectively. *Source:* Zhang et al. [30]. © John Wiley & Sons.

lattice to a cubic-like geometry (c = 2a), the band convergence enhances, leading to a larger N_V, where Γ_{4V} is a nondegenerate band and Γ_{5V} is a doubly degenerate band; Δ_{CF} is the crystal field induced energy split at the top of the Γ_{4V} and Γ_{5V} bands. The pseudocubic structure has a cubic cation framework with $\Delta_{CF} \approx 0$, embedded in non-cubic distorted anion framework.

Physically, the definition of carrier effective mass is $m_b^* = \frac{1}{\hbar^2}\frac{\partial^2 E}{\partial k_i \partial k_j}$, where κ_i and κ_j are the wave vector κ of *i*th and *j*th elements, respectively, and E is the energy. m_b^* is related to the curvature of the bands and can be modified by distorting the band. This can be achieved by increasing the density of states in the vicinity of the Fermi level (DOS(E_F)) [31]. One commonly adopted approach for increasing m_b^* is to introduce resonant levels (RLs), [31a] which refer to the impurity levels that lie in the conduction or valence band of the host material. In addition, a large DOS(E_F) enhances α, according to the Mott relation [32].

11.2.4.2 Carrier Mobility

Most state-of-the-art TE materials are heavily doped semiconductors with the carrier concentrations about 10^{19} – 10^{21} cm^{-3}. The high population of free carriers in these heavily doped semiconductors leads to a decrease of carrier mobility with respect to the lightly doped or undoped ones because of the increased impurity ion scattering.

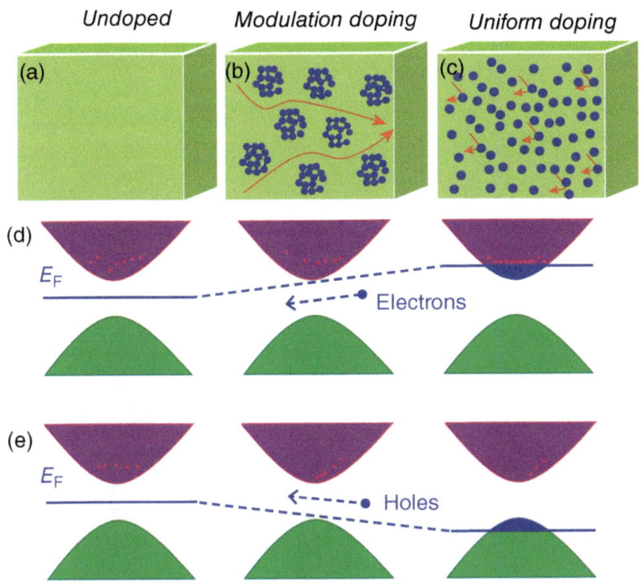

Figure 11.4 Schematic representation of (a) undoped semiconductor and (b) modulation doping where two types of grains consisting of undoped and heavily doped semiconductors are spatially separated. Red arrows show how carriers are scattered when mobilizing across the material. Fermi level positions in undoped, uniformly doped, and modulation-doped semiconductors: (c) heavily doped semiconductor where dopants are uniformly dispersed in the host matrix, and (d) n-type doping and (e) p-type doping. The Fermi level imbalance between the undoped and the heavily doped grains prompts the carriers to diffuse from the latter to the former. *Source:* Tan et al. [1]. © American Chemical Society.

Three-dimensional (3D) modulation doping has been recently demonstrated to be effective in increasing the ZT values of several important TE materials by carrier mobility enhancement. Modulation-doped samples (Fig. 11.4b) actually are two-phase composites composed of undoped (Figure 11.4a) and heavily doped (Figure 11.4c) counterparts. The undoped pristine sample has low carrier concentration but high carrier mobility, while the uniformly and heavily doped samples have high carrier concentration but low carrier mobility. The Fermi level of the undoped sample is usually located at the middle of the energy gap and would move deep into the conduction band (n-type doping, Figure 11.4d) or the valence band (p-type doping, Figure 11.4e) in the case of a heavily doped sample. Because of the imbalance of the Fermi levels between the undoped and the heavily doped phases, the carriers in the modulation-doped sample spill over from the heavily doped region to the undoped region (Figure 11.4d, e). This results in carrier mobility enhancements relative to the uniform doping because of avoidance of impurity ion scattering.

11.3 Low Lattice Thermal Conductivity of Cluster-based Materials

Manipulating thermal-transport properties is important for various applications ranging from integrated circuits and optoelectronics to TE energy conversion [34]. For semiconducting materials, the main contribution to thermal conductivity comes from phonons, i.e.

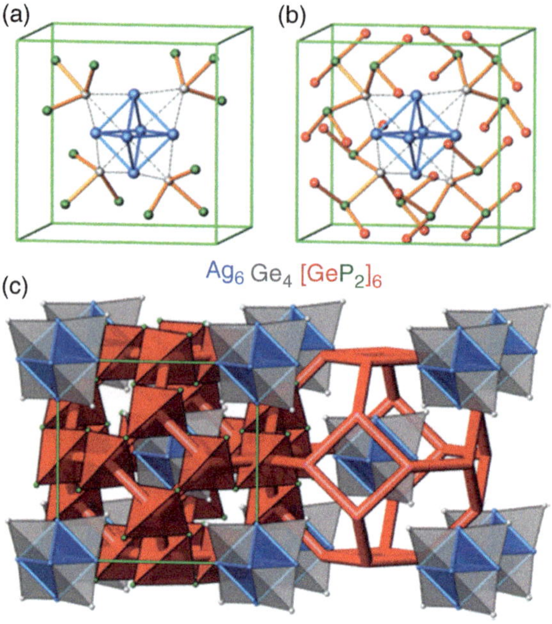

Figure 11.5 Perspective representation of the crystal structure of $Ag_6Ge_{10}P_{12}$, with margins of the unit cell (green). Color code: Ag (blue spheres), Ge(cap)$^{2+}$ (gray spheres), Ge(fr)$^{4+}$ (red spheres), P (green spheres), Ge-P bonds (orange sticks). (a) $Ag_6Ge_4P_{12}$ unit buildup by four $[GeP_3]^{7-}$ Zintl anions coordinated to the Ag_6 octahedron (dotted lines), (b) $Ag_6Ge_4P_{12}$ units bonded to Ge^{4+} ions, and (c) polyhedral representation showing Ag_6Ge_4 units (blue/gray) embedded in a tetrahedral $GeP_{4/2}$ framework (red). *Source:* Nuss et al. [33]. © American Chemical Society.

the lattice vibration modes, which can be described with wave vector **q** in branch *s*. A well-known theoretical model of the phonon dispersion relations is the Debye model, which can successfully reproduce the temperature dependence of specific heat at low temperatures observed in experiments by incorporating the spectral feature of phonon states. The spectral feature of phonons must be taken into account to accurately describe thermal conduction. From the Boltzmann transport equation, the lattice thermal conductivity κ_L can be expressed as

$$\kappa_L = \frac{1}{3}\sum_{q,s} C_{q,s} \upsilon_{q,s}^2 \tau_{q,s} = \frac{1}{3}\sum_{q,s} C_{q,s} v_{q,s} \Lambda_{q,s} \tag{11.12}$$

where C is the specific heat, v is the phonon group velocity, τ is relaxation time, and $\Lambda = v\tau$ is the phonon mean free path (MFP) [35]. Heat transport in TE materials results from the cumulative contributions of phonons with a broad range of MFPs. Different MFPs contribute different amounts to κ_L for a specific material [36]. Generally, heat is mainly carried by the long wavelength acoustic phonons, and the most significant phonon scattering at high temperature is usually phonon-phonon scattering. In cluster-based materials, much effort has been on enhancing phonon-phonon scattering to reduce the lattice thermal conductivity through the mechanisms discussed below.

11.3.1 Crystal Complexity of Cluster-based Materials

Low thermal conductivity is generally associated with crystals containing large, complex unit cells. The number of atoms (N) in the primitive cell indicates the complexity of a crystal structure, which usually includes many differently bonded atoms with different masses.

With increasing N, the ratio of acoustic branches to optical branches can be reduced to $1/(N-1)$. As a simple approximation, crystal structure complexity has the same effect as a significantly reduced population of acoustic phonons, which dominate the lattice thermal conductivity. This would enable an extremely low lattice thermal conductivity inherent to complex structures. For example, complex Zintl-phase materials are polyanionic compounds and appear to be promising TE materials with low lattice thermal conductivity [37]. In general, the Zintl-phase compounds are valence-precise semiconductors with electropositive cations and covalently bonded anionic units or networks formed from more electronegative elements. Zintl ions are naked anionic clusters of main group elements (e.g. Sn_9^{4-} and Sb_7^{3-}), and bimetallic Zintl ions such as $Pt_2@Sn^{144-}$ provide an excellent bridge between large molecular clusters and small metallic nanoparticles. The Zintl-phase compounds often have small bandgaps and benefit from the flexibility of alloying on the cation sites to obtain doping without strongly perturbing the anionic network that often plays the main role in electrical transport. In addition, the heavy atoms, complex structures, and soft lattices of many of these materials may lead to low lattice thermal conductivity.

In addition to the Zintl-phase compounds mentioned above, another type of Zintl compound with weakly coupled anionic clusters and intercalated cations have lower κ_L. For instance, $Na(K)_8SnSb_4$ consists of weakly coupled clusters $SnSb_4$, which are likewise expected to produce soft phonons [38]. This suggests effective coupling between these soft optical phonons and the acoustic phonons, leading to enhanced scattering and hence lower κ_L. In fact, the very soft lattice of this anionic cluster compound and low optic phonon frequencies imply that the thermal conductivity may be very low.

Slack quantitatively showed that the lattice thermal conductivity has an inverse dependence on the number of atoms in the primitive unit cell (namely, the complexity of the crystal structure) [39]. Assuming that the optical phonons do not carry heat, lattice thermal conductivity κ_L can be approximated as,

$$\kappa_L \approx BM\delta\left(\theta_D\right)^3 N^{-2/3} T^{-1} \gamma_2^{-2} \tag{11.13}$$

where B is a constant, M is the mean atomic mass, δ is a length parameter (δ^3 is the volume of the primitive cell), θ_D is the Debye temperature, N is the number of atoms in the primitive cell, and γ_2 is the Gruneisen constant. As discussed by Slack, for similar compounds, N can be considered to be a good measure of the crystal complexity [39]. For example, both YB_{66} and $YB_{44}Si_2$ have complex structures with B_{12}-based icosahedron cluster as the building block. Indeed, N takes relatively large values of 396 and 372 in YB_{66} [40] and $YB_{44}Si_2$ [41], respectively. One can see that a large N is an intrinsic feature of cluster-based materials as it offers flexibility of changing size, shape, and orientation of clusters.

11.3.2 Chemical Bond Hierarchy in Cluster-based Materials

Chemical bond hierarchy (CBH) depicts a unique structural type of solids, in which fractions of the materials are loosely bonded in a relatively rigid framework. The weak bonding in materials with CBH induces special atomic vibrational motions and has a significant influence on the lattice thermal conductivity. Different from well-bonded materials, the caged compounds and filled skutterudites as well as clathrates have loosely bonded ions in their structures with CBH features. Owing to the oversized crystalline cages, these fillers usually show distinctly large atomic displacement parameters (ADPs), compared with the framework atoms [42]. The fillers are sometimes called rattlers because they are considered to be rattling around in their equilibrium sites, and they reduce κ_L. The so-called rattling phenomenon can be interpreted as guest atoms in voids being weakly bound to the host atoms, thus behaving as an Einstein oscillator in scattering acoustic phonons [43] as discussed in $Ag_6Ge_{10}P_{12}$ [33].

In $Ag_6Ge_{10}P_{12}$, the high performance of its TE properties is mainly induced by the very low lattice thermal conductivity [300 K, ~1.2 W/(m K)]. Because of the moderate electrical conductivity, the absolute values can primarily be attributed to lattice contributions. This main property, the low lattice thermal conductivity of $Ag_6Ge_{10}P_{12}$, is largely consistent with the detailed crystal structure analysis. The structure can be described by $Ag_6[GeP_3]_4$ units interconnected by the remaining Ge(fr) atoms (Figure 11.5a, b). The porous Zeolite-like framework provides the space for the Ag_6Ge_4 units (Figure 11.5c).

Analysis of the local bonding properties reveals the coexistence of different types of chemical bonds leading to a strong "local bonding asymmetry" [44]. Interactions vary from strong covalent bonding in the rigid framework of Ge(fr) and P to weak bonding between lone-pair electrons of Ge(cap) and P with the silver atoms. The weak interactions lead to lone-pair induced anharmonic rattling, which reduces thermal and phonon transport. The compound can be described as Ag_6Ge_4 clusters being trapped in a closed and rigid 3D network with weak interactions.

$Ag_6Ge_{10}P_{12}$ clearly shows the feature of CBH with different types of chemical bonds with hierarchical strengths. This results in low-frequency vibrations and thus thermally induced large displacement parameters of mobile silver atoms, causing thermal damping and reduced lattice thermal conductivity, close to the amorphous limit (phonon glass-like behavior).

11.3.3 Structural Disorder in Cluster-based Materials

The structural disorder of the materials enhances phonon scattering and greatly suppresses the lattice thermal conductivity. Consider Ag_9GaS_6 as an example, whose detailed crystal structure has been well studied in the literature [45]. Being similar to many argyrodites, a phase transition was found in $[GaS_4]^-$ cluster-based materials from α-Ag_9GaS_6 with an orthorhombic structure to β-Ag_9GaS_6 with a cubic structure [46], as shown in Figure 11.6a (unit cell) and Figure 11.6b (primitive cell). The Ag atoms are arranged in a disorderly fashion in the anion framework at three different sites: 48*h* (0.232, 0.232, 0.029), 48*h* (−0.169, −0.169, 0.031), and 96*i* (−0.160, 0.073, 0.129), with occupancies of 0.25, 0.25, and 0.125, respectively [47]. The structure consists of $[GaS_4]^-$ and S^{2-} frameworks (Figure 11.6a) and has a total 16 atoms. The crystal structure of Ag_9GaS_6 is much more disordered than conventional thermoelectrics such as PbTe [48], SnTe [49], $CoSb_3$ [50], and half-Heusler alloy [51]. Temperature-dependent total thermal conductivity (κ) and lattice thermal conductivity (κ_L) of $Ag_9Ga(S_{1-x}Se_x)_6$ compounds are shown in Figure 11.6c and d, respectively. κ for all samples is found to be lower than 0.4 W/(m K) in the entire temperature range. The κ_L is calculated by subtracting the electronic thermal conductivity (κ_e) via the Wiedemann–Franz law ($\kappa_e = \sigma LT$) [52] from the total thermal conductivity, where L is the Lorenz factor determined by the single

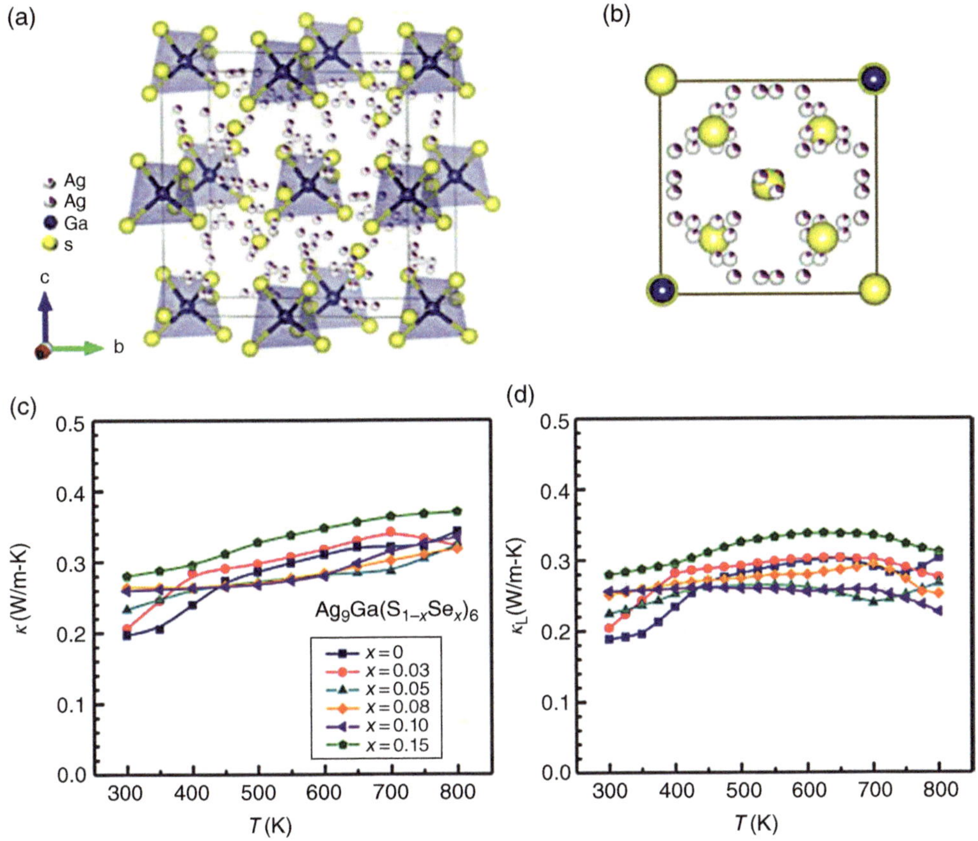

Figure 11.6 (a) Crystal structure and (b) the primitive cell for cubic Ag_9GaS_6, with partial occupancy of Ag at 48*h* and 96*i* sites. Color code: Ag (mixed white and red), Ga (blue), S (yellow). Temperature-dependent thermal conductivity (c) and lattice thermal conductivity (d) for $Ag_9Ga(S_{1-x}Se_x)_6$. *Source:* Lin et al. [10a]. © Elsevier.

parabolic band (SPB) model with acoustic scattering. Owing to the low κ_e, κ largely comes from the lattice component κ_L, which is as low as 0.25 W/(m K) in the entire temperature range in this compound. This is much lower than that of conventional thermoelectrics such as PbTe [48], SnTe [49], GeTe [53], and CoSb3 [50]. Such a low κ_L is mainly attributed to the highly disordered Ag^+ cations. It can be concluded that the $Ag_9Ga(S_{1-x}Se_x)_6$ compounds inherently show an extremely low κ_L, indicating the potential for TE applications.

11.3.4 Orientational Disorder in Cluster-based Materials

No clusters are perfectly spherical in geometry; thus, the orientation can be used as a parameter to tune the functionalities of cluster-based materials [54]. For example, the thermal transport behavior can be modulated through the manipulation of the vibrational landscape and orientational order of the superatoms as found in $Co_6E_8(PEt_3)_6$, $[Co_6E_8(PEt_3)_6][C_{60}]_2$, [55] and some kinds of fullerene assembled films [56].

11.3.4.1 $Co_6E_8(PEt_3)_6$ and $[Co_6E_8(PEt_3)_6][C_{60}]_2$

The molecular structure of the four superatoms and $[Co_6Te_8][C_{60}]_2$ superatomic crystal (SAC) are shown in Figure 11.7a and b. The thermal conductivities of these SACs are measured by frequency domain thermoreflectance (FDTR), which is a noncontact optical pump-probe technique [57]. These results range from 0.13 to 0.25 W/m K at room temperature (Table 11.1). With increasing the cluster core mass from Co_6S_8 to Co_6Se_8 to Co_6Te_8, their lattice thermal conductivities decrease, while the slightly higher thermal conductivity of $[Co_6Se_8][C_{60}]_2$ relative to $[Co_6Se_8]$ results from the strong C–C covalent bonding.

The results of the temperature-dependent thermal conductivity are plotted in Figure 11.8. As the temperature decreases to 260 K, C_{60} undergoes a structural transition from a face-centered cubic (fcc) unit cell with freely rotating C_{60} superatoms to a simple cubic (sc) structure with a basis of four orientationally ordered C_{60} superatoms [58]. Yu et al. suggested that orientational disorder in the fcc phase strongly scatters phonons, leading to temperature-independent thermal conductivity above 260 K [58]. The thermal conductivities of $[Co_6S_8]$ and $[Co_6Se_8]$ are temperature independent, while it decreases slightly for $[Co_6Te_8]$ with decreasing temperature. Below 190 K, $[Co_6Se_8][C_{60}]_2$ is orientationally ordered and behaves like a crystalline solid whose thermal conductivity increases with decreasing temperature. Above 190 K, the orientational disorder strongly scatters phonons, leading to a lower lattice thermal conductivity. By contrast, $[Co_6Te_8]$

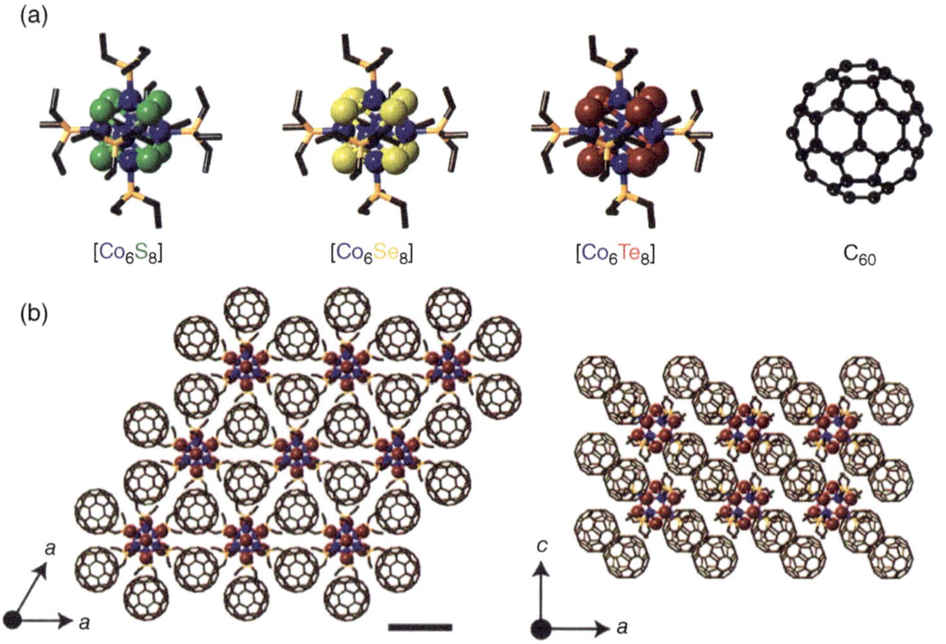

Figure 11.7 (a) Molecular structure of the superatoms $[Co_6E_8]$ and C_{60}. Hydrogens are omitted for clarity. (b) Crystal structure of $[Co_6Te_8][C_{60}]_2$ looking down the c-axis (left) and down the *a*-axis (right). In (a, b), scale bar is 1 nm; C, black; P, orange; Co, blue; Te, red; hydrogen atoms are omitted for clarity. These structures were determined by single crystal X-ray diffractometer (SCXRD) at 100 K. *Source:* Ong et al. [55]. © Springer Nature.

Table 11.1 Thermal and physical properties of SACs at room temperature.

SAC	Superatom composition	Thermal conductivity (W/m k)
$[Co_6S_8]$	$Co_6S_8(PEt_3)_6$	0.22±0.05
$[Co_6Se_8]$	$Co_6Se_8(PEt_3)_6$	0.18±0.04
$[Co_6Te_8]$	$Co_6Te_8(PEt_3)_6$	0.13±0.03
$[Co_6Se_8][C_{60}]_2$	$[Co_6Se_8(PEt_3)_6][C_{60}]_2$	0.25±0.06
$[Co_6Te_8][C_{60}]_2$	$[Co_6Te_8(PEt_3)_6][C_{60}]_2$	0.16±0.04

Source: Ong et al. [55]. © Springer Nature.

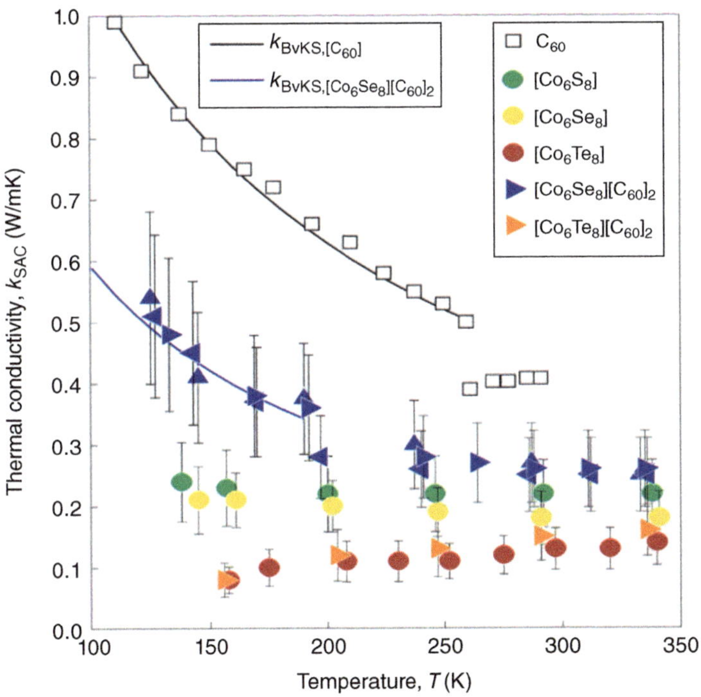

Figure 11.8 Thermal conductivity of SACs as a function of temperature. *Source:* Ong et al. [55]. © Springer Nature.

$[C_{60}]_2$ shows an amorphous thermal conductivity behavior, because the neighboring C_{60} molecules in $[Co_6Te_8][C_{60}]_2$ cannot approach closely enough to experience the short-range anisotropic interactions for orientational ordering [59].

11.3.4.2 Fullerene Assembled Films

Using time domain thermoreflectance (TDTR), Giri et al. [56] recently reported ultralow thermal conductivities of alkyl- and indene-group functionalized fullerene assembled thin films including [6,6]-phenyl C_{61} butyric acid methyl ester (PCBM), [6,6]-phenyl C_{61} butyric

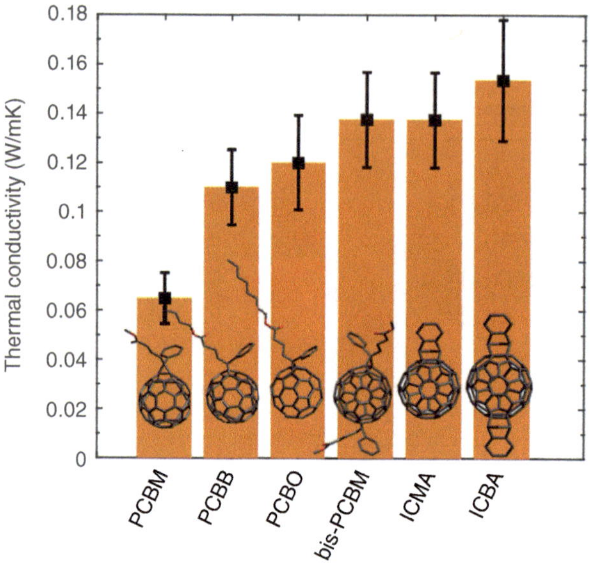

Figure 11.9 Measured thermal conductivities for the fullerene derivative films. *Source:* Giri et al. [56]. © American Physical Society.

acid butyl ester (PCBB), [6,6]-phenyl C_{61} butyric acid octyl ester (PCBO), indene-C_{60} mono-adduct (ICMA), indene-C_{60} bisadduct (ICBA), and bisadduct PCBM (bis-PCBM), as shown in Figure 11.9, where PCBM films show "ultralow" thermal conductivities of ~0.06 W/m K. The thermal conductivity of the PCBB and PCBO films with longer molecular tails are ~70% higher than that of PCBM, while a thermal conductivity of ICBA is about twice as large as that of PCBM, suggesting the flexibility of thermal conductivity via changing the way of assembling clusters.

11.4 Thermoelectric Properties of some Selected Cluster-based Materials

11.4.1 Mo_6 and Mo_9 Cluster-based Selenides

Mo-based cluster compounds containing Mo_6 and Mo_9 cluster units have long been known for their rich chemistry as well as the diversity and complexity of their crystal structures. These compounds are promising materials for high-temperature TE applications due to their extremely low intrinsic thermal conductivities. In this section, we provide a brief overview of the recent results on the TE properties of these materials. Due to the diversity of structures that can be built with Mo-based clusters, we focus only on $Ag_xMo_9Se_{11}$ ($3.4 \leq x \leq 3.9$) [60] with its known electronic and thermal transport properties.

11.4.1.1 Crystal Structures

$Ag_xMo_9Se_{11}$ compounds crystallize in the orthorhombic space group *Cmcm* [61]. The building block is the bioctahedral $Mo_9Se_{11}Se_6$ cluster unit shown in Figure 11.10a, which can be seen as the fusion of two octahedral $Mo_6Se_8Se_6$ cluster units (Figure 11.10b), similar to

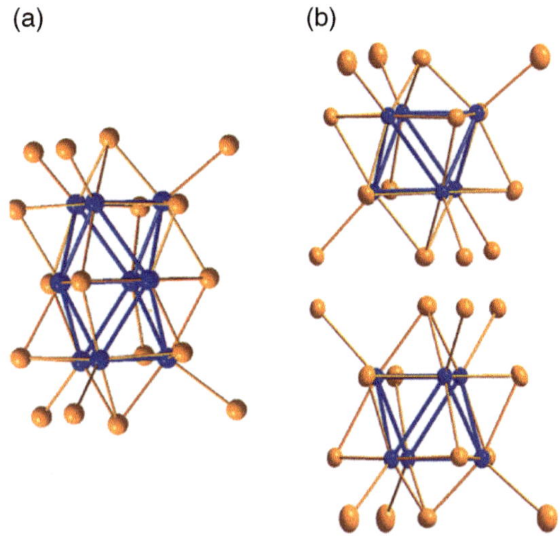

Figure 11.10 Schematic representation of the formation of the $Mo_9Se_{11}Se_6$ cluster unit (a) resulting from the condensation of two octahedral $Mo_6Se_8Se_6$ cluster units (b). The Mo and Se atoms are represented in blue and orange, respectively. *Source:* Candolfi et al. [6]. © Springer Nature.

those found in the Chevrel phases [62]. The 3D network formed by these clusters has empty channels that accommodate Ag atoms, delocalized over four crystallographically independent sites. The Ag nonstoichiometry in these compounds arises essentially from the filling of the Ag_4 site (shown in dark gray in Figure 11.11) located in rhomboid cross-sectional channels running along the a direction. The occupation of this site varies from 10% (in $Ag_{3.4}Mo_9Se_{11}$) up to 25% (in $Ag_4Mo_9Se_{11}$).

11.4.1.2 Electronic Structures

The electronic structure of most of the Mo-based cluster compounds can be deduced from the molecular orbital diagram of the isolated cluster units [63]. This approach proves to be efficient whenever the clusters in the crystal structure are isolated enough from the neighboring clusters. It was found that the interactions between the chalcogen atoms that connect the clusters are not strong enough to perturb the electronic structure of the isolated cluster unit. In such a case, the number of electrons available for metal-metal bonds, called the metallic electron count (MEC), is the critical parameter that determines the metallic or semiconducting nature of the compound. Semiconducting behavior is predicted when the MEC reaches the optimum value for the cluster unit. In the case of the Mo_9Se_{11} clusters, the optimum MEC has been found to be 36 [64]. Assuming ionic interactions between the cations and the clusters and $a+1$ valence state for the former, Ag atom may thus help achieve an optimum MEC for the Mo_9Se_{11} clusters. Varying the concentration of the Ag^+ may also allow for finetuning the electronic properties by adjusting the carrier concentration.

A shift from a p-type metallic-like behavior ($x = 3$) to a p-type heavily doped semiconducting state ($x = 4$) of $Ag_xMo_9Se_{11}$, shown in Figure 11.12, confirms this picture. $Ag_4Mo_9Se_{11}$ exhibits a direct bandgap of 0.7 eV. The rigid-band model could apply when the Ag content varies. This is what can be observed, at least qualitatively, because the band structure of $Ag_3Mo_9Se_{11}$ (Fig. 11.15b) is very similar to that of $Ag_4Mo_9Se_{11}$. The band structure of $Ag_3Mo_9Se_{11}$ demonstrates that it has a p-type metallic-like behavior.

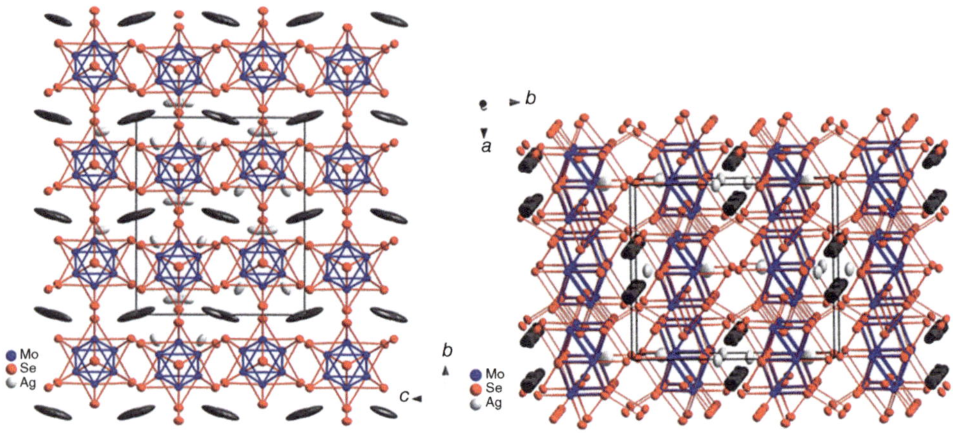

Figure 11.11 Projection of the crystal structure of the $Ag_xMo_9Se_{11}$ compounds (space group *Cmcm*, No. 63) in ellipsoid representation (97% probability level) along the *a*-axis (left) and *c*-axis (right). The Ag, Mo, and Se atoms are in gray, blue, and red, respectively. The Ag_4 atoms are distinguished from the other Ag atoms by their dark-gray color. *Source:* Candolfi et al. [6]. © Springer Nature.

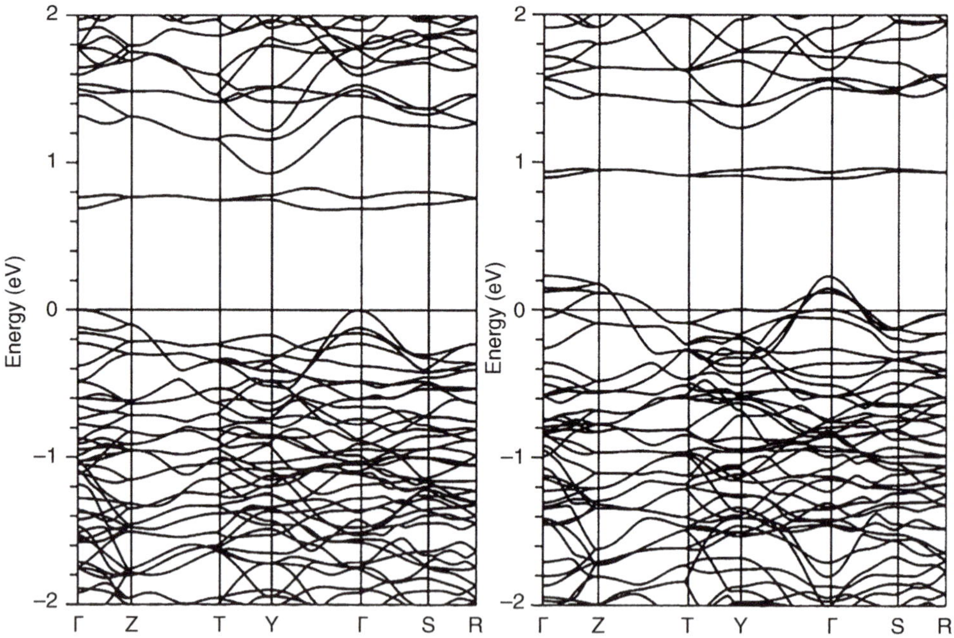

Figure 11.12 DFT dispersion curves of the (a) $Ag_4Mo_9Se_{11}$ and (b) $Ag_3Mo_9Se_{11}$. *Source:* Zhou et al. [60]. © American Chemical Society.

11.4.1.3 Thermal Properties

Figure 11.13a shows the measured temperature-dependent electrical resistivity of $Ag_xMo_9Se_{11}$ with $3.41 \leq x \leq 3.81$ [60]. The sample with $x = 3.41$ shows a metallic-like behavior. Above $x \geq 3.71$, the metallic character remains down to 125 K, where a steep rise in the

ρ values is seen as temperature decreases further. The fact that these samples behave as highly doped semiconductors may explain this phenomenon. Figure 11.13b presents the temperature dependence of the thermopower for various Ag-filled samples. The results show that this metallic nature remains unaffected up to $x \leq 3.5$, before evolving toward a semiconducting state expected to develop at $x = 4.0$. The overall variation in the α values with x is in agreement with their resistivity behavior.

One of the distinguishing features of all these Mo-based materials is their extremely low lattice thermal conductivity κ_L that is typically well below 1 W/(m K), with a value of 0.5 W/(m K) often achieved at high temperatures. Several key characteristics of the crystal structure of these compounds likely concur to yield the observed very low, glass-like lattice thermal conductivity. In addition to the large unit cells that result in low phonon cutoff wave vectors [66], the large, anisotropic ADP values of the cations [67] likely lead to low-lying optical modes that cut the acoustic phonon dispersion. The phase space available for acoustic phonons is thus strongly reduced, which results in a significant reduction of the lattice thermal conductivity. Low-energy optical modes also open an additional channel for Umklapp scattering events, which persist down to low temperatures and suppress the dielectric maximum. This mechanism is similar to what was observed in tetrahedrites, for instance, a class of sulfur-based compounds exhibiting similar thermal transport. In these minerals, the low-energy optical modes have been shown to be strongly anharmonic [68]. Because anharmonicity determines the scattering probability of phonons during phonon–phonon interactions, highly anharmonic bonds favor these scattering events and, hence, help to maintain low lattice thermal conductivity with increasing temperature. Combination of these factors mentioned above leads to the ultralow lattice thermal conductivity of Mo-based cluster compounds.

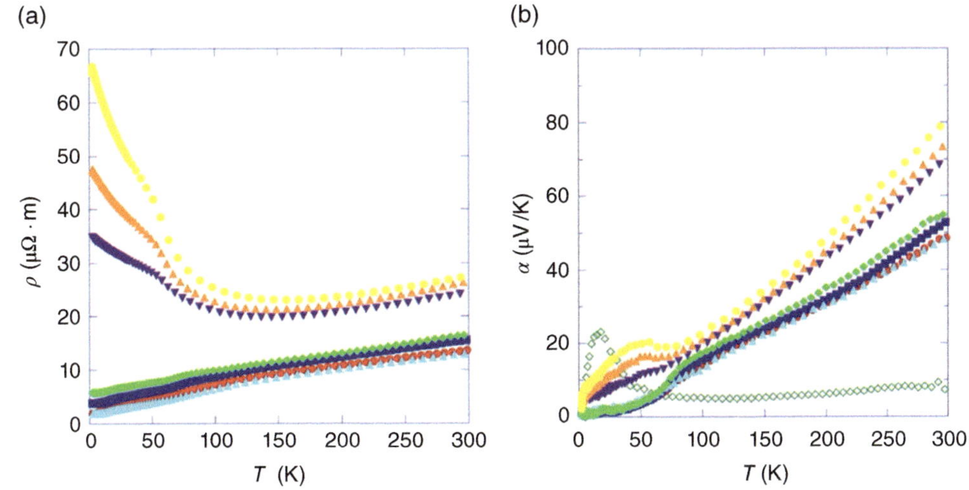

Figure 11.13 (a) Temperature dependence of the electrical resistivity ρ and (b) thermopower α of the $Ag_xMo_9Se_{11}$ for $x = 0.05$ (open green diamonds), $x = 3.41$ (red circles), $x = 3.53$ (blue squares), $x = 3.54$ (light blue triangles), $x = 3.61$ (green diamonds), $x = 3.71$ (orange triangles), $x = 3.73$ (downward-pointing purple triangles), and $x = 3.78$ (yellow circles). *Source:* Zhou et al. [60]. © American Chemical Society.

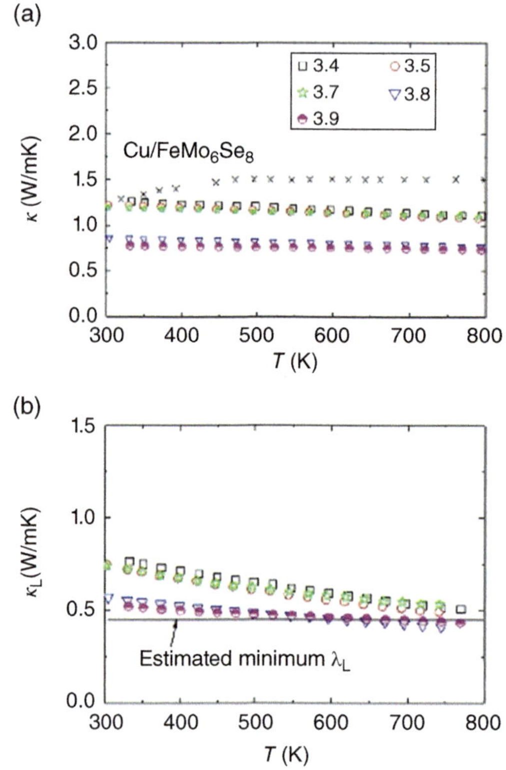

Figure 11.14 (a) The total thermal conductivity κ and (b) the lattice thermal conductivity κ_L of $Ag_xMo_9Se_{11}$. The behavior of the Cu/Fe-filled Chevrel phase is shown for comparison [65]. *Source:* Zhou et al.[60b]. © AIP Publishing.

The temperature dependence of the thermal conductivity of the $Ag_xMo_9Se_{11}$ compounds is plotted in Figure 11.14a and compared to that of $Cu/FeMo_6Se_8$ [65]. One can see that the κ in $Ag_xMo_9Se_{11}$ compounds is lower than that in the Chevrel phase, irrespective of the value of x. At 300 K, the values range from 0.8 to 1.2 W/(m K) for $x = 3.9$ and $x = 3.4$, respectively. The κ value remains nearly constant with temperature, indicating that the thermal conductivity approaches a glass-like limit. The lattice thermal conductivity, κ_L, displayed in Figure 11.14b, is in the range 0.4–0.7 W/(m K). These values are lower than those reported for the best TE materials like the partially filled skutterudites [69].

11.4.1.4 Thermoelectric Figure of Merit ZT

The dimensionless figure of merit ZT as a function of temperature for all the $Ag_xMo_9Se_{11}$ compounds is displayed in Figure 11.15 and compared to that of the $Cu/FeMo_6Se_{8.9}$ compounds. The $Ag_xMo_9Se_{11}$ ternary compounds possess ZT values greater than the best Chevrel phase over the temperature range investigated. ZT reaches the value of 0.65 at 800 K in compounds having the highest Ag nominal content ($x = 3.8$–3.9).

11.4.2 Boron-based Cluster Materials

Among functional materials, compounds with interesting structural features and strong structure–property relationships are attractive because of their potential for applications [70]. Boron cluster-based materials are among these. Experimentally, it was found

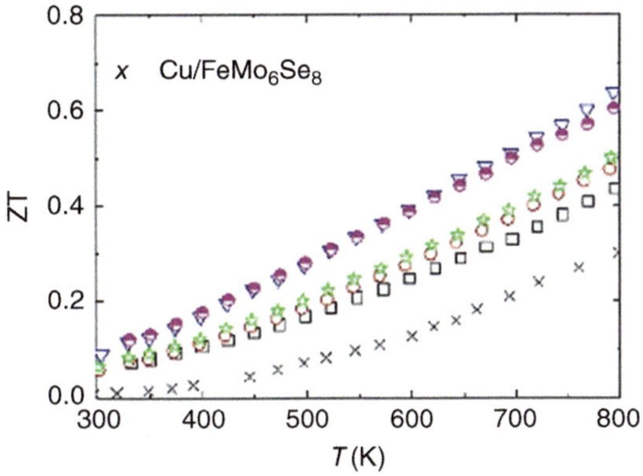

Figure 11.15 Temperature dependence of ZT of the $Ag_xMo_9Se_{11}$ compounds. *Source:* Zhou et al. [60b]. © AIP Publishing.

that the B_{12} icosahedral cluster can significantly enhance and flexibly mediate magnetic coupling in dilute magnetic systems [71]. More interestingly, boron icosahedral compounds can also exhibit large Seebeck coefficients and relatively low thermal conductivity. This section reviews briefly the possible mechanisms proposed to lower the lattice thermal conductivity in boron cluster-based materials and several borides that have been recently found to possess enhanced TE properties.

11.4.2.1 Crystal Structures

Similar to carbon, boron materials show rich structural features. Having one electron less than carbon, boron exhibits interesting chemistry, especially when binding with rare earth elements ("R") that have localized 4*f* orbitals, forming varied metal–boron cluster compounds with high stability as discussed in review papers and book chapters [70, 72]. In particular, among the rare earth binary boron-rich boride structures with atomic ratio [B]/[R] >1, RB_2 has a layered structure with boron graphitic layers sandwiching metal atoms (space group *P6/mmm*). When the boron ratio is increased, the boron structural building block shifts from a $[B]_6$ atomic net to the B_6 octahedron cluster, forming RB_4 (space group *P4/mbm*) and RB_6 (space group *Pm*-3*m*). On the other hand, R_2B_5 apparently forms only for very limited rare earth elements such as Nd, Sm, and Gd [70c]. Further increase of the concentration of B leads to the boron cuboctahedron and RB_{12} (space group *Fm*-3*m*). When the atomic ratio [B]/[R] > 12, the compounds, including even ternary and quaternary compounds, have the B_{12} icosahedron cluster as the building block (Figure 11.16).

11.4.2.2 Thermoelectric Properties

In this section, we discuss the TE properties of RB_{66}-type compounds with B_{12}-icosahedra as the basic building blocks. These materials generally possess high Seebeck coefficients [73] due to their complex crystal structures and disorder, causing intrinsically low thermal conductivity [73b, 74]. Additionally, these compounds have melting temperatures exceeding 2000 K, which is favorable for working in the high-temperature range [75]. These combined attributes make RB_{66}-type compounds promising for applications in high-temperature TE devices.

New interest in RB_{66}-type compounds has been sparked by the surprisingly good TE performance of SmB_{66} phase [76] with ZT reaching 0.13 at 1050 K, that is about 40 times higher than that of the previously reported values in YB_{66} [77] and ErB_{66} [78]. This big

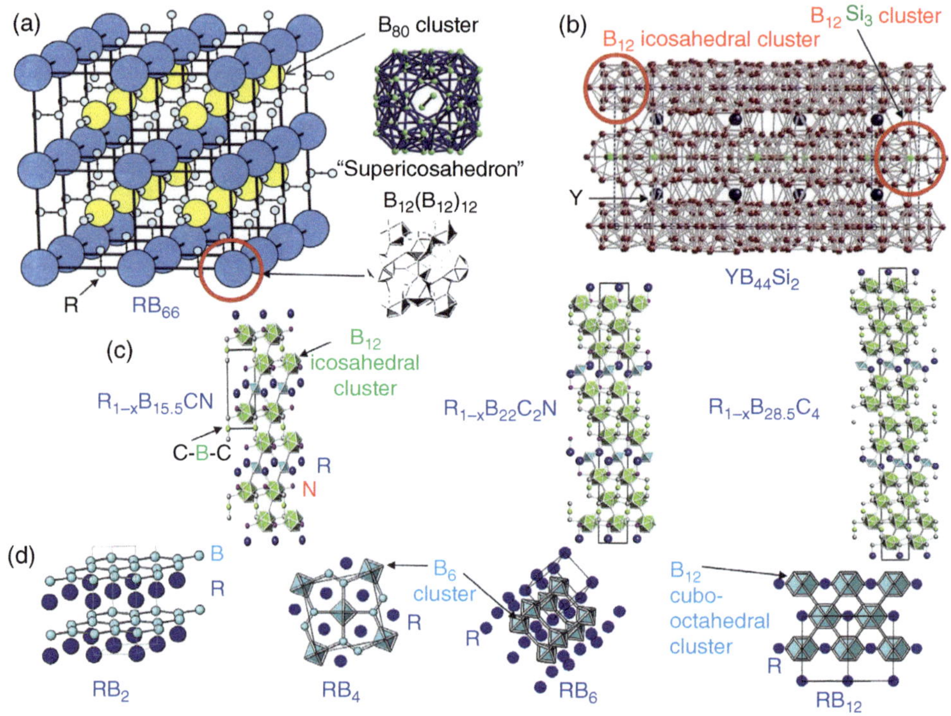

Figure 11.16 Views of crystal structures of (a) RB_{66} [70a], (b) $YB_{44}Si_2$ along the growth direction [70a], (c) R–B–C(N) homologous compounds [70a], and (d) several binary rare earth borides [70c]. *Source:* Mori [8]. © Elsevier.

enhancement is due to a large drop in the electrical resistivity while the Seebeck coefficient decreased only slightly. This effect is in part attributed to a Sm(+2)/Sm(+3) mixed valence state. Note that spin entropy enhanced Seebeck coefficients have been previously observed for some transition metal oxides [79]. The enhancement of the Seebeck coefficient through magnetic interaction and spin fluctuation has also been recently demonstrated [80]. Here, we consider a new phase of RB_{66}-type compounds, YbB_{66} [81], as an example to discuss the TE properties.

The temperature dependence of the electrical resistivity of YbB_{66} is shown in Figure 11.17 and compared to that of SmB_{66}, YB_{66}, YB_{48}, and ErB_{66}. The resistivity of YbB_{66} is roughly two orders of magnitude lower than that of YB_{66} and ErB_{66} throughout the whole temperature range and very close to that of SmB_{66}. The resistivity of YB_{48} is significantly lower at low temperature, but the resistivities of YbB_{66}, SmB_{66}, and YB_{48} seemingly converge at higher temperatures.

For disordered semiconductors, the localization length ξ appears to be an important characterization parameter [83], which can be calculated from the relationship,

$$k_B T_0 = \frac{18.1}{D(E_F).\xi^3} \tag{11.14}$$

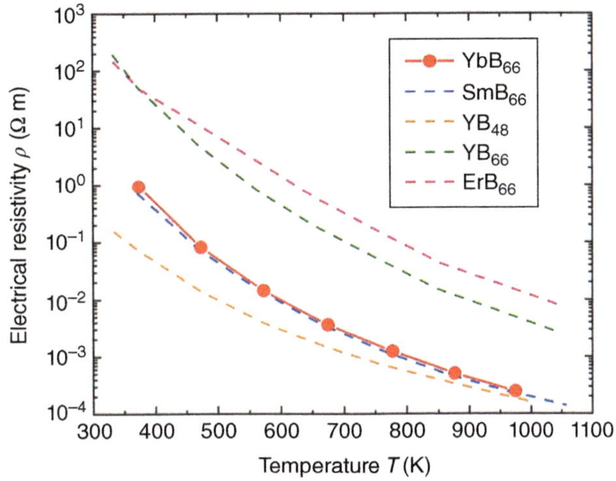

Figure 11.17 Temperature dependence of the electrical resistivity ρ of YbB_{66} with SmB_{66} [76], YB_{66} [74b], YB_{48} [82], and ErB_{66} [78] as reference values. *Source:* Sauerschnig et al. [81]. © Elsevier.

Here, k_B and $D(E_F)$ are the Boltzmann constant and the density of states at the Fermi level, respectively. It is found that $\xi = 0.56$, 0.33, and 2.41 for YB_{66}, SmB_{62}, and YB_{48}, respectively [82], indicating that the carrier wave function in YB_{48} has some delocalization features. Indeed, YB_{48} has the lowest electrical resistivity among these compounds. It has been previously observed that in doped semiconductors carrier compensation can lead to electron delocalization [84]. In this case, metal-rich YB_{48} can be viewed as having electrons injected when compared with the situation in YB_{66}. On the other hand, divalency exists in YbB_{66}, leading to high charge concentration and low electrical resistivity.

Figure 11.18 shows the temperature dependence of the Seebeck coefficient α, in YbB_{66}, which is positive throughout the whole temperature range. At 373 K, the value is 600 μV/K, and it drops monotonically with increasing temperature to roughly 241 μV/K at 973 K. Except for YB_{48}, the Seebeck coefficients of other compounds are above 500 μV/K. The power factor for YbB_{66} is shown in Figure 11.19, which increases steadily with increasing

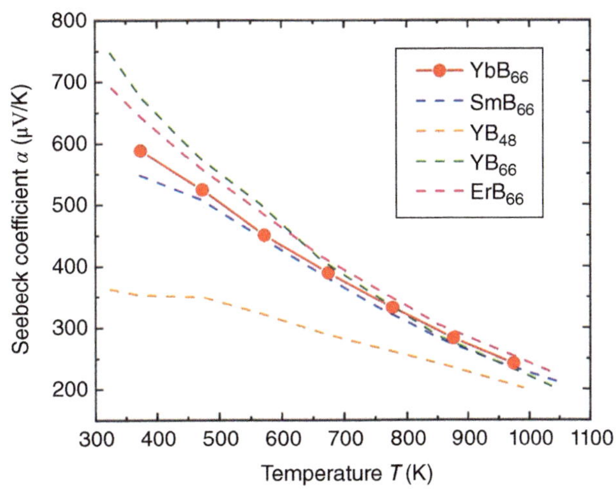

Figure 11.18 Temperature dependence of the Seebeck coefficient of YbB_{66} with SmB_{66} [76], YB_{66} [74b], YB_{48} [82], and ErB_{66} [78] as reference values. *Source:* Sauerschnig et al. [81]. © Elsevier.

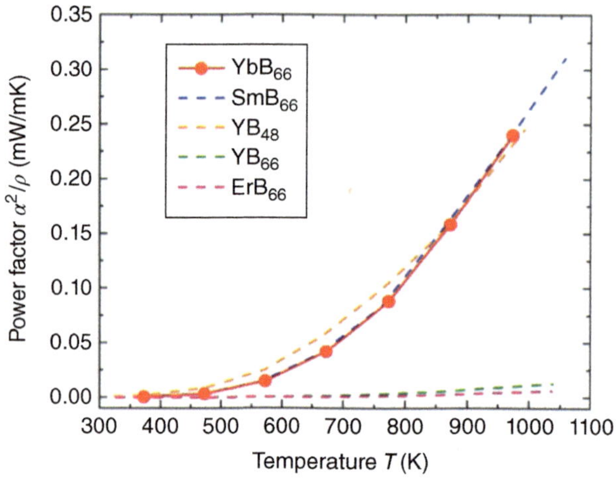

Figure 11.19 Temperature dependence of the power factor of YbB_{66} with SmB_{66} [76], YB_{66} [74b], YB_{48}, [82] and ErB_{66} [78] as reference values. *Source:* Sauerschnig et al. [81]. © Elsevier.

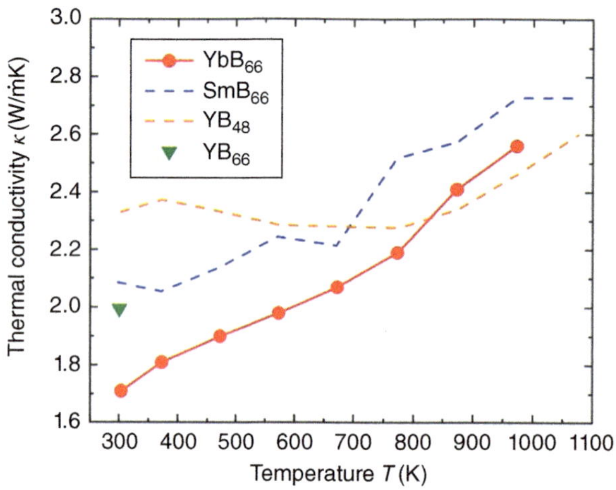

Figure 11.20 Temperature dependence of the thermal conductivity κ of YbB_{66} with SmB_{66} [76], YB_{66} [74b], and YB_{48} [82] as reference values. *Source:* Sauerschnig et al. [81]. © Elsevier.

temperature, reaching a value of 0.24 mW/(m K) at 973 K, which is very close to the values of YB_{48} and SmB_{66}.

The temperature dependence of the thermal conductivity is plotted in Figure 11.20. One can see that the thermal conductivity values for YbB_{66} are slightly lower than that of SmB_{66} throughout the whole temperature range. This can be explained by the higher atomic weight of Yb as compared to Sm. The figure of merit ZT shown in Figure 11.21 for YbB_{66} is similar to that of SmB_{66} and reaches a value close to 0.1 at 973 K.

11.4.3 Silver-based Cluster Materials

Recently, a series of argyrodite silver-based cluster compounds, such as Ag_8SnSe_6 [85], Ag_9GaSe_6 [86], Ag_9AlSe_6 [10b], Ag_8SiTe_6 [87], and Ag_8GeTe_6 [10a, 88], have been developed and demonstrated as promising TE materials with a peak ZT > 1.0. Such a high ZT largely relies on the ultralow intrinsic κ_L in these materials, originating from a high concentration

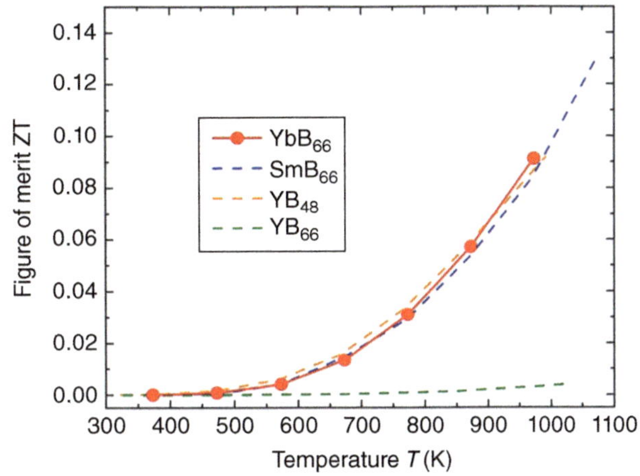

Figure 11.21 Temperature dependence of the dimensionless figure of merit ZT of YbB_{66} with SmB_{66} [76], YB_{66} [74b], and YB_{48} [82] as reference values. *Source:* Sauerschnig et al. [81]. © Elsevier.

of weakly bonded cations, a large number of atoms in the primitive cell, and a strong lattice anharmonicity. Note that the large number of atoms in the primitive cell leads to a significantly reduced acoustic-to-optical fraction of phonons. Note that acoustic phonons are the main contributors to lattice thermal conduction [39, 89]. The existence of highly diffusive Ag/Cu cations results in a weak bonding force for low sound velocity and strong lattice anharmonicity [86]. All of these factors effectively contribute to a low lattice thermal conductivity.

In this section, we first focus on a new argyrodite compound, Ag_9GaS_6 [10a], as a representative of the silver-based cluster compounds for potential TE applications. The carrier concentration is found to be tunable by Se/S substitutions, which leads to an optimization of carrier concentration for enhancing the TE power factor. The sound velocity is found to be as low as 1500 m/s, which is comparable to that of other reported argyrodite compounds. In addition, the strong lattice anharmonicity and the large primitive cell contribute to the intrinsically low κ_L with only ~0.25 W/(m K) in the entire temperature range. As a result, a peak ZT of ~0.6 is realized in $Ag_9Ga(S_{1-x}Se_x)_6$ at 800 K.

The structure and κ_L of $Ag_9Ga(S_{1-x}Se_x)_6$ has been discussed in Section 11.3.3, such a low κ_L is believed to be the result of the following three factors: (i) complex crystal structure for a small population of acoustic phonons, (ii) soft bonding for an overall low sound velocity, and (iii) massive disordering of Ag ions. The electronic transport properties of $Ag_9Ga(S_{1-x}Se_x)_6$ are shown in Figure 11.22. Temperature-dependent Seebeck coefficient and resistivity for $Ag_9Ga(S_{1-x}Se_x)_6$ are shown in Figure 11.22a and b, respectively. All the Seebeck coefficients are negative, indicating n-type conduction. Both Seebeck coefficient and resistivity increase with increasing electron concentration. The power factor (PF) is found to increase with increasing temperature and Se concentration (Figure 11.22c). It should be noted that the PF of $<3\,mW/cm\,K^2$ is much lower than that in conventional thermoelectrics such as PbTe [48], SnTe [49], GeTe [53], and $CoSb_3$ [50]. Importantly, the carrier-concentration-dependent PF can be well understood using the SPB model (Figure 11.22d), which is consistent with other reported TE argyrodites [90].

Temperature-dependent TE figure of merit (ZT) for $Ag_9Ga(S_{1-x}Se_x)_6$ is shown in Figure 11.23 where ZT is found to increase with increasing temperature. With the

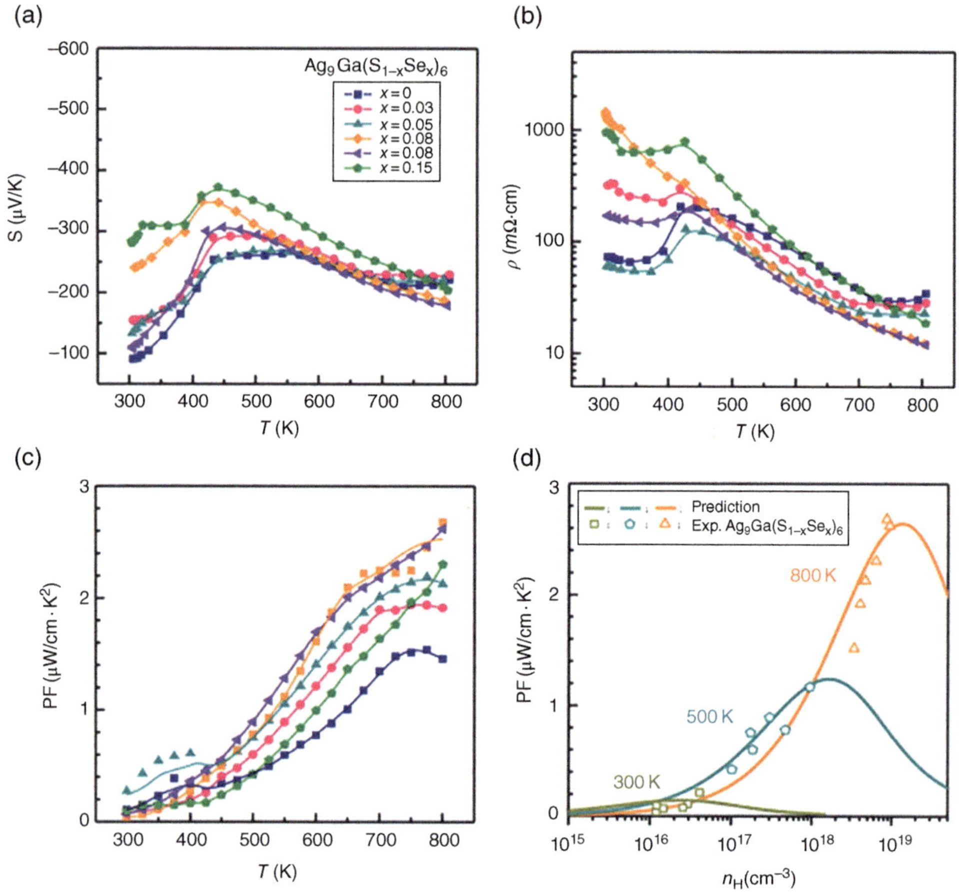

Figure 11.22 Temperature-dependent Seebeck coefficient (a), resistivity (b), power factor (c), and carrier-concentration-dependent power factor at different temperatures with a comparison to SPB model predictions (d) for $Ag_9Ga(S_{1-x}Se_x)_6$. SPB, single parabolic band. *Source:* Lin et al. [10a]. © Elsevier.

optimization of carrier concentration combined with the intrinsically ultralow κ_L, a promising peak ZT value of 0.6 is achieved in $Ag_9Ga(S_{1-x}Se_x)_6$, even with a PF of <3 mW/cm K^{2s}. Thus, Ag_9GaS_6 class of materials can be viewed as a potential sulfide-based TE.

It is worth mentioning another silver-cluster compound, $Ag_6Ge_{10}P_{12}$ [33], an air-stable phosphide, which was synthesized from its elements in gram quantities. The structure of $Ag_6Ge_{10}P_{12}$ is closely related to that of the tetrahedrite $Cu_{12}Sb_4S_{13}$ [92], which are prominent TE materials [93]. The structure and low lattice thermal conductivity have been discussed in Section 11.3.2.

The interesting structural features of this compound are reflected in its physical and in particular in its TE and transport properties. The semiconducting character of this material is seen from band structure calculations. Measurements of the thermal diffusivity exhibit a very low thermal conductivity, $\kappa < 1$ W/(m K), which is close to a phonon glass-like behavior and has its origin in a strong local bonding asymmetry, induced by strong bonding of

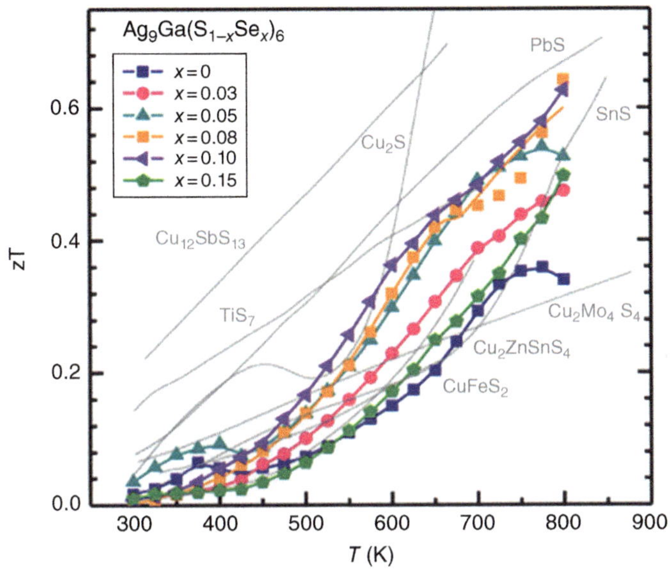

Figure 11.23 Temperature-dependent figure of merit (ZT) for $Ag_9Ga(S_{1-x}Se_x)_6$, compared to other sulfides in the literature [91] and model prediction. *Source:* Lin et al. [10a]. © Elsevier.

the phosphorus–germanium (Ge^{4+}) covalent framework and weak bonding of lone-pair electrons (Ge^{2+}). This CBH creates a pronounced anisotropic behavior of the silver atoms leading to low-frequency vibrations and thermal damping. The moderate electrical resistivity ($\rho \sim 15\,\text{m}\Omega$ cm) together with the high Seebeck coefficient ($\alpha \sim 380\ \mu$V/K) results in a remarkably high figure of merit (ZT) of about 0.6 at 700 K.

11.5 Conclusion

In this chapter we discussed the characteristics of TE materials that can yield high ZT values and how these can be accomplished with clusters as building blocks. Clusters unlike atoms have the advantage that their size, structure, and composition can be tailored to meet specific measurements. One of the primary requirements for efficient TE materials is that their thermal conductivity should be small. Increasing the number of atoms in the unit cell and tailoring the bonding between the structural units can accomplish this goal. Many complex solids exist that can be viewed as cluster-based solids as individual units can be identified as clusters. Thus, the vast knowledge that has been accumulating in cluster science over the past 50 years can be brought into bear for the synthesis of cluster-based solids. The unique advantages of clusters, i.e. their size, symmetry, and composition can meet three basic requirements of anti-Slack criteria for ultralow lattice thermal conductivity – namely, (i) a complex crystal structure, (ii) high anharmonicity, and (iii) weak interatomic bonding. These features endow cluster-based materials with unprecedented potential for novel materials with ultralow lattice thermal conductivity and/or high TE performance. The resulting small group velocity, large Gruneisen parameter, high anharmonic scattering rate, and large three-phonon scattering space lead to low lattice thermal conductivity and high ZT values. The research progress in this area not only expands the application of clusters but also sheds light on new materials design.

References

1 (a) Goldsmid, H. (1964). *Thermoelectric Refrigeration: Applications of Thermoelectric Refrigeration*, 210–223. Springer. (b) Hogan, T., Shih, T., and Rowe, D. (2005). *Thermoelectrics Handbook: Macro To Nano*, vol. *12*. Boca Raton, FL: *CRC*. (c) Tan, G., Zhao, L.-D., and Kanatzidis, M.G. (2016). Rationally designing high-performance bulk thermoelectric materials. *Chemical Reviews 116*: 12123–12149.

2 Kanatzidis, M.G., Hogan, T., and Mahanti, S. (2012). *Chemistry, Physics, and Materials Science of Thermoelectric Materials: Beyond Bismuth Telluride*. Springer Science & Business Media.

3 Tritt, T.M., Böttner, H., and Chen, L. (2008). Thermoelectrics: Direct solar thermal energy conversion. *MRS Bulletin 33*: 366–368.

4 Tritt, T.M. (2011). Thermoelectric phenomena, materials, and applications. *Annual Review of Materials Research 41*: 433–448.

5 Zlatic, V. and Hewson, A.C. (2009). *Properties and Applications of Thermoelectric Materials*. Springer Science & Business Media.

6 Candolfi, C., Gougeon, P., Gall, P. et al. (2019). *Ligated Transition Metal Clusters in Solid-state Chemistry: The legacy of Marcel Sergen: Thermoelectric Properties of Ternary and Quaternary Mo6 and Mo9 Cluster Selenides* (ed. J.-F. Halet), 125–141. Cham: Springer International Publishing.

7 Nuss, J.r., Wedig, U., Xie, W. et al. (2017). Phosphide–tetrahedrite Ag6Ge10P12: thermoelectric performance of a long-Forgotten silver-cluster compound. *Chemistry of Materials 29*: 6956–6965.

8 Mori, T. (2019). Thermoelectric and magnetic properties of rare earth borides: Boron cluster and layered compounds. *Journal of Solid State Chemistry 275*: 70–82.

9 Zhang, L., Du, M.-H., and Singh, D.J. (2010). Zintl-phase compounds with SnSb 4 tetrahedral anions: electronic structure and thermoelectric properties. *Physical Review B 81*: 075117.

10 (a) Lin, S., Li, W., Bu, Z. et al. (2018). Thermoelectric properties of Ag9GaS6 with ultralow lattice thermal conductivity. *Materials Today Physics 6*: 60–67. (b) Li, W., Lin, S., Weiss, M. et al. (2018). Crystal structure induced ultralow lattice thermal conductivity in thermoelectric Ag9AlSe6. *Advanced Energy Materials 8*: 1800030.

11 O'Brien, E.S., Russell, J.C., Bartnof, M. et al. (2018). Spontaneous electronic band formation and switchable behaviors in a phase-rich superatomic crystal. *Journal of the American Chemical Society 140*: 15601–15605.

12 Seebeck, T.J. (1895). *Magnetische Polarisation der Metalle und erze Durch Temperatur-Differenz*. W. Engelmann.

13 Kishore, R.A. and Priya, S. (2018). A review on low-grade thermal energy harvesting: materials, methods and devices. *Materials (Basel) 11*: 1433.

14 Hsu, C.-T., Huang, G.-Y., Chu, H.-S. et al. (2011). An effective Seebeck coefficient obtained by experimental results of a thermoelectric generator module. *Applied Energy 88*: 5173–5179.

15 Spanner, D. (1951). The Peltier effect and its use in the measurement of suction pressure. *Journal of Experimental Botany*: 145–168.

16 Thomson, W. (1857). On a mechanical theory of thermo-electric currents. *Proceedings of the Royal society of Edinburgh 3*: 91–98.

17 Goldsmid, J. (2017). *The Physics of Thermoelectric Energy Conversion*. Morgan & Claypool Publishers.

18 Apertet, Y. and Goupil, C. (2016). On the fundamental aspect of the first Kelvin's relation in thermoelectricity. *International Journal of Thermal Sciences 104*: 225–227.

19 He, J. and Tritt, T.M. (2017). Advances in thermoelectric materials research: Looking back and moving. *Science 357*: eaak9997.

20 (a) Chasmar, R.P. and Stratton, R. (1959). The thermoelectric figure of merit and its relation to thermoelectric generators. *International Journal of Electronics 7*: 52–72. (b) Mahan, G.D. (1997). Good thermoelectrics. *Solid State Physics 51*: 81–157.

21 Wang, H., Pei, Y., Lalonde, A.D., and Snyder, G.J. (2013). *Material Design Considerations Based on Thermoelectric Quality Factor*. Berlin Heidelberg: Springer.

22 Snyder, G.J. and Ursell, T.S. (2003). Thermoelectric efficiency and compatibility. *Physical Review Letters 91*: 148301.

23 Ursell, T.S. and Snyder, G.J. (2002). *International Conference on Thermoelectrics. 2002. Proceedings ICT '02*, 2002, Long Beach, CA, USA (29 August 2002), pp. 412–417.

24 Harman, T., Walsh, M., and Turner, G. (2005). Nanostructured thermoelectric materials. *Journal of Electronic Materials 34*: L19–L22.

25 Rosi, F. (1968). Thermoelectricity and thermoelectric power generation. *Solid-State Electronics 11*: 833–868.

26 Pei, Y., Gibbs, Z.M., Balke, B. et al. (2014). Optimum carrier concentration in n-type PbTe thermoelectrics. *Advanced Energy Materials 4*: 9201–9210.

27 Ioffe, A.F. and Infosearch, L. (1957). *Semiconductor Thermoelements and Thermoelectric Cooling*. London: Infosearch Ltd.

28 Yamini, S.A., Ikeda, T., Lalonde, A. et al. (2013). Rational design of p-type thermoelectric PbTe: temperature dependent sodium solubility. *Journal of Materials Chemistry A 1*: 8725–8730.

29 Pei, Y., Wang, H., and Snyder, G.J. (2012). Thermoelectric materials: band engineering of thermoelectric materials (Adv. Mater. 46/2012). *Advanced Materials 24*: 6124–6124.

30 Zhang, J., Liu, R., Cheng, N. et al. (2014). High-performance pseudocubic thermoelectric materials from non-cubic chalcopyrite compounds. *Advanced Materials 26*: 3848–3853.

31 (a) Heremans, J.P., Wiendlocha, B., and Chamoire, A.M. (2012). Resonant levels in bulk thermoelectric semiconductors. *Energy & Environmental Science 5*: 5510–5530. (b) Heremans, J.P., Jovovic, V., Toberer, E.S. et al. (2008). Enhancement of thermoelectric efficiency in PbTe by distortion of the electronic density of states. *Science 321*: 554–557.

32 Pu, Y., Chiba, D., Matsukura, F. et al. (2008). Mott relation for anomalous Hall and Nernst Effects in Ga 1 − x Mn x As ferromagnetic semiconductors. *Physical Review Letters 101*: 117208.

33 Nuss, J., Wedig, U., Xie, W. et al. (2017). Phosphide-Tetrahedrite Ag6Ge10P12: thermoelectric performance of a Long-forgotten silver-cluster compound. *Chemistry of Materials*: 29.

34 (a) Moore, A.L. and Shi, L. (2014). Emerging challenges and materials for thermal management of electronics. *Materials Today 17*: 163–174. (b) Tian, Z., Lee, S., and Chen,

G. (2013). Heat transfer in thermoelectric materials and devices. *Journal of Heat Transfer*: 135.
35 Srivastava, G.P. and Kresin, V. (1991). The physics of phonons. *Physics Today 44*: 75.
36 Minnich, A.J., Johnson, J.A., Schmidt, A.J. et al. (2011). Thermal conductivity spectroscopy technique to measure phonon mean free paths. *Physical Review Letters 107*: 095901.
37 Kauzlarich, S.M., Brown, S.R., and Snyder, G.J. (2007). Zintl phases for thermoelectric devices. *Dalton Transactions*: 2099–2107.
38 Zhang, L., Du, M.-H., and Singh, D.J. (2010). Zintl-phase compounds with SnSb4 tetrahedral anions: electronic structure and thermoelectric properties. *Physical Review B*: 81.
39 Slack, G.A. (1979). *Solid State Physics: The Thermal Conductivity of Nonmetallic Crystals*, vol. *34* (eds. H. Ehrenreich, F. Seitz and D. Turnbull), 1–71. Elsevier B.V.
40 Richards, S.M. and Kaspar, J.S. (1969). The crystal structure of YB66. *Acta Crystallographica Section B 25*: 237–251.
41 Higashi, I., Tanaka, T., Kobayashi, K. et al. (1997). Crystal Structure of YB41Si1.2. *Journal of Solid State Chemistry 133*: 11–15.
42 Sales, B.C., Mandrus, D., Chakoumakos, B.C. et al. (1997). Filled skutterudite antimonides: Electron crystals and phonon glasses. *Physical Review B 56*: 15081–15089.
43 (a) Nolas, G., Slack, G., Morelli, D. et al. (1996). The effect of rare-earth filling on the lattice thermal conductivity of skutterudites. *Journal of Applied Physics 79*: 4002–4008. (b) Shi, X., Yang, J., Salvador, J.R. et al. (2011). Multiple-filled skutterudites: high thermoelectric figure of merit through separately optimizing electrical and thermal transports. *Journal of the American Chemical Society 133*: 7837–7846.
44 (a) Skoug, E.J. and Morelli, D.T. (2011). Role of lone-pair electrons in producing minimum thermal conductivity in nitrogen-group chalcogenide compounds. *Physical Review Letters 107*: 235901. (b) Lai, W., Wang, Y., Morelli, D.T., and Lu, X. (2015). From bonding asymmetry to anharmonic rattling in Cu12Sb4S13 tetrahedrites: when lone-pair electrons are not so lonely. *Advanced Functional Materials 25*: 3648–3657.
45 (a) Brandt, G. and Krämer, V. (1976). Phase investigations in the silver-gallium-sulphur system. *Materials Research Bulletin 11*: 1381–1388. (b) Hellstrom, E.E. and Huggins, R.A. (1980). Silver ionic and electronic conductivity in Ag9GaS6, Ag9AlS6, AgGaS2, AgAlS2, and AgAl5S8. *Journal of Solid State Chemistry 35*: 207–214.
46 Carcaly, C., Ollitrault-Fichet, R., Houphouët, D. et al. (1979). Les composes Ag7AsS6 et Ag7AsSe6 etude des proprietes thermiques, cristallographiques et electriques. *Materials Research Bulletin 14*: 1497–1503.
47 Deloume, J.-P. and Faure, R. (1981). Un nouveau matériau, Ag9GaSe6: Etude structurale de la phase α. *Journal of Solid State Chemistry 36*: 112–117.
48 Pei, Y., Shi, X., LaLonde, A. et al. (2011). Convergence of electronic bands for high performance bulk thermoelectrics. *Nature 473*: 66–69.
49 Li, W., Zheng, L., Ge, B. et al. (2017). Promoting SnTe as an eco-friendly solution for p-PbTe thermoelectric via band convergence and interstitial defects. *Advanced Materials (Deerfield Beach, Fla.) 29*: 1605887.
50 Tang, Y., Gibbs, Z.M., Agapito, L.A. et al. (2015). Convergence of multi-valley bands as the electronic origin of high thermoelectric performance in CoSb3 skutterudites. *Nature Materials 14*: 1223–1228.

51 Fu, C., Bai, S., Liu, Y. et al. (2015). Realizing high figure of merit in heavy-band p-type half-Heusler thermoelectric materials. *Nature Communications 6*: 8144–8144.

52 Kim, H.S., Gibbs, Z.M., Tang, Y. et al. (2015). Characterization of Lorenz number with Seebeck coefficient measurement. *APL Materials 3*: 041506.

53 Li, J., Zhang, X., Chen, Z. et al. (2018). Low-symmetry rhombohedral GeTe thermoelectrics. *Joule* 2: 976–987.

54 (a) Claridge, S.A., Castleman, A.W., Khanna, S.N. et al. (2009). Cluster-assembled materials. *ACS Nano 3*: 244–255. (b) Roy, X., Lee, C.-H., Crowther, A.C. et al. (2013). Nanoscale atoms in solid-state chemistry. *Science 341*: 157. (c) Baudron, S.A., Batail, P., Coulon, C. et al. (2005). (EDT-TTF-CONH2)6[Re6Se8(CN)6], a metallic Kagome-type organic-inorganic hybrid compound: electronic instability, molecular motion, and charge localization. *Journal of the American Chemical Society 127*: 11785–11797. (d) Tomalia, D.A. and Khanna, S.N. (2016). A systematic framework and nanoperiodic concept for unifying nanoscience: hard/soft nanoelements, superatoms, meta-atoms, new emerging properties, periodic property patterns, and predictive mendeleev-like nanoperiodic tables. *Chemical Reviews 116*: 2705–2774. (e) C. H. Lee, K. I. Choi, M. K. Cho, Y. H. Song, K. C. Park, K. Kim, A novel SONOS structure of SiO2SiNAl2O3 with TaN metal gate for multi-giga bit flash memories International Electron Devices Meeting, 2003, Washington, DC, USA (8–10 December 2003), pp. 613–616; (f) Yoon, B., Luedtke, W.D., Barnett, R.N. et al. (2014). Hydrogen-bonded structure and mechanical chiral response of a silver nanoparticle superlattice. *Nature Materials 13*: 807–811.

55 Ong, W.L., O'Brien, E.S., Dougherty, P.S. et al. (2017). Orientational order controls crystalline and amorphous thermal transport in superatomic crystals. *Nature Materials 16*: 83–88.

56 Giri, A., Chou, S.S., Drury, D.E. et al. (2020). Molecular tail chemistry controls thermal transport in fullerene films. *Physical Review Materials 4*: 065404.

57 Ong, W.-L., Rupich, S.M., Talapin, D.V. et al. (2013). Surface chemistry mediates thermal transport in three-dimensional nanocrystal arrays. *Nature Materials 12*: 410–415.

58 Yu, R.C., Tea, N., Salamon, M.B. et al. (1992). Thermal conductivity of single crystal C60. *Physical Review Letters 68*: 2050–2053.

59 David, W.I.F., Ibberson, R.M., Matthewman, J.C. et al. (1991). Crystal structure and bonding of ordered C60. *Nature 353*: 147–149.

60 (a) Zhou, T., Colin, M., Candolfi, C. et al. (2014). Comprehensive study of the low-temperature transport and thermodynamic properties of the cluster compounds AgxMo9Se11. *Chemistry of Materials 26*: 4765–4775. (b) Zhou, T., Lenoir, B., Colin, M. et al. (2011). Promising thermoelectric properties in Ag x Mo 9 Se 11 compounds ($3.4\leq x\leq 3.9$). *Applied Physics Letters 98*: 162106.

61 (a) Masschelein, P., Candolfi, C., Dauscher, A. et al. (2018). Influence of S and Te substitutions on the thermoelectric properties of the cluster compound Ag3. 8Mo9Se11. *Journal of Alloys and Compounds 739*: 360–367. (b) Colin, M., Zhou, T., Lenoir, B. et al. (2012). Optimization of bulk thermoelectrics: influence of Cu insertion in Ag 3.6 Mo 9 Se 11. *Journal of Electronic Materials 41*: 1360–1364. (c) Gougeon, P., Padiou, J., Le Marouille, J. et al. (1984). Ag3. 6Mo9Se11: Premier compose′ aclusters Mo9 dans des motifs Mo9Se11. *Journal of Solid State Chemistry 51*: 218–226.

62 Chevrel, R., Sergent, M., and Prigent, J. (1971). Sur de nouvelles phases sulfurées ternaires du molybdène. *Journal of Solid State Chemistry 3*: 515–519.

63 (a) Gougeon, P., Gall, P., Al Rahal Al Orabi, R. et al. (2012). Synthesis, crystal and electronic structures, and thermoelectric properties of the novel cluster compound Ag3In2Mo15Se19. *Chemistry of Materials 24*: 2899–2908. (b) Picard, S., Saillard, J.-Y., Gougeon, P. et al. (2000). Rb2n(Mo9S11)(Mo6nS6n+2) (n=1 to 4): a novel family of superconducting molybdenum cluster compounds. *J. Solid State Chem. 155*: 417–426. (c) Gougeon, P., Potel, M., and Gautier, R. (2004). Syntheses and structural, physical, and theoretical studies of the novel isostructural Mo9 cluster compounds Ag2. 6CsMo9Se11, Ag4. 1ClMo9Se11, and h-Mo9Se11 with tunnel structures. *Inorganic Chemistry 43*: 1257–1263.

64 Gautier, R., Gougeon, P., Halet, J.-F. et al. (1997). New experimental and theoretical studies on condensed molybdenum chalcogenide cluster compounds. *Journal of Alloys and Compounds 262–263*: 311–315.

65 Caillat, T., Fleurial, J.P., and Snyder, G.J. (1999). Potential of Chevrel phases for thermoelectric applications. *Solid State Sciences 1*: 535–544.

66 Mori, T. (2017). Novel principles and nanostructuring methods for enhanced thermoelectrics. *Small 13*: 1702013.

67 (a) Gougeon, P., Gall, P., Merdrignac-Conanec, O. et al. (2017). Synthesis, crystal structure, and transport properties of the hexagonal mo 9 cluster compound Ag3RbMo9Se11. *Inorganic Chemistry 56*: 9684. (b) Gougeon, P., Gall, P., Al Rahal, R. et al. (2019). Electronic band structure and transport properties of the cluster compound Ag3Tl2Mo15Se19. *Inorganic Chemistry 58*: 5533–5542.

68 Bouyrie, Y., Candolfi, C., Pailhès, S. et al. (2015). From crystal to glass-like thermal conductivity in crystalline minerals. *Physical Chemistry Chemical Physics 17*: 19751–19758.

69 Puyet, M., Dauscher, A., Lenoir, B. et al. (2007). Influence of Ni on the thermoelectric properties of the partially filled calcium skutterudites CayCo4−xNixSb12. *Physical Review B 75*: 245110.

70 (a) Gschneidner, K.A., Bunzli, J.C.G., and Pecharsky, V.K. (1987). *Handbook on the Physics and Chemistry of Rare Earths 57*: 295–300. (b) Fehlner, T., Halet, J.F., and Saillard, J.Y. (2007). Main-group clusters: geometric and electronic structure. In: *Molecular Clusters: A Bridge to Solid-State Chemistry*, 33–84. Cambridge: Cambridge University Press https://doi.org/10.1017/CBO9780511628887. (c) Etourneau, J. and Hagenmuller, P. (1985). Structure and physical features of the rare-earth borides. *Philosophical Magazine B Physics of Condensed Matter 52*: 589–610.

71 (a) Mori, T., Kuzmych-Ianchuk, I., Yubuta, K. et al. (2012). Direct elucidation of the effect of building defects on the physical properties of alpha-TmAlB4; An AlB2-type analogous "tiling" compound. *Journal of Applied Physics 111*: 105. (b) Mori, T., Kudou, K., Shishido, T., and Okada, S. (2011). f-electron dependence of the physical properties of REAlB4; an AlB2-type analogous "tiling" compound. *Journal of Applied Physics 109*: 07E111. (c) Mori, T., Shishido, T., and Nakajima, K. (2009). Doping effects in rare-earth borides. *Journal of Electronic Materials 38*: 1098–1103.

72 (a) Bauer, J., Halet, J.F., and Saillard, J.Y. (1998). Rare earth metal borocarbides: examples of coordination compounds in solid-state chemistry. *Coordination Chemistry Review 178–180*: 723–753. (b) Chernyak, G.V., Chaban, N.F., and Kuz'Ma, Y.B. (1983). Ternary systems {Dy, Er}FeB. *Soviet Powder Metallurgy & Metal Ceramics 22*: 479–480. (c) Albert, B. and Hillebrecht, H. (2010). Boron: elementary challenge for experimenters and theoreticians. *Angewandte Chemie International Edition 48*: 8640–8668. (d) Oliver, J., Oliver, N., Rainer, P., and Bernard, C. (2016). Cerium intermetallics with TiNiSi-type structure. *Zeitschrift für Naturforschung B 71*: 737–764.

73 (a) Golikova, O.A. (2006). Semiconductors with complex lattice and the amorphization problem. *Physica Status Solidi A 101*: 277–314. (b) Mori, T. (2008). Higher borides. *Handbook on the Physics and Chemistry of Rare Earths 38*: 105–173.

74 (a) Cahill, D.G., Fischer, H.E., Watson, S.K. et al. (1989). Thermal properties of boron and borides. *Physical Review B Condensed Matter 40*: 3254. (b) Tanaka, T., Kamiya, K., Numazawa, T. et al. (2006). The effect of transition metal doping on thermal conductivity of YB66. *Zeitschrift Für Kristallographie 221*: 472–476. (c) Mori, T., Martin, J., and Nolas, G. (2007). Thermal conductivity of YbB44Si2. *Journal of Applied Physics* 102: 073510.

75 Spear, K. and Solov'ev, G. (1972). High boron content rare-earth borides. *National Bureau of Standards (US), Special Publication* 364: 597–604.

76 Sussardi, A., Tanaka, T., Khan, A.U. et al. (2015). Enhanced thermoelectric properties of samarium boride. *Journal of Materiomics 1*: 196–204.

77 Mori, T. and Tanaka, T. (2006). Effect of transition metal doping and carbon doping on thermoelectric properties of YB66 single crystals. *Journal of Solid State Chemistry 179*: 2889–2894.

78 Mori, T. (2005). High temperature thermoelectric properties of B12 icosahedral cluster-containing rare earth boride crystals. *Journal of Applied Physics 97*: 093703.

79 (a) Terasaki, I., Sasago, Y., and Uchinokura, K. (1997). Large thermoelectric power in NaCo2O4 crystals. *Physical Review B 56*: 75397. (b) Terasaki, I. (2013). *Layered Cobalt Oxides: Correlated Electrons for Thermoelectrics*. Berlin Heidelberg: Springer.

80 (a) Ahmed, F., Tsujii, N., and Mori, T. (2017). Thermoelectric properties of CuGa1−xMnxTe2: power factor enhancement by incorporation of magnetic ions. *Journal of Materials Chemistry A* 5: 7545–7554. (b) Vaney, J.-B., Yamini, S.A., Takaki, H. et al. (2019). Magnetism-mediated thermoelectric performance of the Cr-doped bismuth telluride tetradymite. *Materials Today Physics 9*: 100090. (c) Tsujii, N., Nishide, A., Hayakawa, J., and Mori, T. (2019). Observation of enhanced thermopower due to spin fluctuation in weak itinerant ferromagnet. *Science Advances 5*: eaat5935.

81 Sauerschnig, P., Tsuchiya, K., Tanaka, T. et al. (2020). On the thermoelectric and magnetic properties, hardness, and crystal structure of the higher boride YbB66. *Journal of Alloys and Compounds 813*: 152182.

82 Hossain, M.A., Tanaka, I., Tanaka, T. et al. (2015). YB48 the metal rich boundary of YB66; crystal growth and thermoelectric properties. *Journal of Physics and Chemistry of Solids 87*: 221–227.

83 Shklovskiĭ, B.I. and Efros, A.L. (2013). *Electronic Properties of Doped Semiconductors*, vol. 45. Springer Science & Business Media.

84 Mori, T., Shimazu, Y., and Ikehata, S. (1993). NMR study on compensated Si: P. *Journal of the Physical Society of Japan 62*: 3172–3180.

85 Li, W., Lin, S., Ge, B. et al. (2016). Low sound velocity contributing to the high thermoelectric performance of Ag8SnSe6. *Advanced Science 3*: 1600196.

86 Lin, S., Li, W., Li, S. et al. (2017). High thermoelectric performance of Ag9GaSe6 enabled by low cutoff frequency of acoustic phonons. *Joule 1*: 816–830.

87 Charoenphakdee, A., Kurosaki, K., Muta, H. et al. (2009). Ag8SiTe6: a new thermoelectric material with low thermal conductivity. *Japanese Journal of Applied Physics 48*: 011603.

88 (a) Weldert, K.S., Zeier, W.G., Day, T.W. et al. (2014). Thermoelectric transport in Cu7PSe6 with high copper ionic mobility. *Journal of the American Chemical Society 136*: 12035–12040. (b) Charoenphakdee, A., Kurosaki, K., Muta, H. et al. (2008). Reinvestigation of the thermoelectric properties of Ag8GeTe6. *Physica Status Solidi (RRL) – Rapid Research Letters 2*: 65–67.

89 Morelli, D.T. and Slack, G.A. (2006). *High Thermal Conductivity Materials*, 37–68. Springer.

90 (a) Jiang, B., Qiu, P., Eikeland, E. et al. (2017). Cu8GeSe6-based thermoelectric materials with an argyrodite structure. *Journal of Materials Chemistry C 5*: 943–952. (b) Li, W., Lin, S., Zhang, X. et al. (2016). *Chemistry of Materials 28*: 6227–6232.

91 (a) Beaumale, M., Barbier, T., Bréard, Y. et al. (2014). Electron doping and phonon scattering in Ti1+ xS2 thermoelectric compounds. *Acta Materialia 78*: 86–92. (b) Suekuni, K., Tsuruta, K., Kunii, M. et al. (2013). High-performance thermoelectric mineral Cu12− x Ni x Sb4S13 tetrahedrite. *Journal of Applied Physics 113*: 043712. (c) Zhou, B., Li, S., Li, W. et al. (2017). Thermoelectric properties of SnS with Na-doping. *ACS Applied Materials & Interfaces 9*: 34033–34041. (d) Liu, M.L., Huang, F.Q., and Chen, L.D. (2009). A wide-band-gap p-type thermoelectric material based on quaternary chalcogenides of Cu2ZnSnQ4 (Q=S,Se). *I. Chen 94*: 223. (e) Ohta, M., Yamamoto, A., and Obara, H. (2010). Thermoelectric properties of chevrel-phase sulfides M xMo 6S 8 (M: Cr, Mn, Fe, Ni). *Journal of Electronic Materials 39*: 2117–2121. (f) Wang, H., Schechtel, E., Pei, Y., and Snyder, G.J. (2013). High Thermoelectric Efficiency of n-type PbS. *Advanced Energy Materials 3*: 488–495. (g) Berthebaud, D., Lebedev, O.I., and Maignan, A. (2015). Thermoelectric properties of n-type cobalt doped chalcopyrite Cu1−xCoxFeS2 and p-type eskebornite CuFeSe2. *Journal of Materiomics 1*: 68–74. (h) Wang, H., Gibbs, Z.M., Takagiwa, Y., and Snyder, G. (2014). Tuning bands of PbSe for better thermoelectric efficiency. *Energy and Environmental Science 7*: 804–811.

92 Bullett, D.W. and Dawson, W.G. (1986). Bonding relationships in some ternary and quarternary phosphide and tetrahedrite structures: (Ag6M4P12)M6', Cu12+xSb4S13and Cu14-xSb4S13, Ln6Ni6P17. *Journal of Physics C 19*: 5837.

93 (a) Skoug, E.J., Cain, J.D., and Morelli, D.T. (2011). High thermoelectric figure of merit in the Cu3SbSe4-Cu3SbS4 solid solution. *Applied Physics Letters 98*: 261911. (b) Lu, X., Morelli, D.T., Xia, Y. et al. (2013). High Performance Thermoelectricity in Earth-Abundant Compounds Based on Natural Mineral Tetrahedrites. *Advanced Energy Materials 3*: 342–348. (c) Heo, J., Laurita, G., Muir, S. et al. (2014). Enhanced thermoelectric performance of synthetic tetrahedrites. *Chemistry of Materials 26*: 2047–2051.

12

Clusters for CO_2 Activation and Conversion

Haoming Shen[1,2], Qiang Sun[1,2], and Puru Jena[3]

[1] *School of Materials Science and Engineering, Peking University, Beijing, China*
[2] *Center for Applied Physics and Technology, Peking University, Beijing, China*
[3] *Physics Department, Virginia Commonwealth University, Richmond, VA, USA*

12.1 Introduction

It is well-known that CO_2 is a main greenhouse gas, which is primarily responsible for global warming through the absorption and release of infrared radiation. The excessive concentration of CO_2 overheats the atmosphere, increases the Earth's temperature, and perturbs the climate. CO_2 emission mostly comes from petrochemical industries as well as power plants that use coal to generate energy. With the growth of heavy industry and the increasing number of vehicles using fossil fuels, CO_2 emissions have become a huge issue for the environment. Therefore, considerable efforts are being made to study the capture and conversion of CO_2 and to develop new catalysts for this purpose. A good catalyst for CO_2 should possess the following properties: (i) CO_2 adsorption with low reduction overpotentials, (ii) easy dissociation to products or low overpotentials for converting it into other hydrocarbons, (iii) thermodynamical stability, and (iv) limited side reactions like hydrogen evolution reaction (HER).

The linear nonpolar molecule CO_2 has a simple geometry but complex chemistry. The C–O bond in CO_2 is a Π_3^4 bond with three-center-four-electron delocalization, which stabilizes the O–C–O system, resulting in the C–O bond length even shorter than that of some C–O double bonds in other chemicals. CO_2 has an electron affinity of −0.6 eV, and the extra electron in its anionic state is in the σ^* antibonding orbital, changing *sp* hybridization to "quasi"-sp^2. This leads to a bent geometry of CO_2^- with C_{2v} symmetry. To better understand the interactions of CO_2 with materials at atomic level, Bowen and coworkers [1] carried out photoelectron spectroscopy experiment and computational study of CO_2 absorbed on metal ions (M = Cu, Ag, Au). The structures of both anionic and neutral forms were determined using density functional theory (DFT) and Becke, 3-parameter, Lee Yang Parr (B3LYP) hybrid functional for exchange-correlation potential and Stuttgart and Dresden pseudopotentials (SDD) basis set. The geometries with bond length (Å) and O–C–O bond angle (in degrees) are shown in Figure 12.1. For $Au^-(CO_2)$ and $Ag^-(CO_2)$, the M–C bond

Superatoms: Principles, Synthesis and Applications, First Edition. Edited by Purusottam (Puru) Jena and Qiang Sun.
© 2022 John Wiley & Sons Ltd. Published 2022 by John Wiley & Sons Ltd.

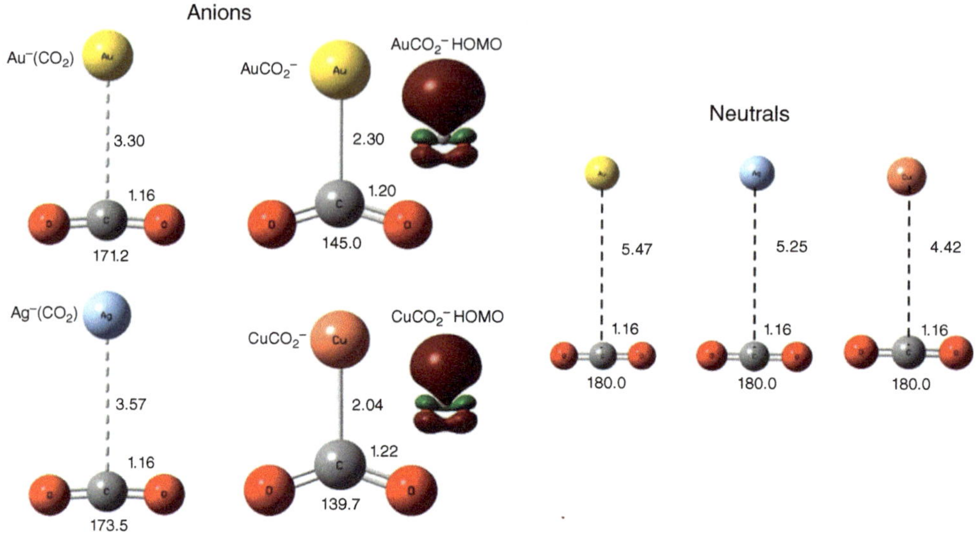

Figure 12.1 Calculated structures of $Au^-(CO_2)$, $AuCO_2^-$, $Ag^-(CO_2)$, $CuCO_2^-$, $AuCO_2$, $AgCO_2$, and $CuCO_2$. HOMOs of $AuCO_2^-$ and $CuCO_2^-$ are also presented. *Source:* Zhang et al. [1]. © AIP Publishing.

lengths are significantly longer than the typical chemical (single) bonds, indicating weak interactions between Au^-/Ag^- and CO_2. Although the C–O bond length is 1.16 Å, close to the value in an isolated CO_2, the O–C–O moiety is slightly bent due to the interaction with metal ions.

The M–C bond lengths in $AuCO_2^-$ and $CuCO_2^-$ are 2.30 and 2.04 Å, respectively, indicative of a single bond between M and C in both cases. The natural population analysis showed that in $AuCO_2^-$, the CO_2 moiety has a negative charge of −0.45 e, whereas in $CuCO_2^-$, the charge on CO_2 is −0.64 e. Therefore, CO_2 has been significantly reduced in both complexes. Compared to $AuCO_2^-$, the CO_2 moiety in $CuCO_2^-$ is more activated, as evidenced from their M–C and C–O bond lengths; the C–O bond (1.22 Å) in $CuCO_2^-$ is longer than that in $AuCO_2^-$ (1.20 Å), both being longer than the C–O bond length in an isolated CO_2 (1.16 Å). This implies that larger is the negative charge on the CO_2 moiety, more activated is the C-O bond. The difference in charges is also consistent with the O–C–O bond angle, which is bent more in $CuCO_2^-$ (139.7°) than that in $AuCO_2^-$ (145.0°). This originates from the difference in electronegativity (EA) between the metal atoms, i.e. EA(Au) > EA(Cu). Like hydrogen atom, gold or copper atom each has only one valence electron. Both $AuCO_2^-$ and $CuCO_2^-$ are isostructural and isoelectronic to the formate anion; they are metalloformates [2]. The highest occupied molecular orbitals (HOMOs) of $AuCO_2^-$ and $CuCO_2^-$ accommodate the excess electron, as plotted in Figure 12.1. In all the three neutral M–CO_2 (M = Cu, Ag, Au) species, the metal atom is far away from CO_2, indicative of physisorption.

Since CO_2 would assume a bent structure when an electron is transferred to it or due to its interaction with the metal atom, [3] one would expect that the stable geometry of M–CO_2^- would depend upon the ionization potential (IP) of the metal atom M, making it easier for an atom with a smaller IP to transfer an electron to CO_2 than the one with a large

IP. Similarly, the Coulomb repulsion from the metal atom would also bend the linear CO_2 structure [3]. The IPs of Cu, Ag, and Au are, respectively, 7.73, 7.58, and 9.23 eV. Bowen and coworkers [1] found the calculated O–C–O angles of $AuCO_2^-$ and $CuCO_2^-$ to be 145.0° and 139.7°, respectively, while the geometry of CO_2 in $Ag(CO_2)^-$ remained linear. However, one would have expected it to be a bent structure as in the case of $CuCO_2^-$ because both Cu and Ag have nearly the same IP. These results suggest that both the electronic structure and the size of the metal atom play a role in electron transfer.

It is known that alkali metal atoms exhibit the lowest IPs (5.39–3.89 eV) [4] among all the elements in the periodic table, making them good electron donors in chemical reactions. The IPs can be further reduced to less than (3.89 eV) [4] in "superalkalis" composed of polyatomic molecules or clusters. Such extraordinary species show exciting prospect of activation and conversion of CO_2. In this chapter we briefly summarize the advances that have been made in this topic.

12.2 Superalkali Catalysts

Initially, Gutsev and Boldyrev [5] designed superalkalis by decorating an electronegative central atom with alkali metal ligands, serving as mononuclear clusters. They proposed a simple formula ML_{k+1} for such superalkalis, where M is an electronegative atom with the maximal formal valence k and L is an alkali metal atom. Most of the commonly seen mononuclear superalkalis are Li-containing clusters such as FLi_2 [6], OLi_3 [7], NLi_4 [5, 8], CLi_5 [9], and $SiLi_5$ [10]. In 2009 [11], Wu and coworkers extended the design of superalkalis to include homologous binuclear species, M_2L_{2k+1} (where L is an alkali metal and k is the maximal formal valence of nonmetal M atom). In such superalkalis, the $2k$ valence electrons from alkali metal ligands will fully close the atomic shells of two nonmetal M atoms. The one excess electron in the network leads to their low IPs. Moreover, some attention was paid to the design of polynuclear superalkali cations YL_{k+1}^+ (where L is an alkali metal atom and k is the valence state of the anionic groups/clusters) by replacing the central M atom in the mononuclear ML_{k+1} with polyatomic group Y [12]. Later on, more superalkalis were proposed with different structural features. In recent years, some of the superalkalis have been applied to CO_2 activation as discussed below.

12.2.1 Li-based Superalkalis for CO_2 Activation

Zhao, Wang, and Jena [13] showed that the number of superalkalis can be vastly increased by using electron-counting rules beyond the octet rule and were the first ones to use superalkalis for CO_2 activation. The electron-counting rules included the jellium shell closure rule valid for clusters of nearly free electron metals, the 18-electron rule valid for clusters containing transition metal atoms, the Wade–Mingos rule valid for compounds containing electron-deficient boron, and Hückel's rule valid for organic molecules. They used Al_3^+, $Mn(B_3N_3H_6)^{2+}$, $B_9C_3H_{12}^+$, and $C_5NH_6^+$ as examples to illustrate these design principles. Furthermore, to demonstrate the use of these superalkalis in chemical applications, they focused on CO_2 activation that is the first step for producing fuels from CO_2. Realizing that a superalkali with an IP less than that of an alkali atom can easily transfer an electron, they studied the reaction of these superalkalis with CO_2 by computing their equilibrium geometry and charge distribution. In all cases, CO_2 was found to be negatively charged, although the amount of charge transfer depends on the nature of the superalkali. However, the O–C–O bond angle is found to be smaller than CO_2^- in the gas phase. The results are shown in Figure 12.2 and suggest that superalkalis can be used as catalysts for CO_2 activation. This has inspired many research groups to work on this subject. In the following, we focus on Li-based superalkalis for CO_2 activation.

Srivastava [14] studied the reduction of CO_2 on Li-based superalkalis by using Møller–Plesset perturbative (MP2) level of theory and focusing on FLi_2, OLi_3, and NLi_4 superalkalis. He found stable superalkali–CO_2 complexes in which the structure of CO_2 is bent due to electron transfer from the superalkalis. The ionization energies of these species are calculated to be 3.85, 3.46, and 3.23 eV for FLi_2, OLi_3, and NLi_4, respectively, which are all less than that of Li (-5 eV). The equilibrium structures of superalkali–CO_2 complexes are displayed in Figure 12.3 along with CO_2^- anion and superalkali cations. The relative energy (ΔE) of superalkali–CO_2 complexes and their isomers are collected in Table 12.1. In all the superalkali–CO_2 complexes, the lowest energy structures are those in which both O atoms of CO_2 interact with two Li atoms of the superalkalis. The bond lengths of Li–O interaction lie between 1.865 and 1.892 Å. The low-lying isomers of FLi_2–CO_2 and OLi_3–CO_2 (interacting via single O atom) are 1.07 and 0.40 eV higher in energy than the bridge-bonded ones as shown in Figure 12.3. In NLi_4–CO_2, no competing isomer obtained was found in the calculations. In all superalkali–CO_2 complexes, CO_2 moiety becomes bent with an angle 133° and the C–O bond lengths become 1.246 Å. These values are comparable to the bond lengths of 1.237 Å and bond angle of 137° in the CO_2^- anion. This suggests that CO_2 can be significantly activated by superalkalis.

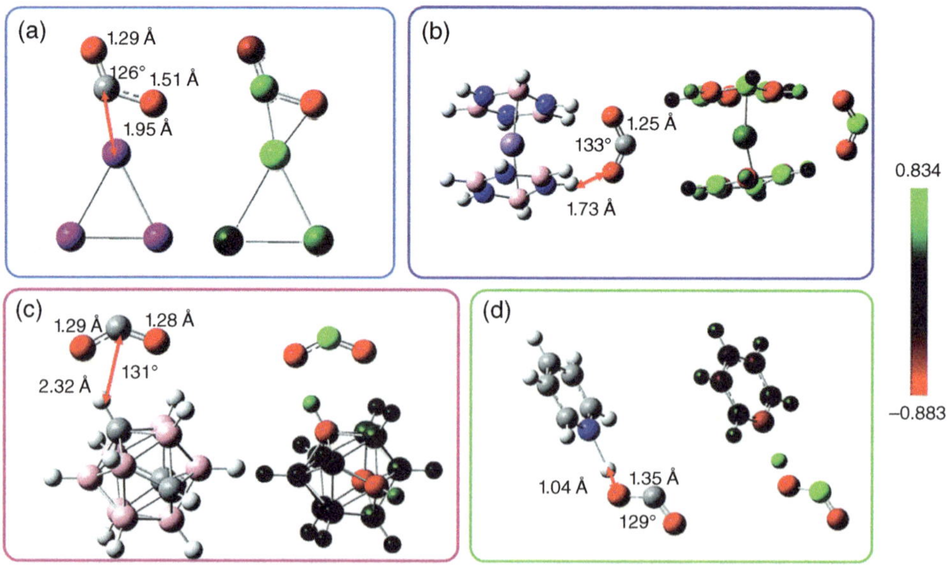

Figure 12.2 Optimized geometries (left panel) and natural bond orbital (NBOO charge distribution (right panel) of four neutral superalkalis: Al_3^+ (a), $Mn(B_3N_3H_6)^{2+}$ (b), $B_9C_3H_{12}^+$ (c), and $C_5NH_6^+$ (d) interacting with the CO_2 molecule. *Source:* Zhao et al. [13]. © Royal Society of Chemistry.

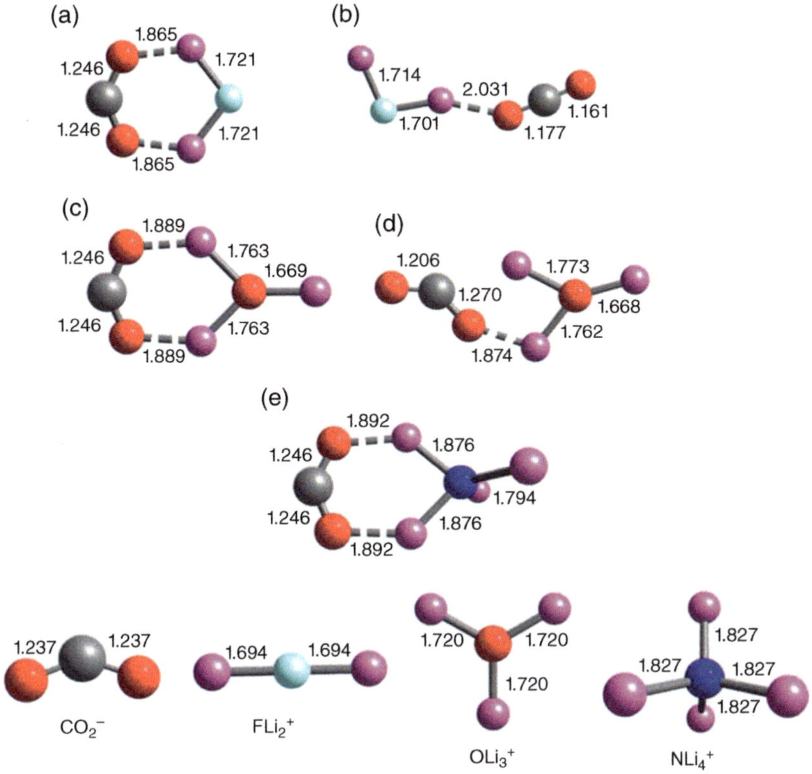

Figure 12.3 Equilibrium structures of superalkali–CO_2 complexes with bond lengths in Å. Equilibrium structures of CO_2^- anion and FLi_2^+, OLi_3^+, and NLi_4^+ superalkali cations are also given. *Source:* Srivastava [14]. © John Wiley & Sons.

Table 12.1 The MP2/6‐311+G(d) calculated relative energy (ΔE), binding energy (BE), charge transferred to CO_2 (Δq), spin density located at CO_2 ($\Delta \rho_s$), and SOMO energy (E_{SOMO}) of superalkali–CO_2 complexes.

Systems	Isomer	ΔE (eV)	BE (eV)	Δq (*e*)	$\Delta \rho_s$ (a.u.)	E_{SOMO} (eV)
FLi_2–CO_2	(a)	0	2.41	−0.90	0.96	−9.74
	(b)	1.07	1.34	−0.17	0.08	−4.35
OLi_3–CO_2	(c)	0	1.23	−0.88	0.96	−8.91
	(d)	0.40	0.83	−0.82	0.89	−8.58
NLi_4–CO_2	(e)		1.17	−0.85	0.96	−6.57

Source: Srivastava [14]. © John Wiley & Sons.

The binding energy (BE) of CO_2 with superalkali decreases in the order BE(FLi_2–CO_2) > BE(OLi_3–CO_2) > BE(NLi_4–CO_2). The interaction between superalkali and CO_2 becomes slightly weaker with the increase in the size of superalkalis. This can be expected due to the fact that larger superalkalis usually possess more delocalized electron cloud [15]. Therefore,

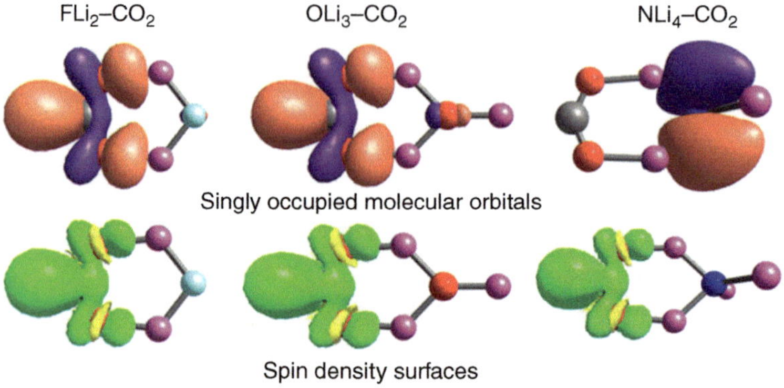

Figure 12.4 The SOMOs (isovalue = 0.02 a.u.) and spin density surfaces (isovalue = 0.002 a.u.) of superalkali–CO_2 complexes. *Source:* Srivastava [14]. © John Wiley & Sons.

charge transfer from larger superalkali such as NLi_4 is not as much effective as in the case of smaller superalkali such as FLi_2.

The singly occupied molecular orbitals (SOMOs) and spin density surfaces are plotted in Figure 12.4. The SOMOs of FLi_2–CO_2 and OLi_3–CO_2 complexes are localized on the CO_2 moiety, which clearly indicates electron transfer from FLi_2 and OLi_3 superalkali moieties, respectively. In the case of NLi_4–CO_2 complex, on the contrary, the SOMO is localized on the N atom in NLi_4. Although there is indeed an electron transfer from NLi_4 to CO_2, the localization of SOMO is due to the lone pair of nitrogen. This fact is further confirmed by spin density surfaces, which suggests that the spin density is localized on CO_2 moiety due to the unpaired electron for all superalkali–CO_2 complexes studied here. The SOMO eigenvalues of superalkali–CO_2 complexes are also listed in Table 12.1. One can note that the SOMO energies are fairly large, which suggests that the unpaired electron is strongly bound. However, these SOMO energy values decrease with the increase in the size of the superalkalis, following the same trend as those of the charge transfer and BEs discussed earlier.

Thus, superalkalis are indeed capable of reducing CO_2. It should, however, be noted that the size of superalkalis is more effective than their ionization energy in CO_2 activation. For example, the size of superalkalis considered is in the order of $FLi_2 < OLi_3 < NLi_4$, while their ionization energies follow a reverse order: $FLi_2 > OLi_3 > NLi_4$. The calculated binding, charge transfer, and SOMO energies of superalkali–CO_2 complexes indicate that FLi_2 is more effective for CO_2 reduction than NLi_4. A similar observation was made earlier by Zhao et al. [13] in the case of some novel superalkalis.

Park and Meloni [16] also studied the reduction of CO_2 on Li superalkalis using complete basis set (CBS)-QB3 method. The focused system is binuclear superalkali Li_3F_2, and N_2 is also analyzed for discussing the Li_3F_2 selectivity of CO_2 from N_2. They found that Li_3F_2 has reduction capability toward CO_2 with low selectivity of N_2. The calculated binding energies and charge transfer data are listed in Table 12.2. The potential energy surfaces of the Li_3F_2 with CO_2 and N_2 reaction are shown in Figure 12.5.

The geometry of neutral Li_3F_2 is planar (C_{2v}) with a F–Li–F bond angle of 129° and bond lengths of 1.71 and 1.68 Å for the medial and terminal Li–F, respectively. $[Li_3F_2]^+$ ($D_{\infty h}$) is

Table 12.2 Adiabatic ionization energies (AIE) in eV, adiabatic electron affinities (AEA) in eV, binding energy (BE) in kJ/mol, charge flows (Δq) in a.u., and HOMO–LUMO gaps in eV obtained using the CBS-QB3 composite model.

Species	AIE	AEA	BE	Δq	HOMO–LUMO gap
Li_3F_2	3.80	0.60	—	—	5.60
CO_2	13.8	−0.55	—	—	16.4
N_2	15.6	−2.01	—	—	20.3
Li_3F_2/CO_2 (**1**)	5.16	1.21	158	0.63	9.95
Li_3F_2/CO_2 (**2**)	5.21	0.36	163	0.78	10.0
Li_3F_2/CO_2 (**3**)	7.43	1.41	106	0.88	10.4
$[Li_3F_2/CO_2]^+$ (**1**), (**2**)	—	—	27	—	15.9
$[Li_3F_2/CO_2]^+$ (**3**)	—	—	18	—	15.2
$[Li_3F_2/CO_2]^-$ (**1**)	—	—	218	—	5.35
$[Li_3F_2/CO_2]^-$ (**2**)	—	—	140	—	3.46
$[Li_3F_2/CO_2]^-$ (**3**)	—	—	184	—	5.51
Li_3F_2/N_2 (**a**)	4.00	−0.38	51	0.60	7.85
Li_3F_2/N_2 (**b**)	4.05	−0.13	35	0.70	8.14
$[Li_3F_2/N_2]^+$ (**a**)	—	—	32	—	15.9
$[Li_3F_2/N_2]^+$ (**b**)	—	—	11	—	16.0
$[Li_3F_2/N_2]^-$ (**a**)	—	—	−42	—	1.07
$[Li_3F_2/N_2]^-$ (**b**)	—	—	−35	—	1.30

Source: Park and Meloni [16]. © Royal Society of Chemistry.

linear upon electron removal from the neutral. The anion of Li_3F_2 (C_{2v}) becomes nearly linear with the Li–F–Li angle of 174° [17]. Visualization of spin density (ρ) migration from the superalkali to the carbon atom can be seen in Figure 12.6. The unpaired electron in Li_3F_2 SOMO is localized at the two terminal lithium atoms. The spin density moves to the carbon atom once the superalkali combines with CO_2. After the migration, the unpaired electrons in the SOMOs of all the three CO_2 clusters look identical to the HOMO of the CO_2 anion. This proves that Li_3F_2 successfully donates the electron to CO_2. The reduction capability of the superalkali toward N_2 can also be explained by the visualized HOMO from Figure 12.6. The HOMOs of clusters (**a**) and (**b**) are identical to the HOMO of the N_2 anion. The unevenly distributed spin density of the two nitrogen atoms in Li_3F_2/N_2 (**a**) lowers the symmetry from C_{2v} to *Cs*. On the other hand, cluster (**b**) shows an even distribution of spin density and retains its symmetry. In this case, the strongest calculated BE is only 51 kJ/mol, which ensures that Li_3F_2 is capable of reducing CO_2 with high selectivity over N_2. The geometric transformation of CO_2 from linear to bent is observed as well; the activated CO_2 can be further converted into other fuels.

Polynuclear superalkali C_6Li_6 has also been studied by Srivastava [18] using long-range dispersion corrected density functional ωB97xD/6–311 + G(d) calculations (Figure 12.7). A planar star-like C_6Li_6 configuration has been previously studied, with its application in

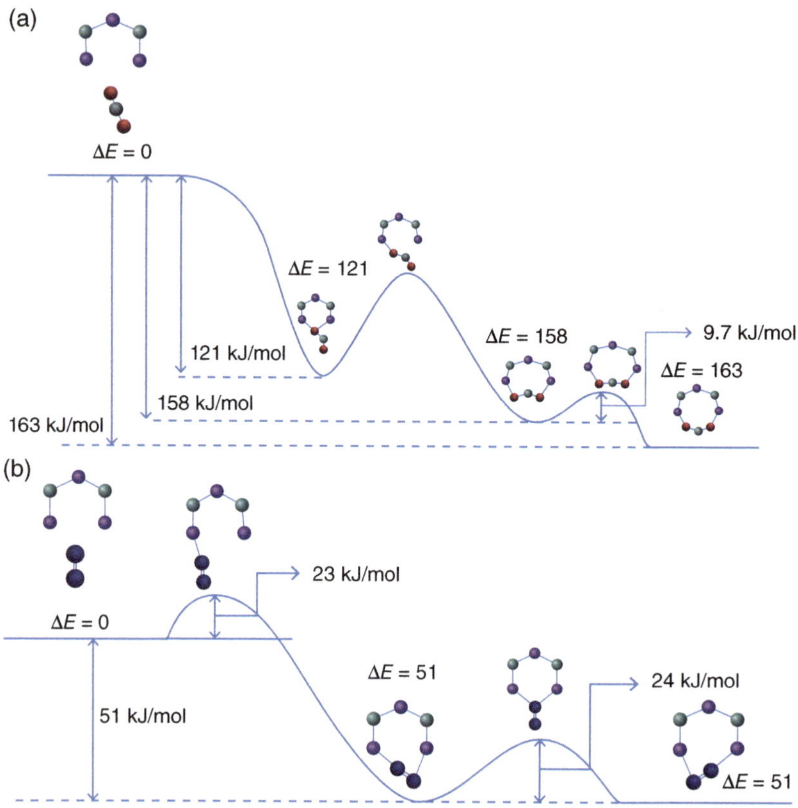

Figure 12.5 Calculated CBS-QB3 potential energy surface of the Li_3F_2 with CO_2 (a) and N_2 (b) reaction. The relative energy values are in kJ mol^{-1}. *Source:* Park and Meloni [16]. © Royal Society of Chemistry.

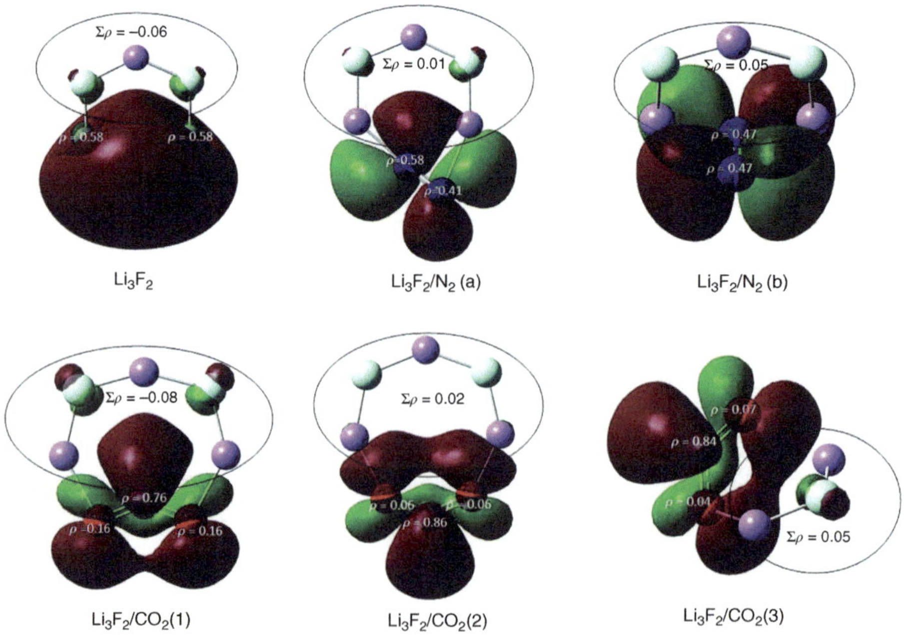

Figure 12.6 The SOMOs for Li_3F_2; Li_3F_2/CO_2 (**1**), (**2**), and (**3**); and Li_3F_2/N_2 (**a**) and (**b**). ρ is the spin density that adds up to 1. Violet = Li; electric blue = F; red = O; gray = C; blue = N. *Source:* Park and Meloni [16]. © Royal Society of Chemistry.

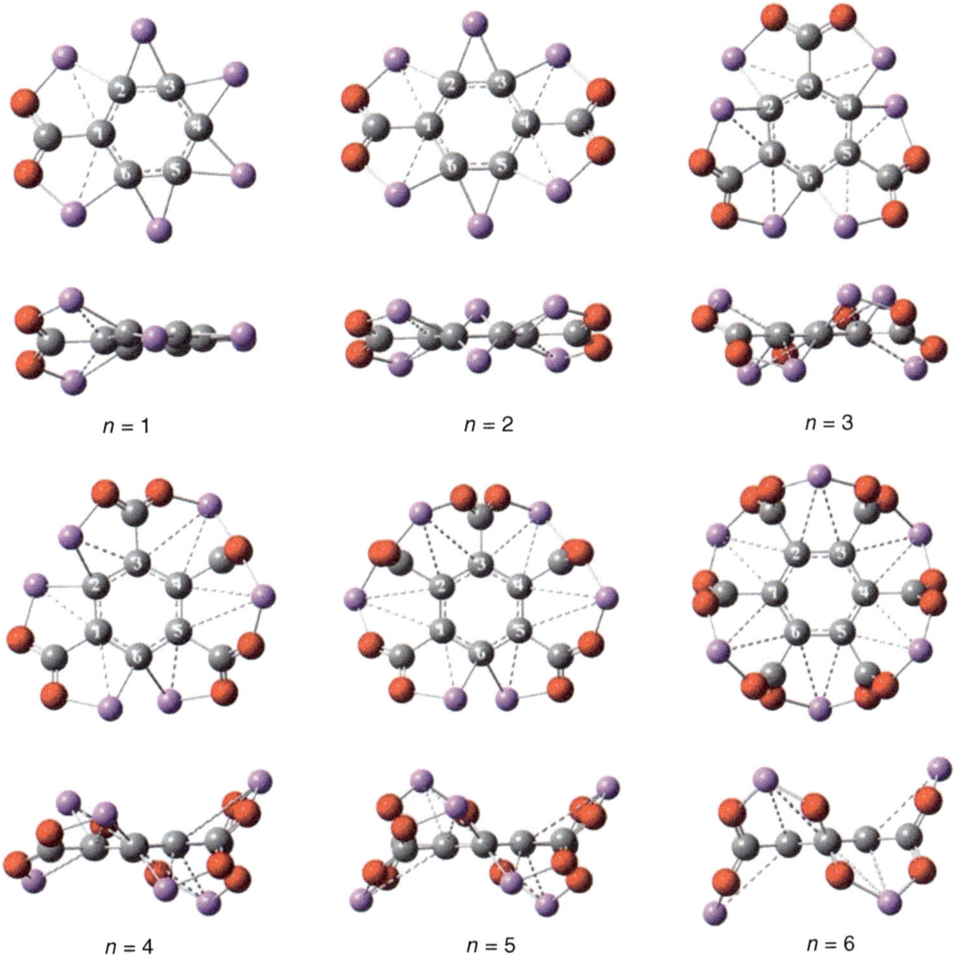

Figure 12.7 Equilibrium structure of $C_6Li_6-nCO_2$ complexes for n = 1–6 obtained at ωB97xD/6-311+G(d) level. Front as well as side views are also displayed. *Source:* Srivastava [18]. © John Wiley & Sons.

hydrogen storage [19]. However, Moreno et al. [20] described the lowest-energy structure as three C_2^{2-} fragments strongly aggregated through lithium bridges. It was noticed that the IP of C_6Li_6 is lower than that of the Li atom [21], and polylithium organic compounds have already been synthesized [22], including C_6Li_6 [23].

The interactions lead to stable $C_6Li_6-nCO_2$ complexes (n = 1–6) in which the structure of CO_2 moieties is bent considerably (122–125°) due to electron transfer from C_6Li_6, whose planarity is distorted only slightly (≤7°). The result suggests that the CO_2 molecules can successfully be activated and captured by C_6Li_6. It has also been noticed that the bond length of CO_2 in $C_6Li_6-nCO_2$ complexes increases monotonically, whereas adsorption energy decreases, ranging between 3.18 and 2.79 eV/CO_2 as the n increases. C_6Li_6 is not only capable of reduction but also effective in the sequential adsorption of six CO_2 molecules.

12.2.2 Supported or Embedded Superalkalis for CO_2 Capture

For practical applications, it is desirable to search for materials where the superalkalis can be supported or embedded without changing their intrinsic properties. Ramondo et al. [24] performed a theoretical study on Li_xF_y superalkalis supported on pyridinic graphene using the CBS-QB3 method. Both Li_3F_2 and Li_5F_4 on graphene can still reduce CO_2 forming cyclic structures as found for the freestanding species (Figure 12.8). Reduction of CO_2 may occur for all the structures with lithium terminal atoms. There is a competition between graphene and CO_2 toward the superalkali; Li_3F_2 shows a higher affinity to graphene than Li_5F_4 but less affinity to CO_2. The study shows that pyridinic sites on graphene surfaces can significantly adsorb hyperlithiated species without changing their reducing ability toward CO_2.

Fang et al. [25] used Li_3O superalkali as the building block to form one-dimensional electride encapsulated in a (4,8) boron-nitride nanotube (BNNT) for activating CO_2. Note that BNNT is hydrophobic. Thus, H_2O molecules, which are often present during the gas conversion process, would not adversely affect the reaction. They found that the ultralow work function of the hierarchical electride would make it easy to transfer electrons to CO_2 molecule for activation by bending its linear ground state structure, as is the case with CO_2^-. The C–O–C angle of CO_2^- is 137° and the C–O bond is stretched to 1.24 Å, much longer than 1.15 Å in its neutral ground state. As shown in Figure 12.9a, the unit cell of Li_3O@BNNT can readily activate a CO_2 molecule to a bent configuration, with an O–C–O angle of 129° and stretched C–O bond of 1.28 Å. About 1.3 electrons (0.32 per Li_3O on average) are transferred from Li_3O@BNNT to CO_2. The BE between Li_3O@BNNT and the CO_2 molecule is 1.1 eV. All these results suggest that the electride can efficiently capture and activate CO_2. On the other hand, the empty BNNT cannot adsorb and activate CO_2.

To understand the activation process, the (partial) electronic DOS of Li_3O@BNNT·CO_2 was calculated as plotted in Figure 12.9b, showing interactions of CO_2 with Li_3O@BNNT.

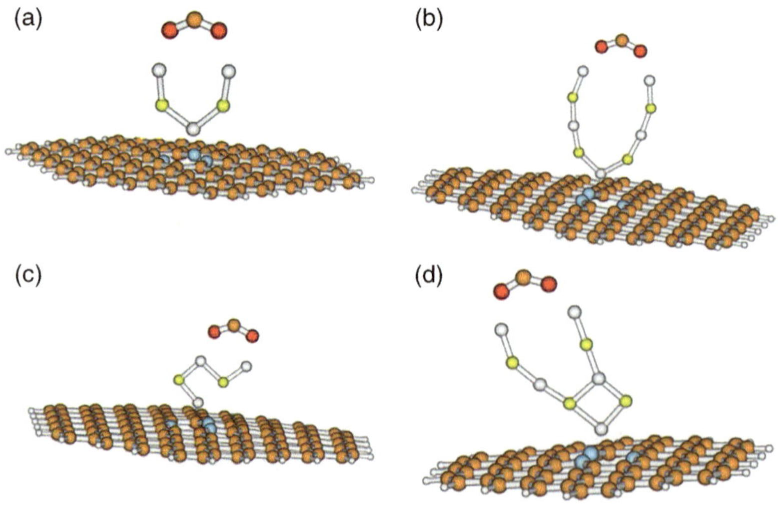

Figure 12.8 Optimized structures of graphene ... Li_3F_2 ... CO_2 (a, c) and graphene ... Li_5F_4 ... CO_2 complexes (b, d). *Source:* Ramondo et al. [24]. © John Wiley & Sons.

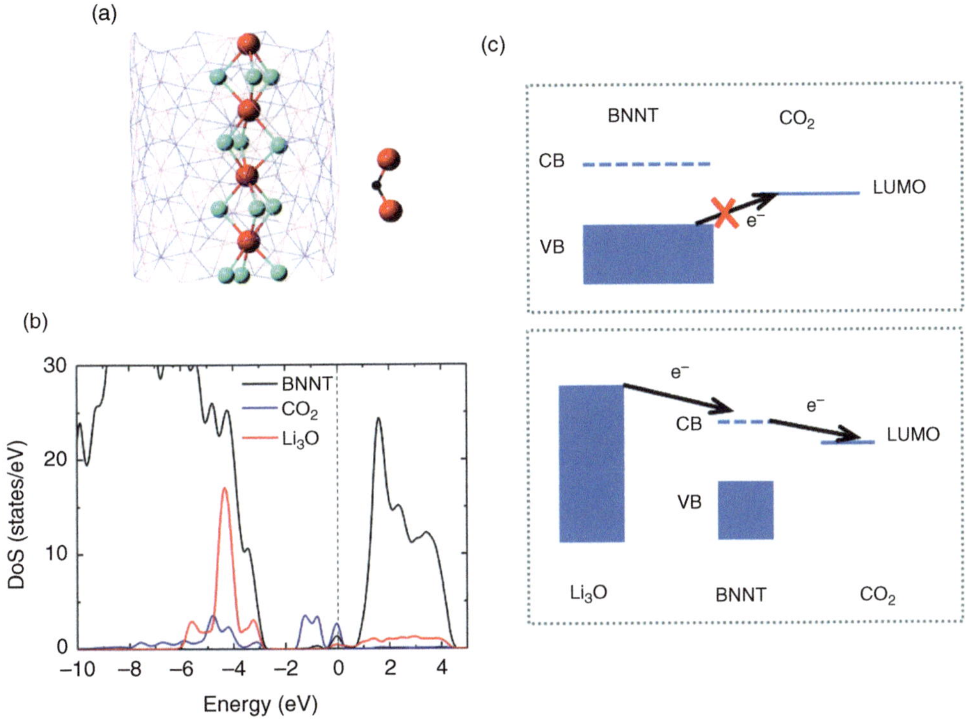

Figure 12.9 (a) Optimized structure of Li_3O@BNNT with an adsorbed CO_2 molecule. Oxygen is in red, lithium in cyan, and carbon in black. (b) Calculated (partial) electronic density of states (DOS), showing the band alignment between Li_3O, BNNT, and CO_2. (c) Schematics of a band model to explain the electron transfer from Li_3O to CO_2 molecule. The upper panel corresponds to case without Li_3O, while the lower panel refers to the case after Li_3O insertion. *Source:* Fang et al. [25]. © Royal Society of Chemistry.

According to the Bader charge analysis, Li_3O loses 2.71 electrons in total, 1.44 of which is taken by BNNT and the rest 1.27 is taken by the CO_2 molecule. Based on these results, a band model can be used to explain the process. Without the inserted Li_3O, as shown in the upper panel of Figure 12.9c, the LUMO of CO_2 molecule is above the valence band (VB) of BNNT, preventing electron transfer from the tube to the molecule. After the insertion of Li_3O, the Fermi level of the metallic chain is above the conduction band (CB) of BNNT and the tube will be readily n-doped. Now, with electrons in the CB above the LUMO of CO_2, electrons can be easily transferred to the molecule as shown in Figure 12.9c. Since the Li_3O chain at the core of the hierarchical structure is metallic, it is possible to replenish the transferred electrons by keeping the chain in contact with an electron reservoir.

12.3 Al-based Clusters for CO_2 Capture

Among aluminum clusters, Al_{13}^- is a well-known magic cluster. According to the jellium model, the 39 valence electrons in a neutral Al_{13} cluster make it one electron deficient than needed for the 40-electron shell closure [26]. It was found that Al_{13} cluster mimics halogen behavior when it chemically reacts with an I_2 molecule [27]. In addition, it was proposed that replacing the central Al atom in an Al_{13} icosahedron can enhance the cluster stability as well as tune their electronic properties. For example, doping P or N atom with five valence electrons results in a superalkali atom containing 41 valence electrons [28].

Sengupta et al. [29] have studied the CO_2 absorption by transition metal atoms from Sc to Zn doped in aluminum clusters (Al_5 and Al_7) and found that Sc- and Ti-doped Al clusters are good absorbers. The maximum and minimum values of heat of adsorption were found for Al_7Sc cluster (−48.4 kcal/mol at M06-2X level of theory) and the Al_7Zn cluster (−10.9 kcal/mol in M06-2X), respectively. The resemblance with the bulk is also noticed in the optimized structures of $Al_nM{-}CO_2$ complexes as presented in Figure 12.10. In all cases, the CO_2 molecule is observed to be deviating from linearity along with a substantial increment of both the C=O bonds. These observations signify the existence of charge transfer from metal centers to the CO_2 moiety. Also, as aluminum is more electropositive than transition metals, the increment of both the C=O bonds is found to be asymmetric. The oxygen atom attached to the aluminum side shows longer C=O bond distance (1.3 Å) than the other C=O bond (1.2 Å), which is bonded with the doped transition metal.

Kagdada et al. [30] reported DFT-based calculations of the structural and electronic properties of transition metal (Tm) Zr- and Hf-doped small aluminum clusters containing 1–7 aluminum atoms. They found that these clusters can adsorb CO_2 and that there are mainly two different configurations of bonding between CO_2 and T_mAl_n clusters [29]. In one configuration, one O atom in CO_2 binds with the transition metal in the cluster, and the other oxygen atom binds with the aluminum atom (see Figure 12.11). It is interesting to note that the exothermicity for $HfAl_6$ is even higher than that of Sc- and Ti-doped aluminum clusters [29], and $HfAl_6$ cluster with the low spin state (singlet) has a larger adsorption energy for CO_2 as compared with other spin states.

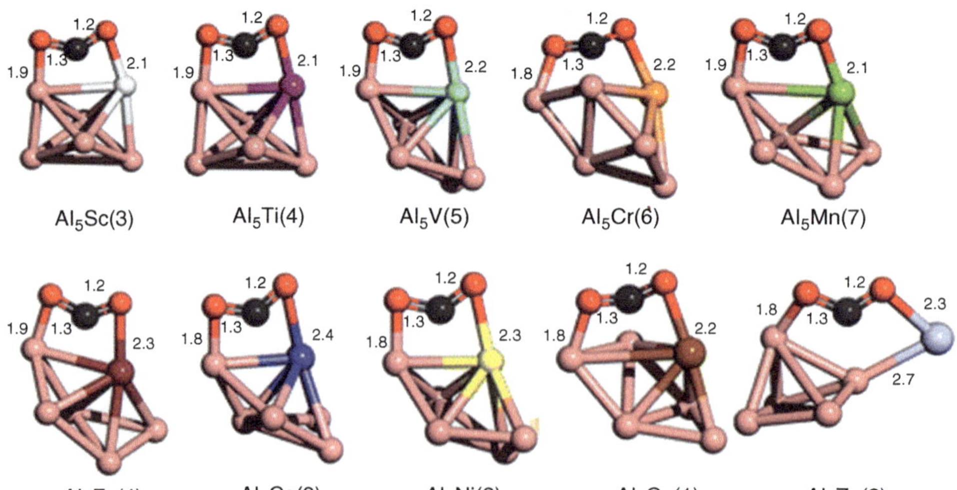

Figure 12.10 Structures of optimized ground state $Al_5M \ldots CO_2$ complexes with spin multiplicity included within the parentheses. *Source:* Sengupta et al. [29]. © American Chemical Society.

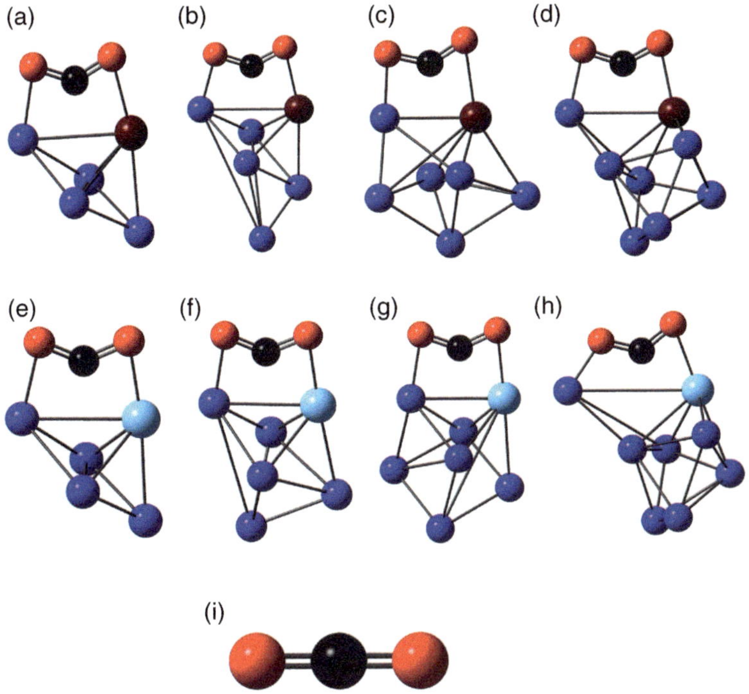

Figure 12.11 Configurations of CO_2 adsorptions on $ZrAl_n$ (a–d), $HfAl_n$ (e–h), n = 4–7. *Source:* Kagdada et al. [30]. © Elsevier.

The BE of the clusters considered increases with an increase in cluster size. The calculations show that the $ZrAl_4$ and $HfAl_4$ are relatively stable structures. It was found that the ZrAl is kinetically more stable with a HOMO–LUMO gap of 2.458 eV, while $HfAl_6$ cluster with the low spin state (singlet) has higher enthalpy difference and binds more strongly to CO_2.

12.4 Ligand-protected Au_{25} Clusters for CO_2 Conversion

Different from the bulk metal surfaces and larger nanoparticles, ligand-protected atomically precise metal clusters exhibit not only well-defined atomic structures but also size-dependent electronic, chemical, and catalytic properties. Recently, synthetic advances have identified a new class of ligand-protected Au clusters with unique chemical and electronic properties [31]. Crystallographic efforts have confirmed that such Au clusters form into atomically precise structures with inherent anionic (negative) charge [32]. Ligand-protected Au_{25} clusters are such examples showing appealing catalytic ability for CO_2 reduction.

Kauffman et al. [33] reported an experimental and computational investigation of Au_{25} clusters for CO_2 adsorption. Figure 12.12a and b present the optical spectra of an initially N_2-purged Au_{25} solution that was subsequently saturated with CO_2. The introduction of CO_2 bleached the a', a, and b absorbance features and increased the absorbance at 2.04 eV. The apparent interaction between Au_{25} and CO_2 was quite different from the behavior of

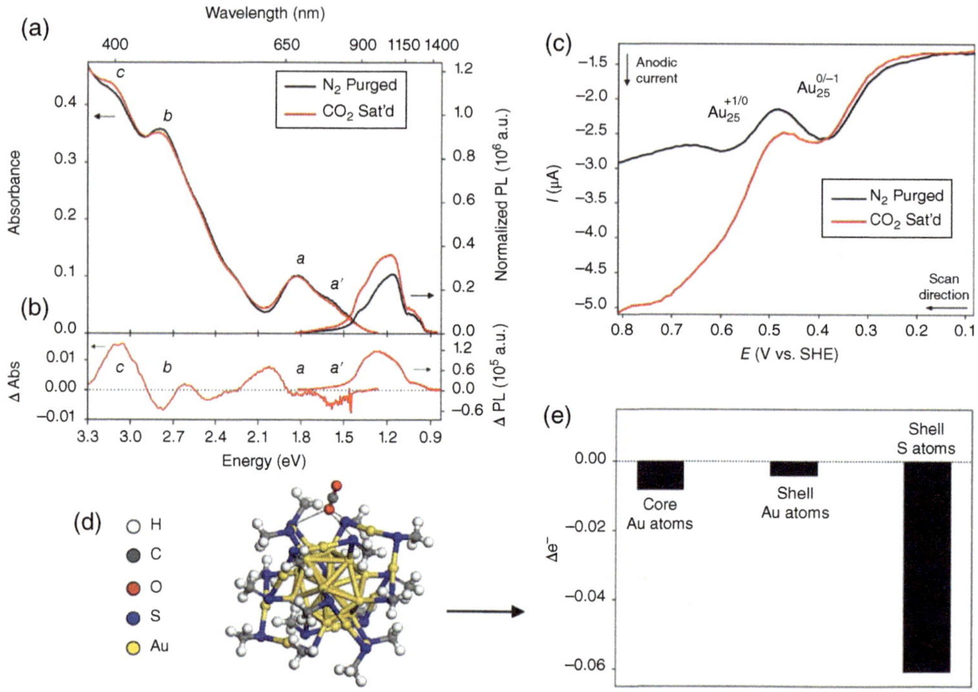

Figure 12.12 Experimental results and DFT modeling of the Au_{25}–CO_2 couple. (a) Optical absorbance and photoluminescence (PL) spectra, and (b) difference spectra (Δabs and ΔPL) of Au_{25} in N_2-purged and CO_2-saturated dimethylformamide (DMF). (c) Square wave voltammetry of Au_{25} in N_2-purged and CO_2-saturated DMF + 0.1 M tetrabutylammonium perchlorate (TBAP). (d) DFT model of stable CO_2 adsorption. (e) Bader charge analysis. *Source:* Kauffman [33]. © American Chemical Society.

CO_2 on traditional gold surfaces [34]. Experiments showed that the CO_2-induced optical changes are reversible by simply purging the solution with N_2, suggesting a comparatively weak interaction between the Au_{25} cluster and CO_2.

The unique interaction between Au_{25} and CO_2 provided an exciting opportunity to test CO_2 reduction with a negatively charged nanocatalyst. Comparing the electrocatalytic activity of Au_{25} (-1 nm) and larger Au catalysts for the electrochemical reduction of CO_2 in aqueous 0.1 M $KHCO_3$, potential-dependent product analysis identified significant and reproducible CO formation at an onset potential of −0.193 V versus reversible hydrogen electrode (RHE) (>95% confidence level [CL], $n = 3$). Electrolysis in N_2-purged $KHCO_3$ ruled out spurious CO evolution from Au_{25}'s organic ligands or the carbon black support. Remarkably, the onset of CO formation was within 90 mV of the $CO_2 \rightarrow CO$ formal potential (−0.103 V versus RHE) [35]. This low overpotential constitutes an approximate 200–300 mV reduction compared to the case of the larger Au catalysts.

Following this study, Zhao et al. [36] reported two types of atomically precise Au_{25} nanoclusters (1 nm) with distinct morphology (i.e. nanosphere and nanorod) for CO_2 reduction catalysis. The Au_{25} nanosphere exhibits higher Faradaic efficiency for CO with higher formation rates as compared with the Au_{25} nanorod. First-principles calculations reveal that the negative charge and the active site generated by removing one ligand from the nanosphere-shaped Au_{25} can better stabilize the important *COOH intermediate in CO_2 electroreduction. To verify the role of shape on CO_2 reduction, Bader charge analysis calculations were performed. In the presence of counterions, the nanosphere-shaped Au_{25} remains negatively charged ($q = -0.77$), whereas the nanorod-shaped Au_{25} is positively charged ($q = +1.94$). This is the reason why the nanosphere-shaped Au_{25} is catalytically more reactive in CO_2 reduction as observed in experiment.

Alfonso et al. [37] performed calculations using DFT and continuum solvation models and found that the fully ligand protected cluster is not active for CO_2 reduction because the formation of the crucial carboxyl intermediate requires very high electrochemical potentials. Instead, the reduction process can feasibly occur on a dethiolated gold site, on which the adsorbed carboxyl intermediate formation was significantly stabilized.

Free energy diagrams were constructed for CO_2 reduction at the singly dethiolated $Au_{25}(SCH_3)_{17}^-$ cluster and the results at 0 V versus RHE are depicted in Figure 12.13. Free energy calculations for the elementary steps I–III predicted that *COOH formation was also the most endoergic step on singly dethiolated $Au_{25}(CH_3)_{17}^-$ nanocluster. The key difference was that *COOH was dramatically stabilized on the singly dethiolated cluster, and *COOH formation free energy was reduced from $\Delta G = 2.04$ eV to $\Delta G = 0.34$ eV. And C-Au bond length also decreases from 2.21 to 2.08 Å, which indicates a stronger

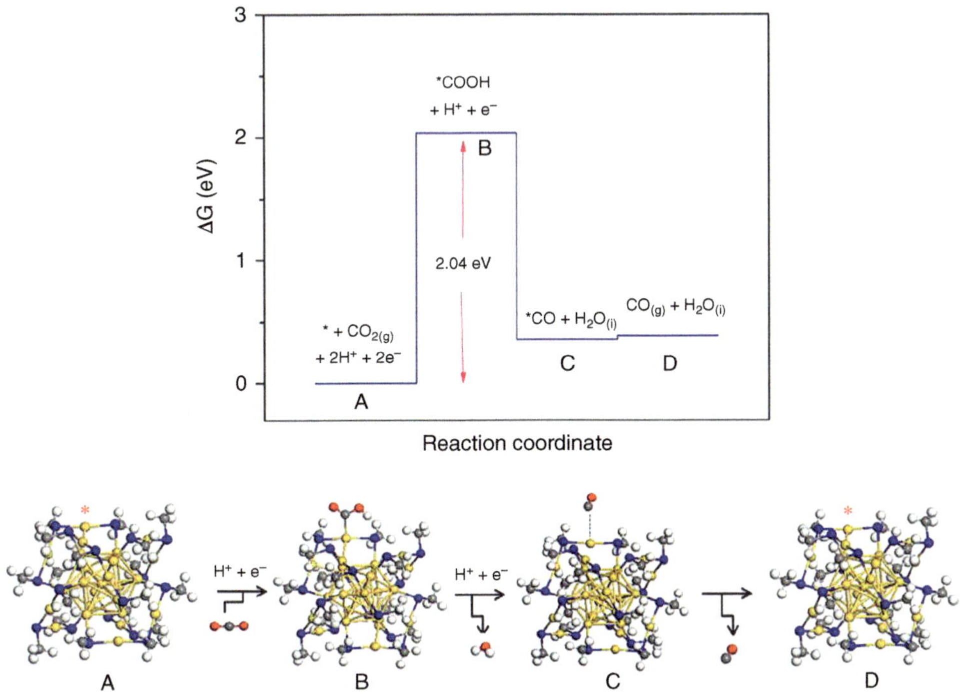

Figure 12.13 Free energy diagram for electrochemical reduction of CO_2 to CO on the singly dethiolated $Au_{25}(SCH_3)_{17}^-$ cluster at $T = 298$ K, pH = 0, and zero applied potential. White, gray, blue, gold, and red balls represent H, C, S, Au atom, and O atoms, respectively. *Source:* Alfonso et al. [37]. © AIP Publishing.

interaction at ligand-free Au sites. Bader charge analysis shows that *COOH behaves as an electronegative adsorbate with gains of 0.33 |e| when adsorbed to singly dethiolated $Au_{25}(SCH_3)_{18}^-$.

Recently, Au_{25} clusters have also been applied in solar energy conversion to produce CO from CO_2. Kim et al. [38] reported that Au_{25} clusters with Au-thiolate staple motifs can initiate electrocatalytic reduction of CO_2 to CO with nearly zero energy loss and achieve a high CO_2RR current density of 540 mA/cm^2 in a gas-phase reactor. Electrochemical kinetic investigations revealed that the high CO_2RR activity of the Au_{25} originates from the strong CO_2 binding affinity, leading to high CO_2 electrolysis performance with 18.0% Solar-to-CO (STC) efficiency over 12 hours in 100% CO_2 and a 15.9% STC efficiency in 10% CO_2.

12.5 M@Ag_{24} Clusters for CO_2 Conversion

Similar to Au, Ag_{25} cluster has a core-shell structure (Ag@Ag_{24}), where an Ag_{13} icosahedral core is encapsulated by an incomplete shell consisting of 12 Ag atoms in 6 extended staples (–S–Ag–S–Ag–S–Ag–) [39]. The Ag atom located at the icosahedral core of the Ag@Ag_{24} cluster can be replaced by a foreign atom (Au, Pd, and Pt) to form Au@Ag_{24}, Pd@Ag_{24}, and Pt@Ag_{24} clusters. It can be anticipated that such single-atom doping into the core of Ag@Ag_{24} clusters could modify the catalytic nature and lead to completely different catalytic behaviors.

Liu et al. [40] reported the catalytic conversion of CO_2 toward carbon–carbon bond formation using the centrally single-atom doped clusters as catalysts and found a unique anion–anion mechanism that promotes the carboxylation reaction. Intriguingly, the atomic charge on the central atom governs the catalytic properties of the doped clusters. The centrally doped clusters exhibit a synergistic enhancement in the catalytic efficiency, as compared to the original Ag@Ag_{24}.

As shown in Figure 12.14, the propiolic acid derivatives were synthesized from CO_2 and terminal alkyne using Au@Ag_{24}, Pd@Ag_{24}, Pt@Ag_{24}, and Ag@Ag_{24} clusters, respectively. Among the four catalysts, Au@Ag_{24} catalyst gave rise to the highest product yield (>92%), while Ag@Ag_{24} exhibited the lowest activity. It is noted that Pd@Ag_{24} and Pt@Ag_{24} have no notable difference in their catalytic performance of the carboxylation of CO_2 with terminal alkyne. The catalytic activity can be presented in a descending sequence of Au@Ag_{24} > Pd@Ag_{24} ≈ Pt@Ag_{24} > Ag@Ag_{24}.

To obtain molecular-level understanding into the enhancement of carboxylation of CO_2 with terminal alkyne by the core-doped clusters, DFT calculations were performed. The results show that despite M@Ag_{24} being anion clusters, the adsorption of $[CH_3C{\equiv}C]^-$ can occur at the outer Ag (Figure 12.15a). At 0 K, the adsorption of $[CH_3C{\equiv}C]^-$ is thermally neutral on Au@Ag_{24}, endothermic by about 35 kcal/mol on Pd@Ag_{24} and Pt@Ag_{24}, and exothermic by 11 kcal mol^{-1} on Ag@Ag_{24}. The weak adsorption of $[CH_3C{\equiv}C]^-$ on the M@Ag_{24} is due to the radial charge distribution of M@Ag_{24} where the core metal is negatively charged, the inner Ag_{12} shell is neutral, and the outer Ag_{12} shell is positive (Figure 12.15b). Therefore, the adsorption can be described by a simple core M-dipole model, where the dipole is formed by the negative α-C of $[CH_3C{\equiv}C]^-$ and the positive outer Ag.

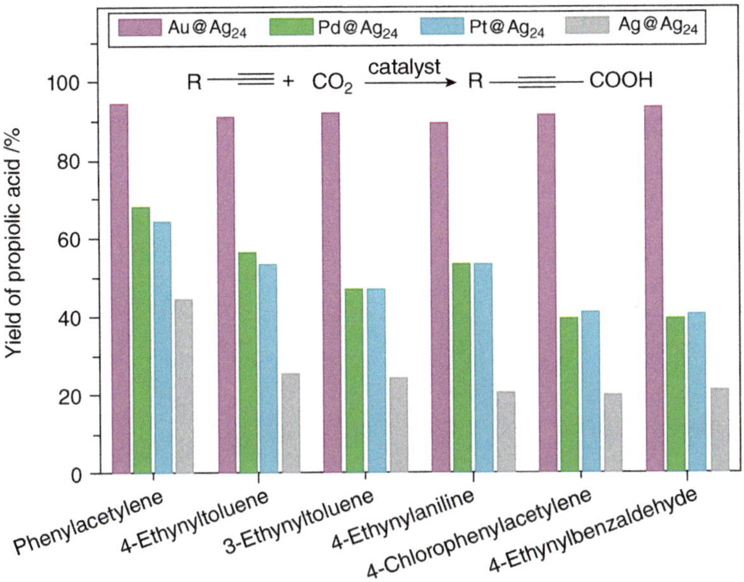

Figure 12.14 The catalytic carboxylation of terminal alkyne with CO_2 over Au@Ag$_{24}$, Pd@Ag$_{24}$, Pt@Ag$_{24}$, and Ag@Ag$_{24}$ clusters. Reaction conditions: catalyst (10 mg), alkyne (1.0 mmol), Cs_2CO_3 (1.2 mmol), CO_2 (1 MPa), 50 °C, 12 hours. *Source:* Liu et al. [40]. © John Wiley & Sons.

The potential energy surface is shown in Figure 12.15c for the reaction of $[CH_3C{\equiv}C]^-$ and CO_2 to form $[CH_3C{\equiv}C\text{-}COO]^-$ (propiolate) in the solutions of the M@Ag$_{24}$ (C brown, H white, O red, S yellow, P black, Cs purple, core Ag dark blue, core Au orange, core Pd dark green, core Pt light green, inner Ag cyan, outer Ag pink). In contrast to the weak anion–anion adsorption of $[CH_3C{\equiv}C]^-$ on M@Ag$_{24}$, the recombination of $[CH_3C{\equiv}C]^-$ and the Cs^+ and $[PPh_4]^+$ cations in the solutions are extremely exothermic. The cations in the solutions serve as $[CH_3C{\equiv}C]^-$ traps and are detrimental to the CO_2 carboxylation reaction. In contrast, the M@Ag$_{24}$ catalysts serve as the temporary hosts for the active $[CH_3C{\equiv}C]^-$ species and facilitate the formation of $[CH_3C{\equiv}C{-}COO]^-$. Furthermore, owing to the excessively strong bonding of $[CH_3C{\equiv}C]^-$ and $[CH_3C{\equiv}C{-}COO]^-$ to the cations, the dissociations of the Cs and PPh_4 propiolates are highly endothermic, which inhibits the formation of the propiolic acid. Overall, the M@Ag$_{24}$ catalysts show unique mechanism in catalyzing the coupling reactions in the presence of bases in the solution.

To understand the influence of the doped atoms, the natural atomic charge analysis was performed. In this reaction, the adsorption energy is contributed not only by the outer Ag attracting α-C but also by the electrostatic interactions between the core M and the dipole. The latter differs for the four clusters, as the negative charges of the core M are Pd/Pt (−2.5) > Au (−1.9) > Ag (−1.7), according to the natural atomic charge analysis (Figure 12.15b, c = core metal M, i = inner Ag$_{12}$ shell, o = outer Ag$_{12}$ shell, m = methyl C, s = S, h = H, α, β, γ = α-, β-, γ-C of $[CH_3C{\equiv}C]^-$). For Ag@Ag$_{24}$, the core Ag is a relatively weak anion, so the attraction between core Ag and outer Ag and the repulsion between the core Ag and the α-C are also weak. Therefore, the attraction between the outer Ag and the

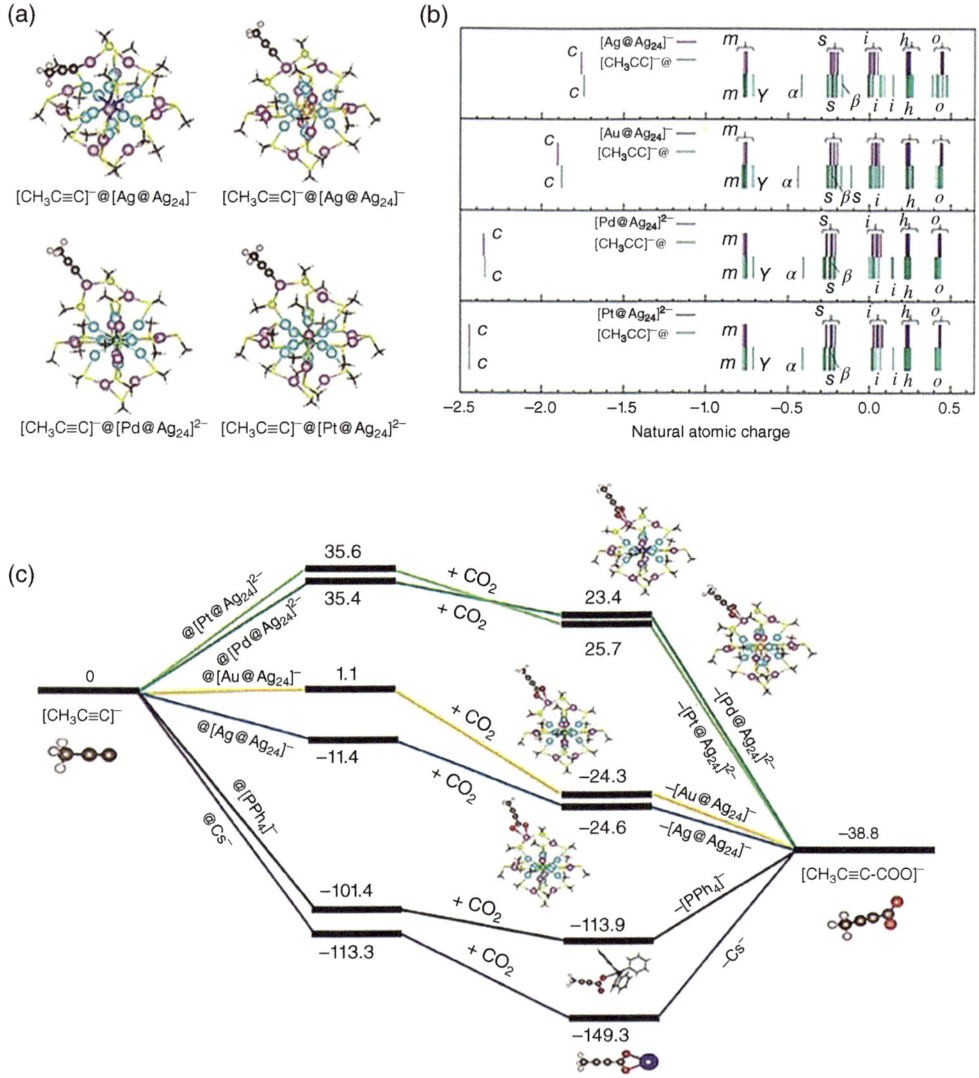

Figure 12.15 (a) DFT-optimized geometries for the adsorption products of $[CH_3C{\equiv}C]^-$ on M@Ag_{24} (M=Au, Pd, Pt, Ag). (b) Natural atomic charge spectra for the adsorption of $[CH_3C{\equiv}C]^-$ on M@Ag_{24}. Purple lines indicate the natural atomic charges for M@Ag_{24}. (c) Potential energy surface (at DFT level in kcal/mol) for $[CH_3C{\equiv}C]^- + CO_2 \rightarrow [CH_3C{\equiv}C\text{-}COO]^-$ reactions in the solutions of the $[Ag_{24}M(SR)_{18}]x^-$ catalysts. *Source:* Liu et al. [40]. © John Wiley & Sons.

α-C can tear the outer Ag-S bond apart. When the core M is heavily negatively charged as found in Pd@Ag_{24} and Pt@Ag_{24}, it can repel the α-C and cause the geometry to deform. When the negative charge on core M is median as found in Au@Ag_{24}, a balance of the electrostatic interactions is reached so that the cluster exhibits high stability and the $[CH_3C{\equiv}C]^-$ has high activity (owing to the weaker outer Ag–α-C interaction). The superior stability of Au@Ag_{24} among the four catalysts under the influence of strong base is also indicated by the natural atomic charge spectra.

12.6 Cu-based Clusters for CO_2 Conversion

Among the experimentally synthesized and structurally precise Cu-based clusters, the neutral $Cu_{32}H_{20}L_{12}$ cluster (L = $S_2P(O^iPr)_2$) [41] is of special interest in CO_2 reduction due to the diverse hydrides with different coordination environments. Tang et al. [42] used the $Cu_{32}H_{20}L_{12}$ cluster as a prototypical Cu–H system to investigate its activity and selectivity for electrochemical reduction of CO_2 using both experimental and theoretical methods. Based on the computed free energy change, the lattice hydride channel (Figure 12.16 right) was found to be more facile, whereby the μ_3-H1 hydride in the $Cu_{32}H_{20}L_{12}$ cluster is moved to the C atom to form adsorbed HCOO* ($Cu_{32}H_{19}L_{12}$-HCOO) with a free energy change of 0.32 eV. In comparison, the proton-reduction mechanism leading to the formation of $Cu_{32}H_{20}L_{12}$-HCOO is highly unfavorable with a high overpotential of about 1.08 eV (Figure 12.16 left). Analysis of the electronic structure suggests that the 2*p*-orbitals of the O atom are highly hybridized with the 3*d*-orbitals of the bonded Cu atoms.

After forming the $Cu_{32}H_{19}L_{12}$-HCOO species, the second step is the addition of another hydride (μ_4-H1) to HCOO*, yielding the final product, HCOOH ($\Delta G = 0.07$ eV; Figure 12.16 right). The loss of two lattice hydrides leads to two hydrogen vacancies in the resulting $Cu_{32}H_{18}L_{12}$ cluster, which could readily proceed via two sequential proton-reduction steps to regenerate the two reacted lattice hydrides. It was found that the first hydrogen goes to the μ_3-H1 vacancy with ΔG = 0.05 eV and the second goes to the μ_4-H1 vacancy with $\Delta G = -0.06$ eV (Figure 12.16 right). In the more favorable lattice-hydride channel, the potential-limiting step for the HCOOH formation is the first hydride transfer for forming the HCOO* intermediate. HCOOH formation and the subsequent hydride regeneration on the $Cu_{32}H_{20}L_{12}$ cluster are summarized in Figure 12.17. One can clearly see that, starting from $Cu_{32}H_{20}L_{12}$ and CO_2, the μ_3-H1 hydride will first add to C to yield $Cu_{32}H_{19}L_{12}$-HCOO

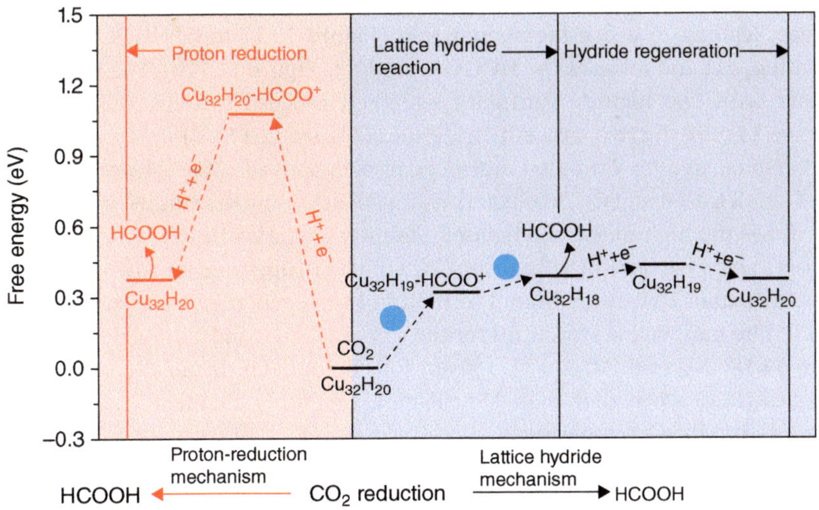

Figure 12.16 Reaction scheme for CO_2 electroreduction on $Cu_{32}H_{20}L_{12}$ to form HCOOH via the proton-reduction channel (to the left) and the lattice-hydride channel (to the right). L_{12} is omitted when labeling the intermediates. *Source:* Tang et al. [42]. © American Chemical Society.

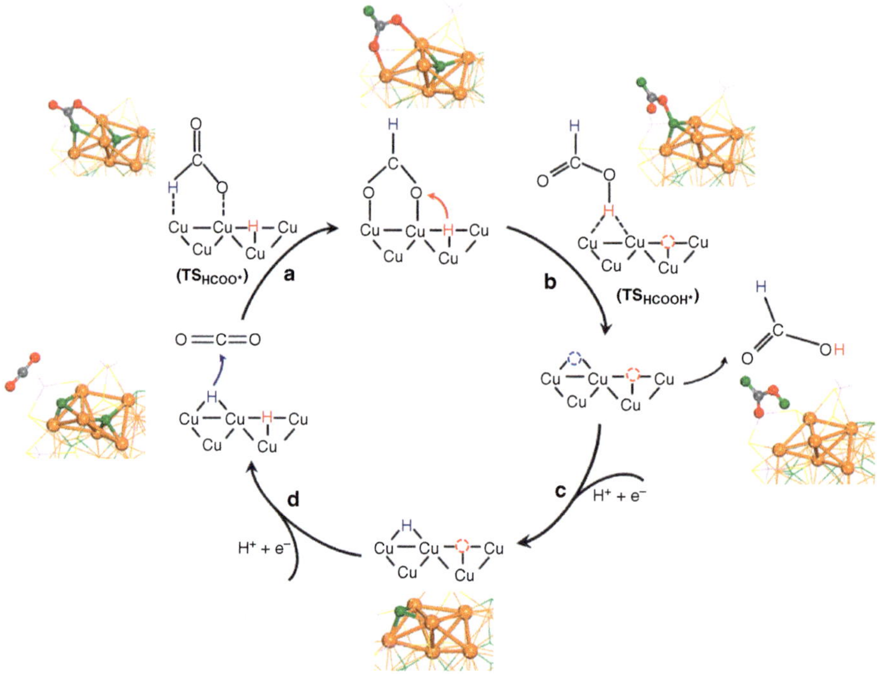

Figure 12.17 Overall mechanism of HCOOH formation from CO_2 reduction on $Cu_{32}H_{20}L_{12}$ via the lattice-hydride channel; the structures of the key intermediates and transition states are shown. Color code: orange, Cu; green, hydride; red, oxygen; gray, carbon; L = S_2PH_2, not shown. *Source:* Tang et al. [42]. © American Chemical Society.

(Figure 12.17a). Notably, the two O atoms in HCOO* intermediate are strongly bonded to Cu on the surface, leading to a five-membered ring (Figure 12.17 top). HCOO* subsequently reacts with μ_4-H1 and releases the HCOOH product (Figure 12.17b). The resulting $Cu_{32}H_{18}L_{12}$ cluster with two hydride vacancies will then proceed through the proton-reduction processes to refill the two lattice-hydride sites (Figure 12.17c, d).

Besides Cu_{32} clusters, smaller Cu-based clusters can also activate CO_2. Alvarez-Garcia et al. [43] studied small Cu-Pd and Cu-Ni clusters using DFT. Among the studied bimetallic systems, Cu_3Pd shows the highest thermodynamic stability (28.82 kcal/mol) for CO_2 binding with the lowest energy barrier (40.91 kcal/mol). Cu_4 cluster supported on Al_2O_3 exhibits good catalytic performance toward methanol formation [44], even at low pressures (near atmospheric) [45]. The most stable structures for the CO_2 on Cu-Ni and Cu-Pd clusters are shown in Figure 12.18. Ni_4 and Pd_4 are also given for comparison. In all the cases, the adsorption energies (E_{ads}) are higher than 15 kcal/mol, signifying chemical interactions between the clusters and the CO_2 molecule.

To model CO_2 dissociation on clusters, the following reaction was considered: $CO_2^* \rightarrow CO^* + O^*$. Figure 12.19 shows the activation energy of transition states (TS) when CO_2^* (R) is dissociated into $CO^* + O^*$ (P) for Cu-Pd alloyed clusters. Results on Cu_4 and Pd_4 are also included for comparison. One can see that the activation energy is between 25.38 and 43.55 kcal/mol. The higher barrier in Pd_4 relative to Cu_3Pd indicates that C–O

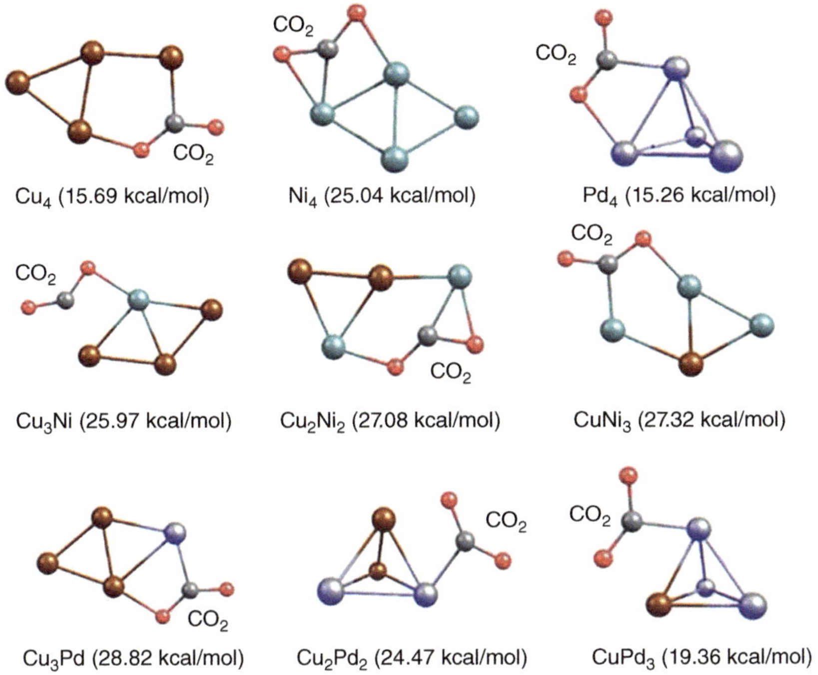

Figure 12.18 Most stable structures and adsorption energies for CO_2 on clusters. *Source:* Alvarez-Garcia et al. [43]. © Elsevier.

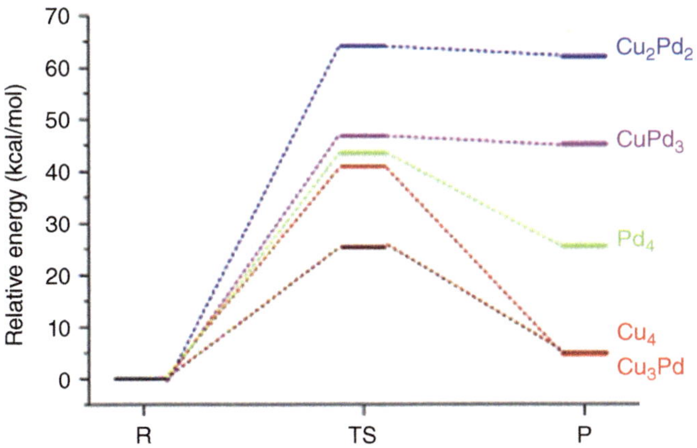

Figure 12.19 Activation energy for the CO_2^* dissociation into $CO^* + O^*$. *Source:* Alvarez-Garcia et al. [43]. © Elsevier.

dissociation is kinetically limited in the palladium cluster, while the barrier on Cu_4 is found to be the lowest. The activation energy for the dissociation of CO_2 ($CO_2^* \rightarrow CO^* + O^*$) on clusters changes in the order $Cu_4 < Cu_3Pd < Pd_4 < CuPd_3 < Cu_2Pd_2$. Therefore, the ideal atomic composition ratio of Cu to Pd in terms of adsorption energy and activation barrier is 3 : 1, which is within the experimentally reported value.

12.7 Metal Encapsulated Silicon Nanocages for CO_2 Conversion

Besides metal clusters, silicon-based clusters are also interesting in catalysis because they are nontoxic, inexpensive, and earth abundant. However, the geometry and properties of pure silicon clusters are hard to control during synthesis [46]. Kumar et al. [47] predicted that silicon cage clusters can be stabilized by encapsulating a transition metal atom in the cage, which was confirmed by experiment for selected cluster size n up to 20 and dopant elements covering most of the 3d, 4d, and 5d transition metal atoms [48]. Latter on, Zhou et al. [49] studied CO_2 hydrogenation on vanadium-doped Si clusters (VSi_n, n = 12–16) using first-principles calculations. They found that VSi_n clusters with n =12–15 are effective catalysts for CO_2 hydrogenation with distinct product selectivity and can utilize the solar energy from the visible to ultraviolet region. The chemisorbed CO_2 molecule is bent with an O–C–O angle of 122–125° since the C atom and one O atom form covalent bonds with Si atoms. The CO_2 molecule gains about 0.82–0.87 electrons from the VSi_n cluster. From physisorption to chemisorption, it has to overcome a moderate barrier of 0.53–0.79 eV. The most active sites for CO_2 adsorption are the Si atoms in the square (VSi_{15}, VSi_{16}), pentagon (VSi_{14}), or hexagon (VSi_{12}, VSi_{13}) faces of the clusters. On the contrary, the Si atoms on the vertex of triangular facets have higher coordination number and thus are inert for CO_2 adsorption.

Generally, the reaction proceeds through either the formate pathway (HCOO*) to form HCOOH or the carboxyl pathway (COOH*) to produce CO, the former path being lower in energy by 0.32–0.72 eV than the latter one. For VSi_{12} and VSi_{14}, the binding strengths of CO* and HCOOH* species are relatively strong, their further hydrogenation is possible to form HCO* species and then to CH_2O, CH_3OH, and CH_4 products as shown in Figure 12.20. On the other hand, VSi_{13} and VSi_{15} weakly absorb CO* and HCOOH* with adsorption energies of −0.31–0.09 eV; they are easy to desorb into gaseous molecules. Therefore, VSi_{13} and VSi_{15} may selectively catalyze CO_2 hydrogenation to form CO and HCOOH products, while other products like CH_2O, CH_3OH, and CH_4 on VSi_{12} or VSi_{14} are also possible.

Pei et al. [50] also found that $TiSi_n$ clusters are capable of photocatalyzing CO_2 hydrogenation to produce various C_1 products with activity and selectivity depending on the cluster size. Figure 12.21 shows the free energy diagram of the most efficient pathway of CO_2 hydrogenation on the $TiSi_{14}$ cluster as a representative, leading to the formation of various C_1 products, including carbon monoxide (CO), formic acid (HCOOH), formaldehyde (CH_2O), methanol (CH_3OH), and methane (CH_4). $TiSi_n$ clusters share similar activities as VSi_n clusters in CO_2 conversion. As the cluster size is reduced, the binding strength of CO_2 molecules on the clusters is enhanced due to the low coordination number and prominent surface charge densities on the silicon cage, which can be correlated to the bonding features of p-orbital of Si atoms.

12.8 Summary and Perspectives

In summary, compared to the conventional bulk materials, clusters show some intrinsic merits for CO_2 activation and conversion. For instance, transition metal Cu is identified as one of the best catalysts in electrocatalysis. However, even with the ability of directly

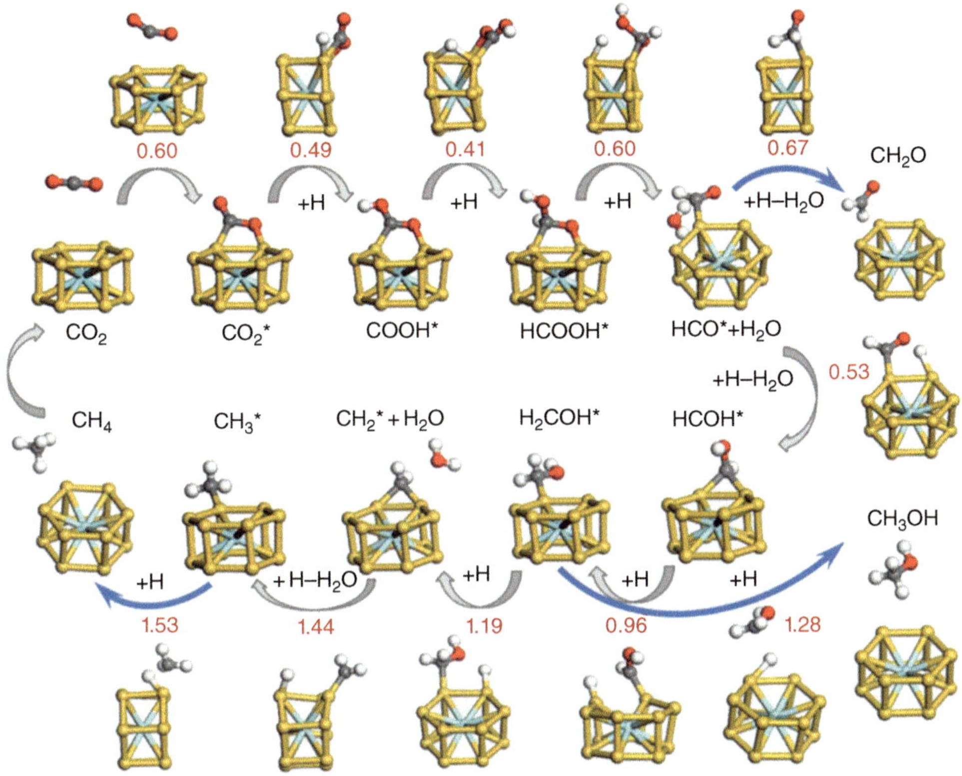

Figure 12.20 Atomic structures of reactants and products for the elemental steps of CO_2 hydrogenation to form CH_2O, CH_3OH, and CH_4 on the VSi_{12} cluster. The red numbers and the structure next to them show the kinetic barrier (in eV) and transition state for each step, respectively. The H, C, O, Si, and V atoms are shown in white, gray, red, yellow, and cyan colors, respectively. *Source:* Zhou et al. [49]. © American Chemical Society

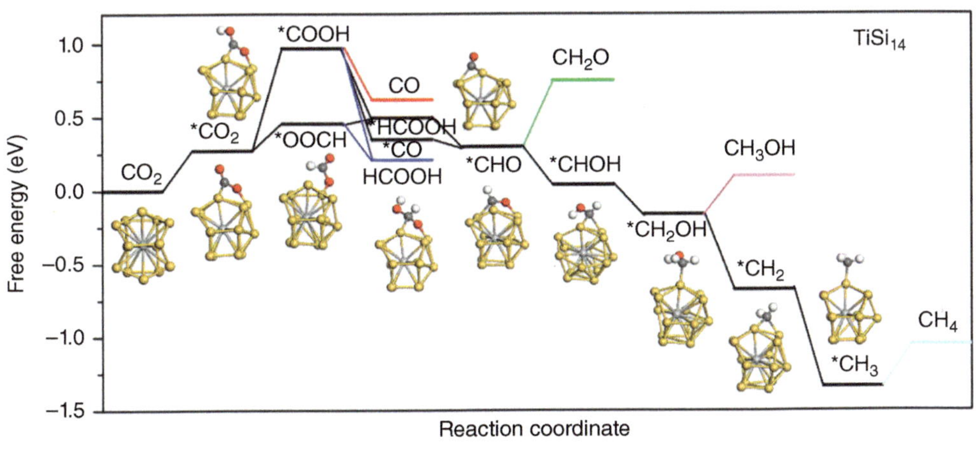

Figure 12.21 Reaction pathway and free energy diagram for CO_2 hydrogenation on the $TiSi_{14}$ cluster. The colored line indicates the formation of various products. The insets show the structures of corresponding reaction intermediates adsorbed on the cluster. The H, C, O, Si, and Ti atoms are shown in white, gray, red, yellow, and sliver colors, respectively. *Source:* Adapted from Pei et al. [50]. © John Wiley & Sons.

producing hydrocarbons and oxygenates at a high current efficiency, the catalytic performance of copper electrode is still impeded by high overpotential (~1 V) [51], although the overpotential can be effectively reduced in Cu clusters by tuning its size and shape. Since charge transfer is vital to CO_2 activation and conversion, clusters behaving as superalkalis have more freedom and flexibility because their reduced IP promotes electron donation and can be tailored by changing the composition, size, geometry, or ligands. However, currently this field remains in its preliminary stage. Future research directions include: (i) assembling clusters to form stable materials while keeping the intrinsic features of the cluster units. (ii) Designing clusters for multicarbon products. Currently, the products of most CO_2 reduction reaction (CO_2RR) are C_1 products like CO, CH_4, CH_3OH, CH_2O, and HCOOH. Two of the most common products are CO and CH_4, which have been widely used in power generation and chemical industry. However, C_1 products have lower volumetric energy densities and are not cost-effective in terms of energy consumption. Several C_2 products are found in CO_2 reduction like C_2H_4, C_2H_5OH, CH_3COOH, and CHOCHO. Compared to CO and CH_4, these C_2 products have higher volumetric energy densities and are more valuable in industrial application [52]. (iii) Designing clusters for high selectivity in CO_2 capture as well as in conversion: in capture, CO_2 selectivity over N_2 is an important issue as both molecules share similar behavior in physical adsorption. Therefore, strategy is needed to selectively capture CO, while in conversion many electrocatalysts are hampered by low Faradic efficiency because of the HER side reaction. A promising electrocatalyst should not only display a low overpotential for CO_2 electroreduction but also effectively suppress the side reactions. Systematic studies of these aspects would promote the development of CO_2 conversion technology.

References

1 Zhang, X., Lim, E., Kim, S.K., and Bowen, K.H. (2015). *J. Chem. Phys. 143*: 174305.

2 Weber, J.M. (2014). *Int. Rev. Phys. Chem. 33*: 489–519.

3 Gutsev, G.L., Bartlett, R.J., and Compton, R.N. (1998). *J. Chem. Phys. 108*: 6756–6762.

4 Lide, D.R. (2004). *CRC Handbook of Chemistry and Physics*, vol. *85*. CRC press.

5 Gutsev, G.L. and Boldyrev, A.I. (1982). *Chem. Phys. Lett. 92*: 262–266.

6 (a) Yokoyama, K., Haketa, N., Tanaka, H. et al. (2000). *Chem. Phys. Lett. 330*: 339–346. (b) Yokoyama, K., Haketa, N., Hashimoto, M. et al. (2000). *Chem. Phys. Lett. 320*: 645–650. (c) Nešković, O.M., Veljković, M.V., Veličković, S.R. et al. (2003). *Rapid Commun. Mass Spectrom. 17*: 212–214. (d) Veličković, S., Djordjević, V., Cvetićanin, J. et al. (2006). *Rapid Commun. Mass Spectrom. 20*: 3151–3153. (e) Veličković, S.R., Koteski, V.J., Belošević Čavor, J.N. et al. (2007). *Chem. Phys. Lett. 448*: 151–155.

7 (a) Kudo, H., Wu, C.H., and Ihle, H.R. (1978). *J. Nucl. Mater. 78*: 380–389. (b) Wu, C.H., Kudoa, H., and Ihle, H.R. (1979). *J. Chem. Phys. 70*: 1815–1820. (c) Schleyer, P.v.R., Wuerthwein, E.U., and Pople, J.A. (1982). *J. Am. Chem. Soc. 104*: 5839–5841. (d) Gutowski, M. and Simons, J. (1994). *J. Phys. Chem. 98*: 8326–8330. (e) Lievens, P., Thoen, P., Bouckaert, S. et al. (1999). *J. Chem. Phys. 110*: 10316–10329. (f) Neukermans, S., Janssens, E., Tanaka, H. et al. (2003). *J. Chem. Phys. 119*: 7206–7213. (g) Wang, D., Graham, J.D., Buytendyk, A.M., and Bowen, K.H. (2011). *J. Chem. Phys. 135*: 164308.

8 Rehm, E., Boldyrev, A.I., and Schleyer, P.v.R. (1992). *Inorg. Chem. 31*: 4834–4842.
9 Schleyer, P.V.R., Wuerthwein, E.U., Kaufmann, E. et al. (1983). *J. Am. Chem. Soc. 105*: 5930–5932.
10 Otten, A. and Meloni, G. (2018). *Chem. Phys. Lett. 692*: 214–223.
11 (a) Tong, J., Li, Y., Wu, D. et al. (2009). *J. Chem. Phys. 131*: 164307. (b) Tong, J., Li, Y., Wu, D. et al. (2011). *J. Phys. Chem. A 115*: 2041–2046.
12 (a) Tong, J., Li, Y., Wu, D., and Wu, Z.-J. (2012). *Inorg. Chem. 51*: 6081–6088. (b) Tong, J., Wu, Z., Li, Y., and Wu, D. (2013). *Dalton Trans. 42*: 577–584.
13 Zhao, T., Wang, Q., and Jena, P. (2017). *Nanoscale 9*: 4891–4897.
14 Srivastava, A.K. (2018). *Int. J. Quantum Chem. 118*: e25598.
15 Yang, H., Li, Y., Wu, D., and Li, Z.-r. (2012). *Int. J. Quantum Chem. 112*: 770–778.
16 Park, H. and Meloni, G. (2017). *Dalton Trans. 46*: 11942–11949.
17 Cochran, E. and Meloni, G. (2014). *J. Chem. Phys. 140*: 204319.
18 Srivastava, A.K. (2019). *Int. J. Quantum Chem. 119*: e25904.
19 (a) Vásquez-Espinal, A., Pino-Rios, R., Fuentealba, P. et al. (2016). *Int. J. Hydrogen Energy 41*: 5709–5715. (b) Giri, S., Lund, F., Núñez, A.S., and Toro-Labbé, A. (2013). *J. Phys. Chem. C 117*: 5544–5551.
20 Moreno, D., Martínez-Guajardo, G., Díaz-Celaya, A. et al. (2013). *Chem-Eur. J. 19*: 12668–12672.
21 Srivastava, A.K. (2018). *Mol. Phys. 116*: 1642–1649.
22 Baran, J.R. and Lagow, R.J. (1990). *J. Am. Chem. Soc. 112*: 9415–9416.
23 Baran, J.R., Hendrickson, C., Laude, D.A., and Lagow, R.J. (1992). *J. Org. Chem. 57*: 3759–3760.
24 Ramondo, F., Leonzi, I., and Meloni, G. (2019). *ChemPhysChem 20*: 3251–3258.
25 Fang, H., Zhou, J., and Jena, P. (2018). *Nanoscale 10*: 22963–22969.
26 de Heer, W.A. (1993). *Rev. Mod. Phys. 65*: 611–676.
27 (a) Bergeron, D.E., Castleman, A.W., Morisato, T., and Khanna, S.N. (2004). *Science 304*: 84. (b) Bergeron, D.E., Roach, P.J., Castleman, A.W. et al. (2005). *Science 307*: 231.
28 (a) Akutsu, M., Koyasu, K., Atobe, J. et al. (2006). *J. Phys. Chem. A 110*: 12073–12076. (b) Wang, B., Zhao, J., Shi, D. et al. (2005). *Phys. Rev. A 72*: 023204.
29 Sengupta, T., Das, S., and Pal, S. (2016). *J. Phys. Chem. C 120*: 10027–10040.
30 Kagdada, H., Dabhi, S., Mankad, V. et al. (2018). *Mater. Chem. Phys. 239*: 122264–122272.
31 (a) Sardar, R., Funston, A.M., Mulvaney, P., and Murray, R.W. (2009). *Langmuir 25*: 13840–13851. (b) Jupally, V.R., Kota, R., Dornshuld, E.V. et al. (2011). *J. Am. Chem. Soc. 133*: 20258–20266. (c) Parker, J.F., Fields-Zinna, C.A., and Murray, R.W. (2010). *Acc. Chem. Res. 43*: 1289–1296. (d) Jin, R. (2010). *Nanoscale 2*: 343–362.
32 (a) Heaven, M.W., Dass, A., White, P.S. et al. (2008). *J. Am. Chem. Soc. 130*: 3754–3755. (b) Zhu, M., Aikens, C.M., Hollander, F.J. et al. (2008). *J. Am. Chem. Soc. 130*: 5883–5885.
33 Kauffman, D.R., Alfonso, D., Matranga, C. et al. (2012). *J. Am. Chem. Soc. 134*: 10237–10243.
34 Freund, H.J. and Roberts, M.W. (1996). *Surf. Sci. Rep. 25*: 225–273.
35 Gattrell, M., Gupta, N., and Co, A. (2006). *J. Electroanal. Chem. 594*: 1–19.
36 Zhao, S., Austin, N., Li, M. et al. (2018). *ACS Catal. 8*: 4996–5001.
37 Alfonso, D.R., Kauffman, D., and Matranga, C. (2016). *J. Chem. Phys. 144*: 184705.
38 Kim, B., Seong, H., Song, J.T. et al. (2020). *ACS Energy Letters 5*: 749–757.

39 Joshi, C.P., Bootharaju, M.S., Alhilaly, M.J., and Bakr, O.M. (2015). *J. Am. Chem. Soc. 137*: 11578–11581.

40 Liu, Y., Chai, X., Xiao, C. et al. (2018). *Angew. Chem., Int. Ed. 57*: 9775–9779.

41 Dhayal, R.S., Liao, J.-H., Kahlal, S. et al. (2015). *Chem-Eur. J. 21*: 8369–8374.

42 Tang, Q., Lee, Y., Li, D.-Y. et al. (2017). *J. Am. Chem. Soc. 139*: 9728–9736.

43 Alvarez-Garcia, A., Flórez, E., Moreno, A., and Jimenez-Orozco, C. (2020). *Mol. Catal. 484*: 110733.

44 Yang, B., Liu, C., Halder, A. et al. (2017). *J. Phys. Chem. C 121*: 10406–10412.

45 Liu, C., Yang, B., Tyo, E. et al. (2015). *J. Am. Chem. Soc. 137*: 8676–8679.

46 Mitas, L., Grossman, J.C., Stich, I., and Tobik, J. (2000). *Phys. Rev. Lett. 84*: 1479–1482.

47 Kumar, V. and Kawazoe, Y. (2001). *Phys. Rev. Lett. 87*: 045503.

48 (a) Hiura, H., Miyazaki, T., and Kanayama, T. (2001). *Phys. Rev. Lett. 86*: 1733–1736. (b) Koyasu, K., Akutsu, M., Mitsui, M., and Nakajima, A. (2005). *J. Am. Chem. Soc. 127*: 4998–4999. (c) Koyasu, K., Atobe, J., Akutsu, M. et al. (2007). *J. Phys. Chem. A 111*: 42–49. (d) Lau, J.T., Hirsch, K., Klar, P. et al. (2009). *Phys. Rev. A 79*: 053201. (e) Claes, P., Janssens, E., Ngan, V.T. et al. (2011). *Phys. Rev. Lett. 107*: 173401.

49 Zhou, S., Yang, X., Pei, W. et al. (2019). *J. Phys. Chem. C 123*: 9973–9980.

50 Pei, W., Zhou, S., and Bai, Y. (2020). *J. Cluster Sci. 31*: 627–635.

51 (a) Mikkelsen, M., Jørgensen, M., and Krebs, F.C. (2010). *Energy Environ. Sci. 3*: 43–81. (b) Li, Y., Chan, S.H., and Sun, Q. (2015). *Nanoscale 7*: 8663–8683. (c) Peterson, A.A., Abild-Pedersen, F., Studt, F. et al. (2010). *Energy Environ. Sci. 3*: 1311–1315.

52 (a) Zheng, Y., Vasileff, A., Zhou, X. et al. (2019). *J. Am. Chem. Soc. 141*: 7646–7659. (b) Garza, A.J., Bell, A.T., and Head-Gordon, M. (2018). *ACS Catal. 8*: 1490–1499. (c) Fan, Q., Zhang, M., Jia, M. et al. (2018). *Mater. Today Energy 10*: 280–301.

13

Conclusions and Future Outlook

Puru Jena[1] *and Qiang Sun*[2,3]

[1]*Department of Physics, Virginia Commonwealth University, Richmond, VA, USA*
[2]*School of Materials Science and Engineering, Peking University, Beijing, China*
[3]*Center for Applied Physics and Technology, Peking University, Beijing, China*

Nearly half a century of research on atomic clusters has revealed many exciting and unexpected properties of this new phase of matter. Not only are the geometries, electronic structure, and properties of atomic clusters very different from those of their bulk materials, but also these properties change abruptly and unpredictably with the addition of a single atom or a single electron. In addition, it has been found that atomic clusters, with a specific size and composition, may mimic the chemistry of an atom in the periodic table, even if it does not contain a single element belonging to that group. These special atomic clusters are termed superatoms. It has been suggested that these superatoms can be used as building blocks of a new class of cluster-assembled materials with properties very different than those where individual atoms form the building blocks. One of the challenges, however, is that these atomic clusters are produced in the gas phase. Once removed from the vacuum environment, they would either react with the environmental gases or coalesce. Either way, they would lose their inherent cluster properties.

There are two ways in which these clusters can be prevented from coalescence. Either they are so stable that they can retain their pristine structure, or they have to be protected by coating them with ligands or by depositing them on noninteracting substrates. In the latter case, the environment will play a role on the structure and properties of these clusters and hence on their assembly. Instead of thinking that these are adverse effects, one can use both ligands and substrate to further tailor the properties of cluster-assembled materials. However, it is important to realize that synthesis of cluster-assembled materials in large quantities is difficult if one starts from their gas-phase origin. Thus, for practical applications, finding alternate synthesis methods for cluster-based materials is essential. To this end, we note that numerous materials with complex architecture, such as perovskites, argyrodites, borane-based materials, and Zintl and Chevrel phase compounds, exist with constituents that can be thought of as clusters. Knowledge gained from the stability and structure of these materials can guide in the solution-based synthesis of new cluster-based materials.

Superatoms: Principles, Synthesis and Applications, First Edition. Edited by Purusottam (Puru) Jena and Qiang Sun.
© 2022 John Wiley & Sons Ltd. Published 2022 by John Wiley & Sons Ltd.

In this book we began by summarizing the recent developments in the rational design of superatoms using different electron-counting rules, applied individually or in combination. Superatomic clusters that satisfy the electronic shell closure rules tend to be stable as neutral species. Reactive clusters can be designed by modifying the composition of the neutral clusters such that they either lack or have excess electrons needed to satisfy electronic shell closure. Stable clusters and their assemblies can also be created by endohedral doping of cage clusters as well by suitable choice of ligands. Two classes of clusters that behave either like halogen or like alkali atoms and supersalts composed of superions are discussed at length. Magnetic superatoms composed of heteroatomic species are designed where localized *d*- or *f*-electrons carry the magnetic moments while *s*- and *p*-electrons provide the chemical stability. One-, two-, and three-dimensional cluster-based materials and their application in energy harvesting (solar cells), energy storage (batteries), energy conversion (thermoelectrics), CO_2 capture and conversion, and ferroelectrics are illustrated in remaining chapters.

Cluster science, once motivated by our desire to learn how properties evolve from clusters to crystals, has entered a new phase where atom-precise clusters are used to build a new class of materials. Consequently, cluster science has now bridged physics, chemistry, and materials science. While it is difficult to predict the future of any discipline, let alone the one that is emerging, we believe that cluster science has a bright future where materials and phenomena commonly unknown in nature will be revealed. In the following, we list a few of the challenges and opportunities in cluster science.

1) Atomically precise design and synthesis of superatoms that go beyond the coinage metal atoms: a lot of superatoms (e.g. superhalogens, superalkalis, superpnictogens, superchalcogens, and endohedrally doped clusters) have been theoretically proposed during the past 20 years, but only few have been synthesized. There is an urgent need to develop advanced technology for synthesis of superatoms in large scale and assembling them in 1D, 2D, and 3D architectures.
2) Controllable assembly of superatoms for materials with desired functions: while much work has been done to illustrate properties of cluster materials, the inverse design should be studied to develop materials with specific functionality. In this regard, precise control of geometric configuration of superatoms could be rewarding.
3) Development of efficient algorithms for geometric structure search of large clusters: unlike crystals, clusters lack translational symmetry constraints. Equally important, the number of isomers increases exponentially with cluster size. For example, there are 487 635 isomers of $C_{60}H_4$, in which 4165 isomers are symmetrically distinct. When one more H is added to form $C_{60}H_5$, the number of isomers increases to 5 461 512, among them 45 010 are symmetrically distinct. This huge increase in the number of isomers just by adding one atom creates an enormous challenge for determining the ground state atomic structure.
4) Development of effective algorithms for property-directed search: in the global structure search, the energy or free energy is the target function for the global search by assuming that an experimentally synthesizable cluster is in the ground state structure if it is kinetically accessible. However, when the energy barrier is too high, the cluster can stay in the local minima with even more attractive properties as it is the case with

diamond versus graphite, the former being energetically metastable. Thus, for the design of cluster-based functional materials, the focus is on the properties instead of the global minima. In this context, property-directed search becomes important.

5) Development of effective algorithms for geometric structure search of cluster-assembled materials: it has been a long-standing dream to design materials with superatoms. However, the complexity of superatoms in composition, symmetry, size, and shape makes it challenging for determining the ground state geometry of 3D-assembled cluster materials. Conventional atom-based global search codes should be extended to cluster-based materials.
6) Applying machine learning for cluster science: for complex superatom-based materials with large unit cell, explicit electronic-structure methods such as density functional theory are not very effective due to the large computational cost. This can be overcome by using machine learning approach that can discover complex rules and find the invisible relationships among multivariables either by "learning" electronic-structure data or by deriving machine learning based interatomic potentials for reaching similar accuracy levels, but orders of magnitude faster.
7) Search algorithms for metastable phases: the current structural search methods use randomly distributed atoms followed by optimization using a variety of techniques. The use of molecular precursors instead of atoms may lead to metastable materials that otherwise will be missed by the current search techniques. These metastable phases possess unexpected properties.

It is our hope that this book will motivate both the development of tools and studies of superatoms and their assemblies using the synergy between experiment and theory.

Index

Superatoms: Principles, Synthesis and Applications, First Edition. Edited by Purusottam (Puru) Jena and Qiang Sun.
© 2022 John Wiley & Sons Ltd. Published 2022 by John Wiley & Sons Ltd.

Printed in the USA/Agawam, MA
September 16, 2026

910661.050